Applied Mathematics and Fractional Calculus II

Applied Mathematics and Fractional Calculus II

Editors

Francisco Martínez González
Mohammed K. A. Kaabar

MDPI

Basel • Beijing • Wuhan • Barcelona • Belgrade • Novi Sad • Cluj • Manchester

Editors
Francisco Martínez González
Universidad Politécnica de Cartagena,
Cartagena, Spain

Mohammed K. A. Kaabar
Samarkand International University of Technology,
Samarkand, Uzbekistan

Editorial Office
MDPI
St. Alban-Anlage 66
4052 Basel, Switzerland

This is a reprint of articles from the Special Issue published online in the open access journal *Symmetry* (ISSN 2073-8994) (available at: https://www.mdpi.com/journal/symmetry/special_issues/Applied_Mathematics_Fractional_Calculus_II).

For citation purposes, cite each article independently as indicated on the article page online and as indicated below:

Lastname, A.A.; Lastname, B.B. Article Title. *Journal Name* **Year**, *Volume Number*, Page Range.

ISBN 978-3-0365-9424-8 (Hbk)
ISBN 978-3-0365-9425-5 (PDF)
doi.org/10.3390/books978-3-0365-9425-5

Contents

About the Editors

Francisco Martínez González

Francisco Martínez González is a tenured professor at the Universidad Politécnica de Cartagena, Spain. He received a Ph.D. degree in Physics from Universidad de Murcia in 1992. His research interests include nonlinear dynamics methods and their applications, fractional calculus, fractional differential equations, multivariate calculus or special functions, and the divulgation of mathematics.

Mohammed K. A. Kaabar

Mohammed K. A. Kaabar is a Full Professor of Mathematics at the Samarkand International University of Technology, listed as one of the 2023 World's Top 2% Scientists by Stanford University, and he received his BSc, MSc, and PhD degrees in Mathematics from Washington State University, Pullman, Washington, USA and Universiti Malaya, Kuala Lumpur, Malaysia, respectively. He has a global, diverse experience in teaching/research. He has worked as a Professor of Mathematics, a Math Lab Instructor, and a Lecturer at various US institutions, such as Moreno Valley College, California, USA, Washington State University, Washington, USA, and Colorado Early Colleges, Colorado, USA. He is an aspiring educator, a researcher, a keynote speaker at conferences held in various countries, and a Director of the International Engineering and Technology Institute (IETI), headquarters in Malaysia.

Preface

In the last three decades, fractional calculus has broken into the field of mathematical analysis, both at the theoretical level and the level of its applications. In essence, the fractional calculus theory is a mathematical analysis tool applied to studying integrals and derivatives of arbitrary order, which unifies and generalizes the classical notions of differentiation and integration. These fractional and derivative integrals, which until a few years ago had been used in purely mathematical contexts, have been revealed as instruments with great potential to model problems in various scientific fields, such as fluid mechanics, viscoelasticity, physics, biology, chemistry, dynamical systems, signal processing, and entropy theory. Since fractional order's differential and integral operators are nonlinear operators, fractional calculus theory provides a tool for modeling physical processes, which in many cases is more useful than classical formulations; this is why applying fractional calculus theory has become a focus of international academic research. This Special Issue, "Applied Mathematics and Fractional Calculus II," has published excellent research studies in the field of applied mathematics and fractional calculus, authored by many well-known mathematicians and scientists from diverse countries worldwide, such as the USA, Ireland, Romania, Bulgaria, Türkiye, China, Pakistan, Iran, Egypt, India, Iraq, and Saudi Arabia.

Francisco Martínez González and Mohammed K. A. Kaabar
Editors

symmetry MDPI

Article

Mittag-Leffler Type Stability of Delay Generalized Proportional Caputo Fractional Differential Equations: Cases of Non-Instantaneous Impulses, Instantaneous Impulses and without Impulses

Ravi P. Agarwal [1], Snezhana Hristova [2,*] and Donal O'Regan [3]

1 Department of Mathematics, Texas A&M University-Kingsville, Kingsville, TX 78363, USA
2 Faculty of Mathematics and Informatics, Plovdiv University "P. Hilendarski", 4000 Plovdiv, Bulgaria
3 School of Mathematical and Statistical Sciences, National University of Ireland, H91 TK33 Galway, Ireland
* Correspondence: snehri@uni-plovdiv.bg

Abstract: In this paper, nonlinear differential equations with a generalized proportional Caputo fractional derivative and finite delay are studied in this paper. The eventual presence of impulses in the equations is considered, and the statement of initial value problems in three cases is defined: namely non-instantaneous impulses, instantaneous impulses and no impulses. The relations between these three cases are discussed. Additionally, some stability properties are investigated. We apply the Mittag–Leffler function which plays a vital role and which gives well-known bounds on the norm of the solutions. The symmetry of this function about a line and the bounds is a property that plays an important role in stability. Several sufficient conditions are presented via appropriate new comparison results and the modified Razumikhin method. The results generalize several known results in the literature.

Keywords: generalized proportional fractional derivatives; delays; non-instantaneous impulses; instantaneous impulses; Mittag–Leffler stability; Razumikhin method; Lyapunov functions

MSC: 34A34; 34K45; 34A08; 34D20

Citation: Agarwal, R. P.; Hristova, S.; O'Regan, D. Mittag-Leffler Type Stability of Delay Generalized Proportional Caputo Fractional Differential Equations: Cases of Non-Instantaneous Impulses, Instantaneous Impulses and without Impulses. *Symmetry* **2022**, *14*, 2290. https://doi.org/10.3390/sym14112290

Academic Editors: Francisco Martínez González and Mohammed K. A. Kaabar

Received: 10 October 2022
Accepted: 24 October 2022
Published: 1 November 2022

1. Introduction

Fractional calculus in real world phenomena is very applicable because of some typical properties such as memory. Various types of kernels in fractional integrals and fractional derivatives are applied (for example, in [1,2] the fourth-order time-fractional integro-differential equation with various types of kernels are studied numerically). A very general type of kernel was studied in [3] and called a general fractional integral/derivative. These general fractional integrals and derivatives were systematically studied by Y. Luchko [4,5] in appropriate function spaces in the framework of fractional calculus. Luchko also studied some qualitative properties of solutions of various types of differential equations with general fractional derivatives (see, [5]). In this paper, we focus on stability for a particular kernel (to be described in Section 3). Stability properties for fractional differential equations were studied by many authors (see, for example, [6,7]). As mentioned in [8], the generalized energy of a system does not have to decay exponentially for the system to be stable in the sense of Lyapunov, and recently the Mittag–Leffler stability and the fractional Lyapunov direct method were introduced for various types of fractional differential equations (see, for example, [9–12]) and applied in fractional models ([13–17]).

Many real processes are characterized by rapid changes in their state, and they are adequately modeled by differential equations with impulses. The acting time of these changes could be short relative to the duration of the whole process and they could be modeled as *instantaneous impulses* (see, for example, the classical book for ordinary differential equations [18] and the cited references therein). In some processes, the duration of changes might not be negligible, i.e., they start at arbitrary fixed points and remain active on finite time intervals. These types of changes could be modeled by *non-instantaneous impulses* (see, the overview given in the book [19]).

Even though fractional derivatives have memory, often various types of delays are involved in the fractional differential equations to represent some dynamics of the corresponding processes. When one studies fractional differential equations with delays and any type of impulse, there are a number of technical and theoretical difficulties.

In this paper, we study nonlinear differential equations with finite delay and with a generalized proportional Caputo fractional derivative. We consider three main cases: the case when there are non-instantaneous impulses in the equation, the case when there are instantaneous impulses in the equation and the case without any impulses. In all of these cases, we set up the initial value problem and we discuss the relation between them. The appropriate Mittag–Leffler type stability is defined, and several sufficient conditions are obtained. Our study is based on the Razumikhin method and its appropriate modifications. Some of the obtained results are generalizations of results known in the literature for the case of Caputo fractional differential equations.

Our contributions in this paper include:

1. The statement of the initial value problem for nonlinear systems of generalized proportional Caputo fractional differential equations with finite delays, and we consider three cases:
   - With non-instantaneous impulses;
   - With instantaneous impulses;
   - Without impulses.
2. An appropriate interpretation and connection between the three cases are provided.
3. Generalized proportional Mittag–Leffler stability of the three types of systems is defined.
4. The appropriate modifications of the Razumikhin method are applied in the three cases.
5. Some extensions of the comparison principle are provided.
6. Sufficient conditions for the Mittag–Leffler-type stability are obtained.

The paper is organized as follows. In Section 2, we recall some basic definitions about generalized proportional fractional integrals and Caputo-type derivatives, and some basic results are presented. In Section 3, we discuss the statements of fractional order delay systems in our three cases, and the relationships between them is provided. In Section 4, in the three cases, the generalized proportional Mittag–Leffler stability is defined, some comparison results are proved and several sufficient conditions are obtained with the help of appropriate modifications of the Razumikhin method.

2. Preliminary Notes and Results

We will give some basic notations used in this paper.

Let $u : [0,b] \to \mathbb{R}^n$, $b > 0$, $b \le \infty$ and $\tau \in (0,b)$. Then, we will use the following notations $u(\tau) = u(\tau - 0) = \lim_{t\uparrow\tau} u(t)$ and $u(\tau+0) = \lim_{t\downarrow\tau} u(t)$.

Let $r > 0$ be a given number and consider the set $E = \{\phi : [-r,0] \to \mathbb{R}^n$ is continuous everywhere except at a finite number of points $\tau_j \in (-r,0) : \phi(\tau_j - 0) = \phi(\tau_j), \phi(\tau_j + 0) < \infty\}$ with a norm $||\phi||_0 = \sup_{s\in[-r,0]} ||\phi(s)||$, where $||.||$ is a norm in $\mathbb{R}^n$.

Let two sequences of points $\{t_i\}_{i=1}^{\infty}$ and $\{s_i\}_{i=0}^{\infty}$ be given such that $0 < s_{i-1} \le t_i < s_i < t_{i+1}$, $i = 1,2,\ldots$, and $\lim_{k\to\infty} s_k = \infty$. Denote $t_0 = 0$.

Let $J \subset [0,\infty)$ be a given interval. Consider the following classes of functions:

$$\begin{aligned} NPC(J,\mathbb{R}^n) = \{u : J \to \mathbb{R}^n :\ & u \in C[J \cap (\bigcup \cup_{k=0}^{\infty}(t_k,s_k]),\mathbb{R}^n] : \\ & u(s_k) = u(s_k - 0) = \lim_{t\uparrow s_k} u(t) < \infty, \\ & u(s_k + 0) = \lim_{t\downarrow s_k} u(t) < \infty,\ k :\ s_k \in J\}, \end{aligned}$$

and

$$PC(J,\mathbb{R}^n) = \{v : J \to \mathbb{R}^n : \ v \in C[J \cap ([0,\infty)/\{t_k\}_{k=1}^{\infty}), \mathbb{R}^n] :$$
$$v(t_k) = v(t_k - 0) = \lim_{t \uparrow t_k} v(t) < \infty,$$
$$v(t_k + 0) = \lim_{t \downarrow t_k} v(t) < \infty, \ k : \ t_k \in J\},$$

We will give a brief overview of the literature on fractional integrals and derivatives with general kernels. In [4], Luchko described what was known in the literature on general fractional integrals (GFI) and general fractional derivatives (GFD) and studied GFI and GFD with the Sonine kernel. In [5], Luchko studied some analytical properties of initial-value problems for single and multi-term fractional differential equations with GFD with a Sonine kernel that possess integrable singularities of power function-type at the point zero. Luchko introduced the set of Sonine kernels $\mathbb{S}_{-1}$ and he considered GFI with a kernel $\kappa \in \mathbb{S}_{-1}$ (Definition 3.2 [5]):

$$(\mathbb{I}_{(\kappa)} f)(t) = \int_0^t \kappa(t-\tau) f(\tau) d\tau, \ t > 0, \tag{1}$$

GFD of Riemann–Liouville type (Definition 3.3 [5]):

$$(\mathbb{D}_{(\kappa)} f)(t) = \frac{d}{dt} \int_0^t \kappa(t-\tau) f(\tau) d\tau, \ t > 0, \tag{2}$$

and GFD of Caputo-type (Definition 3.3 [5]):

$$({}_*\mathbb{D}_{(\kappa)} f)(t) = (\mathbb{D}_{(\kappa)} f)(t) - f(0)\kappa(t), \ t > 0. \tag{3}$$

In [5], the first fundamental theorem of fractional calculus for the GFD (Theorem 3.1 [5]) and the second fundamental theorem of FC for the GFD (Theorem 3.2 [5]) are proved. Additionally, an explicit form of the solution of the initial value problem (IVP) for the linear fractional differential equation with Caputo type GFD is obtained. This formula significantly depends on the kernel $\kappa \in \mathbb{S}_{-1}$. Since the main goal of this paper is the study of fractional generalization of exponential stability, i.e., so-called Mittag–Leffler-type of stability, we will use a spacial type of the kernel $\kappa \in \mathbb{S}_{-1}$:

$$\kappa(t;\alpha,\rho) = \frac{\rho^{\alpha-1} \, t^{-\alpha}}{\Gamma(1-\alpha)} e^{\frac{\rho-1}{\rho} t} \in \mathbb{S}_{-1}, \ \alpha \in (0,1), \rho \in (0,1], \ t \geq 0. \tag{4}$$

Then, the definitions of GFI and GFD given by (1)–(3) are reduced:

$$\begin{aligned}
&(\mathcal{I}^{1-\alpha,\rho} f)(t) = (\mathbb{I}_{(\kappa(t,1-\alpha,\rho))} f)(t) = \int_0^t \frac{\rho^{-\alpha} \, (t-s)^{\alpha-1}}{\Gamma(\alpha)} e^{\frac{\rho-1}{\rho}(t-s)} f(s) ds, \\
&\quad \alpha > 0, \ \rho \in (0,1], \\
&({}^{RL}\mathcal{D}^{\alpha,\rho} f)(t) = (\mathbb{D}_{(\kappa(t;\alpha,\rho))} f)(t) \\
&\quad = \frac{1}{\rho^{1-\alpha}\Gamma(1-\alpha)} \frac{d}{dt} \int_0^t (t-s)^{-\alpha} e^{\frac{\rho-1}{\rho}(t-s)} f(s) ds, \quad \alpha \in (0,1), \ \rho \in (0,1], \\
&({}^{C}\mathcal{D}^{\alpha,\rho} f)(t) = ({}_*\mathbb{D}_{(\kappa(t;\alpha,\rho))} f)(t) \\
&\quad = \frac{\rho^{\alpha-1}}{\Gamma(1-\alpha)} \frac{d}{dt} \int_0^t (t-s)^{-\alpha} e^{\frac{\rho-1}{\rho}(t-s)} f(s) ds - f(0) \frac{\rho^{\alpha-1} \, t^{-\alpha}}{\Gamma(1-\alpha)} e^{\frac{\rho-1}{\rho} t}, \\
&\quad \text{for } t > 0, \ \ \alpha \in (0,1), \ \ \rho \in (0,1].
\end{aligned} \tag{5}$$

Remark 1. *The fractional integral $(\mathcal{I}^{1-\alpha,\rho} f)(t)$, the fractional derivatives $({}^{RL}\mathcal{D}^{\alpha,\rho} f)(t)$ and $({}^{C}\mathcal{D}^{\alpha,\rho} f)(t)$ are called generalized proportional fractional integral, generalized proportional Rieman–*

Liouville fractional integral and generalized proportional Caputo fractional derive, respectively, and they are studied in [20,21].

Remark 2. *(see Remark 3.2 [20]) If $\alpha \in (0,1)$ and $\rho \in (0,1]$ then the relation $({}_a^C\mathcal{D}^{\alpha,\rho} e^{\frac{\rho-1}{\rho}(.)})(t) = 0$ for $t > a$ holds. At the same time $({}_a^C\mathcal{D}^{\alpha,\rho} K)(t) \neq 0$ for $K \in \mathbb{R},\ K \neq 0$.*

We recall some results about generalized proportional Caputo fractional derivatives and their applications in differential equations, which will be applied in the main result in the paper.

Lemma 1. *(Proposition 5.2 [20]) For $\rho \in (0,1]$ and $\alpha \in (0,1)$ we have*

$$({}_a^C\mathcal{D}^{\alpha,\rho}(e^{\frac{\rho-1}{\rho}t}(t-a)^{\beta-1})(t) = \frac{\rho^\alpha \Gamma(\beta)}{\Gamma(\beta-\alpha)} e^{\frac{\rho-1}{\rho}t}(t-a)^{\beta-1-\alpha}, \quad \beta > 0.$$

Lemma 2. *(Lemma 3.2 [22]) Let $u \in C^1([a,b],\mathbb{R})$ with $a, b \in \mathbb{R},\ b \leq \infty$ (if $b = \infty$ then the interval is half open), and $q \in (0,1),\ \rho \in (0,1]$ be two reals. Then,*

$$({}_a^C\mathcal{D}^{\alpha,\rho} u^2)(t) \leq 2u(t)({}_a^C\mathcal{D}^{\alpha,\rho} u)(t), \quad t \in (a,b].$$

Lemma 3. *(Lemma 5 [23]) Let $u \in C([t_0, T, \mathbb{R}),\ T > t_0$, and there exists a point $t^* \in (t_0, T]$ such that $u(t^*) = 0$, and $u(t) < 0$, for $t_0 \leq t < t^*$. Then, if the generalized proportional Caputo fractional derivative of u exists for $t = t^*$, then the inequality $({}_{t_0}^{\,c}\mathcal{D}^{\alpha,\rho} u)(t)|_{t=t^*} > 0$ holds.*

Lemma 4. *(Example 5.7 [20]) The scalar linear generalized proportional Caputo fractional initial value problem*

$$({}_a^C\mathcal{D}^{\alpha,\rho} u)(t) = \lambda u(t), \quad u(a) = u_0, \quad \alpha \in (0,1), \ \rho \in (0,1]$$

has a solution

$$u(t) = u_0 e^{\frac{\rho-1}{\rho}(t-a)} E_\alpha(\lambda(\frac{t-a}{\rho})^\alpha), \quad t > a,$$

where $\lambda \in \mathbb{R}$, $E_\alpha(z) = \sum_{i=0}^{\infty} \frac{z^i}{\Gamma(i\alpha+1)}$ is the Mittag–Leffler function of one parameter.

Lemma 5. *Let $\alpha \in (0,1)$ and $\rho \in (0,1]$. Then*

$$({}_a^C\mathcal{D}^{\alpha,\rho}\left(e^{\frac{\rho-1}{\rho}(t-a)} E_\alpha\left(\lambda\left(\frac{(t-a)}{\rho}\right)^\alpha\right)\right) = \lambda e^{\frac{\rho-1}{\rho}(t-a)} E_\alpha\left(\lambda\left(\frac{(t-a)}{\rho}\right)^\alpha\right).$$

Proof. From Lemma 1 and the definition of Mittag–Leffler function with one parameter, we obtain

$$({}_a^C\mathcal{D}^{\alpha,\rho}(E_\alpha(\lambda(\frac{t-a}{\rho}^\alpha)))e^{\frac{\rho-1}{\rho}(t-a)}) = \sum_{i=0}^{\infty} \frac{({}_a^C\mathcal{D}^{\alpha,\rho}(e^{\frac{\rho-1}{\rho}(t-a)})(\lambda(\frac{t-a}{\rho})^\alpha)^i}{\Gamma(i\alpha+1)}$$

$$= \sum_{i=1}^{\infty} \frac{\lambda^i \rho^\alpha \Gamma(\alpha i+1) e^{\frac{=\rho-1}{\rho}(t-a)})(t-a)^{\alpha i-\alpha}}{\rho^{\alpha i}\Gamma(\alpha i+1-\alpha)\Gamma(i\alpha+1)}$$

$$= \lambda e^{\frac{\rho-1}{\rho}(t-a)} \sum_{i=1}^{\infty} \frac{\lambda^{i-1}(t-a)^{\alpha(i-1)}}{\rho^{\alpha(i-1)}\Gamma(\alpha(i-1)+1)} = \lambda e^{\frac{\rho-1}{\rho}(t-a)} E_\alpha\left(\lambda\left(\frac{(t-a)}{\rho}\right)^\alpha\right).$$

□

3. Statement of the Problems

In this paper, we will consider three cases: non-instantaneous impulses, instantaneous impulses and without impulse,s and we give the relations between them.

3.1. Non-Instantaneous Impulses

Let two sequences of points $\{t_i\}_{i=1}^{\infty}$ and $\{s_i\}_{i=0}^{\infty}$ be given such that $0 < s_{i-1} \le t_i < s_i < t_{i+1}$, $i = 1, 2, \ldots$, and $\lim_{k\to\infty} s_k = \infty$. Let $t_0 \ge 0$ be the given fixed initial time. Without loss of generality, we will assume $0 \le t_0 < s_0 < t_1$.

Remark 3. *The intervals $(s_k, t_{k+1}]$, $k = 0, 1, 2, \ldots$ are called intervals of non-instantaneous impulses.*

Let $J \subset \mathbb{R}$ be a given interval. Consider the following class of functions:

$$NPC^{\alpha,\rho}(J, \mathbb{R}^n) = \{u : J \to \mathbb{R}^n : \ u \in NPC(J, \mathbb{R}^n) : \text{ for any } k = 0, 1, 2, \cdots : \ t_k \in J,$$
$$({}^{C}_{t_k}\mathcal{D}^{\alpha,\rho}u)(t) \ \text{ exists for } \ t \in (t_k, s_k] \cap J\},$$

Consider the system of *non-instantaneous impulsive delay differential equations* (NIDDE) with the generalized proportional Caputo fractional derivative

$$\begin{aligned} ({}^{C}_{t_k}\mathcal{D}^{\alpha,\rho}x)(t) &= f(t, x_t) \text{ for } \ t \in (t_k, s_k], \ k = 0, 1, 2, \ldots \\ x(t) &= \Phi_k(t, x(s_k - 0)) \quad \text{for } t \in (s_k, t_{k+1}], \ k = 0, 1, 2, \ldots, \end{aligned} \tag{6}$$

with initial condition

$$x(t + t_0) = \phi(t) \ \text{ for } t \in [-r, 0], \tag{7}$$

where $f : [t_0, s_0] \bigcup \cup_{k=1}^{\infty} [t_k, s_k] \times \mathbb{R}^n \to \mathbb{R}^n$, $\Phi_i : [s_i, t_{i+1}] \times \mathbb{R}^n \to \mathbb{R}^n$, $(i = 0, 1, 2, 3, \ldots)$, $r > 0$ is a given number, $\phi : [-r, 0] \to \mathbb{R}^n$ and $x_t = x(t + s), s \in [-r, 0]$.

Remark 4. *The functions $\Phi_k(t, x)$, $k = 1, 2, \ldots$, are called non-instantaneous impulsive functions.*

Remark 5. *For some detailed explanations about non-instantaneous impulses in generalized proportional Caputo fractional differential equations without delays, see [24].*

We will introduce the following conditions:
(A 1.1.) The function $f \in C(\cup_{k=0}^{\infty}[t_k, s_k] \times \mathbb{R}^n, \mathbb{R}^n)$.
(A 1.2.) For any natural number k the functions $\Phi_k \in C([s_k, t_k] \times \mathbb{R}^n, \mathbb{R}^n)$, $k = 1, 2, \ldots$.

Remark 6. *We will assume that for any initial function $\phi \in E$ the IVP for the system of NIDDE (6) and (7) has a solution $x(t; t_0, \phi) \in NPC^{\alpha,\rho}([t_0, \infty), \mathbb{R}^n)$.*

We now give a brief description of the solution of IVP for NIDDE (6) and (7). The solution $x(t; t_0, \phi)$ of (6) and (7) is given by

$$x(t; t_0, \phi) = \begin{cases} X_k(t), & \text{for } t \in (t_k, s_k], \ k = 0, 1, 2, \ldots, \\ \Phi_k(t, X_k(s_k - 0)), & \text{for } \ t \in (s_k, t_{k+1}] \ \ k = 1, 2, \ldots \end{cases} \tag{8}$$

where

- On the interval $[t_0 - r, t_0]$, the solution satisfies the initial condition (7);
- On the interval $[t_0, s_0]$, the solution coincides with $X_0(t)$ which is the solution of $({}^{C}_{t_0}\mathcal{D}^{\alpha,\rho}x)(t) = f(t, x_t)$, $t \in (t_0, s_0]$ with initial condition (7);
- On the interval $(s_0, t_1]$, the solution $x(t; t_0, \phi)$ satisfies the equation

$$x(t; t_0, \phi) = \Phi_0(t, X_0(s_0 - 0));$$

- On the interval $(t_1, s_1]$, the solution coincides with $X_1(t)$ which is the solution of $({}^C_{t_1}\mathcal{D}^{\alpha,\rho}x)(t) = f(t, x_t),\ t \in (t_1, s_1]$ and initial condition $x(t + t_1) = \tilde{\phi}(t),\ t \in [-r, 0]$ with
$$\tilde{\phi}(t) = \begin{cases} \Phi_0(t_1, X_0(s_0 - 0)) & t = 0 \\ x(t - t_1; t_0, \phi) & t \in [-r, 0); \end{cases}$$
- On the interval $(s_1, t_2]$, the solution $x(t; t_0, \phi)$ satisfies the equation
$$x(t; t_0, x_0) = \Phi_1(t, X_1(s_1 - 0));$$

and so on.

In connection with the study of the stability properties of zero solutions, we introduce the following assumption:

(A 1.3.) The equalities $f(t, 0) = 0$ and $\Phi_k(t, 0) \equiv 0,\ k = 0, 1, 2, \ldots,$ hold.

3.2. Instantaneous Impulses

Let the sequence of points $\{t_i\}_{i=1}^{\infty}$ be given such that $0 < t_i \leq t_{i+1}$, $i = 1, 2, \ldots,$ and $\lim_{k\to\infty} t_k = \infty$. Let $t_0 \geq 0$ be the given fixed initial time. Without loss of generality we will assume $0 \leq t_0 < t_1$.

Remark 7. *The points $t_k,\ k = 0, 1, 2, \ldots$ are called points of impulses.*

Let $J \subset \mathbb{R}$ be a given interval. Consider the following class of functions
$$PC^{\alpha,\rho}(J, \mathbb{R}^n) = \{v : J \to \mathbb{R}^n :\ v \in PC(J, \mathbb{R}^n) :\ \text{for any } t_k \in J,\ k = 0, 1, 2, \cdots : \\ ({}^C_{t_k}\mathcal{D}^{\alpha,\rho}v)(t) \text{ exists for } t \in (t_k, t_{k+1}] \cap J\}.$$

Consider the system of *instantaneous impulsive delay differential equations* (IDDE) with the generalized proportional Caputo fractional derivative
$$\begin{aligned} &({}^C_{t_k}\mathcal{D}^{\alpha,\rho}x)(t) = f(t, x_t) \text{ for } t \in (t_k, t_{k+1}],\ k = 0, 1, 2, \ldots \\ &x(t_k + 0) = \Psi_k(x(t_k - 0)) \quad \text{for } k = 1, 2, \ldots, \end{aligned} \tag{9}$$
with initial condition (7), where $f : [t_0, \infty) \times \mathbb{R}^n \to \mathbb{R}^n$, $\Psi_i : \mathbb{R}^n \to \mathbb{R}^n$, $(i = 1, 2, 3, \ldots)$.

Remark 8. *The functions $\Psi_k(y),\ k = 1, 2, \ldots,$ are called impulsive functions.*

Remark 9. *In the case in Section 3.1 that both sequences coincide, i.e., $s_i = t_{i+1},\ i = 0, 1, 2, \ldots$, the system (6) is reduced to the system (9) with $\Phi_k(t, u) = \Psi_k(u),\ k = 0, 1, 2, \ldots,$ i.e., the case of non-instantaneous impulses could be considered as a generalization of the case of instantaneous impulses.*

We will introduce the following conditions:

(A 2.1.) The function $f \in C([t_0, t_1] \bigcup_{k=1}^{\infty}(t_k, t_{k+1}] \times \mathbb{R}^n, \mathbb{R}^n)$.

(A 2.2.) The functions $\Phi_k \in C(\mathbb{R}^n, \mathbb{R}^n),\ k = 1, 2, \ldots.$

(A 2.3.) The function $f(t, 0) = 0,\ t \geq t_0$ and the functions $\Psi_k(0) = 0,\ k = 1, 2, \ldots.$

If condition (A 2.3) is satisfied, then for the zero initial function, the IVP for IDDE (7) and (9) has a zero solution.

Remark 10. *We will assume that for any initial function $\phi \in E$ the IVP for the system of IDDE (7) and (9) has a solution $x(t; t_0, \phi) \in PC^{\alpha,\rho}([t_0, \infty), \mathbb{R}^n)$*

3.3. No Impulses

Consider the system of *delay differential equations* (DDE) with the generalized proportional fractional derivative

$$({}_{t_0}^{C}\mathcal{D}^{\alpha,\rho}x)(t) = f(t, x_t) \text{ for } t > t_0 \tag{10}$$

with initial condition (7), where $f : [t_0, \infty) \times \mathbb{R}^n \to \mathbb{R}^n$.

Remark 11. *The system (10) could be considered as a partial case of (9) in the case when there are no impulses, i.e., in Section 3.2* $t_i = t_0,\ i = 1, 2, \ldots,$ *i.e., the case of instantaneous impulses could be considered as a generalization of the case of without impulses.*

Let $J \subset \mathbb{R}$ be a given interval. Consider the following classes of functions

$$\begin{aligned} C^{\alpha,\rho}(J, \mathbb{R}^n) \quad &= \quad \{u : J \to \mathbb{R}^n : \ u \in C(J \cap [a, \infty), \mathbb{R}^n) : \\ & \qquad ({}_{a}^{C}\mathcal{D}^{\alpha,\rho}u)(t) \ \text{ exists for } \ t \in [a, \infty) \cap J\}. \end{aligned}$$

We will introduce the following conditions:
(A 3.1.) The function $f \in C([t_0, \infty) \times \mathbb{R}^n, \mathbb{R}^n)$.
(A 3.2.) The function $f(t, 0) = 0, \ t \geq t_0$.

Remark 12. *We will assume that for any initial function* $\phi \in E$*, the IVP for the system of DDE (7) and (10) has a solution* $x(t; t_0, \phi) \in C^{\alpha,\rho}([t_0, \infty), \mathbb{R}^n)$

4. Mittag–Leffer-Type Stability Properties

We will study the Mittag–Leffler-type stability for NIDDE (6), IDDE (9) and DDE (10) by Lyapunov functions and an appropriate modification of the Razumikhin method.

4.1. Non-Instantaneous Impulses

Definition 1. *The zero solution of the system NIDDE (6) and (7) is said to be* generalized proportional Mittag–Leffler stable *if there exist constants* $\beta, \gamma, C, \lambda > 0$ *such that the inequality*

$$||x(t; t_0, \phi)|| \\ \leq \begin{cases} C||\phi||_0^\beta \Big(\Big(\prod_{i=0}^{k-1} e^{\frac{\rho-1}{\rho}(s_i - t_i)} E_\alpha(-\lambda(\frac{s_i - t_i}{\rho})^\alpha) \Big) e^{\frac{\rho-1}{\rho}(t-t_k)} E_\alpha(-\lambda(\frac{t-t_k}{\rho})^\alpha) \Big)^\gamma, \\ \qquad t \in (t_k, s_k], \ k = 0, 1, \ldots, \\ C||\phi||_0^\beta \Big(\prod_{i=0}^{k} e^{\frac{\rho-1}{\rho}(s_i - t_i)} E_\alpha(-\lambda(\frac{s_i - t_i}{\rho})^\alpha) \Big)^\gamma, \\ \qquad t \in (s_k, t_{k+1}], \ k = 0, 1, 2, \ldots \end{cases} \tag{11}$$

holds, where $x(t; t_0, \phi)$ *is a solution of the IVP for NIDDE (6) and (7) (with an arbitrary initial function* $\phi \in E$*).*

Remark 13. *The definition for generalized proportional Mittag–Leffler stability for NIDDE (6) and (7) depends significantly on the type of intervals—the intervals of differential equations and the intervals of non-instantaneous impulses (see, the first and the second line, respectively, in (11)).*

We will use the following class of Lyapunov-like functions (for more details, see the book [19]):

Definition 2. *Let* $a < b \leq \infty$ *be given numbers ,* $\Omega \subset \mathbb{R}^n$*,* $0 \in \Omega$*. Then, the function* $V : [a - r, b] \times \Omega \to [0, \infty)$ *is from the class* $N\Lambda([a - r, b], \Omega)$ *if:*

- $V \in C([a, b]/\{s_k\} \times \Omega, [0, \infty))$ *and it is Lipschitz with respect to the second argument;*
- *For any* $s_k \in (a, b),\ x \in \Omega$*, there exist finite limits* $V(s_k - 0, x) = \lim_{t \uparrow s_k} V(t, x)$ *and* $V(s_k + 0, x) = \lim_{t \downarrow s_k} V(t, x)$.

We will consider the following scalar non-instantaneous impulsive differential equation (NIDE) as a comparison equation

$$\begin{aligned} &({}_{t_k}^{C}\mathcal{D}^{\alpha,\rho}u)(t) = -\lambda u(t), \quad \text{for } t \in (t_k, s_k],\ k = 0,1,2,\ldots, \\ &u(t) = \Xi_k(t, u(s_k - 0)) \text{ for } t \in (s_k, t_{k+1}],\ \ k = 0,1,2,\ldots, \\ &u(t_0) = u_0. \end{aligned} \tag{12}$$

According to Lemma 4, the solution of the IVP for NIDE (12) is given by

$$u(t) = \begin{cases} u_0 e^{\frac{\rho-1}{\rho}(t-t_0)} E_\alpha(-\lambda(\frac{t-t_0}{\rho})^\alpha), & t \in [t_0, s_0] \\ \Xi_k(t, u(s_k - 0)), & t \in (s_k, t_{k+1}],\ k = 0,1,2,\ldots, \\ \Xi_{k-1}(t_k, u(s_{k-1} - 0)) e^{\frac{\rho-1}{\rho}(t-t_k)} E_\alpha(-\lambda(\frac{t-t_k}{\rho})^\alpha), & t \in (t_k, s_k],\ k = 1,2,\ldots. \end{cases}$$

Applying the scalar NIDE (12) as a comparison equation, we will obtain the following comparison result for NIDDE (6).

Lemma 6. *Suppose:*

1. *The function $x^*(t) = x(t; t_0, \phi) \in NPC^{\alpha,\rho}([t_0, \infty), \Delta)$ is a solution of the NIDDE (6) and (7), where $\Delta \subset \mathbb{R}^n$.*
2. *The functions $\Xi_k \in C([s_k, t_{k+1}] \times \mathbb{R}, \mathbb{R})$ and $\Xi_k(t, u) \leq u$ for $t \in [s_k, t_{k+1}]$, $u \geq 0$, $k = 0,1,2,\ldots$.*
3. *The function $V \in N\Lambda([t_0 - r, \infty), \Delta)$ and*
   (i) for any $t \in (t_k, s_k]$ with $k = 0,1,\ldots$ such that

$$V(t, x^*(t)) \frac{e^{\frac{1-\rho}{\rho}(t-t_k)}}{E_\alpha\Big(-\lambda\Big(\frac{(t-t_k)}{\rho}\Big)^\alpha\Big)} \geq \sup_{s \in [t-r,t] \cap [t_k, t]} \frac{e^{\frac{1-\rho}{\rho}(s-t_k)}}{E_\alpha\Big(-\lambda\Big(\frac{(s-t_k)}{\rho}\Big)^\alpha\Big)} V(s, x^*(s)) \tag{13}$$

   the inequality

$${}_{t_k}^{C}\mathcal{D}^{\alpha,\rho} V(t, x^*(t)) \leq -\lambda V(t, x^*(t))$$

   holds where $\lambda > 0$ is a given number.
   (ii) For any $k = 0,1,\ldots$ the inequalities

$$V(t, \Phi_k(t, x^*(s_k - 0))) \leq \Xi_k(t, V(s_k - 0, x^*(s_k - 0))) \ \text{for } t \in (s_k, t_{k+1}].$$

   hold.

Then, the inequality

$$V(t, x^*(t)) \leq \begin{cases} M\Big(\prod_{i=0}^{k-1} e^{\frac{\rho-1}{\rho}(s_i - t_i)} E_\alpha(-\lambda(\frac{s_i - t_i}{\rho})^\alpha)\Big) e^{\frac{\rho-1}{\rho}(t-t_k)} E_\alpha(-\lambda(\frac{t-t_k}{\rho})^\alpha), \\ \qquad t \in (t_k, s_k],\ k = 0,1,\ldots, \\ M\Big(\prod_{i=0}^{k} e^{\frac{\rho-1}{\rho}(s_i - t_i)} E_\alpha(-\lambda(\frac{s_i - t_i}{\rho})^\alpha)\Big), \quad t \in (s_k, t_{k+1}], k = 0,1,2,\ldots, \end{cases} \tag{14}$$

holds where $M = \max_{s \in [-r,0]} V(t_0 + s, \phi(s))$.

Proof. *Case 1.* Let $t \in [t_0, s_0]$. Define the function $m(t) = V(t, x^*(t))$ for $t \in [t_0 - r, s_0]$. Then, the function $m(t) \in C^{\alpha,\rho}([t_0, s_0], \mathbb{R}_+)$ and the inequality $m(t_0) = V(t_0, \phi(0)) \le \sup_{s\in[-r,0]} V(t_0 + s, \phi(s)) = M$ hold. We will prove that

$$m(t) < Me^{\frac{\rho-1}{\rho}(t-t_0)} E_\alpha\left(-\lambda\left(\frac{t-t_0}{\rho}\right)^\alpha\right) + \varepsilon e^{\frac{\rho-1}{\rho}(t-t_0)}, \quad t \in [t_0, s_0], \tag{15}$$

where $\varepsilon > 0$ is a small enough number. Note for $t = t_0$ inequality (15) holds. Assume (15) is not true on $(t_0, s_0]$. Therefore, there exists $t^* \in (t_0, s_0]$ such that

$$\begin{aligned} m(t) &< Me^{\frac{\rho-1}{\rho}(t-t_0)} E_\alpha\left(-\lambda\left(\frac{t-t_0}{\rho}\right)^\alpha\right) + \varepsilon e^{\frac{\rho-1}{\rho}(t-t_0)}, \quad t \in [t_0, t^*), \\ m(t^*) &= e^{\frac{\rho-1}{\rho}(t^*-t_0)} E_\alpha\left(-\lambda\left(\frac{t^*-t_0}{\rho}\right)^\alpha\right) + \varepsilon e^{\frac{\rho-1}{\rho}(t^*-t_0)}. \end{aligned} \tag{16}$$

Consider the function $\xi(t) = m(t) - Me^{\frac{\rho-1}{\rho}(t-t_0)} E_\alpha\left(-\lambda\left(\frac{t-t_0}{\rho}\right)^\alpha\right) - \varepsilon e^{\frac{\rho-1}{\rho}(t-t_0)}$ for $t \in [t_0, s_0]$. According to Lemma 3 with $u(t) \equiv \xi(t)$ the inequality $({}_{t_0}^{C}\mathcal{D}^{\alpha,\rho}\xi)(t)|_{t=t^*} > 0$ holds. Therefore, according to Lemma 5 and Remark 2, we obtain

$$({}_{t_0}^{C}\mathcal{D}^{\alpha,\rho}m)(t)|_{t=t^*} > -\lambda Me^{\frac{\rho-1}{\rho}(t^*-t_0)} E_\alpha\left(-\lambda\left(\frac{t^*-t_0}{\rho}\right)^\alpha\right). \tag{17}$$

Case 1.1. Let $r < t^* - t_0$. Then, $t^* - r > t_0$ and $[t^* - r, t^*] \subset (t_0, t^*]$, i.e., $[t^* - r, t^*] \cap [t_0, t^*] = [t^* - r, t^*]$. Therefore, since the function $E_\alpha(-\lambda t)$ is decreasing for $t \in (t_0, t^*]$, i.e., $\frac{1}{E_\alpha\left(-\lambda\left(\frac{(t-t_0)}{\rho}\right)^\alpha\right)} \le \frac{1}{E_\alpha\left(-\lambda\left(\frac{(t^*-t_0)}{\rho}\right)^\alpha\right)}$ for $t \in [t^* - r, t^*]$ by (16), we obtain

$$\begin{aligned} m(t)\frac{e^{\frac{1-\rho}{\rho}(t-t_0)}}{E_\alpha\left(-\lambda\left(\frac{t-t_0}{\rho}\right)^\alpha\right)} &< M + \varepsilon\frac{1}{E_\alpha\left(-\lambda\left(\frac{t-t_0}{\rho}\right)^\alpha\right)} \\ &\le M + \varepsilon\frac{1}{E_\alpha\left(-\lambda\left(\frac{t^*-t_0}{\rho}\right)^\alpha\right)} \\ &= m(t^*)\frac{e^{\frac{1-\rho}{\rho}(t^*-t_0)}}{E_\alpha\left(-\lambda\left(\frac{t^*-t_0}{\rho}\right)^\alpha\right)}, \quad t \in [t^* - r, t^*], \end{aligned} \tag{18}$$

i.e., inequality (13) is satisfied for $t = t^*$.

According to condition 3(i) the inequality

$$\begin{aligned} &({}_{t_0}^{C}\mathcal{D}^{\alpha,\rho}m)(t)|_{t=t^*} = ({}_{t_0}^{C}\mathcal{D}^{\alpha,\rho}V(t, x^*(t)))|_{t=t^*} \le -\lambda V(t^*, x^*(t^*)) \\ &= -\lambda m(t^*) = -\lambda Me^{\frac{\rho-1}{\rho}(t^*-t_0)} E_\alpha\left(-\lambda\left(\frac{t^*-t_0}{\rho}\right)^\alpha\right) - \lambda\varepsilon e^{\frac{\rho-1}{\rho}(t^*-t_0)} \end{aligned} \tag{19}$$

holds.

From inequalities (17) and (19), it follows that $-\lambda\varepsilon e^{\frac{\rho-1}{\rho}(t^*-t_0)} > 0$. The obtained contradiction proves the inequality (15) on $[t_0, s_0]$.

Case 1.2. Let $r \ge t^* - t_0$. Then, $t^* - r \le t_0$ and $[t^* - r, t^*] \cap [t_0, t^*] = [t_0, t^*] = \{t_0\} \cup (t_0, t^*]$. Similar to the proof in Case 1.1, we obtain the inequality

$$m(t)\frac{e^{\frac{1-\rho}{\rho}(t-t_0)}}{E_\alpha\left(-\lambda\left(\frac{t-t_0}{\rho}\right)^\alpha\right)} \le m(t^*)\frac{e^{\frac{1-\rho}{\rho}(t^*-t_0)}}{E_\alpha\left(-\lambda\left(\frac{t^*-t_0}{\rho}\right)^\alpha\right)}, \quad t \in (t_0, t^*].$$

For $t = t_0$, apply (16), $E_\alpha(0) = 1$ and obtain $m(t^*)\frac{e^{\frac{1-\rho}{\rho}(t^*-t_0)}}{E_\alpha\left(-\lambda\left(\frac{t^*-t_0}{\rho}\right)^\alpha\right)} > M \geq m(t_0)$.

Therefore, inequality (13) holds for $t = t^*$.

Thus, condition 3(i) is applicable and as in Case 1.1 we obtain a contradiction.

The contradiction proves inequality (15). From inequality (15) as $\varepsilon \to 0$ follows the validity of (14) on $[t_0, s_0]$.

Case 2. Let $t \in (s_0, t_1]$. Then, $x^*(t) = \Phi_1(t, x^*(s_0 - 0))$. From conditions 2, 3(ii) for $k = 0$ and Case 1, we obtain

$$
\begin{aligned}
V(t, x^*(t)) &= V(t, \Phi_0(t, x^*(s_0 - 0))) \leq \Xi_0(t, V(s_0 - 0, x^*(s_0 - 0))) \\
&\leq V(s_0 - 0, x^*(s_0 - 0)) \\
&\leq M e^{\frac{\rho-1}{\rho}(s_0 - t_0)} E_\alpha\left(-\lambda\left(\frac{s_0 - t_0}{\rho}\right)^\alpha\right), \quad t \in (s_0, t_1].
\end{aligned}
$$

Therefore, inequality (14) holds on $(s_0, t_1]$.

Case 3. Let $t \in (t_1, s_2]$. Define the function

$$
m_1(t) = \begin{cases} V(t_1, x^*(t_1)) & \text{for } t \in [t_1 - r, t_1], \\ V(t, x^*(t)) & \text{for } t \in (t_1, s_1]. \end{cases}
$$

Then, the function $m_1(t) \in C^{\alpha,\rho}([t_1, s_1], \mathbb{R}_+)$. Denote $M_1 = V(t_1, x^*(t_1))$. Then,

$$
\left(\max_{s \in [-r, 0]} m_1(t_1 + s)\right) = M_1
$$

and according to Case 2, the inequality

$$
M_1 < M e^{\frac{\rho-1}{\rho}(s_0 - t_0)} E_\alpha\left(-\lambda\left(\frac{s_0 - t_0}{\rho}\right)^\alpha\right)
$$

holds.

Similar to the proof of inequality (15) in Case 1, we have the validity of the inequality

$$
m_1(t) < M_1 e^{\frac{\rho-1}{\rho}(t - t_1)} E_\alpha\left(-\lambda\left(\frac{t - t_1}{\rho}\right)^\alpha\right) + \varepsilon e^{\frac{\rho-1}{\rho}(t - t_1)}. \; t \in [t_1, s_1].
$$

Thus,

$$
\begin{aligned}
m_1(t) &< M e^{\frac{\rho-1}{\rho}(s_0 - t_0)} E_\alpha\left(-\lambda\left(\frac{s_0 - t_0}{\rho}\right)^\alpha\right) e^{\frac{\rho-1}{\rho}(t - t_1)} E_\alpha\left(-\lambda\left(\frac{t - t_1}{\rho}\right)^\alpha\right) \\
&+ \varepsilon e^{\frac{\rho-1}{\rho}(t - t_1)}, \quad t \in (t_1, s_1].
\end{aligned} \tag{20}
$$

Taking the limit in (20) as $\varepsilon \to 0$ we obtain the claim of Lemma 6 on $(t_1, s_1]$.

Continue this process and an induction argument proves the claim in Lemma 6. □

Remark 14. *The condition (13) is a modified Razumikhin condition applied in connection with generalized proportional fractional derivatives.*

Remark 15. *The inequality (13) in condition 3(i) of Lemma 6 could be replaced by*

$$
V(t, x^*(t)) \geq \sup_{s \in [t-r, t] \cap [t_k, t]} \frac{e^{\frac{1-\rho}{\rho}(s - t_k)}}{E_\alpha\left(-\lambda\left(\frac{(s - t_k)}{\rho}\right)^\alpha\right)} V(s, x^*(s)) \tag{21}
$$

Note that if (21) holds, then inequality (13) is also satisfied.

Remark 16. *If the condition (21) is satisfied, then the classical Razumikhin condition* $V(t, x^*(t)) \geq \sup_{s\in[t-r,t]\cap[t_k,t]} V(s, x^*(s))$ *holds.*

Remark 17. *The condition 3(i) is satisfied only at some particular points of t from the studied interval.*

We study the generalized Mittag–Leffler stability properties of the zero solution of NIDDE (6).

Theorem 1. *Suppose :*

1. *Conditions (A 1.1)–(A 1.3) are satisfied.*
2. *There exists a function* $V \in N\Lambda([t_0 - r, \infty), \mathbb{R}^n)$ *such that*
   - *(i) There exist positive constants* A, B, a, b *such that the inequalities* $A||x||^a \leq V(t,x) \leq B||x||^{ab},\ t \geq t_0,\ \ x \in \mathbb{R}^n$ *hold.*
   - *(ii) For any point* $t \in (t_k, s_k]$ *with* $k = 0,1,2,\ldots$ *and any function* $\psi \in C^{\alpha,\rho}(t_k, [t - r, t], \mathbb{R}^n)$ *such that* $\left({}_{t_k}^{C}\mathcal{D}^{\alpha,\rho}\psi\right)(t) = f(t, \psi_t)$ *and*

$$V(t, \psi(t))\frac{e^{\frac{1-\rho}{\rho}(t-t_k)}}{E_\alpha\left(-D\left(\frac{(t-t_k)}{\rho}\right)^\alpha\right)} \geq \sup_{s\in[t-r,t]\cap[t_k,t]} \frac{e^{\frac{1-\rho}{\rho}(s-t_k)}}{E_\alpha\left(-D\left(\frac{(s-t_k)}{\rho}\right)^\alpha\right)} V(s, \psi(s)) \tag{22}$$

     the inequality

$$\left({}_{t_k}^{C}\mathcal{D}^{\alpha,\rho} V(t, \psi(t))\right) \leq -DV(t, \psi(t)) \tag{23}$$

     holds where $D > 0$ *is a given number.*
   - *(iii) For any* $k = 0,1,\ldots$ *and* $u \in \mathbb{R}^n$*, the inequalities*

$$V(t, \Phi_k(t,u)) \leq C||u||^a \ \ \text{for } t \in (s_k, t_{k+1}].$$

     hold where $C \in (0, A]$.

Then, the zero solution of NIDDE (6) with the zero initial function is generalized proportional Mittag–Leffler stable with $C = \sqrt[a]{\frac{B}{A}}, \beta = b,\ \lambda = D,\ \gamma = \frac{1}{a}$.

Proof. Let $\phi \in E$ be an arbitrary initial function and now let $x(t) = x(t; t_0, \phi) \in NPC^{\alpha,\rho}([t_0, \infty), \mathbb{R}^n)$ be the solution of the IVP for NIDDE (6) and (7). Let $t^* \in (t_k, s_k]$ with k a non-negative integer, be such that the inequality (22) holds with $\psi(t) = x(t)$. Note that $x \in C^{\alpha,\rho}(t_k, [t^* - r, t^*], \mathbb{R}^n)$ and $\left({}_{t_k}^{C}\mathcal{D}^{\alpha,\rho}x\right)(t)|_{t=t^*} = f(t^*, x_{t^*})$. Then, according to condition 2(ii) of Theorem 1, the inequality (23) holds, i.e., we have

$$\left({}_{t_k}^{C}\mathcal{D}^{\alpha,\rho} V(t, x(t))\right)|_{t=t^*} \leq -DV(t^*, x(t^*)),$$

i.e., the condition 3(i) of Lemma 6 is satisfied with $\lambda = D$.

Let $k = 0,1,\ldots$ be an arbitrary number. Then, from conditions 2(i) and 2(iii) of Theorem 1, we obtain $V(t, \Phi_k(t, x(s_k - 0))) \leq C||x(s_k - 0)||^a \leq \frac{C}{A}V(s_k - 0, x(s_k - 0))$, i.e., condition 3(ii) of Lemma 6 is satisfied with $\Xi_k(t,u) = \frac{C}{A}u \leq u$ according to the choice of the constants A, C.

According to Lemma 6, the inequality

$$
\begin{aligned}
&V(t,x(t))\\
&\le \begin{cases} M\Big(\prod_{i=0}^{k-1} e^{\frac{\rho-1}{\rho}(s_i-t_i)}E_\alpha(-D(\frac{s_i-t_i}{\rho})^\alpha)\Big)e^{\frac{\rho-1}{\rho}(t-t_k)}E_\alpha(-D(\frac{t-t_k}{\rho})^\alpha), \\ \qquad t\in(t_k,s_k],\ k=0,1,\dots, \\ M\Big(\prod_{i=0}^{k} e^{\frac{\rho-1}{\rho}(s_i-t_i)}E_\alpha(-D(\frac{s_i-t_i}{\rho})^\alpha)\Big),\ t\in(s_k,t_{k+1}],\ k=0,1,2,\dots. \end{cases}
\end{aligned}
\tag{24}
$$

holds where $M\le B||\phi||_0^{ab}$.

Thus, from condition 2(i) of Theorem 1, we obtain

$$
\begin{aligned}
&||x(t)||\\
&\le \begin{cases} \sqrt[a]{\frac{B}{A}}||\phi||_0^b\Big(\Big(\prod_{i=0}^{k-1} e^{\frac{\rho-1}{\rho}(s_i-t_i)}E_\alpha(-D(\frac{s_i-t_i}{\rho})^\alpha)\Big)e^{\frac{\rho-1}{\rho}(t-t_k)}E_\alpha(-D(\frac{t-t_k}{\rho})^\alpha)\Big)^{\frac{1}{a}}, \\ \qquad t\in(t_k,s_k],\ k=0,1,\dots, \\ \sqrt[a]{\frac{B}{A}}||\phi||_0^b\Big(\prod_{i=0}^{k} e^{\frac{\rho-1}{\rho}(s_i-t_i)}E_\alpha(-D(\frac{s_i-t_i}{\rho})^\alpha)\Big)^{\frac{1}{a}}, \\ \qquad t\in(s_k,t_{k+1}],\ k=0,1,2,\dots. \end{cases}
\end{aligned}
\tag{25}
$$

Thus, the zero solution of (6) is generalized Mittag-Leffler stable with $C=\sqrt[a]{\frac{B}{A}},\beta=b,\ \lambda=D,\ \gamma=\frac{1}{a}$. □

Corollary 1. *Let the conditions of Theorem 1 be satisfied where the inequality (22) is replaced by*

$$
V(t,\psi(t))\ge \sup_{s\in[t-r,t]\cap[t_k,t]}\frac{e^{\frac{1-\rho}{\rho}(s-t_k)}}{E_\alpha\Big(-D\Big(\frac{(s-t_k)}{\rho}\Big)^\alpha\Big)}V(s,\psi(s))
\tag{26}
$$

Then, the zero solution of NIDDE (6) with the zero initial function is generalized proportional Mittag–Leffler stable.

Proof. If the inequality (26) is satisfied for the point t, then we obtain

$$
V(t,\psi(t))\frac{e^{\frac{1-\rho}{\rho}(t-t_k)}}{E_\alpha\Big(-D\Big(\frac{(t-t_k)}{\rho}\Big)^\alpha\Big)}\ge V(t,\psi(t)),
$$

i.e., inequality (22) is satisfied. □

Corollary 2. *Let the conditions of Theorem1 be satisfied where the condition 2(ii) is replaced by 2(ii)* for any point $t\in(t_k,s_k]$ with $k=0,1,2,\dots$ and any function $\psi\in C^{\alpha,\rho}(t_k,[t-r,t],\mathbb{R}^n)$ such that $\left({}^{C}_{t_k}\mathcal{D}^{\alpha,\rho}\psi\right)(t)=f(t,\psi_t)$ and*

$$
V(t,\psi(t))\ge \sup_{s\in[t-r,t]\cap[t_k,t]}\frac{e^{\frac{1-\rho}{\rho}(s-t_k)}}{E_\alpha\Big(-D\Big(\frac{(s-t_k)}{\rho}\Big)^\alpha\Big)}V(s,\psi(s))
\tag{27}
$$

the inequality

$$
\left({}^{C}_{t_k}\mathcal{D}^{\alpha,\rho}V(t,\psi(t))\right)\le -D\sup_{s\in[t-r,t]\cap[t_k,t]}||\psi(s)||^{ab}
\tag{28}
$$

holds where $D>0$ is a given number.

Then, the zero solution of NIDDE (6) with the zero initial function is generalized proportional Mittag–Leffler stable.

Proof. From condition 2(iii) of Theorem 1 and inequality (27), we have that $||\psi(s)||^{ab} \geq V(s,\psi(s)),\ s \in [t-r,t] \cap [t_k,t]$, i.e.,

$$-D\left(\sup_{s\in[t-r,t]\cap[t_k,t]} ||\psi(s)||^{ab}\right) \leq -D\left(\sup_{s\in[t-r,t]\cap[t_k,t]} V(s,\psi(s))\right) = -DV(t,\psi(t)).$$

Thus, from inequality (28) we have inequality (23). □

Corollary 3. *Let the conditions of Theorem1 be satisfied where the inequality (23) is replaced by*

$$^{C}_{t_k}\mathcal{D}^{\alpha,\rho}V(t,\psi(t)) \leq 0, \tag{29}$$

and condition 2(i) is changed by

2(i) There exist positive constants A, B such that the inequalities $A||x|| \leq V(t,x) \leq B||x||,\ t \geq t_0,\ x \in \mathbb{R}^n$ hold.*

Then, the zero solution of NIDDE (6) with the zero initial function is stable.

Proof. Inequality (29) is a partial case of (23) with $D = 0$, then use $E_\alpha(0) = 1$ and inequality (25) and we obtain $||x(t)|| \leq \frac{B}{A}||\phi||_0$ for $t \geq t_0$, which proves the stability of the solution. □

Example 1. . *Consider the scalar IVP for NIDDE*

$$\begin{aligned} &\left(^{C}_{t_k}\mathcal{D}^{\alpha,\rho}x\right)(t) = -\frac{2+t}{t+1}(x(t) - 0.5x_t^{(k)}), \quad \text{for } t \in (t_k,s_k],\ k = 0,1,2,\dots, \\ &x(t) = 0.5(\sin t)x(s_k - 0) \text{ for } t \in (s_k,t_{k+1}],\ k = 0,1,2,\dots, \\ &x(t_0+s) = \phi(s),\ s \in [-r,0], \end{aligned} \tag{30}$$

where for any $t \in (t_k,s_k]$ we denote $x_t^{(k)}(s) = x(t+s), s \in [max\{-r,t_k-t\},0]$.

The scalar IVP for NIDDE (30) with $\phi(s) \equiv 0$ has a zero solution.

Consider the Lyapunov function $V(t,x) = x^2$. Then, condition 2(i) of Theorem 1 is satisfied with $A = 0.25,\ B = 1, a = 2, b = 1$. Let k be a whole number and the point $t \in (t_k,s_k]$ and the function $\psi \in C^{\alpha,\rho}(t_k,[t-r,t],\mathbb{R})$ be such that $\left(^{C}_{t_k}\mathcal{D}^{\alpha,\rho}\psi\right)(t) = -\frac{2+t}{t+1}(\psi(t) - 0.5\psi_t^{(k)})$ and

$$\psi^2(t) \geq \sup_{s\in[t-r,t]\cap[t_k,t]} \frac{e^{\frac{1-\rho}{\rho}(s-t_k)}}{E_\alpha\left(-\left(\frac{(s-t_k)}{\rho}\right)^\alpha\right)}\psi^2(s). \tag{31}$$

Then applying $\sup_{s\in[t-r,t]\cap[t_k,t]} \frac{e^{\frac{1-\rho}{\rho}(s-t_k)}}{E_\alpha\left(-\left(\frac{(s-t_k)}{\rho}\right)^\alpha\right)}\psi^2(s) \geq \sup_{s\in[t-r,t]\cap[t_k,t]} \psi^2(s)$ we obtain

$$\begin{aligned} &\left(^{C}_{t_k}\mathcal{D}^{\alpha,\rho}\psi^2\right)(t) \leq 2\psi(t)\left(^{C}_{t_k}\mathcal{D}^{\alpha,\rho}\psi\right)(t) \\ &= -2\frac{2+t}{t+1}(\psi^2(t) - 0.5\psi(t)\psi_t^{(k)}) \\ &\leq \frac{2+t}{t+1}(-2\psi^2(t) + 0.5\psi^2(t) + 0.5(\psi_t^{(k)})^2) \\ &\leq \frac{2+t}{t+1}(-2\psi^2(t) + 0.5\psi^2(t) + 0.5\sup_{s\in[t-r,t]\cap[t_k,t]} \psi^2(s)) \\ &\leq \frac{2+t}{t+1}(-1.5\psi^2(t) + 0.5\psi^2(t)) = -\frac{2+t}{t+1}\psi^2(t) \\ &< -V(t,\psi(t)). \end{aligned} \tag{32}$$

Let $t \in (s_k,t_{k+1}]$ where $k = 0,1,2,\dots$. Then, $(0.5\sin t\ u)^2 \leq 0.25u^2 = 0.25|u|^2$.

Therefore, the conditions of Corollary 1 are satisfied with $D = 1, C = A = 0.25, B = 1, a = 2, b = 1$. According to Corollary 1 the zero solution of the scalar NIDDE (30) is generalized proportional Mittag–Leffler stable with $C = \sqrt{4} = 2, \beta = 1,\ \lambda = 1,\ \gamma = 0.5$, i.e., the inequality

$$||x(t)|| \le \begin{cases} 2||\phi||_0 \sqrt{\left(\Pi_{i=0}^{k-1} e^{\frac{\rho-1}{\rho}(s_i-t_i)} E_\alpha(-(\frac{s_i-t_i}{\rho})^\alpha)\right) e^{\frac{\rho-1}{\rho}(t-t_k)} E_\alpha(-(\frac{t-t_k}{\rho})^\alpha)}, \\ \qquad t \in (t_k, s_k],\ k = 0,1,\dots, \\ 2||\phi||_0 \sqrt{\Pi_{i=0}^{k} e^{\frac{\rho-1}{\rho}(s_i-t_i)} E_\alpha(-(\frac{s_i-t_i}{\rho})^\alpha)}, \\ \qquad t \in (s_k, t_{k+1}],\ k = 0,1,2,\dots. \end{cases}$$

holds.

Remark 18. *The Mittag–Leffler type stability for the Caputo fractional differential equations (with $\rho = 1$) is studied in [25].*

4.2. Instantaneous Impulses

As mentioned in Remark 9, the case of non-instantaneous impulses could be considered as a generalization of the case of instantaneous impulses. That is why we can translate the results from the previous section to instantaneous impulses.

Definition 3. *The zero solution of the system IDDE (7) and (9) (with $\phi \equiv 0$) is said to be* generalized proportional Mittag–Leffler stable *if there exist constants $\beta, \gamma, C, \lambda > 0$ such that the inequality*

$$||x(t;t_0,\phi)|| \le C||\phi||_0^\beta \left(e^{\frac{\rho-1}{\rho}(t-t_k)} E_\alpha(-\lambda(\frac{t-t_k}{\rho})^\alpha)\right)^\gamma, \quad t \in (t_k, t_{k+1}],\ k = 0,1,\dots, \tag{33}$$

holds, where $x(t;t_0,\phi)$ is a solution on the IVP for IDDE (7) and (9) with an arbitrary initial function $\phi \in E$.

We will use some comparison results for IDDE (9) by applying piecewise continuous Lyapunov functions and we introduce a class of Lyapunov-like functions:

Definition 4. *Let $a < b \le \infty$ be given numbers , $\Omega \subset \mathbb{R}^n$, $0 \in \Omega$. Then, the function $V : [a-r, b] \times \Omega \to [0,\infty)$ is from the class $P\Lambda([a-r,b],\Omega)$ if:*

- *$V \in C([a,b]/\{t_k\} \times \Omega, [0,\infty))$ and it is Lipschitz with respect to the second argument;*
- *For any $t_k \in (a,b)$, $x \in \Omega$, there exist finite limits $V(t_k - 0, x) = \lim_{t\uparrow t_k} V(t,x)$ and $V(t_k + 0, x) = \lim_{t\downarrow t_k} V(t,x)$.*

The comparison scalar equation (IDE) is

$$\begin{aligned} &({}_{t_k}^{C}\mathcal{D}^{\alpha,\rho}u)(t) = -\lambda u(t), \quad \text{for } t \in (t_k, t_{k+1}],\ k = 0,1,2,\dots, \\ &u(t) = \Xi_k(u(t_k - 0)) \text{ for } k = 1,2,\dots, \\ &u(t_0) = u_0. \end{aligned} \tag{34}$$

According to Lemma 4, the solution of the IVP for IDE (34) is given by

$$u(t) = \begin{cases} u_0 e^{\frac{\rho-1}{\rho}(t-t_0)} E_\alpha(-\lambda(\frac{t-t_0}{\rho})^\alpha) & t \in [t_0, t_1] \\ \Xi_k(u(t_k-0)) e^{\frac{\rho-1}{\rho}(t-t_k)} E_\alpha(-\lambda(\frac{t-t_k}{\rho})^\alpha) & t \in (t_k, t_{k+1}],\ k = 1,2,\dots. \end{cases}$$

The auxiliary Lemma, corresponding to Lemma 6, reduces to

Lemma 7. *Suppose:*

1. *The function* $x^*(t) = x(t; t_0, \phi) \in PC^{\alpha,\rho}([t_0, \infty), \Delta)$ *is a solution of the IDDE* (7) *and* (9) *where* $\Delta \subset \mathbb{R}^n$.
2. *The functions* $\Xi_k \in C(\mathbb{R}, \mathbb{R})$ *and* $\Xi_k(u) \leq u$ *for* $u \geq 0$, $k = 1, 2, \ldots$.
3. *The function* $V \in P\Lambda([t_0 - r, \infty), \Delta)$ *and*
   (i) *For any* $t \in (t_k, t_{k+1}]$ *with* $k = 0, 1, \ldots,$ *such that*

$$V(t, x^*(t)) \frac{e^{\frac{1-\rho}{\rho}(t-t_k)}}{E_\alpha\Big(-\lambda\Big(\frac{(t-t_k)}{\rho}\Big)^\alpha\Big)} \geq \sup_{s\in[t-r,t]\cap[t_k,t]} \frac{e^{\frac{1-\rho}{\rho}(s-t_k)}}{E_\alpha\Big(-\lambda\Big(\frac{(s-t_k)}{\rho}\Big)^\alpha\Big)} V(s, x^*(s)) \tag{35}$$

   the inequality

$${}^{C}_{t_k}\mathcal{D}^{\alpha,\rho} V(t, x^*(t)) \leq -\lambda V(t, x^*(t))$$

   holds where $\lambda > 0$ *is a given number.*
   (ii) *For any* $k = 1, \ldots,$ *the inequalities*

$$V(t_k - 0, \Psi_k(x^*(t_k - 0))) \leq \Xi_k(V(t_k - 0, x^*(t_k - 0))),$$

   hold.

Then, the inequality

$$V(t, x^*(t)) \leq \Big(\max_{s\in[-r,0]} V(t_0 + s, \phi(s))\Big) e^{\frac{\rho-1}{\rho}(t-t_k)} E_\alpha(-\lambda(\frac{t-t_k}{\rho})^\alpha), \tag{36}$$

$t \in (t_k, t_{k+1}]$, $k = 0, 1, \ldots,$

holds.

Remark 19. *The comparison scalar Equation* (34) *is chosen such that its explicit solution is known and condition 3(i) will be satisfied for the Lyapunov function.*

Theorem 2. *Suppose:*

1. *Conditions (A 2.1)–(A 2.3) are satisfied.*
2. *There exists a function* $V \in P\Lambda([t_0 - r, \infty), \mathbb{R}^n)$ *such that*
   (i) *There exist positive constants* A, B, a, b *such that the inequalities* $A||x||^a \leq V(t, x) \leq B||x||^{ab}$, $t \geq t_0$, $x \in \mathbb{R}^n$ *hold.*
   (ii) *For any point* $t \in (t_k, t_{k+1}]$ *with* $k = 0, 1, 2, \ldots$ *and any function* $\psi \in C^{\alpha,\rho}(t_k, [t - r, t], \mathbb{R}^n)$ *such that* $\Big({}^{C}_{t_k}\mathcal{D}^{\alpha,\rho}\psi\Big)(t) = f(t, \psi_t)$ *and*

$$V(t, \psi(t)) \frac{e^{\frac{1-\rho}{\rho}(t-t_k)}}{E_\alpha\Big(-D\Big(\frac{(t-t_k)}{\rho}\Big)^\alpha\Big)} \geq \sup_{s\in[t-r,t]\cap[t_k,t]} \frac{e^{\frac{1-\rho}{\rho}(s-t_k)}}{E_\alpha\Big(-D\Big(\frac{(s-t_k)}{\rho}\Big)^\alpha\Big)} V(s, \psi(s)) \tag{37}$$

the inequality

$$ {}_{t_k}^{C}\mathcal{D}^{\alpha,\rho}V(t,\psi(t)) \le -DV(t,\psi(t))r \tag{38}$$

holds where $D > 0$ is a given number.

(iii) For any $k = 1,2,\dots$ and $u \in \mathbb{R}^n$ the inequalities

$$V(t,\Psi_k(u)) \le C||u||^a \ \ for\ t \in (t_k, t_{k+1}].$$

hold where $C \in (0, A]$.

Then, the zero solution of IDDE (9) with the zero initial function is generalized proportional Mittag–Leffler stable with $C = \sqrt[a]{\frac{B}{A}}, \beta = b,\ \lambda = D,\ \gamma = \frac{1}{a}$.

Now we will provide an example illustrating the application of the given above sufficient conditions. To be able to compare both cases about non-instantaneous impulses and instantaneous impulses we will consider the scalar IVP for NIDDE (30) with appropriate changes.

Example 2. . *Consider the scalar IVP for IDDE*

$$\begin{aligned}
&\left({}_{t_k}^{C}\mathcal{D}^{\alpha,\rho}x\right)(t) = -\frac{2+t}{t+1}(x(t) - 0.5x_t^{(k)}) \quad for\ t \in (t_k, t_{k+1}],\ k = 0,1,2,\dots,\\
&x(t_k+0) = 0.5(\sin t_k)x(t_k-0) \quad for\ \ k = 1,2,\dots,\\
&x(t_0+s) = \phi(s),\ s \in [-r,0].
\end{aligned} \tag{39}$$

The scalar IVP for IDDE (39) with $\phi(s) \equiv 0$ has a zero solution.

Let $V(t,x) = x^2$. Thus, the condition 2(i) of Theorem 2 is satisfied with $A = 0.25, B = 1, a = 2, b = 1$.

Let k be a given natural number and $t \in (t_k, t_k+1)$, and the function $\psi \in C^{\alpha,\rho}(t_k,[t-r,t],\mathbb{R})$ be such that

$$\left({}_{t_k}^{C}\mathcal{D}^{\alpha,\rho}\psi\right)(t) = -\frac{2+t}{t+1}(\psi(t) - 0.5\psi_t^{(k)})$$

and

$$\psi^2(t) \ge \sup_{s\in[t-r,t]\cap[t_k,t]} \frac{e^{\frac{1-\rho}{\rho}(s-t_k)}}{E_\alpha\left(-\left(\frac{(s-t_k)}{\rho}\right)^\alpha\right)}\psi^2(s).$$

Then, we obtain $\left({}_{t_k}^{C}\mathcal{D}^{\alpha,\rho}\psi^2\right)(t) < -V(t,\psi(t))$ (see (32)), i.e., condition 2(ii) of Theorem 2 is satisfied with $D = 1$.

For any $k = 1,2,\dots$ we obtain $(0.5\sin t_k\ u)^2 \le 0.25u^2 = 0.25|u|^2$, i.e., the condition 2(iii) of Theorem 2 is satisfied with $C = 0.25$.

According to Theorem 2, the zero solution of the scalar IDDE (39) is a generalized proportional Mittag–Leffler stable with $C = 2, \beta = 1,\ \lambda = 1,\ \gamma = 0.5$, i.e., the inequality

$$||x(t;t_0,\phi)|| \le 2||\phi||_0\sqrt{e^{\frac{\rho-1}{\rho}(t-t_i)}E_\alpha(-(\frac{t-t_k}{\rho})^\alpha)},\ \ t \in (t_k, t_{k+1}],\ k = 0,1,\dots$$

holds (compare with the special case $t_{k+1} = s_k,\ k = 0,1,2,\dots$ of Example 1) .

4.3. No Impulses

As mentioned in Remark 11 the case of instantaneous impulses could be considered as a generalization of the case of no impulses, i.e., the system (10) could be considered as a partial case of (9) with $t_i = t_0,\ i = 1,2,\dots$. That is why we can translate the results from the previous section to the case without impulses.

Definition 5. *The zero solution of the system DDE* (10) *(with* $\phi \equiv 0$*) is said to be* generalized proportional Mittag-Leffler stable *if there exist constants* $\beta, \gamma, C, \lambda > 0$ *such that the inequality*

$$||x(t;t_0,\phi)|| \leq C||\phi||_0^\beta \left(e^{\frac{\rho-1}{\rho}(t-t_0)}E_\alpha(-\lambda(\frac{t-t_0}{\rho})^\alpha)\right)^\gamma, \ t \geq t_0, \tag{40}$$

holds, where $x(t;t_0,\phi)$ *is a solution on the IVP for DDE* (7) *and* (10).

Remark 20. *In the case* $\rho = 1$, *Definition* 5 *is the same as in* [26].

We will use some comparison results for DDE (10) by applying Lyapunov functions:

Definition 6. *Let* $a < b \leq \infty$ *be given numbers* , $\Omega \subset \mathbb{R}^n$, $0 \in \Omega$. *Then, the function* $V : [a-r,b] \times \Omega \to [0,\infty)$ *is from the class* $\Lambda([a-r,b],\Omega)$ *if* $V \in C([a,b]/\{t_k\} \times \Omega, [0,\infty))$ *and it is Lipschitz with respect to the second argument.*

The comparison scalar equation (DE) is

$$\begin{aligned} ({}_{t_0}^C\mathcal{D}^{\alpha,\rho}u)(t) &= -\lambda u(t), \quad \text{for } t > t_0, \\ u(t_0) &= u_0. \end{aligned} \tag{41}$$

According to Lemma 4, the solution of the IVP for DE (41) is given by $u(t) = u_0 e^{\frac{\rho-1}{\rho}(t-t_0)} E_\alpha(-\lambda(\frac{t-t_0}{\rho})^\alpha)$. $t \geq t_0$.

The auxiliary Lemma, corresponding to Lemma 6 reduces to

Lemma 8. *Suppose:*

1. *The function* $x^*(t) = x(t;t_0,\phi) \in C^{\alpha,\rho}([t_0,\infty),\Delta)$ *is a solution of the DDE* (7) *and* (10), *where* $\Delta \subset \mathbb{R}^n$.
2. *The function* $V \in C\Lambda([t_0-r,\infty),\Delta)$ *and for any point* $t > t_0$ *such that*

$$\begin{aligned} &V(t,x^*(t))\frac{e^{\frac{1-\rho}{\rho}(t-t_0)}}{E_\alpha\left(-\lambda\left(\frac{(t-t_0)}{\rho}\right)^\alpha\right)} \\ &\geq \sup_{s\in[t-r,t]\cap[t_0,t]} \frac{e^{\frac{1-\rho}{\rho}(s-t_0)}}{E_\alpha\left(-\lambda\left(\frac{(s-t_0)}{\rho}\right)^\alpha\right)}V(s,x^*(s)) \end{aligned} \tag{42}$$

the inequality

$$\left({}_{t_0}^C\mathcal{D}^{\alpha,\rho}V(t,x^*(t))\right) \leq -\lambda V(t,x^*(t))$$

holds where $\lambda > 0$ *is a given number.*

Then, the inequality

$$V(t,x^*(t)) \leq \max_{s\in[-r,0]} V(t_0+s,\phi(s))e^{\frac{\rho-1}{\rho}(t-t_0)}E_\alpha(-\lambda(\frac{t-t_0}{\rho})^\alpha), \ t > t_0$$

holds.

Theorem 3. *Suppose:*

1. *Conditions (A 3.1), (A 3.2) are satisfied.*
2. *There exists a function* $V \in \Lambda([t_0-r,\infty),\mathbb{R}^n)$ *such that*
   (i) *There exist positive constants* A, B, a, b *such that* $C \leq A$ *and the inequalities* $A||x||^a \leq V(t,x) \leq B||x||^{ab}$, $t \geq t_0$, $x \in \mathbb{R}^n$ *hold.*

(ii) *For any point $t > t_0$ and any function $\psi \in C^{\alpha,\rho}(t_0,[t-r,t],\mathbb{R}^n)$ such that $\left({}_{t_0}^{C}\mathcal{D}^{\alpha,\rho}\psi\right)(t) = f(t,\psi_t)$ and*

$$V(t,\psi(t))\frac{e^{\frac{1-\rho}{\rho}(t-t_0)}}{E_\alpha\left(-\lambda\left(\frac{(t-t_0)}{\rho}\right)^\alpha\right)} \geq \sup_{s\in[t-r,t]\cap[t_0,t]} \frac{e^{\frac{1-\rho}{\rho}(s-t_k)}}{E_\alpha\left(-\lambda\left(\frac{(s-t_0)}{\rho}\right)^\alpha\right)} V(s,\psi(s)) \tag{43}$$

the inequality

$${}_{t_0}^{C}\mathcal{D}^{\alpha,\rho}V(t,\psi(t)) \leq -DV(t,\ \psi(t)) \tag{44}$$

holds where $D > 0$ is a given number.

Then, the zero solution of DDE (10) with the zero initial function is generalized proportional Mittag–Leffler stable with constants $C = \sqrt[a]{\frac{B}{A}}, \beta = b,\ \lambda = D,\ \gamma = \frac{1}{a}$.

Example 3. *Consider the scalar IVP for DDE*

$$\begin{aligned}\left({}_{t_0}^{C}\mathcal{D}^{\alpha,\rho}x\right)(t) &= -\frac{2+t}{t+1}(x(t) - 0.5\sup_{s\in[-r,0]} x(t+s)),\ \ t > t_0,\\ x(t_0+s) &= \phi(s),\ s\in[-r,0].\end{aligned} \tag{45}$$

The scalar IVP for DDE (45) with $\phi(s) \equiv 0$ has a zero solution.

Let $V(t,x) = x^2$. Thus, the condition 2(i) of Theorem 3 is satisfied with $A = 0.25, B = 1, a = 2, b = 1$.

Let $t > t_0$ and the function $\psi \in C^{\alpha,\rho}(t_0,[t-r,t],\mathbb{R})$ be such that $\left({}_{t_0}^{C}\mathcal{D}^{\alpha,\rho}\psi\right)(t) = -\frac{2+t}{t+1}(\psi(t) - 0.5\sup_{s\in[-r,0]}\psi(t+s)$ and $\psi^2(t) \geq \sup_{s\in[t-r,t]\cap[t_0,t]} \frac{e^{\frac{1-\rho}{\rho}(s-t_0)}}{E_\alpha\left(-\left(\frac{(s-t_0)}{\rho}\right)^\alpha\right)}\psi^2(s)$. Then, we obtain

$$\left({}_{t_0}^{C}\mathcal{D}^{\alpha,\rho}\psi^2\right)(t) < -V(t,\psi(t))$$

(see (32)), i.e., condition 2(ii) of Theorem 3 is satisfied with $D = 1$.

According to Theorem 3, the zero solution of the scalar DDE (45) is generalized proportional Mittag–Leffler stable with $C = 2, \beta = 1,\ \lambda = 1,\ \gamma = 0$, i.e., the inequality

$$||x(t;t_0,\phi)|| \leq 2||\phi||_0\sqrt{e^{\frac{\rho-1}{\rho}(t-t_0)}E_\alpha(-(\frac{t-t_0}{\rho})^\alpha)},\ \ t \geq t_0.$$

holds (compare with the special case of $t_0 = t_k,\ k = 1,2,\dots$ of Example 2).

5. Conclusions

In this paper, a system of nonlinear differential equations with finite delay and with a generalized proportional Caputo fractional derivative is studied. The basic cases are presented: the case when there are non-instantaneous impulses in the equations, the case when there are instantaneous impulses in the equations, and the case without any impulses in all equations. The appropriate initial value problem is set up in all these cases, and the relation between them is discussed. It is shown that the case of non-instantaneous impulses is a generalization of the case of instantaneous impulses, and the case of instantaneous impulses could be considered as a generalization of the case without any impulses. These statements could be applied to study various qualitative properties of the solutions. In this paper, based on the application of Lyapunov functions and an appropriate modification of the Razumikhin method, the Mittag–Leffler type stability is investigated.

Author Contributions: Conceptualization, R.P.A., S.H. and D.O.; Formal analysis, R.P.A., S.H. and D.O.; Investigation, R.P.A., S.H. and D.O.; Methodology, R.P.A., S.H. and D.O.; Supervision, R.P.A., S.H. and D.O.; Writing—original draft, R.P.A., S.H. and D.O.; Writing—review and editing, R.P.A., S.H. and D.O.. All authors have read and agreed to the published version of the manuscript.

Funding: This research was funded by the Bulgarian National Science Fund under Project KP-06-N32/7.

Data Availability Statement: Not applicable.

Conflicts of Interest: The authors declare no conflict of interest.

References

1. Qiu, W.; Xu, D.; Guo, J. Numerical solution of the fourth-order partial integro-differential equation with multi-term kernels by the Sinc-collocation method based on the double exponential transformation. *Appl. Math. Comput.* **2021**, *392*, 125693. [CrossRef]
2. Xu, D.; Qiu, W.; Guo, J. A compact finite difference scheme for the fourth-order time-fractional integro-differential equation with a weakly singular kernel. *Numer. Meth. Part. Differ. Equ.* **2020**, *36*, 439–458. [CrossRef]
3. Kochubei, A.N. General fractional calculus, evolution equations, and renewal processes. *Integr. Equ. Oper. Theory* **2011**, *71*, 583–600. [CrossRef]
4. Luchko, Y. General Fractional Integrals and Derivatives with the Sonine Kernels. *Mathematics* **2021**, *9*, 594. [CrossRef]
5. Luchko, Y. Operational calculus for the general fractional derivatives with the Sonine kernels. *Fract. Calc. Appl. Anal.* **2021**, *24*, 338–375. [CrossRef]
6. Momani, S.; Hadid, S. Lyapunov stability solutions of fractional integrodifferential equations. *Intern. J. Math. Math. Sci.* **2004**, *47*, 2503–2507. [CrossRef]
7. Zhang, L.G.; Li, J.M.; Chen, G.P. Extension of Lyapunov second method by fractional calculus. *Pure Appl. Math.* **2005**, *21*, 291–294.
8. Chen, Y.Q. Ubiquitous fractional order controls? *IFAC Proc. Vol.* **2006**, *39*, 481–492. [CrossRef]
9. Li, Y.; Chen, Y.Q.; Podlubny, I. Mittag-Leffler stability of fractional order nonlinear dynamic systems. *Automatica* **2009**, *45*, 2965–2969. [CrossRef]
10. Abed-Elhameed, T.M.; Aboelenen, T. Mittag–Leffler stability, control, and synchronization for chaotic generalized fractional-order systems. *Adv. Cont. Discr. Mod.* **2022**, *50*. [CrossRef]
11. Stamova, I. Mittag-Leffler stability of impulsive differential equations of fractional order. *Quart. Appl. Math.* **2015**, *73*, 525–535. [CrossRef]
12. Yang, X.; Li, C.; Huang, T.; Song, Q. Mittag–Leffler stability analysis of nonlinear fractional-order systems with impulses. *Appl. Math. Comput.* **2017**, *293*, 416–422. [CrossRef]
13. Gao, Y. Mittag-Leffler stability for a new coupled system of fractional-order differential equations on network. *Adv. Differ. Equ.* **2018**, *121*. [CrossRef]
14. Sene, N. Mittag-Leffler input stability of fractional differential equations and its applications.*Discr. Cont. Dynam. Syst.-S* **2020**, *13*, 867–880. [CrossRef]
15. Tatar, N. Mittag—Leffler stability for a fractional Euler—Bernoulli problem. *Chaos Solitons Fractals* **2021**, *149*, 1110777. [CrossRef]
16. Moharramnia, A.; Eghbali, N.; Rassias, J.M. Mittag-Leffler-Hyers-Ulam stability of Prabhakar fractional integral equation. *Int. J. Nonlinear Anal. Appl.* **2021**, *12*, 25–33.
17. Wu, A.; Liu, L.; Huang, T.; Zeng, Z. Mittag-Leffler stability of fractional-order neural networks in the presence of generalized piecewise constant arguments. *Neural Netw.* **2017**, *85*, 118–127. [CrossRef]
18. Lakshmikantham, V.; Bainov, D.D.; Simeonov, P.S. *Theory of Impulsive Differential Equations*; World Scientific: Singapore, 1989.
19. Agarwal, R.; Hristova, S.; O'Regan, D. *Non-Instantaneous Impulses in Differential Equations*; Springer: Berlin/Heidelberg, Germany, 2017.
20. Jarad, F.; Abdeljawad, T.; Alzabut, J. Generalized fractional derivatives generated by a class of local proportional derivatives. *Eur. Phys. J. Spec. Top.* **2017**, *226*, 3457–3471. [CrossRef]
21. Jarad, F.; Abdeljawad, T. Generalized fractional derivatives and Laplace transform. *Discret. Contin. Dyn. Syst. Ser. S* **2020**, *13*, 709–722. [CrossRef]
22. Bohner, M.; Hristova, S. Stability for generalized Caputo proportional fractional delay integro-differential equations. *Bound Value Probl.* **2022**, *14*, 14. [CrossRef]
23. Almeida, R.; Agarwal, R.P.; Hristova, S.; O'Regan, D. Quadratic Lyapunov functions for stability of generalized proportional fractional differential equations with applications to neural networks. *Axioms* **2021**, *10*, 322. [CrossRef]
24. Agarwal, R.; Hristova, S.; O'Regan, D. Generalized Proportional Caputo Fractional Differential Equations with Noninstantaneous Impulses: Concepts, Integral Representations, and Ulam-Type Stability. *Mathematics* **2022**, *10*, 2315. [CrossRef]
25. Agarwal, R.; Hristova, S.; O'Regan, D. Mittag-Leffler stability for non-instantaneous impulsive Caputo fractional differential equations with delays. *Math. Slovaca* **2019**, *69*, 583–598. [CrossRef]
26. Sadati, S.J.; Baleanu, D.; Ranjbar, A.; Ghaderi, R.; Abdeljawad, T. Mittag–Leffler stability theorem for fractional nonlinear systems with delay. *Abstract Appli. Anal.* **2010**, *2010*, 108651. [CrossRef]

symmetry MDPI

Article

Fractional Integrals Associated with the One-Dimensional Dunkl Operator in Generalized Lizorkin Space

Fethi Bouzeffour

Department of Mathematics, College of Sciences, King Saud University, P.O. Box 2455, Riyadh 11451, Saudi Arabia; fbouzaffour@ksu.edu.sa

Abstract: This paper explores the realm of fractional integral calculus in connection with the one-dimensional Dunkl operator on the space of tempered functions and Lizorkin type space. The primary objective is to construct fractional integral operators within this framework. By establishing the analogous counterparts of well-known operators, including the Riesz fractional integral, Feller fractional integral, and Riemann–Liouville fractional integral operators, we demonstrate their applicability in this setting. Moreover, we show that familiar properties of fractional integrals can be derived from the obtained results, further reinforcing their significance. This investigation sheds light on the utilization of Dunkl operators in fractional calculus and provides valuable insights into the connections between different types of fractional integrals. The findings presented in this paper contribute to the broader field of fractional calculus and advance our understanding of the study of Dunkl operators in this context.

Keywords: Dunkl theory; fractional Integral; Bessel functions

MSC: 42B30; 33C52; 33C67; 33D67; 33D80; 35K08; 42B25; 42C05

Citation: Bouzeffour, F. Fractional Integrals Associated with the One-Dimensional Dunkl Operator in Generalized Lizorkin Space. *Symmetry* **2023**, *15*, 1725. https://doi.org/10.3390/sym15091725

Academic Editors: Francisco Martínez González and Mohammed K. A. Kaabar

Received: 30 August 2023
Revised: 5 September 2023
Accepted: 6 September 2023
Published: 8 September 2023

1. Introduction

On the real line, for a positive real number κ, the Dunkl operator $\mathscr{D}_\kappa$ provides a one-parameter deformation of the ordinary derivative $\frac{d}{dx}$. It is defined as:

$$\mathscr{D}_\kappa := \frac{d}{dx} + \frac{\kappa}{x}(1-s), \tag{1}$$

where s is the reflection operator acting on a function $f(x)$ of a real variable x as $sf(x) := f(-x)$. The Dunkl operator incorporates the additional term $\frac{\kappa}{x}(1-s)$, which accounts for reflection symmetry and introduces a dependence on the parameter κ. This operator plays a fundamental role in generalizing various classical results in harmonic analysis and approximation theory, as explored in the works of Dunkl [1,2] Trimeche [3], de Jeu [4], Rosler [5–7], and others.

Fractional calculus [8–15] has gained significant importance in recent decades as a powerful tool for developing advanced mathematical models involving fractional differential and integral operators. When applied to the Dunkl operator, fractional calculus offers a fresh perspective by incorporating the effects of reflection and asymmetry within the underlying space.

A notable feature of the Dunkl setting is the existence of a natural Riesz transform, which shares similarities with classical singular integrals. In the multidimensional case, S. Thangavelu and Y. Xu [16,17] established the L^p-boundedness of the associated Riesz transform. This study was further extended by Amri and Sifi [18], who considered the general case for $1 < p < \infty$. Additionally, investigations into singular integrals and multipliers were carried out in [18–22]. These contributions have significantly enriched our understanding of the Dunkl operator and its associated Riesz transform.

In this study, our main focus is on the comprehensive exploration of the one-dimensional fractional Dunkl integral within Lizorkin type spaces [10–12], with a specific emphasis on analytic continuation techniques. The obtained operators go beyond the conventional Riesz fractional integral [9] and Feller fractional integral [8,11], as they are specifically tailored to operate within the Dunkl setting. By extending the applicability of these operators to the Dunkl context, we aim to unlock new possibilities and gain deeper insights into the realm of fractional calculus.

To address the challenges posed by the divergence of fractional Dunkl operators, we adopt a unique approach that incorporates the regularization technique for divergent integrals, inspired by the work described in the book by Samko [11,12]. Our methodology involves utilizing specific segments of the Taylor formula associated with the Dunkl operator, as originally formulated by Mourou [23]. This regularization technique plays a pivotal role in extending the fractional integral operators to the domain of $\Re(\alpha) > 0$. As a result, we introduce an alternative normalization scheme for tempered power functions, offering a fresh and insightful perspective on fractional calculus within the Dunkl setting. It is important to note that while Soltani [24] relies on the conventional Taylor series, our approach, based on the Taylor formula of Mourou [23], better suits the specific requirements of the Dunkl operator.

Our paper is organized as follows: In Section 2, we begin by collecting some essential facts about the Dunkl operator and the Lizorkin space. Section 3 focuses on studying the generalized power function and its analytic continuation. Moving on to Section 4, we dedicate that section to the study of extensions of well-known fractional integrals such as the Riesz fractional integral, the Feller fractional integral, and the Weyl fractional integral.

2. Preliminaries

In this section, we introduce some notations and gather some facts about the one-dimensional Dunkl operator.

2.1. The One-Dimensional Dunkl Operator

Let $\kappa \geq 0$, and f be a differentiable function on $\mathbb{R}$. The Dunkl derivative $\mathscr{D}_\kappa f(x)$ is defined by

$$\mathscr{D}_\kappa f(x) = \begin{cases} f'(x) + \kappa \frac{f(x)-f(-x)}{x}, & \text{if} \quad x \neq 0, \\ (2\kappa+1) f'(0), & \text{if} \quad x = 0. \end{cases} \tag{2}$$

We denote by $L^p_\kappa(\mathbb{R})$ $(1 \leq p)$, the Lebesgue space associated with the measure

$$\sigma_\kappa(dx) = \frac{|x|^{2\kappa}}{2^{\kappa+1/2}\Gamma(\kappa+1/2)} dx \tag{3}$$

and by $\|f\|_{\kappa,p}$ the usual norm given by

$$\|f\|_{\kappa,p} = \left(\int_{\mathbb{R}} |f(x)|^p \, \sigma_\kappa(dx) \right)^{1/p}. \tag{4}$$

Now, consider the so-called *nonsymmetric Bessel function*, also called *Dunkl type Bessel function*, in the rank one case (see [25]) [§10.22(v)]:

$$\mathscr{E}_\kappa(x) := \mathscr{J}_{\kappa-1/2}(ix) + \frac{x}{2\kappa+1} \mathscr{J}_{\kappa+1/2}(ix). \tag{5}$$

where the normalized Bessel functions is defined by

$$\begin{aligned} \mathscr{J}_\kappa(x) &:= \Gamma(\kappa+1)\,(2/x)^\kappa J_\kappa(x) \\ &= \sum_{n=0}^{\infty} \frac{(-1)^n}{n!\Gamma(\kappa+n+1)} (\frac{x}{2})^{2n+\kappa}, \quad x > 0. \end{aligned}$$

It is evident to the reader that the Dunkl kernel $\mathcal{E}_\kappa(i\lambda x)$ coincides with the exponential function when the parameter κ is equal to zero, i.e., $\mathcal{E}_0(i\lambda x) = e^{i\lambda x}$. This function also has a close connection with the Wright function.

$$\mathcal{E}_\kappa(x) = \Gamma(\kappa + 1/2)\Big[W_{1,\kappa+1/2}(\frac{x^2}{4}) + \frac{x}{2}\,W_{1,\kappa+3/2}(\frac{x^2}{4})\Big], \tag{6}$$

where the Wright function is defined by the series representation, valid in the whole complex plane [26]

$$W_{\alpha,\beta}(z) := \sum_{n=0}^{\infty} \frac{z^n}{n!\Gamma(\alpha n + \beta)}, \quad \alpha > -1, \quad \beta \in \mathbb{C}. \tag{7}$$

The Wright function provides a powerful tool for dealing with fractional calculus problems, as it allows for the analysis of fractional differential and integral equations in a unified framework, see [26,27].

The function $\mathscr{E}_k(i\xi x)$ satisfies the following eigenvalue problem

$$\mathscr{D}_\kappa(\mathscr{E}_k(i\xi x)) = i\,\xi\,\mathscr{E}_\kappa(i\xi x), \quad \mathcal{E}_k(0)) = 1 \tag{8}$$

and has the *Laplace* representation

$$\mathscr{E}_\kappa(ix) = \frac{\Gamma(\kappa+1/2)}{\Gamma(1/2)\Gamma(\kappa)} \int_{-1}^{1} e^{tx}(1-t)^{\kappa-1}(1+t)^{\kappa}\,dt. \tag{9}$$

The Dunkl transform is defined by [1,3,4]

$$(\mathscr{F}_\kappa f)(\lambda) := \int_{-\infty}^{\infty} f(x)\,\mathcal{E}_\kappa(-i\lambda x)\,\sigma_\kappa(dx). \tag{10}$$

The *Dunkl* transform can be extended to an isometry of $L^2_\kappa(\mathbb{R})$, that is,

$$\int_{\mathbb{R}} |f(x)|^2\,\sigma_\kappa(dx) = \int_{\mathbb{R}} |\widehat{f}_\kappa(\lambda)|^2\,\sigma_\kappa(d\lambda). \tag{11}$$

For any $f \in L^1_\kappa(\mathbb{R}) \cap L^2_\kappa(\mathbb{R})$, the inverse is given by

$$f(x) = \int_{\mathbb{R}} \widehat{f}_\kappa(\lambda)\,\mathcal{E}_\kappa(i\lambda x)\,\sigma_\kappa(d\lambda). \tag{12}$$

As in the classical case, a generalized translation operator was defined in the Dunkl setting side on $L^2_\kappa(\mathbb{R})$ by Trimèche [3]

$$\mathcal{F}_\kappa\{\tau^y f(x);\xi\} := \mathcal{E}_\kappa(i\xi y)\mathcal{F}_\kappa\{f(x);\xi\}, \quad y,\,\xi \in \mathbb{R}. \tag{13}$$

We also define the Dunkl convolution product for suitable functions f and g by

$$f * g(x) = \int_{\mathbb{R}} \tau^{-x} f(y) g(y) \sigma_\kappa(dy).$$

Explicitly, the generalized translation $\tau^x f(y)$ takes the explicit form (see [28] Theorem 6.3.7):

$$\begin{aligned}\tau^x f(y) :=& \frac{1}{2}\int_{-1}^{1} f(\sqrt{x^2+y^2-2xyt})(1+\frac{x-y}{\sqrt{x^2+y^2-2xyt}})h_k(t)dt \\ +& \frac{1}{2}\int_{-1}^{1} f(-\sqrt{x^2+y^2-2xyt})(1-\frac{x-y}{\sqrt{x^2+y^2-2xyt}})\,h_k(t)\,dt,\end{aligned} \tag{14}$$

where

$$h_\kappa(t) = \frac{\Gamma(\kappa+1/2)}{2^{2\kappa}\sqrt{\pi}\Gamma(\kappa)}(1+t)(1-t^2)^{\kappa-1}.$$

2.2. The Generalized Lizorkin Space

For a comprehensive treatment of the standard Lizorkin space, we recommend referring to the book [12] §2, where the authors provide a detailed and in-depth analysis of this topic. Additionally, the study of the generalized Lizorkin space has been carried out by Soltani [24]. While we cannot provide a detailed overview of the entire subject here, we can highlight some important points for clarity.

We denote by $S(\mathbb{R})$ the Schwartz space, which is the space of C^∞-functions on $\mathbb{R}$ which are rapidly decreasing as well as their derivatives, endowed with the topology defined by the seminorms

$$\|f\|_{n,m} = \sup_{x\in\mathbb{R},\, j\le m} (1+x^2)^n \mathscr{D}_\kappa^j \varphi(x), \quad n,\, m \in \mathbb{N},$$

It is not difficult to check that

$$\mathscr{D}f(x) = f'(x) + \kappa \int_{-1}^{1} f'(xt)dt.$$

From this representation, we see that the operator $\mathscr{D}$ leaves $S(\mathbb{R})$ invariant.

In the context of distribution theory, the space $S'(\mathbb{R})$ denotes the topological dual of $S(\mathbb{R})$, which consists of generalized functions, also known as tempered distributions. The value of a generalized function f as a functional on a test function $\varphi \in S(\mathbb{R})$ is denoted by (f, φ).

A generalized function is said to be κ-regular if there exists a locally integrable function f with respect to the measure $\sigma_\kappa(dx)$, such that the integral $\int_{\mathbb{R}} f(x)\varphi(x)\sigma_\kappa(dx)$ is finite for every $\varphi \in S(\mathbb{R})$. The action of the κ-regular generalized function f on a test function φ is denoted as (f, φ) or equivalently $\langle f, \varphi\rangle_\kappa$. Here, the integral on the right-hand side of the equation is denoted by $\langle f, \varphi\rangle_\kappa$. It is important to note that the measure $\sigma_\kappa(dx)$ depends on the specific context and properties of the Dunkl operators. The notation and definitions provided above establish a general framework for understanding κ-regular generalized functions and their evaluation on test functions.

The Dunkl transform is a powerful mathematical tool that acts as a topological isomorphism between the Schwartz space $S(\mathbb{R})$ and itself. This transform extends naturally to generalized functions by considering the Dunkl transform of a generalized function $f \in S'(\mathbb{R})$. The definition of the Dunkl transform for generalized functions can be expressed using duality as follows: for any $\varphi \in S(\mathbb{R})$, the pairing between the Dunkl transform of f and φ is given by

$$(\mathscr{F}_\kappa f,\, \varphi) = (f,\, \mathscr{F}_\kappa \varphi), \quad \varphi \in S(\mathbb{R}).$$

In terms of integral notation, it can be written as:

$$\int_{\mathbb{R}} (\mathscr{F}_\kappa f)(x)\varphi(x)\sigma\kappa(dx) = \int_{\mathbb{R}} f(x)\mathscr{F}_\kappa\varphi(x)\sigma\kappa(dx), \quad \varphi \in S(\mathbb{R}), \tag{15}$$

provided f and $\mathscr{F}_\kappa f$ are κ-regular.

The space $S(\mathbb{R})$ itself is not invariant under multiplication by power functions. However, we can define an invariant subspace by utilizing the Dunkl transforms. This leads us to the set $\Psi_\kappa(\mathbb{R})$ consisting of functions $\varphi \in S(\mathbb{R})$ that satisfy the conditions:

$$\mathscr{D}_\kappa^n \varphi(0) = 0, \quad \text{for } n = 0, 1, 2, \ldots,$$

where $\mathscr{D}_\kappa^n \varphi$ denotes the nth order Dunkl transform of φ. In other words, φ belongs to $\Psi_\kappa(\mathbb{R})$ if all the Dunkl transforms of φ evaluated at the origin are zero. By imposing these conditions, we construct a space of functions that possess certain transformation properties with respect to the Dunkl operators. The generalized Lizorkin space $\Phi_\kappa(\mathbb{R})$ is introduced as the *Dunkl* transform preimage of the space $\Psi_\kappa(\mathbb{R})$ in the space $S(\mathbb{R})$,

$$\Phi_\kappa(\mathbb{R}) = \Big\{\varphi \in S(\mathbb{R}) \,:\, \varphi = \mathscr{F}_\kappa(\psi),\ \psi \in \Psi_\kappa(\mathbb{R})\Big\}. \tag{16}$$

According to this definition, any function $\varphi \in \Phi_\kappa(\mathbb{R})$ satisfies the orthogonality conditions

$$\int_{\mathbb{R}} x^n\, \varphi(x)\sigma_\kappa(dx) = 0, \quad n = 0, 1, 2, \dots . \tag{17}$$

3. Regularization of Integrals with Power Singularity

In this section, we examine two types of power functions defined on the entire real line

- Even, $|x|^\alpha$;
- Odd, $\mathrm{sgn}(x)\,|x|^\alpha$; where

$$\mathrm{sgn}(x) := \begin{cases} 1, & \text{if } \ x > 0 \\ -1, & \text{if } \ x < 0. \end{cases}$$

Other types of tempered power functions can be defined as follows

$$x_\pm^\alpha = \frac{1}{2}\big[|x|^\alpha \pm |x|^\alpha \mathrm{sgn}(x)\big],$$
$$(\pm i\,x)^\alpha = |x|^\alpha\big(\cos(\pi\alpha/2) \pm i\,\mathrm{sgn}(x)\sin(\pi\alpha/2)\big).$$

These tempered power functions capture different aspects of fractional calculus and are used to generalize the concept of differentiation and integration to noninteger orders.

3.1. *Taylor–Dunkl Formula*

To facilitate the forthcoming discussion on analytic continuation, we begin by presenting an additional formula that proves to be valuable in the process.

Let $f \in C^\infty(\mathbb{R})$; for every $n \in \mathbb{N}$, we have [19]

$$\tau^y f(x) = \sum_{j=0}^{n-1} b_j(x)\mathscr{D}_\kappa^j f(x) + r_n(x, y; f), \tag{18}$$

where

$$\begin{cases} r_{j+1}(x,y;f) = \int_{-|y|}^{|y|} \big(\frac{\mathrm{sgn}(y)}{2|y|^{2\kappa}} + \frac{\mathrm{sgn}(u)}{2|u|^{2\kappa}}\big) r_j(x,u;\mathscr{D}_\kappa f)\,|u|^{2\kappa} du, \\ r_1(x,y;f) = \tau^y f(x) - f(x) \end{cases}$$

and

$$b_{j+1}(x) = \int_{-|y|}^{|y|} \big(\frac{\mathrm{sgn}(y)}{2|y|^{2\kappa}} + \frac{\mathrm{sgn}(u)}{2|u|^{2\kappa}}\big) b_j(u)\,|u|^{2\kappa} du, \quad b_0(x) = 1. \tag{19}$$

Then,

$$b_{2s}(x) = \frac{\Gamma(\kappa+1/2)}{\Gamma(\kappa+s+1/2)}\frac{x^{2s}}{s!}, \quad b_{2s+1}(x) = \frac{\Gamma(\kappa+1/2)}{\Gamma(\kappa+s+3/2)}\frac{x^{2s+1}}{s!}, \ s = 0, 1, 2, \dots .$$

From the work of Mourou [23], we can extract the following proposition, which provides a complete asymptotic expansion for $\tau_\kappa f(x)$ as x approaches a.

Lemma 1. *Let $f \in C^{\infty}(\mathbb{R})$ and $a \in \mathbb{R}$; then, one has*

$$\tau_{\kappa}^{a} f(x) \sim \sum_{s=0}^{\infty} b_s(x) \mathscr{D}_{\kappa}^{s} f(a), \quad as \quad x \to a, \tag{20}$$

3.2. Generalized Power Functions

By considering $|x|^{-\alpha}$ and $\text{sign}(x)|x|^{-\alpha}$ as elements of $\Psi'_{\kappa}(\mathbb{R})$, we recognize them as κ-regular generalized functions for all $\alpha \in \mathbb{C}$, that is,

$$\langle |x|^{-\alpha}, \varphi \rangle_{\kappa} = \int_{\mathbb{R}} \frac{1}{|x|^{\alpha}} \varphi(x) \sigma_{\kappa}(dx), \tag{21}$$

$$\langle \text{sgn}(x)|x|^{-\alpha}, \varphi \rangle_{\kappa} = \int_{\mathbb{R}} \frac{\text{sgn}(x)}{|x|^{\alpha}} \varphi(x) \sigma_{\kappa}(dx). \tag{22}$$

When considering the functions $|x|^{-\alpha}$ and $\text{sign}(x)|x|^{-\alpha}$ as elements of $S'(\mathbb{R})$ or $\Phi'_{\kappa}(\mathbb{R})$, they are not κ-regular if $\Re(\alpha) \geq 2\kappa + 1$. To handle these generalized functions, let $\alpha \in \mathbb{C}$ such that $\alpha \neq 2\kappa + 2s + 1$ for $s = 0, 1, 2, \ldots$. For $\varphi \in S(\mathbb{R})$, we can define the generalized power function $|x|^{-\alpha}$ as follows:

$$\begin{aligned}(|x|^{-\alpha}, \varphi) = & \int_{|x|<1} \frac{1}{|x|^{\alpha}} \left[\varphi(x) - \sum_{s=0}^{m} b_s(x) \mathscr{D}_{\kappa}^{s} \varphi(0) \right] \sigma_{\kappa}(dx) \\ & + \sum_{s=0}^{[\frac{m}{2}]} \frac{\mathscr{D}_{\kappa}^{2s} \varphi(0)}{2^{\kappa - 1/2} \Gamma(\kappa + s + 1/2), s!} \frac{1}{2\kappa + 2s + 1 - \alpha} \\ & + \int_{|x| \geq 1} \frac{\varphi(x)}{|x|^{\alpha}} \sigma_{\kappa}(dx),\end{aligned} \tag{23}$$

where $m > \text{Re}(\alpha) - 2\kappa - 1$. It is important to note that the right-hand side of Equation (23) does not depend on the choice of m as long as $m > \Re(\alpha) - 2\kappa - 1$. Since $\varphi \in S(\mathbb{R})$, Lemma 1 guarantees that

$$\varphi(x) - \sum_{s=0}^{m} b_s(x) \mathscr{D}_{\kappa}^{s} \varphi(0) = \mathcal{O}(x^{m+1}) \quad (\text{as} \quad x \to 0).$$

This property ensures the well-definedness of the expression. The mapping $\alpha \to (|x|^{-\alpha}, \varphi)$ from $\mathbb{C}$ to $S'(\mathbb{R})$ can be extended to a holomorphic function on $\mathbb{C} - \{2\kappa + 2s + 1 : s = 0, 1, 2, \ldots\}$, with simple poles at $\alpha = 2\kappa + 2s + 1$. The residues of the function at these poles are given by

$$\text{Res}\big((|x|^{-\alpha}, \varphi); 2\kappa + 2s + 1\big) = -\frac{2^{-\kappa + 1/2} \mathscr{D}_{\kappa}^{2s} \varphi(0)}{\Gamma(\kappa + s + 1/2)\, s!}. \tag{24}$$

When $\alpha = 2\kappa + 2s + 1$ with $s = 0, 1, 2, \ldots$, we define the even, tempered power function $|x|^{-2\kappa - 2s - 1}$ as

$$(|x|^{-2\kappa - 2s - 1}, \varphi) = \lim_{\alpha \to 2\kappa + 2s + 1} \left\{ (|x|^{-\alpha}, \varphi) + \frac{\mathscr{D}_{\kappa}^{2s} \varphi(0)}{2^{\kappa - 1/2} \Gamma(\kappa + s + 1/2)\, s!} \frac{1}{\alpha - 2\kappa - 2n - 1} \right\}. \tag{25}$$

This provides a definition for the even, tempered power $|x|^{-\alpha}$ for all $\alpha \in \mathbb{C}$.

Similarly, for $\alpha \in \mathbb{C}$ such that $\alpha \neq 2\kappa + 2s + 2$ with $s = 0,\ 1,\ 2 \ldots$, we define the odd tempered power function $|x|^{-\alpha}\text{sgn}(x)$ by

$$(\frac{\mathrm{sgn}(x)}{|x|^{\alpha}},\varphi)=\int_{|x|<1}\frac{\mathrm{sgn}(x)}{|x|^{\alpha}}\big[\varphi(x)-\sum_{s=0}^{m}b_s(x)\mathscr{D}_{\kappa}^{s}\varphi(0)\big]\,\sigma_{\kappa}(dx) \quad (26)$$
$$+\sum_{s=0}^{[\frac{m-1}{2}]}\frac{\mathscr{D}_{\kappa}^{2s+1}\varphi(0)}{2^{\kappa-1/2}\Gamma(\kappa+s+3/2)\,s!}\frac{1}{2\kappa+2s+2-\alpha}$$
$$+\int_{|x|\geq 1}\frac{\mathrm{sgn}(x)}{|x|^{\alpha}}\varphi(x)\,\sigma_{\kappa}(dx)\quad (m>\Re(\alpha)-2\kappa-2).$$

It follows that the mapping $\alpha\to(|x|^{-\alpha}\mathrm{sgn}(x),\,\varphi)$ is analytic on $\mathbb{C}-\{2\kappa+2s+2,$ $s=0,1,2,\dots\}$, with simple poles at $\alpha=2\kappa+2s+2$ and

$$\mathrm{Res}\big((|x|^{-\alpha}\mathrm{sgn}(x),\,\varphi);2\kappa+2s+2\big)=-\frac{2^{-\kappa+1/2}\mathscr{D}_{\kappa}^{2s+1}\varphi(0)}{\Gamma(\kappa+s+3/2)\,s!}.$$

For $\alpha=2\kappa+2s+2$, with $s=0,\,1,\,2,\,\dots$, we define the odd, tempered powers function $\mathrm{sgn}(x)|x|^{-2\kappa-2s-2}$ as

$$(\mathrm{sgn}(x)|x|^{-2\kappa-2s-2},\,\varphi)=\lim_{\alpha\to 2\kappa+2s+2}\big\{(\mathrm{sgn}(x)|x|^{-\alpha},\,\varphi)+\frac{\mathscr{D}_{\kappa}^{2s+1}\varphi(0)}{2^{\kappa-1/2}\Gamma(\kappa+s+3/2)\,s!}\frac{1}{\alpha-2\kappa-2s-2}\big\}. \quad (27)$$

4. Fractional-Type Integral and Derivative for the Dunkl Operator

In this section, we embark on a comprehensive exploration of fractional-type integral operators associated with the Dunkl operator. These operators transcend the conventional Riesz fractional integral, Feller fractional integral, and Liouville fractional integral, as they are specifically designed to operate within the Dunkl setting.

4.1. The Riesz–Dunkl Fractional Integral

In this section, our focus lies on extending the Riesz fractional integral to any arbitrary value of $\Re(\alpha)>0$. As a reminder, the Riesz fractional integral $I^{\alpha}f$ is defined by

$$(I^{\alpha}f)(x)=\frac{1}{\gamma(\alpha)}\int_{\mathbb{R}}k_{\alpha}(x-y)f(y)dy, \quad (28)$$

where $k_{\alpha}(x)$ is defined as:

$$k_{\alpha}(x)=\begin{cases}|x|^{\alpha-1}, & \alpha\neq 1,3,5,\dots,\\ -|x|^{\alpha-1}\ln|x|, & \alpha=1,3,5,\dots.\end{cases} \quad (29)$$

The normalization factor $\gamma(\alpha)$ depends on the value of α and is given by:

$$\gamma(\alpha)=\begin{cases}\frac{2^{\alpha-1/2}\pi^{1/2}\Gamma(\frac{\alpha}{2})}{\Gamma(\frac{1-\alpha}{2})}, & \alpha\neq 2s+1,\ s=0,2,\dots,\\ (-1)^{s}s!\pi^{1/2}2^{2s}\Gamma(s+1/2), & \alpha=2s+1,\ s=0,2,\dots.\end{cases} \quad (30)$$

Lemma 2. *Let $\kappa<\alpha<2\kappa+1$. Then, the Dunkl transform of $|x|^{\alpha-2\kappa-1}$ exists in the usual sense, and it is given by*

$$\mathscr{F}_{\kappa}^{-1}(|x|^{-\alpha})=\frac{\Gamma(\kappa+\frac{1-\alpha}{2})}{2^{\alpha-\kappa-1/2}\Gamma(\frac{\alpha}{2})}\,|x|^{\alpha-2\kappa-1}.$$

Proof. By using (5), we obtain

$$\mathscr{F}_{\kappa}^{-1}(|x|^{-\alpha})(x)=\int_{-\infty}^{\infty}|u|^{-\alpha}\mathcal{E}_{\kappa}(iux)\sigma_{\kappa}(du)$$
$$=\frac{2}{2^{\kappa+1/2}\Gamma(\kappa+\frac{1}{2})}\int_{0}^{\infty}\mathscr{J}_{\kappa-1/2}(|x|u)u^{-\alpha+2\kappa}\,du.$$

Making the substitution $t = |x|u$ yields

$$\mathscr{F}_\kappa^{-1}(|x|^{-\alpha})(x) = |x|^{\alpha-2\kappa-1}\int_0^\infty \frac{J_{\kappa-1/2}(u)}{u^{\alpha-\kappa-1/2}}\,du.$$

The result follows from the following Weber formula [29] §13.24:

$$\int_0^\infty \frac{J_\nu(t)}{t^{\nu-\mu+1}}\,dt = \frac{1}{2^{\nu-\mu+1}}\frac{\Gamma(\frac{\mu}{2})}{\Gamma(\nu-\frac{\mu}{2}+1)}, \quad 0<\Re(\mu)<\Re(\nu)+\frac{3}{2}. \tag{31}$$

□

Proposition 1. *The Dunkl transform of* $|x|^{-\alpha}\in\Psi'_\kappa(\mathbb{R})$ *is given by*

$$\mathscr{F}_\kappa^{-1}(|x|^{-\alpha}) = \frac{1}{\gamma_\kappa(\alpha)}\begin{cases} |x|^{\alpha-2\kappa-1}, & \alpha\neq -2s,\ \alpha\neq 2\kappa+2s+1,\ s\in\mathbb{N}_0, \\ |x|^{\alpha-2\kappa-1}\ln\frac{1}{|x|}, & \alpha = 2\kappa+2s+1,\ s\in\mathbb{N}_0, \\ (-1)^s\mathscr{D}_\kappa^{2s}\delta, & \alpha=-2s,\ s\in\mathbb{N}_0, \end{cases}$$

where

$$\gamma_\kappa(\alpha) = \begin{cases} \frac{2^{\alpha-\kappa-1/2}\Gamma(\frac{\alpha}{2})}{\Gamma(\kappa+\frac{1-\alpha}{2})} & \alpha\neq-2s,\ \alpha\neq 2\kappa+2s+1, \\ (-1)^s s!\,2^{\kappa+2s+1/2}\Gamma(\kappa+s+1/2), & \alpha=2\kappa+2s+1, \\ 1, & \alpha=-2s \end{cases}$$

and δ *is the Dirac delta distribution.*

Proof. From Lemma 2, it is evident that by analytic continuation, for $\alpha\in\mathbb{C}$ such that $\alpha\neq 2\kappa+2s+1$ and $\alpha\neq -2s$ for $s=0,1,2,\ldots$, we have:

$$\frac{1}{|x|^\alpha} = \frac{\Gamma(\kappa+\frac{1-\alpha}{2})}{2^{\alpha-\kappa-1/2}\Gamma(\frac{\alpha}{2})}\mathscr{F}_\kappa(|x|^{\alpha-2\kappa-1}). \tag{32}$$

The case $\alpha=-2s$ for $s=0,1,2,\ldots$ follows from the fact that

$$(\mathscr{F}_\kappa\mathscr{D}_\kappa^{2s}\varphi)(x) = (-1)^s|x|^{2s}(\mathscr{F}_\kappa\varphi)(x), \quad \varphi\in S(\mathbb{R}).$$

It remains to consider the case $\alpha=\alpha_s=2\kappa+2s+1$ for $s\in\mathbb{N}_0$. From Equation (32), we have

$$\frac{\partial}{\partial\alpha}((\alpha-\alpha_s)(|x|^{-\alpha},\mathscr{F}_\kappa\varphi)) = \frac{\partial}{\partial\alpha}\Big(\eta(\alpha)(|x|^{\alpha-2\kappa-1},\varphi)\Big), \quad \eta(\alpha)=\frac{\alpha-\alpha_s}{\gamma_\kappa(\alpha)}. \tag{33}$$

By considering (23) and (25), the limit as $\alpha\to\alpha_s$ of the left-hand side of (33) can be evaluated as follows:

$$\lim_{\alpha\to\alpha_s}\frac{\partial}{\partial\alpha}\Big((\alpha-\alpha_k)(|x|^{-\alpha},\mathscr{F}_\kappa\varphi)\Big) = (|x|^{-2\kappa-2s-1},\mathscr{F}_\kappa\varphi).$$

The limit of the right-hand side of Equation (33) as $\alpha\to\alpha_s$ can be evaluated as follows:

$$\lim_{\alpha\to\alpha_s}\frac{\partial}{\partial\alpha}\Big(\eta(\alpha)\,(|x|^{\alpha-2\kappa-2},\varphi)\Big) = \lim_{\alpha\to\alpha_s}((\eta'(\alpha)+\eta(\alpha)\ln|x|)|x|^{\alpha-2\kappa-1},\varphi).$$

A straightforward computation shows that

$$\lim_{\alpha\to\alpha_s}\eta(\alpha)=\frac{(-1)^{s+1}}{s!2^{\kappa+2s-1/2}\Gamma(\kappa+s+1/2)}. \quad (34)$$

Taking into account Equation (17), in the limit as α approaches α_s, we obtain the following expression:

$$(|x|^{-2\kappa-2s-1},\mathscr{F}_\kappa\varphi)=\frac{(-1)^s}{s!2^{\kappa+2s-1/2}\Gamma(\kappa+s+1/2)}(|x|^{2s}\ln\frac{1}{|x|},\varphi). \quad (35)$$

□

Definition 1. *For $\Re(\alpha)>0$, we define the Riesz–Dunkl fractional integral $\mathscr{I}_\kappa^\alpha f$ of $f\in\Phi_\kappa(\mathbb{R})$ as:*

$$(\mathscr{I}_\kappa^\alpha f)(x)=\int_{\mathbb{R}}\tau^{-y}\mathscr{K}_{\kappa,\alpha}(x)\,f(y)\sigma_\kappa(dy) \quad (36)$$

where

$$\mathscr{K}_{\kappa,\alpha}(x)=\frac{1}{\gamma_\kappa(\alpha)}\begin{cases}|x|^{\alpha-2\kappa-1}, & \alpha\neq-2s,\ \alpha\neq2\kappa+2s+1\\ \ln(\frac{1}{|x|})\,|x|^{\alpha-2\kappa-1}, & \alpha=2\kappa+2s+1.\end{cases} \quad (37)$$

The following theorem states that the space $\Phi_\kappa(\mathbb{R})$ is closed under the action of the operator $\mathscr{I}_\kappa^\alpha$. This result ensures the consistency and coherence of the space $\Phi_\kappa(\mathbb{R})$ under the Riesz–Dunkl fractional integral. Moreover, the proposition establishes the relationship between the Dunkl transform $\mathscr{F}_\kappa$ and the fractional integral operator $\mathscr{I}_\kappa^\alpha$ and shows the compatibility of the fractional integral operators $\mathscr{I}_\kappa^\alpha$ under composition.

Theorem 1. *The space $\Phi_\kappa(\mathbb{R})$ is invariant under the operator $\mathscr{I}_\kappa^\alpha$, i.e.,*

$$f\in\Phi_\kappa(\mathbb{R})\quad\Rightarrow\quad\mathscr{I}_\kappa^\alpha f\in\Phi_\kappa(\mathbb{R}).$$

Furthermore,

$$(\mathscr{F}_\kappa\mathscr{I}_\kappa^\alpha f)=\frac{1}{|x|^\alpha}\mathscr{F}_\kappa f,$$

and

$$\mathscr{I}_\kappa^\alpha\mathscr{I}_\kappa^\beta=\mathscr{I}_\kappa^{\alpha+\beta},\quad\Re(\alpha),\Re(\delta)>0.$$

The proof of this theorem is omitted, but it can be established by utilizing Lemma 2 and Proposition 1 mentioned earlier, which provide the necessary tools and results to derive these conclusions.

Utilizing the reflection formula for the gamma function, we have:

$$\Gamma(z)\Gamma(1-z)=\frac{\pi z}{\sin\pi z},\quad z\notin\mathbb{Z}.$$

In the limit when $\kappa\downarrow0$, we retrieve the classical *Riesz* and *Feller fractional integral* (see, [11]) §12.1

$$\lim_{\kappa\downarrow0}\mathscr{I}_\kappa^\alpha f(x)=\frac{1}{2\Gamma(\alpha)\cos(\pi\alpha/2)}\int_{-\infty}^{\infty}\frac{1}{|x-y|^{1-\alpha}}f(y)dy. \quad (38)$$

4.2. Feller–Dunkl Fractional Integral

In this section, we aim to establish an analogous version of the classical Feller fractional integral within the framework of Dunkl operators. The Feller fractional integral, denoted as $J_\kappa^\alpha f(x)$, is defined as follows:

$$J_\kappa^\alpha f(x) = \frac{1}{2\Gamma(\alpha)\sin(\pi\alpha/2)} \int_{-\infty}^{\infty} \frac{\text{sgn}(x-y)}{|x-y|^{1-\alpha}} f(y)dy. \tag{39}$$

The following lemmas play a crucial role in establishing an extension of the Feller integral within the framework of the Dunkl operator.

Lemma 3. *Let $\kappa < \alpha < 2\kappa + 2$. Then, the Dunkl transform of $sgn(x)\,|x|^{-\alpha}$ exists in the usual sense, and it is given by*

$$\mathscr{F}_\kappa^{-1}(sgn(x)|x|^{-\alpha}) = i\,\frac{\Gamma(\kappa + \frac{2-\alpha}{2})}{2^{\alpha-\kappa-1/2}\Gamma(\frac{1+\alpha}{2})}\, sgn(x)\,|x|^{\alpha-2\kappa-1}.$$

Proof. Using (5), we have

$$\begin{aligned}
\mathscr{F}_\kappa^{-1}(\text{sgn}(x)|x|^{-\alpha})(x) &= \int_{-\infty}^{\infty} \text{sgn}(u)|u|^{-\alpha}\mathcal{E}_\kappa(iux)\sigma_\kappa(du) \\
&= \frac{2ix}{(2\kappa+1)2^{\kappa+1/2}\Gamma(\kappa+\frac{1}{2})} \int_0^\infty \mathscr{J}_{\kappa+1/2}(xu)\, u^{-\alpha+2\kappa+1}\, du \\
&= i\,\text{sgn}(x)\,|x|^{\alpha-2\kappa-1} \int_0^\infty \frac{J_{\kappa+1/2}(t)}{t^{\alpha-\kappa-1/2}}\, dt.
\end{aligned}$$

The Weber Formula (31) achieves the result. □

Lemma 4. *The following holds: for $\alpha \neq 2\kappa + s + 1$ with $s \in \mathbb{Z}_-$, we have*

$$\mathscr{D}_\kappa |x|^{-\alpha} = -\alpha\,|x|^{-\alpha-1} sgn(x).$$

Proof. Let $\kappa < \Re(\alpha) < 2\kappa + 1$ and $\varphi \in S(\mathbb{R})$, we have

$$\begin{aligned}
< \mathscr{D}_\kappa |x|^{-\alpha}, \varphi >_\kappa &= - < |x|^{-\alpha}, \mathscr{D}_\kappa \varphi >_\kappa \\
&= -\int_{\mathbb{R}} |x|^{-\alpha} \mathscr{D}_\kappa \varphi(x)\sigma_\kappa(dx) \\
&= -\alpha \int_{\mathbb{R}} |x|^{-\alpha-1}\text{sgn}(x)\varphi(x)\sigma_\kappa(dx) \\
&= -\alpha < |x|^{-\alpha-1}\text{sgn}(x), \varphi >_\kappa .
\end{aligned}$$

By analytic continuation for $\alpha \in \mathbb{C}$ such that $\alpha \neq 2\kappa + s + 1,\ s \in \mathbb{N}$, we have

$$\mathscr{D}_\kappa |x|^{-\alpha} = -\alpha\,|x|^{-\alpha-1}\text{sgn}(x),$$

which is the required result.

□

Proposition 2. *The Dunkl transform of $sgn(x)\,|x|^{-\alpha} \in \Psi'_\kappa(\mathbb{R})$ is given by*

$$\mathscr{F}_\kappa^{-1}(-i\,|x|^{-\alpha} sgn(x)) = \frac{1}{\delta_\kappa(\alpha)} \begin{cases} sgn(x)\,|x|^{\alpha-2\kappa-1}, & \alpha \neq -2s-1,\ \alpha \neq 2\kappa+2s+2,\ s \in \mathbb{N}_0, \\ -|x|^{2s+1}\ln|x|, & \alpha = 2\kappa+2s+2,\ s \in \mathbb{N}_0, \\ (-1)^s \mathscr{D}_\kappa^{2s+1}\delta, & \alpha = -2s-1,\ s \in \mathbb{N}_0. \end{cases}$$

where

$$\delta_\kappa(\alpha) = \begin{cases} \frac{2^{\alpha-\kappa-1/2}\Gamma(\frac{\alpha+1}{2})}{\Gamma(\kappa+\frac{2-\alpha}{2})} & \alpha \neq -2s-1,\ \alpha \neq 2\kappa+2s+2, \\ (-1)^s s! 2^{\kappa+2s+3/2}\Gamma(\kappa+s+3/2), & \alpha = 2\kappa+2s+2, \\ 1, \quad \alpha = -2s-1. \end{cases}$$

Proof. The proof of the proposition can be achieved by utilizing the above lemmas. □

Definition 2. *For $\Re(\alpha) > 0$, we define the Riesz–Dunkl fractional integral $\mathscr{J}_\kappa^\alpha f$ of $f \in \Phi_\kappa(\mathbb{R})$ as:*

$$(\mathscr{J}_\kappa^\alpha f)(x) = \int_{\mathbb{R}} \tau^{-y}\mathscr{G}_{\kappa,\alpha}(x) f(y)\sigma_\kappa(dy) \tag{40}$$

where

$$\mathscr{G}_{\kappa,\alpha}(x) = \frac{1}{\delta_\kappa(\alpha)}\begin{cases} sgn(x)\,|x|^{\alpha-2\kappa-1}, & \alpha \neq 2\kappa+2s+2 \\ sgn(x)\,\ln(\frac{1}{|x|})|x|^{\alpha-2\kappa-1}, & \alpha = 2\kappa+2s+2. \end{cases} \tag{41}$$

In the limit when $\alpha \downarrow 0$, we obtain

$$\lim_{\alpha\downarrow 0}(\mathscr{J}_\kappa^\alpha f) := \mathscr{H}_\kappa f(x) := \frac{\Gamma(\kappa+1)}{\sqrt{\pi}\Gamma(\kappa+1/2)} \lim_{\varepsilon\downarrow 0}\int_{|y|\geq\varepsilon} \tau_\kappa^{-y} f(x)\,\frac{dy}{y}, \tag{42}$$

and

$$(\mathscr{F}_\kappa\mathscr{H}_\kappa f)(x) = -i\,\mathrm{sgn}(x)(\mathscr{F}_\kappa f)(x), \quad f \in \Phi_\kappa(\mathbb{R}).$$

For the special case of $\kappa = 0$ and $\alpha = 0$, the Feller–Dunkl fractional integral coincides with the Hilbert transform. The Hilbert transform is a well-known operator in harmonic analysis and signal processing. It acts as a multiplier with the symbol $-isign(x)$.

It can be easily seen from Propositions 1 and 2 that the operators $\mathscr{I}_\kappa^\alpha$ and $\mathscr{J}_\kappa^\alpha$ are connected by

$$\mathscr{I}_\kappa^\alpha = \mathscr{H}_\kappa \mathscr{J}_\kappa^\alpha.$$

4.3. Riemann–Liouville–Dunkl fractional integrals

The Riemann–Liouville fractional integrals are given by [12] formulas (5.1) and (5.2)

$$I_+^\alpha f(x) := \frac{1}{\Gamma(\alpha)}\int_{-\infty}^x (x-y)^{\alpha-1} f(y)dy \tag{43}$$

and

$$I_-^\alpha f(x) := \frac{1}{\Gamma(\alpha)}\int_x^\infty (y-x)^{\alpha-1} f(y)dy \tag{44}$$

They are related to the Riesz fractional integral I^α and its conjugate J^α by

$$I^\alpha f(x) = \frac{I_+^\alpha f(x) + I_-^\alpha f(x)}{2\cos(\frac{\pi\alpha}{2})},$$
$$J^\alpha f(x) = \frac{I_+^\alpha f(x) - I_-^\alpha f(x)}{2\sin(\frac{\pi\alpha}{2})}.$$

Similarly, the correspondent definition of the Riemann–Liouville–Dunkl fractional integral can be given as follows:

$$\mathscr{I}_{\kappa,+}^\alpha f(x) := \cos(\alpha\pi/2)\mathscr{I}_\kappa^\alpha f(x) + \sin(\alpha\pi/2\mathscr{J}_\kappa^\alpha f(x),$$
$$\mathscr{I}_{\kappa,-}^\alpha f(x) := \cos(\alpha\pi/2)\mathscr{I}_\kappa^\alpha f(x) - \sin(\alpha\pi/2\mathscr{J}_\kappa^\alpha f(x).$$

Proposition 3. *The following holds:*

(1) *For $f \in \Phi$, we have*

$$(\mathscr{F}_\kappa \mathscr{I}^\alpha_{\kappa,\pm} f) = (\mp ix)^{-\alpha} (\mathscr{F}_\kappa f)(x).$$

(2) *For $f \in \Phi$ and $\Re(\alpha), \Re(\beta) > 0$, we have*

$$\mathscr{I}^\alpha_{\kappa,\pm} \mathscr{I}^\beta_{\kappa,\pm} = \mathscr{I}^{\alpha+\beta}_{\kappa,\pm}.$$

(3) *Integration by parts:*

$$\int_{\mathbb{R}} \mathscr{I}^\alpha_{\kappa,+} f(x)\, g(x) \sigma_\kappa(dx) = \int_{\mathbb{R}} f(x)\, \mathscr{I}^\alpha_{\kappa,-} g(x) \sigma_\kappa(dx), \quad f, g \in \Phi_\kappa(\mathbb{R}).$$

Funding: This research was funded by the Researchers Supporting Project number (RSPD2023R974), King Saud University, Riyadh, Saudi Arabia.

Data Availability Statement: Not applicable.

Acknowledgments: The author extends his appreciation to the Researchers Supporting Project number (RSPD2023R974), King Saud University, Riyadh, Saudi Arabia.

Conflicts of Interest: The author declares no conflict of interest.

References

1. Dunkl, C.F. Hankel transforms associated to finite reflections groups. *Contemp. Math.* **1992**, *138*, 123–138.
2. Dunkl, C.F. Differential-difference operators associated with reflections groups. *Trans. Amer. Math. Soc.* **1989**, *311*, 167–183. [CrossRef]
3. Trimèche, K. Paley-Wiener Theorems for the Dunkl transform and Dunkl translation operators. *Integral Transform. Spec. Funct.* **2002**, *13*, 17–38. [CrossRef]
4. De Jeu, M.F.E. The Dunkl transform. *Invent. Math.* **1993**, *113*, 147–162. [CrossRef]
5. Rösler, M. Positivity of Dunkl's intertwinning operator. *Duke Math. J.* **1999**, *98*, 445–463. [CrossRef]
6. Rösler, M. *Bessel-Type Signed Hypergroup on $\mathbb{R}$, in Probability Measures on Groups and Related Structures XI*; Heyer, H., Mukherjea, A., Eds.; World Scientific: Singapore, 1995; pp. 292–304.
7. Rösler, M. *Dunkl Operators. Theory and Applications, in Orthogonal Polynomials and Special Functions*; Lecture Notes in Mathematics; Springer: Berlin/Heidelberg, Germany, 2003; Volume 1817, pp. 93–135.
8. Feller, W. *On a Generalization of Marcel Riesz's Potentials and the Semi-Groups Generated by Them*; Gleerup: Lund, Sweden, 1962; pp. 73–81.
9. Stein, E.M.; Weiss, G. *Fractional Integrals on n-Dimensional Euclidean Space*; United States Air Force, Office of Scientific Research: Arlington, VA, USA, 1958; pp. 503–514.
10. Lizorkin, P.I. Generalized Liouville differentiation and function spaces $L^r_p(E_n)$. Embedding theorems. *Mat. Sb.* **1963**, *102*, 325–353.
11. Samko, S.G. *Hypersingular Integrals and Their Applications*; Series Analytical Methods and Special Functions 5; Taylor Francis Group: New York, NY, USA, 2005.
12. Samko, S. Best Constant in the Weighted Hardy Inequality: The Spatial and Spherical Version. *Fract. Calc. Appl. Anal.* **2005**, *8*, 39–52.
13. Mainardi, F. *Fractional Calculus and Waves in Linear Viscoelasticity*, 2nd ed.; World Scientific: Singapore, 2022.
14. Kiryakova, V. A guide to special functions in fractional calculus. *Mathematics* **2021**, *9*, 106. [CrossRef]
15. Riesz, M. L'integrale de Riemann-Liouville et le probleme de Cauchy. *Acta Math.* **1949**, *98*, 1–222. [CrossRef]
16. Thangavelu, S.; Xu, Y. Convolution operator and maximal function for Dunkl transform. *J. d'Analyse Mathématique* **2005**, *97*, 25–55. [CrossRef]
17. Thangavelu, S.; Xu, Y. Riesz transform and Riesz potentials for Dunkl transform *J. Comput. Appl. Math.* **2007**, *199*, 181–195. [CrossRef]
18. Amri, B.; Anker, J.P.; Sifi, M. Three results in Dunkl analysis. *Colloq. Math.* **2010**, *118*, 299–312. [CrossRef]
19. Abdelkefi, C.; Rachdi, M. Some properties of the Riesz potentials in Dunkl analysis. *Ric. Mat.* **2015**, *64*, 195–215. [CrossRef]
20. Abdelkefi, C.; Anker, J.P.; Sassi, F.; Sifi, M. Besov-type spaces on Rd and integrability for the Dunkl transform. *Symmetry Integr. Geom. Methods Appl.* **2009**, *5*, 19.
21. Gorbachev, D.V.; Ivanov, V.I.; Tikhonov, S.Y. L^p-bounded Dunkl-type generalized translation operator and its applications. *Constr. Approx.* **2019**, *49*, 555–605. [CrossRef]
22. Sallam Hassani, S.M.; Sifi, M. Riesz potentials and fractional maximal function for the Dunkl transform. *J. Lie Theory* **2009**, *19*, 725–734.

23. Mourou, M.A. Taylor series associated with a differential-difference operator on the real line. *J. Comp. Appl. Math.* **2003**, *153*, 343–354. [CrossRef]
24. Soltani, F. Sonine Transform assocated to the Dunkl kernel on the real line. *Symmetry Integr. Geom. Methods Appl. SIGMA* **2008**, *4*, 92.
25. Dunkl, C.F.; Xu, Y. *Orthogonal Polynomials of Several Variables*; Cambridge University Press: Cambridge, UK, 2001.
26. Mainardi, F.; Consiglio, A. The Wright functions of the second kind in Mathematical Physics. *Mathematics* **2020**, *8*, 884. [CrossRef]
27. Garra, R.; Mainardi, F. Some aspects of Wright functions in fractional differential equations, *Rep. Math. Phys.* **2021**, *87*, 265–273.
28. Xu, Y. Dunkl operators: Funk–Hecke formula for orthogonal polynomials on spheres and on balls. *Bull. Lond. Math. Soc.* **2000**, 32, 447–457. [CrossRef]
29. Watson, G.N. *A Treatise on the Theory of Bessel Functions*; Cambridge University Press: Cambridge, UK, 1922; ISBN 9780521483919.

symmetry

Article

Artificial Neural Network Solution for a Fractional-Order Human Skull Model Using a Hybrid Cuckoo Search Algorithm

Waseem [1,*], Sabir Ali [2], Shahzad Khattak [2,3], Asad Ullah [4,5], Muhammad Ayaz [6], Fuad A. Awwad [7] and Emad A. A. Ismail [7]

1 School of Mechanical Engineering, Jiangsu University, 301, Xuefu Road, Jingkou District, Zhenjiang 212013, China
2 Department of Mathematics, National University of Modern Languages, Islamabad 44000, Punjab, Pakistan
3 Department of Mathematics, Jiangsu University, 301, Xuefu Road, Jingkou District, Zhenjiang 212013, China
4 School of Finance and Economics, Jiangsu University, 301, Xuefu Road, Jingkou District, Zhenjiang 212013, China; asad@ujs.edu.cn
5 Department of Mathematical Sciences, University of Lakki Marwat, Lakki Marwat 28420, Khyber Pakhtunkhwa, Pakistan
6 Department of Mathematics, Abdul Wali Khan University Mardan, Mardan 23200, Khyber Pakhtunkhwa, Pakistan; mayazmath@awkum.edu.pk
7 Department of Quantitative Analysis, College of Business Administration, King Saud University, P.O. Box 71115, Riyadh 11587, Saudi Arabia
* Correspondence: waseem@ujs.edu.cn

Abstract: In this study, a new fractional-order model for human skull heat conduction is tackled by using a neural network, and the results were further modified by using the hybrid cuckoo search algorithm. In order to understand the temperature distribution, we introduced memory effects into our model by using fractional time derivatives. The objective function was constructed in such a way that the L_2−error remained at a minimum. The fractional order equation was then calculated by using the proposed biogeography-based hybrid cuckoo search (BHCS) algorithm to approximate the solution. When compared to earlier simulations based on integer-order models, this method enabled us to examine the fractional-order (FO) cases, as well as the integer order. The results are presented in the form of figures and tables for the different case studies. The results obtained for the various parameters were validated numerically against the available literature, where our proposed methodology showed better performance when compared to the least squares method (LSM).

Keywords: boundary value problems; fractional derivatives; heat conduction; BHCS algorithm; Cuckoo search; numerical method; human head

Citation: Waseem; Ali, S.; Khattak, S.; Ullah, A.; Ayaz, M.; Awwad, F.A.; Ismail, E.A.A. Artificial Neural Network Solution for a Fractional-Order Human Skull Model Using a Hybrid Cuckoo Search Algorithm. *Symmetry* **2023**, *15*, 1722. https://doi.org/10.3390/sym15091722

Academic Editors: Francisco Martínez González, Mohammed K. A. Kaabar and Junesang Choi

Received: 12 July 2023
Revised: 26 August 2023
Accepted: 5 September 2023
Published: 8 September 2023

1. Introduction

The use of electronic devices is increasing day by day. One reason behind this is the advancement of technology and its applications in various sectors. This advancement in technological equipment has some side effects, especially when it crosses some limit in its use. Some of the devices and systems are Bluetooth, mobile phones, and other headphone-like devices. These devices produce thermal waves that pass through the skin and enter the human head, damaging various tissues, including the brain. The use of these electronic devices produces brain damage and other neural disorders, as explained in the references [1–3]. Furthermore, the analysis of the effects of the thermal and non-thermal waves are numerically and experimentally analyzed by Bernardi et al. [4]. The flow symmetry of heat is widely analyzed by many researchers due to its experimental and theoretical applications [5]. The energy transfer from the electronic object to the human head follows the one-sided symmetry in the skull. The facial, brain, and skull symmetries are well explained by Ratajczak et al. [6].

The analysis of heat in a human head became famous after the work of Flesch [7]. In this work, Flesch used a differential equation for the analysis of heat. This work was further examined by Gray [8] in 1980, which provides a theoretical approach to heat transfer analysis in terms of the human head. The human skull produces more heat on the outer layer than the center. On the other hand, when the surrounding temperature is reduced, the heat production is more peripheral to the skull. Simply put, the temperature outside and the radial distance from the skull's center affect how much heat is produced inside the human skull. Anderson and Arthurs [9] used the complementary extremum approach to analyze this famous problem. Makinde [10] analyzed the human skull problem by using the non-perturbation approach. Raja et al. [11] implemented the stochastic approach to study the human skull problem. Abdelhakem and Youssri [12] analyzed the Lane-Emden and Bratu equations by using the spectral Legendre's algorithm. Youssri et al. [13] used the wavelets approach for a solution of the Lane–Emden equations. A more brief analysis using the numerical approaches to the solution of D.Eqs. can be found in the literature [14,15].

The methodology and the model modifications are both points of interest to researchers. In recent years, the use of fractional derivatives in differential equations has been widely implemented [16,17]. The applications of fractional derivatives are well explained by Podlubny [18]. The use of fractional derivatives in the field of differential equations is explained by Aleksandrovich et al. [19]. Wang [20] studied the febrifuge effect for analyzing the fractional-order human skull problem. The concept of Caputo-type derivatives was introduced by Kumar et al. Kumar et al. [21] for use in differential equations. At the same time, the concept of Caputo-type derivatives for delay-type differential equations was introduced by Odibat et al. [22]. The concept of fractional derivatives in applications can be found in ecology [23,24], psychology [25], chemistry [26], epidemiology [27–30], and physics [31]. Yavuz and Sene [32] examined how various parameters affect fractional-order second-grade fluid flow. Hammouch et al. [33] numerically simulated a fractional chaotic system with changing order. Yavuz [34] studied the classic and generalized Mittag-Leffler kernels used in the fractional derivative definition in the European option pricing model.

The applications of fractional derivatives are not limited to a single definition. The solution to Cauchy and Dirichlet problems are studied by Avci et al. [35] by using the Caputo-Fabrizio derivative definition. Erturk et al. [36] developed a unique Caputo fractional derivative for the corneal shape model of the human eye. The recent literature shows the application of fractional calculus, which can be found in the following references [37–39].

The following is the integer-order model for temperature distribution in the human skull:

$$\begin{aligned} T''(r) + \tfrac{2T'(r)}{r} + \lambda.exp(-mT) = 0, \\ T'(0) = 0, \ \ T'(1) = N_B(1-T). \end{aligned} \tag{1}$$

Here, T, r, N_B, m, and λ denote the temperature, radial distance, Biot number, metabolic thermogenesis slope, and thermogenesis heat production, respectively.

We introduce the Riemann–Liouville definition of the fractional-order derivative of the function $\Im \in C^d_{-1}$ below [18]:

$$D_t^{\mu}\Im(t) = \begin{cases} \frac{d^{\gamma}\Im(t)}{d\psi^{\gamma}}, & \mu = \gamma \in \mathbb{Z}, \\ \frac{1}{\gamma(\gamma-\mu)} \int_0^t (t-\sigma)^{\gamma-\mu-1}\Im^{\gamma}(\sigma)d\sigma, & \gamma - 1 < \mu < \gamma, \gamma \in \mathbb{Z}. \end{cases} \tag{2}$$

In light of Equation (2), the suggested classical BVP (1) is transformed into a fractional-order generalized form:

$$\begin{aligned} \tfrac{2}{r}.{}^cD_0^{\nu}T(r) +^c D_0^{\mu}T(r) + \lambda.exp(-mT) = 0, \\ T'(0) = 0, \ \ T'(1) = N_B(1-T). \end{aligned} \tag{3}$$

Here, $r \in [0.1, 1]$ and the derivatives of the function $T(r)$ with fractional orders $0 < \nu \leq 1$ and $1 < \mu \leq 2$, are represented by the symbols ${}^cD_0^{\nu}$ and ${}^cD_0^{\mu}$ respectively.

The graphical abstract of the proposed methodology is given in Figure 1, whereas the structure of the neural network is presented in Figure 2.

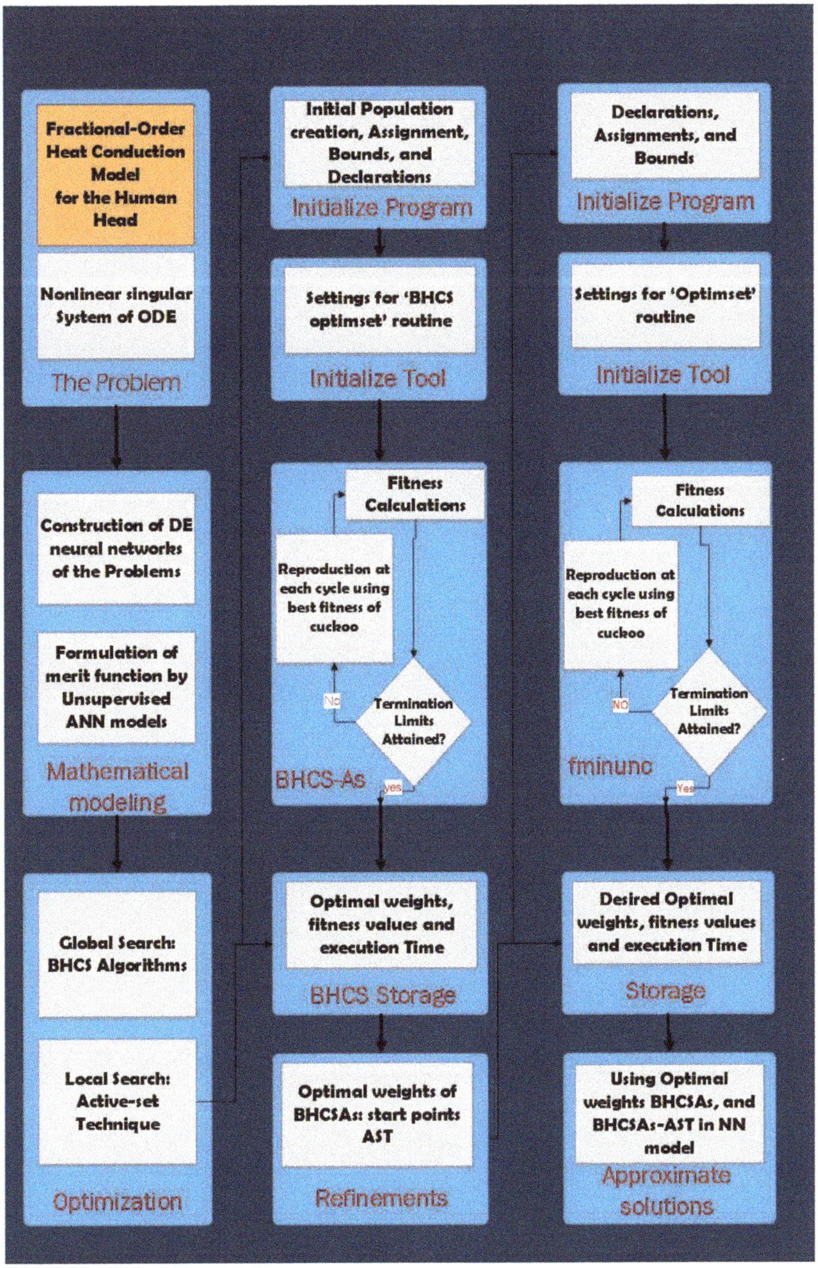

Figure 1. Graphical abstract of the given model.

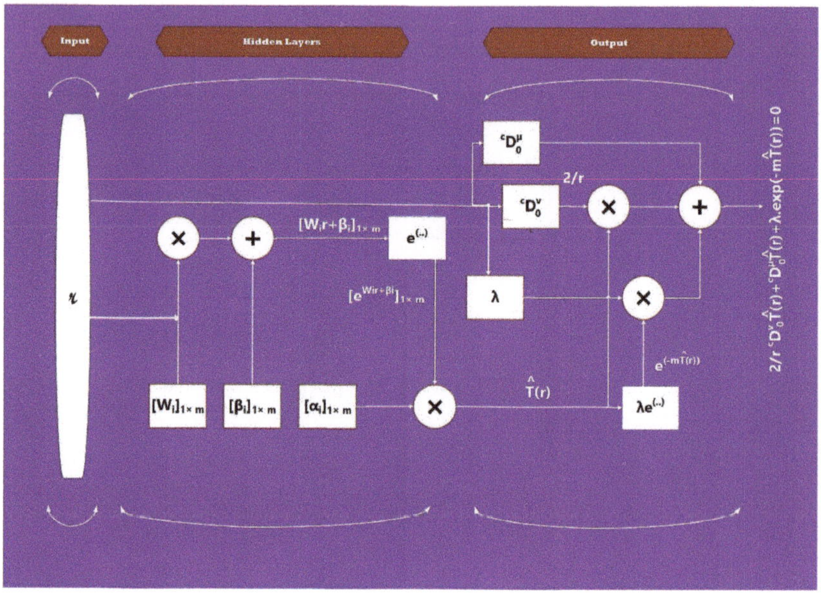

Figure 2. Graphical presentation of the ANN structure for the given model.

Our Contribution

The primary objective of this research study is to develop an approximate solution for a fractional-order human head heat conduction model by using the BHCS algorithm. More specifically, we summarize our contributions as follows:

- To the best of our knowledge, the proposed problem is, for the first time, transformed into a fractional order by using the Riemann–Liouville definition of fractional-order derivatives;
- A new optimal approach has been designed to approximate the solution to the transformed equations;
- We investigated the impacts of radial distance on the dynamics of the temperature curve for various fractional-order values (ν, μ), for which the results are displayed through graphs and tables and were validated against the available literature [40].

In this article, the proposed methodology of BHCS is discussed in Section 2, and the numerical step-up and various proposed cases (with graphs and tables) are presented in Section 3 and are discussed in detail. At the end, a conclusion is provided in Section 4 of the article.

2. The Proposed Methodology

The approximate solution ($\hat{T}(r)$) of the fractional-order model for temperature in the human head by using feed-forward neural networks with the help of an exponential function is given as

$$\hat{T}(r) = \sum_{i=1}^{m} \alpha_i e^{\omega_i r+\beta_i}, \tag{4}$$

$$\frac{d^\mu \hat{T}(r)}{dr^\mu} = \sum_{i=1}^{m} \alpha_i r^{-\mu} e^{\beta_i} E_{1,1-\mu}(\omega_i r), \tag{5}$$

$$\frac{d^\nu \hat{T}(r)}{dr^\nu} = \sum_{i=1}^{m} \alpha_i r^{-\nu} e^{\beta_i} E_{1,1-\nu}(\omega_i r), \tag{6}$$

where α_i, ω_i and β_i are the weights given as $\alpha = [\alpha_1, \alpha_2, \ldots, \alpha_m]$, $\omega = [\omega_1, \omega_2, \ldots, \omega_m]$, $\beta = [\beta_1, \beta_2, \ldots, \beta_m]$, and m represents the number of neurons. Moreover, $\frac{d^{\nu}\hat{T}(r)}{dr^{\nu}}$ and $\frac{d^{\mu}\hat{T}(r)}{dr^{\mu}}$ are the fractional derivatives of the series solution.

2.1. Fitness Function

In the fitness function, we compute the absolute error and make an optimization process to minimize the error ϵ, *i.e.*, when $\epsilon \to 0$, then $\hat{T}(r) \to T(r)$.

The fitness function for the transformed equations is given as

$$\epsilon = \epsilon_1 + \epsilon_2, \tag{7}$$

where ϵ_1 represents the mean squared error for a given differential equation (DE), and ϵ_2 represents the conditions on it. Therefore, we have

$$\epsilon_1 = \frac{1}{k}\sum_{k=1}^{K}\left(\frac{2}{r_k}.{}^{c}D_0^{\nu}\hat{T}_k + {}^{c}D_0^{\mu}\hat{T}_k + \lambda.exp(-m\hat{T}_k)\right)^2, \tag{8}$$

and

$$\epsilon_2 = \frac{1}{2}\left((\hat{T}_0')^2 + (\hat{T}_1' - N_B(1-\hat{T}))^2\right), \tag{9}$$

where $h = \frac{u}{k}$, $\hat{T}_k = \hat{T}(r_k)$, and $r_k = kh$.

2.2. Cuckoo Search (CS) Technique

The cuckoo search (CS) algorithm follows the breeding behavior of the cuckoo bird [41]. In this algorithm, other birds give their eggs to others' nearest nests. When the host bird finds it, she adopts two methods: either to remove the eggs or to find a new nearest nest to lay their own eggs. In this process, the host bird's eggs indicate a solution, and the cuckoo bird's eggs display a fresh potential resolution [42].

This procedure is explained in [43]:

- Each cuckoo bird lays a single egg in the nest of its host;
- Those nests containing eggs of superior quality will be passed on to the next generation;
- The number of hosts' nests is set, and the host bird has a specific chance of discovering an alien egg.

We assume, $y_i = y_{i1}, y_{i2}, y_{i3}, \ldots, y_{iD}$ as i^{th} egg positions. The egg is defined as a solution, and Lévy flights update the new solution y_i^{new} as follows:

$$y_i^{new} = y_i^{old} + \alpha(y_l - x_g) \oplus \text{Levy}(\beta), \tag{10}$$

$$y_i^{new} = y_i^{old} + \frac{0.01u}{|v|^{\frac{1}{\beta}}}(y_i - y_g), \tag{11}$$

where $\oplus$ is the entry-wise product, β indicates the Lévy flight exponent, $\alpha > 0$ is the cuckoo's step size, x_g is the optimal sample, and u and v are random numbers. Furthermore,

$$v \sim N(0, \sigma_v^2), \qquad u \sim N(0, \sigma_u^2), \tag{12}$$

$$\sigma_u = \left[\frac{sin\frac{\pi\beta}{2} \cdot \Gamma(1+\beta)}{2^{\frac{\beta-1}{2}}\beta \cdot \Gamma(\frac{1+\beta}{2})}\right]^{\frac{1}{\beta}}, \qquad \sigma_v = 1. \tag{13}$$

Here, the function σ_u is controlled by the parameter β and the Γ function. In CS, the discovered nests are replaced using a discovery operator that takes the probability p_a into account. Thus, we have the updated solution given as follows:

$$y_{ij}^{\text{new}} = \begin{cases} y_{ij}^{\text{old}} + \text{rand} \cdot (y_{r1}j(k) - y_r 2j(k)) & \text{if } P > p_a, \\ y_{ij}^{\text{old}}(k) & \text{else.} \end{cases} \quad (14)$$

Here, y_{ij}^{new} represents the jth component of the ith solution y_i^{new}, $y_{r1,j}$ is the jth element of the solution y_{r1}, and $y_{r2,j}$ is jth element of the solution y_{r2}. Moreover, $r1$ and $r2$ are two distinct integers within in $[1, NP]$, where NP denotes the size of the population, and p_a is the discovery denoting probability.

2.3. *Biogeography-Based Optimization*

An evolutionary biogeography-based optimization method (BBO) was motivated by several traits of animals found on islands. In BBO, NP habitats (solutions) are used to randomly initialize the population. Each generation ranks the population from best to worst.

Here, we define $\hat{\lambda}$ and $\hat{\mu}$ as the particular habitat's immigration and emigration rates, as given in [44]:

$$\begin{cases} \hat{\lambda}_i = I\left(1 - \frac{\hat{s}_i}{NP}\right), \\ \hat{\mu}_i = E\frac{\hat{s}_i}{NP}, \end{cases} \quad (15)$$

where $E = I = 1$ are the immigration rates, $\hat{s}_i$ is a species of a certain population, which is defined as $\{\hat{s}_i = NP - i\}, i \in N$. The changing parameter updates the corresponding solution, and BBO also utilizes the mutation operator to update the solution accordingly.

2.4. *Hybrid Cuckoo Search*

In order to further improve the best nests obtained from the CS, we applied BBO. Both exploration and exploitation were employed alternatively. By combining exploration and exploitation, the BBO-based heterogeneous cuckoo search (BHCS) method was designed as a hybrid meta-heuristic. The proposed BHCS algorithm comprises two primary steps: heterogeneous CS and biogeography-based discovery.

The Methodology of Heterogeneous CS

The first component of BHCS employs the Lévy flights and a quantum mechanism-based heterogeneous CS. Heterogeneous CS is based on quantum mechanics [42,45].

$$y_i^{\text{new}} = \begin{cases} y_i^{\text{old}} + \alpha \cdot (y_i - y_g) \oplus \text{Lévy}(\beta) & \frac{2}{3} < s_r \leqslant 1, \\ \bar{y} + L \cdot (\bar{y} - y_i^{\text{old}}) & \frac{1}{3} < s_r \leqslant \frac{2}{3}, \\ y_i^{\text{old}} + \varepsilon \cdot (y_g - y_i^{\text{old}}) & \text{else.} \end{cases} \quad (16)$$

Here, the terms $L = \ln\left(\frac{1}{\eta}\right)$, $\epsilon = \delta e^{\eta}$, and y_g refer to the iteration's best solution, s_r is the number $\eta \in [0,1]$, and $\bar{y} = \frac{1}{NP}\sum_{i=1}^{NP} y_i$ is the average of the solutions. Equation (16) demonstrates that three equations are used in a heterogeneous cuckoo search to update the answers with the exact probabilities. The 1st equation is derived from the Lévy flights in the original CS, and the 2nd and 3rd equations are used to update the results by using the quantum-based algorithm. The search space is diversified by updating the solutions with heterogeneous rules, which move toward the actual global region.

3. Results and Discussion

In this section, we calculate four individual case studies and compute the results using our proposed BHCS as a global search technique. The case studies are shown in Table 1. A total of 100 independent runs were performed for each case study by taking the domain $r \in [0.1, 1]$ with 0.1 step size.

Table 1. Different cases with the variation of fractional derivative parameters.

Case Study 1	Case Study 2	Case Study 3	Case Study 4
$\mu = 2, \nu = 1$	$\mu = 1.70, \nu = 0.70$	$\mu = 1.80, \nu = 0.80$	$\mu = 1.90, \nu = 0.90$

The formulation for these case studies is given below:

3.1. Case Study 1

For this case, the fitness function with the boundary conditions is given as

$$\epsilon_1 = \frac{1}{11}\sum_{k=1}^{11}\left(\frac{2}{r_k}.^{c}D_0^{1}\hat{T}_k +^{c} D_0^{2}\hat{T}_k + \lambda.exp(-m\hat{T}_k)\right)^2, \tag{17}$$

$$\epsilon_2 = \frac{1}{2}\left((\hat{T}_0')^2 + (\hat{T}_1' - N_B(1-\hat{T}))^2\right). \tag{18}$$

We considered the parameters $\lambda = 1, m = 1, N_b = 1$. By using BHCS, the best weights for this case are given in the following equation.

$$\hat{T}_{c_1} = \begin{cases} 1.4025e^{(-0.2030r-0.7130)} - 0.0856e^{(-1.4281r+0.5922)} \\ +0.2617e^{(-0.0189r-0.1831)} - 0.7062e^{(0.5749r-1.6112)} \\ -2.4658e^{(-0.5143r-1.7698)} + 1.7948e^{(-0.0388r-0.7866)} \\ -0.1858e^{(-0.5612r-0.5207)} + 0.7182e^{(-1.3911r-1.2173)} \\ +0.9790e^{(0.3547r-2.6460)} - 0.0509e^{(-1.6107r-1.0547)}. \end{cases} \tag{19}$$

Here, Equation (19) is a series solution for case study 1.

3.2. Case Study 2

The fitness function for the current case study is formulated below:

$$\epsilon_1 = \frac{1}{11}\sum_{k=1}^{11}\left(\frac{2}{r_k}.^{c}D_0^{0.70}\hat{T}_k +^{c} D_0^{1.70}\hat{T}_k + \lambda.exp(-m\hat{T}_k)\right)^2, \tag{20}$$

$$\epsilon_2 = \frac{1}{2}\left((\hat{T}_0')^2 + (\hat{T}_1' - N_B(1-\hat{T}))^2\right). \tag{21}$$

Here, we take the parameters $\lambda = 1, m = 1, N_b = 1$. The best weights for this case are given as

$$\hat{T}_{c_2} = \begin{cases} -0.7538e^{(-0.8125r+0.7228)} + 1.3636e^{(-0.3377r-0.4151)} \\ +0.8281e^{(-0.2094r-1.3968)} + 0.4684e^{(-0.4236r-0.2732)} \\ +0.9507e^{(-1.3841r-1.4493)} + 1.1338e^{(-1.1852r-0.9906)} \\ -0.6161e^{(-5.0000r-4.1270)} - 0.1282e^{(-0.7593r-1.3885)} \\ +1.9087e^{(-0.3513r-0.8274)} + 0.0553e^{(0.1238r-0.1236)}. \end{cases} \tag{22}$$

Here, the above equation is the corresponding series solution for this special case 2. The absolute errors (AE) are presented in Table 2. The approximate solutions of the given fractional model, which were obtained by using the series solution of Equation (19), are illustrated in Figure 3. The best fitness values are evaluated by considering the above conditions, and the results are displayed in Figure 4.

Table 2. Minimum Absolute Errors (ϵ).

	Case Study 1	Case Study 2	Case Study 3	Case Study 4
r	$\mu = 2, \nu = 1$	$\mu = 1.70, \nu = 0.70$	$\mu = 1.80, \nu = 0.80$	$\mu = 1.90, \nu = 0.90$
0.1	9.21E−15	4.30E−07	7.63E−08	2.42E−08
0.2	1.27E−11	1.44E−06	3.96E−07	6.66E−08
0.3	2.38E−12	3.64E−08	2.86E−11	3.95E−09
0.4	1.91E−11	2.46E−07	1.07E−07	6.60E−09
0.5	1.22E−11	2.81E−07	7.29E−08	1.20E−08
0.6	2.05E−15	2.94E−08	3.86E−10	4.24E−09
0.7	1.73E−11	6.22E−08	4.86E−08	4.94E−17
0.8	3.83E−11	2.16E−07	9.32E−08	2.16E−09
0.9	1.07E−11	8.58E−08	2.23E−08	2.39E−09
1.0	5.47E−11	1.42E−07	8.26E−08	6.05E−11

We analyzed the FO heat conduction model for the human head given in Equation (3) by using an interval of $[0.1, 1]$. Four different cases were considered by choosing varying values of μ, ν for the fixed parameters $\lambda = 1, N_b = 1$. The approximate solutions for both cases, 1 and 2, are presented in Figure 3. These graphs show that when we decrease the order of the fractional parameters from an integer order to a non-integer, the temperature profile jumps to 1.17 from 1.16. A decreasing trend is observed when $r \longrightarrow 1$. This trend is more sharp in the integer order when compared to the fractional order. The fitness of the functions is shown in Figure 4, where the red lines show the mean. Almost all the values are bounded by the box, and the distance from the mean positions is displayed on the vertical line.

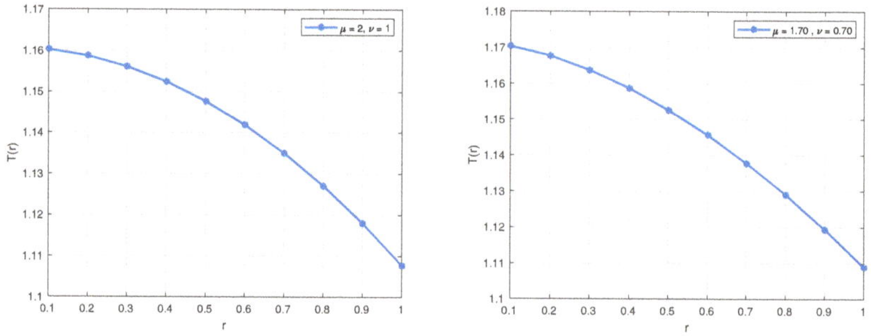

(a) (b)

Figure 3. Graphical representation of the solutions for (**a**) Case study 1 and (**b**) Case study 2.

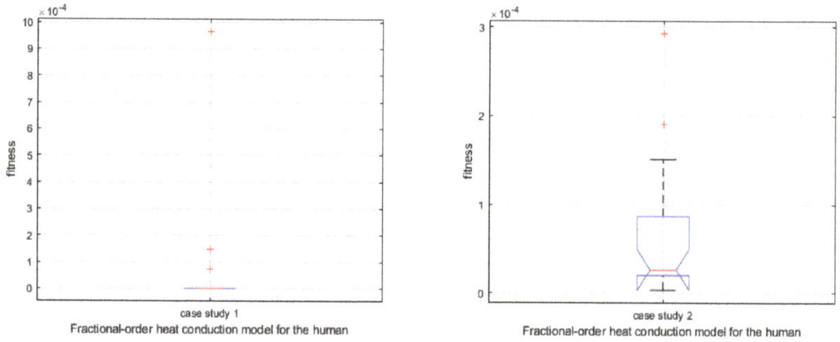

(a) (b)

Figure 4. Graphical representation of fitness functions for (**a**) Case study 1 and (**b**) Case study 2.

3.3. Case Study 3

In case 3, we considered $\mu = 1.80$ and $\nu = 0.80$ by choosing $\lambda = 1, m = 1, N_b = 1$. The fitness functions for this case are given by

$$\epsilon_1 = \frac{1}{11}\sum_{k=1}^{11}\left(\frac{2}{r_k}.^{c}D_0^{0.80}\hat{T}_k +^{c} D_0^{1.80}\hat{T}_k + \lambda.exp(-m\hat{T}_k)\right)^2, \quad (23)$$

$$\epsilon_2 = \frac{1}{2}\left((\hat{T}_0')^2 + (\hat{T}_1' - N_B(1-\hat{T}))^2\right). \quad (24)$$

The corresponding best weights are given below:

$$\hat{T}_{c_3} = \begin{cases} 0.3527e^{(0.7055r-2.1888)} - 1.1520e^{(-0.7899r-1.8388)} \\ +1.4121e^{(-3.4830r-1.8823)} + 1.3158e^{(-0.4966r-0.9646)} \\ +0.1036e^{(0.3872r-0.0793)} - 1.0951e^{(-0.9004r-0.8300)} \\ -0.2317e^{(-3.6303r-0.1507)} + 0.1977e^{(-1.2789r-0.2822)} \\ +0.8166e^{(-0.0468r+0.4276)} - 0.5767e^{(0.5119r-0.9304)}. \end{cases} \quad (25)$$

3.4. Case Study 4

For this case study, we took $\mu = 1.90$ and $\nu = 0.90$. So, the corresponding fitness functions take the following form:

$$\epsilon_1 = \frac{1}{11}\sum_{k=1}^{11}\left(\frac{2}{r_k}.^{c}D_0^{0.90}\hat{T}_k +^{c} D_0^{1.90}\hat{T}_k + \lambda.exp(-m\hat{T}_k)\right)^2, \quad (26)$$

$$\epsilon_2 = \frac{1}{2}\left((\hat{T}_0')^2 + (\hat{T}_1' - N_B(1-\hat{T}))^2\right). \quad (27)$$

By choosing $\lambda = 1, m = 1, N_b = 1$, we have

$$\hat{T}_{c_4} = \begin{cases} -0.5186e^{(-0.8417r-1.3387)} - 0.0069e^{(-4.9937r-1.3860)} \\ +0.1558e^{(-1.3456r-1.2083)} - 0.9332e^{(0.3765r-0.7752)} \\ +1.3882e^{(-0.1069r+0.1492)} - 0.3297e^{(-0.4075r-1.2062)} \\ -0.4607e^{(-0.3663r-0.3495)} + 0.5638e^{(0.5029r-3.9098)} \\ -0.1666e^{(0.4195r-0.8498)} + 2.0192e^{(0.2491r-1.2911)}. \end{cases} \quad (28)$$

Here, $\hat{T}_{c_3}$ and $\hat{T}_{c_4}$ are the series solutions for cases 3 and 4. The AE are plotted in Table 2, whereas the approximate solutions are presented in Figure 5. The fitness values are given in Figure 6. The values of AE ϵ for all the case studies are presented in Table 2.

Similarly, in cases 3 and 4, when we increase the fractional parameters that nearly approach the integer, the solution profiles fall from 1.65 to 1.61. As a result, the suggested fractional-order graph, which takes into account the radial distance (r), Biot number (N_B), metabolic thermogenesis slope parameter (m), and thermogenesis heat production parameter (λ), provides a more accurate representation of the distribution of temperature within the human skull.

The fitness functions for cases 3 and 4 are displayed in Figure 6. The horizontal red line shows the mean, and the red addition symbols show the positions of the results from this point. In both cases, the results are in the range of 10^{-5} and 10^{-4}, respectively. This further recommends that the fitness functions remain as minimal as possible.

In Table 2, the results for the minimum values of the absolute error are displayed numerically. These cases are chosen in such a way that the deviations from the integer order to a fractional order are clearly visible as time varies. First, the decreasing trend from the integer order is observed for various fractional parameters. In the last two columns, the increasing trend towards the integer order is displayed. If we compare the results of the second case and the fourth, we see that the results initially go toward the worst and then beat the integer order at $r = 0.7$. Similarly, the BHCS results are compared against the available literature in Table 3. In case 1, the results for the integer order are almost the same as the LSM. In cases 2, 3, and 4, the results of BHCS are nearer the integer-order solution when compared to LSM. This proves the validity and efficiency of our proposed method, BHCS.

Table 3. Comparison of the approximate solutions of the BHCS neural network using the least squares method (LSM) [40].

	Case study 1		Case Study 2		Case Study 3		Case Study 4	
r	$\mu = 2, \nu = 1$		$\mu = 1.70, \nu = 0.70$		$\mu = 1.80, \nu = 0.80$		$\mu = 1.90, \nu = 0.90$	
	BHCS	**LSM**	**BHCS**	**LSM**	**BHCS**	**LSM**	**BHCS**	**LSM**
0.1	1.1603	1.1603	1.1704	1.1688	1.1648	1.1770	1.1616	1.1848
0.2	1.1587	1.1587	1.1677	1.1669	1.1626	1.1747	1.1598	1.1819
0.3	1.1561	1.1561	1.1637	1.1638	1.1592	1.1710	1.1568	1.1776
0.4	1.1524	1.1524	1.1587	1.1596	1.1547	1.1662	1.1528	1.1722
0.5	1.1477	1.1477	1.1526	1.1543	1.1492	1.1603	11.1477	1.1656
0.6	1.1419	1.1419	1.1457	1.1479	1.1427	1.1534	1.1416	1.1581
0.7	1.1350	1.1350	1.1378	1.1405	1.1353	1.1454	1.1345	1.1496
0.8	1.1271	1.1271	1.1290	1.1320	1.1269	1.1364	1.1263	1.1401
0.9	1.1180	1.1180	1.1194	1.1224	1.1176	1.1264	1.1172	1.1297
1.0	1.1078	1.1078	1.1089	1.1118	1.1073	1.1154	1.1070	1.1183

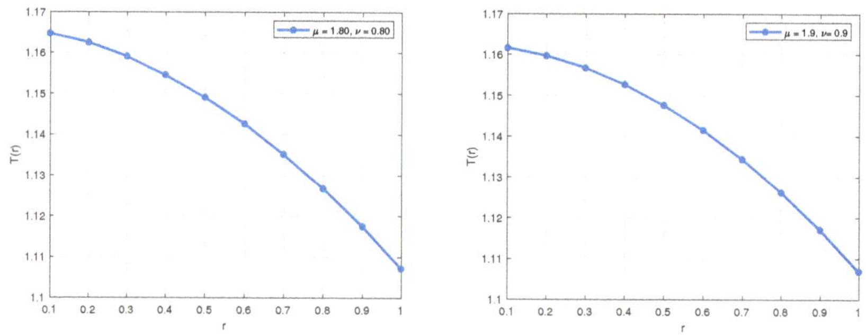

(a) (b)

Figure 5. Graphical representation of the solutions for (**a**) Case study 3 and (**b**) Case study 4.

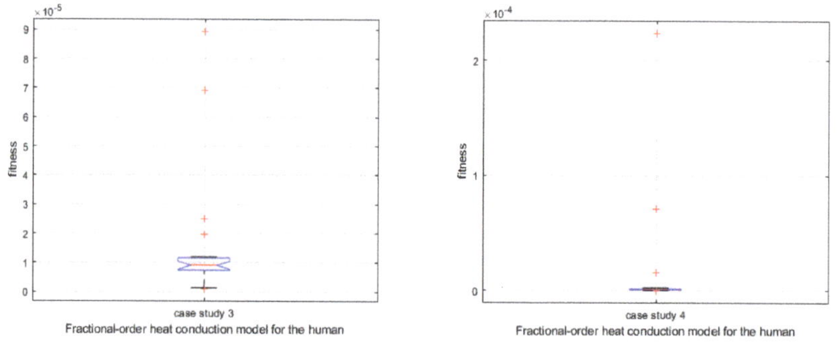

(a) (b)

Figure 6. Graphical representation of fitness function for (**a**) Case study 3 and (**b**) Case study 4.

4. Conclusions

In this article, we discussed the distribution of temperature within the human skull by considering the fractional derivative. In order to solve the proposed model, we utilized the biogeography-based hybrid cuckoo search (BHCS) algorithm to then be used on the transformed fractional-order equation. The following are the main features obtained based on our analysis.

- The proposed problem was tackled by using the Riemann-Liouville definition of fractional-order derivatives for briefly analyzing the transfer of heat at the integer and non-integer points;
- The suggested fractional-order graphs that explain the parameters (r, N_B, m, λ) provide a more accurate representation of the distribution of temperature within the human skull;
- A new type of BHCS algorithm was applied to reduce the L_2-norm for the fitness function;
- On the basis of the L_2-error, we observed that the case obtained extraordinary results that beat the integer order: $r = 0.7$;
- The results were validated against the available literature [40], as per Table 3.

Author Contributions: Conceptualization, W. and S.A.; methodology, S.K.; software, W. and A.U.; validation, M.A., E.A.A.I., S.K., F.A.A. and A.U.; formal analysis, W. and E.A.A.I.; investigation, S.A.; resources, A.U.; data curation, W.; writing—original draft preparation, M.A.; visualization, A.U.; supervision, W.; review writing, E.A.A.I., F.A.A., W., S.A. and A.U.; funding, E.A.A.I. and F.A.A. All authors have read and agreed to the published version of the manuscript.

Funding: This project was funded by King Saud University, Riyadh, Saudi Arabia.

Data Availability Statement: No data are used in this study.

Acknowledgments: Researchers' supporting project number: (RSPD2023R1060), King Saud University, Riyadh, Saudi Arabia.

Conflicts of Interest: The authors declare no conflict of interest.

References

1. Ferreri, F.; Curcio, G.; Pasqualetti, P.; De Gennaro, L.; Fini, R.; Rossini, P.M. Mobile phone emissions and human brain excitability. *Ann. Neurol. Off. J. Am. Neurol. Assoc. Child Neurol. Soc.* **2006**, *60*, 188–196. [CrossRef]
2. Hossmann, K.A.; Hermann, D. Effects of electromagnetic radiation of mobile phones on the central nervous system. *Bioelectromagn. J. Bioelectromagn. Soc. Soc. Phys. Regul. Biol. Med. Eur. Bioelectromagn. Assoc.* **2003**, *24*, 49–62. [CrossRef] [PubMed]
3. Keetley, V.; Wood, A.W.; Spong, J.; Stough, C. Neuropsychological sequelae of digital mobile phone exposure in humans. *Neuropsychologia* **2006**, *44*, 1843–1848. [CrossRef]
4. Bernardi, P.; Cavagnaro, M.; Pisa, S.; Piuzzi, E. Specific absorption rate and temperature increases in the head of a cellular-phone user. *IEEE Trans. Microw. Theory Tech.* **2000**, *48*, 1118–1126. [CrossRef]
5. Sengupta, S.; Guha, A. The fluid dynamics of symmetry and momentum transfer in microchannels within co-rotating discs with discrete multiple inflows. *Phys. Fluids* **2017**, *29*, 093604. [CrossRef]
6. Ratajczak, M.; Ptak, M.; Kwiatkowski, A.; Kubicki, K.; Fernandes, F.A.; Wilhelm, J.; Dymek, M.; Sawicki, M.; Żółkiewski, S. Symmetry of the Human Head—Are Symmetrical Models More Applicable in Numerical Analysis? *Symmetry* **2021**, *13*, 1252. [CrossRef]
7. Flesch, U. The distribution of heat sources in the human head: A theoretical consideration. *J. Theor. Biol.* **1975**, *54*, 285–287. [CrossRef] [PubMed]
8. Gray, B. The distribution of heat sources in the human head—Theoretical considerations. *J. Theor. Biol.* **1980**, *82*, 473–476. [CrossRef]
9. Anderson, N.; Arthurs, A. Complementary extremum principles for a nonlinear model of heat conduction in the human head. *Bull. Math. Biol.* **1981**, *43*, 341–346. [CrossRef] [PubMed]
10. Makinde, O.D. Non-perturbative solutions of a nonlinear heat conduction model of the human head. *Sci. Res. Essays* **2010**, *5*, 529–532.
11. Raja, M.A.Z.; Umar, M.; Sabir, Z.; Khan, J.A.; Baleanu, D. A new stochastic computing paradigm for the dynamics of nonlinear singular heat conduction model of the human head. *Eur. Phys. J. Plus* **2018**, *133*, 364.

12. Abdelhakem, M.; Youssri, Y. Two spectral Legendre's derivative algorithms for Lane-Emden, Bratu equations, and singular perturbed problems. *Appl. Numer. Math.* **2021**, *169*, 243–255. [CrossRef]
13. Youssri, Y.; Abd-Elhameed, W.; Doha, E. Ultraspherical wavelets method for solving Lane–Emden type equations. *Rom. J. Phys* **2015**, *60*, 1298–1314.
14. Abd-Elhameed, W.M.; Youssri, Y.; Doha, E.H. New solutions for singular Lane-Emden equations arising in astrophysics based on shifted ultraspherical operational matrices of derivatives. *Comput. Methods Differ. Equ.* **2014**, *2*, 171–185.
15. Doha, E.; Abd-Elhameed, W.; Youssri, Y. Second kind Chebyshev operational matrix algorithm for solving differential equations of Lane–Emden type. *New Astron.* **2013**, *23*, 113–117.
16. Odibat, Z.; Baleanu, D. Numerical simulation of initial value problems with generalized Caputo-type fractional derivatives. *Appl. Numer. Math.* **2020**, *156*, 94–105. [CrossRef]
17. Caputo, M.; Fabrizio, M. A new definition of fractional derivative without singular kernel. *Prog. Fract. Differ. Appl.* **2015**, *1*, 73–85.
18. Podlubny, I. An introduction to fractional derivatives, fractional differential equations, to methods of their solution and some of their applications. *Math. Sci. Eng* **1999**, *198*, 340.
19. Aleksandrovich, K.; Srivastava, H.M.; Trujillo, J.J. *Theory and Applications of Fractional Differential Equations*; Elsevier: Amsterdam, The Netherlands, 2006; Volume 204.
20. Wang, K.J. A new fractional nonlinear singular heat conduction model for the human head considering the effect of febrifuge. *Eur. Phys. J. Plus* **2020**, *135*, 871.
21. Kumar, P.; Erturk, V.S.; Kumar, A. A new technique to solve generalized Caputo type fractional differential equations with the example of computer virus model. *J. Math. Ext.* **2021**, *15*. [CrossRef]
22. Odibat, Z.; Erturk, V.S.; Kumar, P.; Govindaraj, V. Dynamics of generalized Caputo type delay fractional differential equations using a modified Predictor-Corrector scheme. *Phys. Scr.* **2021**, *96*, 125213.
23. Kumar, P.; Erturk, V.S. Environmental persistence influences infection dynamics for a butterfly pathogen via new generalised Caputo type fractional derivative. *Chaos Solitons Fractals* **2021**, *144*, 110672.
24. Kumar, P.; Erturk, V.S.; Banerjee, R.; Yavuz, M.; Govindaraj, V. Fractional modeling of plankton-oxygen dynamics under climate change by the application of a recent numerical algorithm. *Phys. Scr.* **2021**, *96*, 124044.
25. Kumar, P.; Erturk, V.S.; Murillo-Arcila, M. A complex fractional mathematical modeling for the love story of Layla and Majnun. *Chaos Solitons Fractals* **2021**, *150*, 111091.
26. Kumar, P.; Govindaraj, V.; Erturk, V.S.; Abdellattif, M.H. A study on the dynamics of alkali–silica chemical reaction by using Caputo fractional derivative. *Pramana* **2022**, *96*, 128. [CrossRef]
27. Kumar, P.; Govindaraj, V.; Erturk, V.S. A novel mathematical model to describe the transmission dynamics of tooth cavity in the human population. *Chaos Solitons Fractals* **2022**, *161*, 112370.
28. Nabi, K.N.; Abboubakar, H.; Kumar, P. Forecasting of COVID-19 pandemic: From integer derivatives to fractional derivatives. *Chaos Solitons Fractals* **2020**, *141*, 110283. [PubMed]
29. Nisar, K.S.; Ahmad, S.; Ullah, A.; Shah, K.; Alrabaiah, H.; Arfan, M. Mathematical analysis of SIRD model of COVID-19 with Caputo fractional derivative based on real data. *Results Phys.* **2021**, *21*, 103772. [PubMed]
30. Özköse, F.; Yavuz, M.; Şenel, M.T.; Habbireeh, R. Fractional order modelling of omicron SARS-CoV-2 variant containing heart attack effect using real data from the United Kingdom. *Chaos Solitons Fractals* **2022**, *157*, 111954. [CrossRef] [PubMed]
31. Erturk, V.; Godwe, E.; Baleanu, D.; Kumar, P.; Asad, J.; Jajarmi, A. Novel fractional-order Lagrangian to describe motion of beam on nanowire. *Acta Phys. Pol. A* **2021**, *140*, 265–272.
32. Yavuz, M.; Sene, N.; Yıldız, M. Analysis of the influences of parameters in the fractional second-grade fluid dynamics. *Mathematics* **2022**, *10*, 1125. [CrossRef]
33. Hammouch, Z.; Yavuz, M.; Özdemir, N. Numerical solutions and synchronization of a variable-order fractional chaotic system. *Math. Model. Numer. Simul. Appl.* **2021**, *1*, 11–23. [CrossRef]
34. Yavuz, M. European option pricing models described by fractional operators with classical and generalized Mittag-Leffler kernels. *Numer. Methods Partial. Differ. Equ.* **2022**, *38*, 434–456.
35. Avci, D.; Yavuz, M.; Ozdemir, N. Fundamental solutions to the Cauchy and Dirichlet problems for a heat conduction equation equipped with the Caputo-Fabrizio differentiation. In *Heat Conduction: Methods, Applications and Research*; Nova Science Publishers, Inc.: Hauppauge, NY, USA, 2019.
36. Erturk, V.S.; Ahmadkhanlu, A.; Kumar, P.; Govindaraj, V. Some novel mathematical analysis on a corneal shape model by using Caputo fractional derivative. *Optik* **2022**, *261*, 169086. [CrossRef]
37. Rezk, H.; Albalawi, W.; Abd El-Hamid, H.; Saied, A.I.; Bazighifan, O.; Mohamed, M.S.; Zakarya, M. Hardy-Leindler-Type Inequalities via Conformable Delta Fractional Calculus. *J. Funct. Spaces* **2022**, *2022*, 2399182. [CrossRef]
38. Refaai, D.; El-Sheikh, M.M.; Ismail, G.A.; Zakarya, M.; AlNemer, G.; Rezk, H.M. Stability of Nonlinear Fractional Delay Differential Equations. *Symmetry* **2022**, *14*, 1606. [CrossRef]
39. AlNemer, G.; Kenawy, M.; Zakarya, M.; Cesarano, C.; Rezk, H.M. Generalizations of Hardy's type inequalities via conformable calculus. *Symmetry* **2021**, *13*, 242. [CrossRef]
40. Kumar, P.; Erturk, V.S.; Harley, C. A novel study on a fractional-order heat conduction model for the human head by using the least-squares method. *Int. J. Dyn. Control* **2022**, *11*, 1040–1049. [CrossRef]

41. Yang, X.S.; Deb, S. Cuckoo search via Lévy flights. In Proceedings of the 2009 World Congress on Nature & Biologically Inspired Computing (NaBIC), IEEE, Coimbatore, India, 9–11 December 2009; pp. 210–214.
42. Ding, X.; Xu, Z.; Cheung, N.J.; Liu, X. Parameter estimation of Takagi–Sugeno fuzzy system using heterogeneous cuckoo search algorithm. *Neurocomputing* **2015**, *151*, 1332–1342. [CrossRef]
43. Yang, X.S.; Deb, S. Engineering optimisation by cuckoo search. *arXiv* **2010**, arXiv:1005.2908.
44. Simon, D. Biogeography-based optimization. *IEEE Trans. Evol. Comput.* **2008**, *12*, 702–713. [CrossRef]
45. Cheung, N.J.; Ding, X.M.; Shen, H.B. A nonhomogeneous cuckoo search algorithm based on quantum mechanism for real parameter optimization. *IEEE Trans. Cybern.* **2016**, *47*, 391–402. [CrossRef] [PubMed]

symmetry

MDPI

Article

Averaging Principle for ψ-Capuo Fractional Stochastic Delay Differential Equations with Poisson Jumps

Dandan Yang, Jingfeng Wang and Chuanzhi Bai *

Department of Mathematics, Huaiyin Normal University, Huaian 223300, China; ydd@hytc.edu.cn (D.Y.); jfwang@hytc.edu.cn (J.W.)
* Correspondence: czbai@hytc.edu.cn

Abstract: In this paper, we study the averaging principle for ψ-Capuo fractional stochastic delay differential equations (FSDDEs) with Poisson jumps. Based on fractional calculus, Burkholder-Davis-Gundy's inequality, Doob's martingale inequality, and the Hölder inequality, we prove that the solution of the averaged FSDDEs converges to that of the standard FSDDEs in the sense of L^p. Our result extends some known results in the literature. Finally, an example and simulation is performed to show the effectiveness of our result.

Keywords: averaging principle; ψ-Capuo fractional stochastic delay differential equations; Poisson jumps; L^p convergence

MSC: 34K50; 26A33; 60J75

Citation: Yang, D.; Wang, J.; Bai, C. Averaging Principle for ψ-Capuo Fractional Stochastic Delay Differential Equations with Poisson Jumps. *Symmetry* **2023**, *15*, 1346. https://doi.org/10.3390/sym15071346

Academic Editor: Mohammed K. A. Kaabar

Received: 28 April 2023
Revised: 25 June 2023
Accepted: 27 June 2023
Published: 1 July 2023

1. Introduction

Many systems exhibit natural symmetry, such as chemical, physical, and biological systems. It is well known that stochastic differential equations play an important role in explaining some symmetry phenomena (see [1–3]). Additionally, we know that stochastic differential equations are mathematical tools widely used to simulate and model stochastic processes. Recently, more in-depth research has been conducted on the theory and application aspects of these equations to adapt to more complex systems, such as chemical reaction networks, atmospheric environments, and financial markets; readers can refer to the papers [4–7] for more information.

In 1968, Khasminskii [8] extended the averaging principles for ODEs to the case of stochastic differential equations (SDEs). Since then, the averaging principles for SDEs have found applications in many areas of science and engineering, including fluid dynamics, control theory, and climate modeling. Many people have devoted their efforts to the study of averaging principles for SDEs, for example, see [9–11].

As we all know, compared with integer-order derivatives, fractional-order derivatives provide a magnificent approach to describe the memory and hereditary properties of various processes. Thus, fractional differential equations are more accurate and convenient than integer-order ones. The numerical solution of fractional-order nonlinear systems is an active research area with ongoing developments and improvements in the different numerical algorithms and techniques used [12–14].

With the development of fractional calculus, the averaging principles for fractional stochastic differential equations (FSDEs) have become a widespread concern [15–17]. One notable approach of research is the fractional averaging principle, which extends the classical averaging principle to FSDEs. Another approach of research is the stochastic averaging principle, which combines averaging methods with stochastic calculus. Overall, research into averaging principles for FSDEs is still ongoing, and there is much to be explored in terms of developing new methods and exploring their applications.

Recently, Wang and Lin [18] extended the averaging principle of the following fractional stochastic differential equations (FSDEs)

$$\begin{cases} {}^{C}D_{0}^{\alpha}[x(t) - h(t, x(t)] = f(t, x(t)) + g(t, x(t))\frac{dB_t}{dt}, & t \in J = [0, T], \\ x(0) = x_0, \end{cases} \tag{1}$$

in the sense of mean square (L^2 convergence) to L^p convergence ($p \geq 2$), which generated some works on the averaging principle for FSDES [19–21].

The periodic averaging method for impulsive conformable fractional stochastic differential equations with Poisson jumps are discussed in [22] by Ahmed. For some recent works on Hilfer fractional order stochastic differential systems, we refer to [23–26]. In [27], Ahmed and Zhu investigated the averaging principle for the following Hilfer fractional stochastic delay differential equation with Poisson jumps in the sense of mean square

$$\begin{cases} D_{0}^{\aleph,\hbar}x(t) = \Re(t, x(t), x(t-\tau)) + \sigma(t, x(t), x(t-\tau))\frac{dB}{dt}, \\ \qquad + \int_V h(t, x(t), x(t-\tau), v)\tilde{N}(dt, dv), \quad t \in J = (0, T], \\ x(t) = \phi(t), \quad -\tau \leq t \leq 0, \\ I_{0+}^{(1-\aleph)(1-\hbar)}x(0) = \phi(0). \end{cases} \tag{2}$$

In [28], Almeida generalized the definition of the Caputo fractional derivative by considering the Caputo fractional derivative of a function with respect to another function ψ. Since then, there have been so many papers involving the ψ-Caputo fractional derivative, see [29–32]. Recently, there have been many works on SDEs with Poisson jumps, see, for example, [33–35] and the references therein. However, to the best of our knowledge, the averaging principle for the ψ-Capuo fractional stochastic delay differential equation with Poisson jumps in the sense of L^p convergence has not yet been researched in the literature. In the present paper, motivated by the above-mentioned works, we study the following ψ-Caputo fractional stochastic delay differential equation with Poisson jumps

$$\begin{cases} {}^{C}D_{0}^{\alpha,\psi}[x(t) - h(t, x(t)] = f(t, x(t), x(t-\tau)) + \sigma(t, x(t), x(t-\tau))\frac{dB_t}{dt}, \\ \qquad + \int_V g(t, x(t), x(t-\tau), v)\tilde{N}(dt, dv), \quad t \in J = (0, T], \\ x(t) = \phi(t), \quad -\tau \leq t \leq 0, \end{cases} \tag{3}$$

where ${}^{C}D_{0}^{\alpha,\psi}$ is the left ψ-Caputo fractional derivative with $0 < \alpha < 1$ and $\psi \in C^1([a, b])$ is an increasing function with $\psi'(t) \neq 0$ for all $t \in [0, T]$, $J = (0, T]$, $x \in \mathbb{R}^n$ is a stochastic process, $h, f : J \times \mathbb{R}^n \times \mathbb{R}^n \to \mathbb{R}^n$, $\sigma : J \times \mathbb{R}^n \times \mathbb{R}^n \to \mathbb{R}^{n\times m}$, and $g : J \times \mathbb{R}^n \times \mathbb{R}^n \times V \to \mathbb{R}^n$. Let $(\Omega, \mathcal{F}, P)$ be a complete probability space equipped with a filtration $(\mathcal{F}_t)_{t\geq 0}$ satisfying the usual condition. Here, B_t is an m-dimensional Brownian motion on the probability space $(\Omega, \mathcal{F}, P)$ adapted to the filtration $(\mathcal{F}_t)_{t\geq 0}$. Let $(V, \Phi, \lambda(dv))$ be a σ-finite measurable space, given the stationary Poisson point process $(p_t)_{t\geq 0}$, which is defined on $(\Omega, \mathcal{F}, P)$ with values in V and with the characteristic measure λ. We denote by $N(t, dv)$ the counting measure of p_t such that $\tilde{N}(t, \Theta) := \mathbb{E}(N(t, \Theta)) = t\lambda(\Theta)$ for $\Theta \in \Phi$. Define $\tilde{N}(t, dv) := N(t, dv) - t\lambda(dv)$, and the Poisson martingale measure is generated by p_t.

In this paper, we prove that the solution of the averaged neutral SFDDEs with Poisson random measure converges to that of the standard one in L^p sense. The main contributions and advantages of this paper are as follows:

(i) For the first time in the literature, the averaging principle for ψ-Capuo fractional stochastic delay differential equations with Poisson jumps is investigated.

(ii) The fractional calculus, stochastic inequality, and Hölder inequality are effectively used to establish our result.

(iii) Our work in this paper is novel and more technical. Our result has greatly promoted and extended the main result of [18].

This paper will be organized as follows. In Section 2, we will briefly recall some definitions and preliminaries. Section 3 is devoted to obtaining an averaging principle for the solution of the considered system (3). Additionally, a numerical simulation example is provided to illustrate our main result. Finally, the paper is concluded in Section 4.

2. Preliminaries

In this section, we recall some basic definitions and lemmas, which are used in the sequel.

Definition 1 ([36]). *Let $\alpha > 0$, f be an integrable function defined on $[a,b]$ and $\psi \in C^1([a,b])$ be an increasing function with $\psi'(t) \neq 0$ for all $t \in [a,b]$. The left ψ-Riemann-Liouville fractional integral operator of order α of a function f is defined by*

$$ {}_aI_t^{\alpha,\psi} f(t) = \frac{1}{\Gamma(\alpha)} \int_a^t \psi'(s)(\psi(t) - \psi(s))^{\alpha-1} f(s) ds. \tag{4} $$

Definition 2 ([28,36]). *Let $n-1 < \alpha < n$, $f \in C^n([a,b])$ and $\psi \in C^n([a,b])$ be an increasing function with $\psi'(t) \neq 0$ for all $t \in [a,b]$. The left ψ-Caputo fractional derivative of order α of a function f is defined by*

$$ \begin{aligned} {}_a^CD_t^{\alpha,\psi} f(t) &= ({}_aI_t^{n-\alpha,\psi} f^{[n]})(t) \\ &= \frac{1}{\Gamma(n-\alpha)} \int_a^t (\psi(t) - \psi(s))^{n-\alpha-1} f^{[n]}(s) \psi'(s) ds, \end{aligned} \tag{5} $$

where $n = [\alpha] + 1$ and $f^{[n]}(t) := \left(\frac{1}{\psi'(t)}\frac{d}{dt}\right)^n f(t)$ on $[a,b]$.

In the following, we will give some properties of the combinations of the fractional integral and the fractional derivatives of a function with respect to another function.

Lemma 1 ([28]). *Let $f \in C^n([a,b])$ and $n-1 < \alpha < n$. Then, we have*

(1) ${}_a^CD_t^{\alpha,\psi} {}_aI_t^{\alpha,\psi} f(t) = f(t)$;

(2) $I_t^{\alpha,\psi} {}_a^CD_t^{\alpha,\psi} f(t) = f(t) - \sum_{k=0}^{n-1} \frac{f^{[k]}(a^+)}{\Gamma(k-\alpha)} (\psi(t) - \psi(a))^k$.

In particular, given $\alpha \in (0,1)$, one has

$$ I_t^{\alpha,\psi} {}_a^CD_t^{\alpha,\psi} = f(t) - f(a). $$

To study the averaging method of Equation (3), we impose the following conditions on data of the problem.

(H1) If $|h(0,\phi(0))| < \infty$, $t \in [0,T]$ and for all $x,y \in R^n$, a constant $C_1 \in (0,1)$ exists such that

$$ |h(t,x) - h(t,y)| \leq C_1|x-y|. $$

(H2) For any $x_1, x_2, y_1, y_2 \in R^n$ and $t \in J$, two constants $C_2, C_3 > 0$ exist such that

$$ \begin{aligned} &|f(t,x_1,y_1) - f(t,x_2,y_2)|^p \vee |\sigma(t,x_1,y_1) - \sigma(t,x_2,y_2)|^p \\ &\vee \int_V |g(t,x_1,y_1,v) - g(t,x_2,y_2,v)|^p \lambda(dv) \leq C_2^p(|x_1 - x_2|^p + |y_1 - y_2|^p), \end{aligned} $$

and

$$ |f(t,x_1,y_1)|^p \vee |\sigma(t,x_1,y_1)|^p \vee \int_V |g(t,x_1,y_1,v)|^p \lambda(dv) \leq C_3^p(1 + |x_1|^p + |y_1|^p). $$

According to Lemma 1 and [37], an R^n-value stochastic process $\{x(t), -\tau \leq t \leq T\}$ is called a unique solution of Equation (3) if $x(t)$ satisfies the following :

$$x(t) = \begin{cases} \phi_0 - h(0,\phi_0) + h(t,x(t)) + \frac{1}{\Gamma(\alpha)}\int_0^t (\psi(t)-\psi(s))^{\alpha-1}\psi'(s) f(s,x(s),x(s-\tau))ds \\ +\frac{1}{\Gamma(\alpha)}\int_0^t (\psi(t)-\psi(s))^{\alpha-1}\psi'(s)\sigma(s,x(s),x(s-\tau))dB_s \\ +\frac{1}{\Gamma(\alpha)}\int_0^t (\psi(t)-\psi(s))^{\alpha-1}\psi'(s)\int_V g(s,x(s),x(s-\tau),v)\bar{N}(ds,dv), \quad t\in J, \\ \phi(t), \quad t\in[-r,0], \end{cases} \tag{6}$$

where $\phi_0 = \phi(0)$.

For each $t \in J$, we consider the standard form of Equation (6)

$$x_\epsilon(t) = \phi_0 - h(0,\phi_0) + h(t,x_\epsilon(t)) + \frac{\epsilon}{\Gamma(\alpha)}\int_0^t (\psi(t)-\psi(s))^{\alpha-1}\psi'(s) f(s,x_\epsilon(s),x_\epsilon(s-\tau))ds$$
$$+\frac{\sqrt{\epsilon}}{\Gamma(\alpha)}\int_0^t (\psi(t)-\psi(s))^{\alpha-1}\psi'(s)\sigma(s,x_\epsilon(s),x_\epsilon(s-\tau))dB_s$$
$$+\frac{\sqrt{\epsilon}}{\Gamma(\alpha)}\int_0^t (\psi(t)-\psi(s))^{\alpha-1}\psi'(s)\int_V g(s,x_\epsilon(s),x_\epsilon(s-\tau),v)\bar{N}(ds,dv), \quad t\in J, \tag{7}$$

where $\epsilon \in (0,\epsilon_0]$ is a positive small parameter with ϵ_0 being a fixed number.

Consider the averaged form, which corresponds to the standard form (7) as follows:

$$y_\epsilon(t) = \phi_0 - h(0,\phi_0) + h(t,y_\epsilon(t)) + \frac{\epsilon}{\Gamma(\alpha)}\int_0^t (\psi(t)-\psi(s))^{\alpha-1}\psi'(s) \bar{f}(y_\epsilon(s),y_\epsilon(s-\tau))ds$$
$$+\frac{\sqrt{\epsilon}}{\Gamma(\alpha)}\int_0^t (\psi(t)-\psi(s))^{\alpha-1}\psi'(s)\bar{\sigma}(y_\epsilon(s),y_\epsilon(s-\tau))dB_s$$
$$+\frac{\sqrt{\epsilon}}{\Gamma(\alpha)}\int_0^t (\psi(t)-\psi(s))^{\alpha-1}\psi'(s)\int_V \bar{g}(y_\epsilon(s),y_\epsilon(s-\tau),v)\bar{N}(ds,dv), \quad t\in J, \tag{8}$$

where $\bar{f}: R^n \times R^n \to R^n, \bar{\sigma}: R^n \times R^n \to R^{n\times m}$, and $\bar{g}: R^n \times R^n \times V \to R^n$ satisfying the following averaging condition :

(H3) For any $T_1 \in [0,T]$, $x,y \in R^n$ and $p \geq 2$, a positive bounded function $\beta(\cdot)$ exists such that

$$\frac{1}{T_1}\int_0^{T_1} |f(t,x,y)-\bar{f}(x,y)|^p dt \vee \frac{1}{T_1}\int_0^{T_1} |\sigma(t,x,y)-\bar{\sigma}(x,y)|^p dt$$
$$\vee \frac{1}{T_1}\int_0^{T_1}\left(\int_V |g(t,x,y,v)-\bar{g}(x,y,v)|^p \lambda(dv)\right)dt \leq \beta(T_1)(1+|x|^p+|y|^p),$$

and $\lim_{T_1\to\infty}\beta(T_1) = 0$.

Lemma 2. Suppose that (H2) and (H3) hold. Then, for $T_1 \in (0,T]$ we have

$$|\bar{\sigma}(x,y)|^p \leq C_4(1+|x|^p+|y|^p) \quad \text{and} \quad \int_V |\bar{g}(x,y,v)|^p \lambda(dv) \leq C_4(1+|x|^p+|y|^p),$$

where $C_4 = 2^{p-1}(\beta(T_1)+C_3^p)$.

Proof. Using (H2), (H3) and Jensen's inequality, we obtain

$$|\bar{\sigma}(x,y)|^p \leq \frac{2^{p-1}}{T_1}\int_0^{T_1} |\bar{\sigma}(x,y)-\sigma(t,x,y)|^p dt + \frac{2^{p-1}}{T_1}\int_0^{T_1} |\sigma(t,x,y)|^p dt$$
$$\leq 2^{p-1}\beta(T_1)(1+|x|^p+|y|^p) + 2^{p-1}C_3^p(1+|x|^p+|y|^p)$$

$$= 2^{p-1}(\beta(T_1) + C_3^p)(1 + |x|^p + |y|^p).$$

Similarly, we can prove that

$$\int_V |\bar{g}(x, y, v)|^p \lambda(dv) \le 2^{p-1}(\beta(T_1) + C_3^p)(1 + |x|^p + |y|^p).$$

□

Lemma 3 ([38]). If $p \ge 2$ and $a, b \in \mathbb{R}^n$, then for any $k \in (0, 1)$, one has

$$|a + b|^p \le \frac{|a|^p}{k^{p-1}} + \frac{|b|^p}{(1-k)^{p-1}}.$$

Lemma 4 ([39,40]). Let $\phi : R_+ \times V \to R^n$ and assume that

$$\int_0^t \int_V |\phi(s, v)|^p \lambda(dv) ds < \infty, \quad p \ge 2.$$

Then, $D_p > 0$ exists such that

$$\mathbb{E}\left(\sup_{0 \le t \le u} \left| \int_0^t \int_V \phi(s, v) \tilde{N}(ds, dv) \right|^p \right)$$
$$\le D_p \left\{ \mathbb{E}\left(\int_0^u \int_V |\phi(s, v)|^2 \lambda(dv) ds \right)^{\frac{p}{2}} + \mathbb{E} \int_0^u \int_V |\phi(s, v)|^p \lambda(dv) ds \right\}.$$

Lemma 5 ([41]). *Let u, v be two integrable functions and g be continuous defined on domain $[a, b]$. Let $\psi \in C^1[a, b]$ be an increasing function such that $\psi'(t) \ne 0, \forall t \in [a, b]$. Moreover, assume that*

(1) u and v are nonnegative, and v is nondecreasing;

(2) g is nonnegative and nondecreasing.

If

$$u(t) \le v(t) + g(t) \int_a^t \psi'(\tau)(\psi(t) - \psi(\tau))^{\alpha-1} u(\tau) d\tau,$$

then

$$u(t) \le v(t) E_\alpha(g(t)\Gamma(\alpha)(\psi(t) - \psi(a))^\alpha), \quad \forall t \in [a, b],$$

where E_α is the Mittag–Leffler function.

3. Main Results

Theorem 1. *Assume that (H1)–(H3) are satisfied. Then, for a given arbitrary small number $\delta > 0$, $p = 2$, $\frac{1}{2} < \alpha < 1$, or $p > 2$ and $\max\left\{\frac{p-1}{p}, \frac{p+2}{2p}\right\} < \alpha < 1$, $M > 0$, $\epsilon_1 \in (0, \epsilon_0]$ and $\gamma \in (0, 1)$ exist such that*

$$\mathbb{E}\left(\sup_{t \in [-\tau, M\epsilon^{-\gamma}]} |x_\epsilon(t) - y_\epsilon(t)|^p \right) \le \delta, \tag{9}$$

for all $\epsilon \in (0, \epsilon_1]$.

Proof. If $p = 2$, it is easy to prove that (9) holds by using the similar method as in [27]. In the following, we will only consider the case $p > 2$. From Equations (7) and (8), we obtain

$$x_\epsilon(t) - y_\epsilon(t) = h(t, x_\epsilon(t)) - h(t, y_\epsilon(t))$$

$$+\frac{\epsilon}{\Gamma(\alpha)}\int_0^t(\psi(t)-\psi(s))^{\alpha-1}\psi'(s)[f(s,x_\epsilon(s),x_\epsilon(s-\tau))-\bar{f}(y_\epsilon(s),y_\epsilon(s-\tau))]ds$$
$$+\frac{\sqrt{\epsilon}}{\Gamma(\alpha)}\int_0^t(\psi(t)-\psi(s))^{\alpha-1}\psi'(s)[\sigma(s,x_\epsilon(s),x_\epsilon(s-\tau))-\bar{\sigma}(y_\epsilon(s),y_\epsilon(s-\tau))]dB_s$$
$$+\frac{\sqrt{\epsilon}}{\Gamma(\alpha)}\int_0^t(\psi(t)-\psi(s))^{\alpha-1}\psi'(s)\int_V[g(s,x_\epsilon(s),x_\epsilon(s-\tau),v))$$
$$-\bar{g}(x_\epsilon(s),x_\epsilon(s-\tau),v))]\tilde{N}(ds,dv).$$

Choosing $k=C_1$ in Lemma 3, using (H1) and the following elementary inequalities:

$$|a+b|^p\le 2^{p-1}(|a|^p+|b|^p),\quad |a+b+c|^p\le 3^{p-1}(|a|^p+|b|^p+|c|^p),\tag{10}$$

we obtain

$$|x_\epsilon(t)-y_\epsilon(t)|^p\le C_1|x_\epsilon(t)-y_\epsilon(t)|^p$$
$$+\frac{3^{p-1}\epsilon^p}{(1-C_1)^{p-1}\Gamma(\alpha)^p}\left|\int_0^t(\psi(t)-\psi(s))^{\alpha-1}\psi'(s)[f(s,x_\epsilon(s),x_\epsilon(s-\tau))-\bar{f}(y_\epsilon(s),y_\epsilon(s-\tau))]ds\right|^p$$
$$+\frac{3^{p-1}\epsilon^{\frac{p}{2}}}{(1-C_1)^{p-1}\Gamma(\alpha)^p}\left|\int_0^t(\psi(t)-\psi(s))^{\alpha-1}\psi'(s)[\sigma(s,x_\epsilon(s),x_\epsilon(s-\tau))-\bar{\sigma}(y_\epsilon(s),y_\epsilon(s-\tau))]dB_s\right|^p$$
$$+\frac{3^{p-1}\epsilon^{\frac{p}{2}}}{(1-C_1)^{p-1}\Gamma(\alpha)^p}\left|\int_0^t(\psi(t)-\psi(s))^{\alpha-1}\psi'(s)\int_V[g(s,x_\epsilon(s),x_\epsilon(s-\tau),v))\right.$$
$$\left.-\bar{g}(x_\epsilon(s),x_\epsilon(s-\tau),v))]\tilde{N}(ds,dv)\right|^p.\tag{11}$$

For any $t\in[0,u]\subset[0,T]$, taking the expectation on both sides Equation (11), we have

$$\mathbb{E}\left(\sup_{0\le t\le u}|x_\epsilon(t)-y_\epsilon(t)|^p\right)$$
$$\le\frac{3^{p-1}\epsilon^p}{(1-C_1)^p\Gamma(\alpha)^p}\mathbb{E}\left(\sup_{0\le t\le u}\left|\int_0^t(\psi(t)-\psi(s))^{\alpha-1}\psi'(s)[f(s,x_\epsilon(s),x_\epsilon(s-\tau))-\bar{f}(y_\epsilon(s),y_\epsilon(s-\tau))]ds\right|^p\right)$$
$$+\frac{3^{p-1}\epsilon^{\frac{p}{2}}}{(1-C_1)^p\Gamma(\alpha)^p}\mathbb{E}\left(\sup_{0\le t\le u}\left|\int_0^t(\psi(t)-\psi(s))^{\alpha-1}\psi'(s)[\sigma(s,x_\epsilon(s),x_\epsilon(s-\tau))-\bar{\sigma}(y_\epsilon(s),y_\epsilon(s-\tau))]dB_s\right|^p\right)$$
$$+\frac{3^{p-1}\epsilon^{\frac{p}{2}}}{(1-C_1)^p\Gamma(\alpha)^p}\mathbb{E}\left(\sup_{0\le t\le u}\left|\int_0^t(\psi(t)-\psi(s))^{\alpha-1}\psi'(s)\int_V[g(s,x_\epsilon(s),x_\epsilon(s-\tau),v))\right.\right.$$
$$\left.\left.-\bar{g}(x_\epsilon(s),x_\epsilon(s-\tau),v))]\tilde{N}(ds,dv)\right|^p\right).$$
$$=I_1+I_2+I_3.\tag{12}$$

Applying Jensen inequality, we obtain

$$I_1\le\frac{6^{p-1}\epsilon^p}{(1-C_1)^p\Gamma(\alpha)^p}\mathbb{E}\left(\sup_{0\le t\le u}\left|\int_0^t(\psi(t)-\psi(s))^{\alpha-1}\psi'(s)[f(s,x_\epsilon(s),x_\epsilon(s-\tau))-f(s,y_\epsilon(s),y_\epsilon(s-\tau))]ds\right|^p\right)$$
$$+\frac{6^{p-1}\epsilon^p}{(1-C_1)^p\Gamma(\alpha)^p}\mathbb{E}\left(\sup_{0\le t\le u}\left|\int_0^t(\psi(t)-\psi(s))^{\alpha-1}\psi'(s)[f(s,y_\epsilon(s),y_\epsilon(s-\tau))-\bar{f}(y_\epsilon(s),y_\epsilon(s-\tau))]ds\right|^p\right)$$

$$= I_{11} + I_{12}. \tag{13}$$

Thanks to the Hölder inequality and (H2), we obtain

$$\begin{aligned}
I_{11} \leq & \frac{6^{p-1}\epsilon^p}{(1-C_1)^p\Gamma(\alpha)^p}\left(\int_0^u 1ds\right)^{p-1} \\
& \cdot\mathbb{E}\left(\sup_{0\leq t\leq u}\int_0^t (\psi(u)-\psi(s))^{p(\alpha-1)}\psi'(s)^p|f(s,x_\epsilon(s),x_\epsilon(s-\tau))-f(s,y_\epsilon(s),y_\epsilon(s-\tau))|^p ds\right) \\
\leq & \frac{6^{p-1}\epsilon^p}{(1-C_1)^p\Gamma(\alpha)^p}u^{p-1}K^{p-1}C_2^p \\
& \cdot\mathbb{E}\left(\sup_{0\leq t\leq u}\int_0^t (\psi(u)-\psi(s))^{p(\alpha-1)}\psi'(s)[|x_\epsilon(s)-y_\epsilon(s)|^p+|x_\epsilon(s-\tau))-y_\epsilon(s-\tau))|^p]ds\right) \\
\leq & A_{11}\epsilon^p u^{p-1}\int_0^u (\psi(u)-\psi(s))^{p(\alpha-1)}\psi'(s)\left[\mathbb{E}\left(\sup_{0\leq\theta\leq s}|x_\epsilon(\theta)-y_\epsilon(\theta)|^p\right)\right. \\
& \left.+\mathbb{E}\left(\sup_{0\leq\theta\leq s}|x_\epsilon(\theta-\tau)-y_\epsilon(\theta-\tau)|^p\right)\right]ds,
\end{aligned} \tag{14}$$

where $A_{11} = \frac{6^{p-1}C_2^p K^{p-1}}{(1-C_1)^p\Gamma(\alpha)^p}$ and $K = \sup_{t\in[0,T]}\psi'(t)$.

Applying the Hölder inequality, we obtain

$$\begin{aligned}
I_{12} \leq & \frac{6^{p-1}\epsilon^p}{(1-C_1)^p\Gamma(\alpha)^p}\left(\int_0^u (\psi(u)-\psi(s))^{\frac{(\alpha-1)p}{p-1}}\psi'(s)^{\frac{p}{p-1}}ds\right)^{p-1} \\
& \cdot\mathbb{E}\left(\sup_{0\leq t\leq u}\int_0^t |f(s,y_\epsilon(s),y_\epsilon(s-\tau))-\bar{f}(y_\epsilon(s),y_\epsilon(s-\tau))|^p ds\right).
\end{aligned} \tag{15}$$

Since

$$\begin{aligned}
\int_0^u (\psi(u)-\psi(s))^{\frac{(\alpha-1)p}{p-1}}\psi'(s)^{\frac{p}{p-1}}ds &= \int_0^u (\psi(u)-\psi(s))^{\frac{(\alpha-1)p}{p-1}}\psi'(s)\cdot\psi'(s)^{\frac{1}{p-1}}ds \\
&\leq K^{\frac{1}{p-1}}\int_0^u (\psi(u)-\psi(s))^{\frac{(\alpha-1)p}{p-1}}\psi'(s)ds \\
&= K^{\frac{1}{p-1}}\frac{p-1}{\alpha p-1}(\psi(u)-\psi(0))^{\frac{\alpha p-1}{p-1}},
\end{aligned} \tag{16}$$

we have by (15), (16), and (H3) that

$$I_{12} \leq A_{12}\epsilon^p(\psi(u)-\psi(0))^{\alpha p-1}u, \tag{17}$$

where,

$$A_{12} = \frac{6^{p-1}K}{(1-C_1)^p\Gamma(\alpha)^p}\left(\frac{p-1}{\alpha p-1}\right)^{p-1}\|\beta\|_{L^\infty([0,u])}\left[1+\mathbb{E}\left(\sup_{0\leq t\leq u}|y_\epsilon(t)|^p\right)+\mathbb{E}\left(\sup_{0\leq t\leq u}|y_\epsilon(t-\tau)|^p\right)\right],$$

here, $\|\beta\|_{L^\infty([0,u])} = \sup_{t\in[0,u]}|\beta(t)|$.

For the second term I_2, we have

$$I_2 \leq \frac{6^{p-1}\epsilon^{\frac{p}{2}}}{(1-C_1)^p\Gamma(\alpha)^p}\mathbb{E}\left(\sup_{0\leq t\leq u}\left|\int_0^t (\psi(t)-\psi(s))^{\alpha-1}\psi'(s)[\sigma(s,x_\epsilon(s),x_\epsilon(s-\tau))-\sigma(s,y_\epsilon(s),y_\epsilon(s-\tau))]dB_s\right|^p\right)$$
$$+\frac{6^{p-1}\epsilon^{\frac{p}{2}}}{(1-C_1)^p\Gamma(\alpha)^p}\mathbb{E}\left(\sup_{0\leq t\leq u}\left|\int_0^t (\psi(t)-\psi(s))^{\alpha-1}\psi'(s)[\sigma(s,y_\epsilon(s),y_\epsilon(s-\tau))-\bar{\sigma}(y_\epsilon(s),y_\epsilon(s-\tau))]dB_s\right|^p\right)$$
$$= I_{21}+I_{22}. \tag{18}$$

In view of the Burkholder–Davis–Gundy's inequality, Hölder's inequality, and Doob's martingale inequality, a constant $C_p > 0$ exists such that

$$I_{21} \leq \frac{6^{p-1}\epsilon^{\frac{p}{2}}C_p}{(1-C_1)^p\Gamma(\alpha)^p}\mathbb{E}\left(\int_0^u (\psi(u)-\psi(s))^{2\alpha-2}\psi'(s)^2|\sigma(s,x_\epsilon(s),x_\epsilon(s-\tau))-\sigma(s,y_\epsilon(s),y_\epsilon(s-\tau))|^2ds\right)^{\frac{p}{2}}$$
$$\leq \frac{6^{p-1}C_p}{(1-C_1)^p\Gamma(\alpha)^p}\epsilon^{\frac{p}{2}}u^{\frac{p}{2}-1}\mathbb{E}\left(\int_0^u (\psi(u)-\psi(s))^{(\alpha-1)p}\psi'(s)^p\right.$$
$$\left.\cdot|\sigma(s,x_\epsilon(s),x_\epsilon(s-\tau))-\sigma(s,y_\epsilon(s),y_\epsilon(s-\tau))|^pds\right)$$
$$\leq \frac{6^{p-1}C_p}{(1-C_1)^p\Gamma(\alpha)^p}\epsilon^{\frac{p}{2}}u^{\frac{p}{2}-1}K^{p-1}C_2^p\cdot\mathbb{E}\left(\int_0^u (\psi(u)-\psi(s))^{(\alpha-1)p}\psi'(s)\right.$$
$$\left.\cdot[|x_\epsilon(s)-y_\epsilon(s)|^p+|x_\epsilon(s-\tau)-y_\epsilon(s-\tau)|^p]ds\right)$$
$$\leq A_{21}\epsilon^{\frac{p}{2}}u^{\frac{p}{2}-1}\int_0^u (\psi(u)-\psi(s))^{(\alpha-1)p}\psi'(s)\left[\mathbb{E}\left(\sup_{0\leq\theta\leq s}|x_\epsilon(\theta)-y_\epsilon(\theta)|^p\right)\right.$$
$$\left.+\mathbb{E}\left(\sup_{0\leq\theta\leq s}|x_\epsilon(\theta-\tau)-y_\epsilon(\theta-\tau)|^p\right)\right]ds, \tag{19}$$

where $A_{21} = \frac{6^{p-1}C_pK^{p-1}C_2^p}{(1-C_1)^p\Gamma(\alpha)^p}$.

Since $\alpha > \frac{p-1}{p}$, we have $\alpha p - p + 1 > 0$. Applying Lemma 2 and an estimation method similar to Equation (19), we obtain

$$I_{22} \leq \frac{12^{p-1}C_pK^{p-1}}{(1-C_1)^p\Gamma(\alpha)^p}\epsilon^{\frac{p}{2}}u^{\frac{p}{2}-1}\cdot\mathbb{E}\left(\int_0^u (\psi(u)-\psi(s))^{(\alpha-1)p}\psi'(s)\right.$$
$$\left.\cdot(|\sigma(s,y_\epsilon(s),y_\epsilon(s-\tau))|^p+|\bar{\sigma}(y_\epsilon(s),y_\epsilon(s-\tau))|^p)ds\right)$$
$$\leq A_{22}\epsilon^{\frac{p}{2}}u^{\frac{p}{2}-1}(\psi(u)-\psi(0))^{(\alpha-1)p+1}, \tag{20}$$

where

$$A_{22} = \frac{12^{p-1}C_pK^{p-1}(C_3^p+C_4)}{(1-C_1)^p\Gamma(\alpha)^p(\alpha p-p+1)}\left[1+\mathbb{E}\left(\sup_{0\leq t\leq u}|y_\epsilon(t)|^p\right)+\mathbb{E}\left(\sup_{0\leq t\leq u}|y_\epsilon(t-\tau)|^p\right)\right].$$

Next, we deal with the third term. Similar to the method used in (18), we have

$$I_3 \leq \frac{6^{p-1}\epsilon^{\frac{p}{2}}}{(1-C_1)^p\Gamma(\alpha)^p}\mathbb{E}\left(\sup_{0\leq t\leq u}\left|\int_0^t (\psi(t)-\psi(s))^{\alpha-1}\psi'(s)\int_V [g(s,x_\epsilon(s),x_\epsilon(s-\tau),v)\right.\right.$$

$$-g(s,y_\epsilon(s),y_\epsilon(s-\tau),v)]\tilde{N}(ds,dv)\bigg|^p\Bigg)$$

$$+\frac{6^{p-1}\epsilon^{\frac{p}{2}}}{(1-C_1)^p\Gamma(\alpha)^p}\mathbb{E}\Bigg(\sup_{0\le t\le u}\bigg|\int_0^t(\psi(t)-\psi(s))^{\alpha-1}\psi'(s)\int_V[g(s,y_\epsilon(s),y_\epsilon(s-\tau),v)$$

$$-\bar{g}(y_\epsilon(s),y_\epsilon(s-\tau),v)]\tilde{N}(ds,dv)\bigg|^p\Bigg)$$

$$=I_{31}+I_{32}. \tag{21}$$

From Lemma 4, one has

$$I_{31}\le\frac{6^{p-1}\epsilon^{\frac{p}{2}}}{(1-C_1)^p\Gamma(\alpha)^p}D_p\Bigg\{\mathbb{E}\Bigg(\int_0^u(\psi(u)-\psi(s))^{2\alpha-2}\psi'(s)^2\int_V|g(s,x_\epsilon(s),x_\epsilon(s-\tau),v)$$

$$-g(s,y_\epsilon(s),y_\epsilon(s-\tau),v)|^2\lambda(dv)ds\Bigg)^{\frac{p}{2}}$$

$$+\mathbb{E}\Bigg(\int_0^u(\psi(u)-\psi(s))^{p(\alpha-1)}\psi'(s)^p\int_V|g(s,x_\epsilon(s),x_\epsilon(s-\tau),v)$$

$$-g(s,y_\epsilon(s),y_\epsilon(s-\tau),v)|^p\lambda(dv)ds\Bigg)\Bigg\}. \tag{22}$$

By using the Hölder inequality and (H2), we obtain

$$\mathbb{E}\Bigg(\int_0^u(\psi(u)-\psi(s))^{2\alpha-2}\psi'(s)^2\int_V|g(s,x_\epsilon(s),x_\epsilon(s-\tau),v)-g(s,y_\epsilon(s),y_\epsilon(s-\tau),v)|^2\lambda(dv)ds\Bigg)^{\frac{p}{2}}$$

$$\le(u\lambda(V))^{\frac{p-2}{2}}\mathbb{E}\Bigg(\int_0^u\int_V(\psi(u)-\psi(s))^{p(\alpha-1)}\psi'(s)^p|g(s,x_\epsilon(s),x_\epsilon(s-\tau),v)$$

$$-g(s,y_\epsilon(s),y_\epsilon(s-\tau),v)|^p\lambda(dv)ds\Bigg)$$

$$\le(u\lambda(V))^{\frac{p-2}{2}}K^{p-1}C_2^p\mathbb{E}\Bigg(\int_0^u(\psi(u)-\psi(s))^{p(\alpha-1)}\psi'(s)[|x_\epsilon(s)-y_\epsilon(s)|^p+|x_\epsilon(s-\tau)-y_\epsilon(s-\tau)|^p]ds\Bigg)$$

$$\le K^{p-1}C_2^p\lambda(V)^{\frac{p-2}{2}}u^{\frac{p-2}{2}}\int_0^u(\psi(u)-\psi(s))^{p(\alpha-1)}\psi'(s)\Bigg[\mathbb{E}\Bigg(\sup_{0\le\theta\le s}|x_\epsilon(\theta)-y_\epsilon(\theta)|^p\Bigg)$$

$$+\mathbb{E}\Bigg(\sup_{0\le\theta\le s}|x_\epsilon(\theta-\tau)-y_\epsilon(\theta-\tau)|^p\Bigg)\Bigg]ds, \tag{23}$$

and

$$\mathbb{E}\Bigg(\int_0^u(\psi(u)-\psi(s))^{p(\alpha-1)}\psi'(s)^p\int_V|g(s,x_\epsilon(s),x_\epsilon(s-\tau),v)-g(s,y_\epsilon(s),y_\epsilon(s-\tau),v)|^p\lambda(dv)ds\Bigg)$$

$$\le C_2^p\mathbb{E}\Bigg(\int_0^u(\psi(u)-\psi(s))^{p(\alpha-1)}\psi'(s)^p[|x_\epsilon(s)-y_\epsilon(s)|^p+|x_\epsilon(s-\tau)-y_\epsilon(s-\tau)|^p]ds\Bigg)$$

$$\le C_2^pK^{p-1}\int_0^u(\psi(u)-\psi(s))^{p(\alpha-1)}\psi'(s)\Bigg[\mathbb{E}\Bigg(\sup_{0\le\theta\le s}|x_\epsilon(\theta)-y_\epsilon(\theta)|^p\Bigg)$$

$$+\mathbb{E}\Bigg(\sup_{0\le\theta\le s}|x_\epsilon(\theta-\tau)-y_\epsilon(\theta-\tau)|^p\Bigg)\Bigg]ds. \tag{24}$$

Plugging (23) and (24) into (22), we obtain

$$I_{31} \le A_{31}\epsilon^{\frac{p}{2}}\left(1+\lambda(V)^{\frac{p-2}{2}}u^{\frac{p-2}{2}}\right)\int_0^u (\psi(u)-\psi(s))^{p(\alpha-1)}\psi'(s)\left[\mathbb{E}\left(\sup_{0\le\theta\le s}|x_\epsilon(\theta)-y_\epsilon(\theta)|^p\right)\right.$$

$$\left.+\mathbb{E}\left(\sup_{0\le\theta\le s}|x_\epsilon(\theta-\tau)-y_\epsilon(\theta-\tau)|^p\right)\right]ds, \tag{25}$$

where $A_{31} = \frac{6^{p-1}}{(1-C_1)^p\Gamma(\alpha)^p}D_pC_2^pK^{p-1}$. We also have

$$I_{32} \le \frac{6^{p-1}\epsilon^{\frac{p}{2}}}{(1-C_1)^p\Gamma(\alpha)^p}D_p\cdot\left\{\mathbb{E}\left(\int_0^u(\psi(u)-\psi(s))^{2\alpha-2}\psi'(s)^2\right.\right.$$

$$\left.\cdot\int_V|g(s,y_\epsilon(s),y_\epsilon(s-\tau),v)-\bar{g}(y_\epsilon(s),y_\epsilon(s-\tau),v)|^2\lambda(dv)ds\right)^{\frac{p}{2}}$$

$$\left.+\mathbb{E}\left(\int_0^u(\psi(u)-\psi(s))^{p(\alpha-1)}\psi'(s)^p\int_V|g(s,y_\epsilon(s),y_\epsilon(s-\tau),v)-\bar{g}(y_\epsilon(s),y_\epsilon(s-\tau),v)|^p\lambda(dv)ds\right)\right\}. \tag{26}$$

Since $\alpha > \frac{p+2}{2p}$, we have $2p\alpha-p-2>0$. By using the Hölder inequality, (10), (H2), and (H3), we obtain

$$\mathbb{E}\left(\int_0^u(\psi(u)-\psi(s))^{p(\alpha-1)}\psi'(s)^p\int_V|g(s,y_\epsilon(s),y_\epsilon(s-\tau),v)-\bar{g}(y_\epsilon(s),y_\epsilon(s-\tau),v)|^p\lambda(dv)ds\right)$$

$$\le 2^{p-1}\mathbb{E}\left(\int_0^u(\psi(u)-\psi(s))^{p(\alpha-1)}\psi'(s)^p\left[\int_V(|g(s,y_\epsilon(s),y_\epsilon(s-\tau),v)|^p\right.\right.$$

$$\left.\left.+|\bar{g}(y_\epsilon(s),y_\epsilon(s-\tau),v))|^p)\lambda(dv)ds\right]\right)$$

$$\le 2^{p-1}(C_3^p+C_4)K^{p-1}\mathbb{E}\left(\int_0^u(\psi(u)-\psi(s))^{p(\alpha-1)}\psi'(s)(1+|y_\epsilon(s)|^p+|y_\epsilon(s-\tau)|^p)ds\right)$$

$$\le \frac{2^{p-1}(C_3^p+C_4)K^{p-1}}{p(\alpha-1)+1}(\psi(u)-\psi(0))^{p(\alpha-1)+1}\left[1+\mathbb{E}\left(\sup_{0\le t\le u}|y_\epsilon(t)|^p\right)+\mathbb{E}\left(\sup_{0\le t\le u}|y_\epsilon(t-\tau)|^p\right)\right], \tag{27}$$

and

$$\mathbb{E}\left(\int_0^u(\psi(u)-\psi(s))^{2\alpha-2}\psi'(s)^2\int_V|g(s,y_\epsilon(s),y_\epsilon(s-\tau),v)-\bar{g}(y_\epsilon(s),y_\epsilon(s-\tau),v)|^2\lambda(dv)ds\right)^{\frac{p}{2}}$$

$$\le\mathbb{E}\left[\left(\int_0^u\int_V(\psi(u)-\psi(s))^{\frac{2p(\alpha-1)}{p-2}}\psi'(s)^{\frac{2p}{p-2}}\lambda(dv)ds\right)^{\frac{p-2}{2}}\right.$$

$$\left.\cdot\left(\int_0^u\int_V|g(s,y_\epsilon(s),y_\epsilon(s-\tau),v)-\bar{g}(y_\epsilon(s),y_\epsilon(s-\tau),v)|^p\lambda(dv)ds\lambda(dv)ds\right)\right]$$

$$\le K^{\frac{p+2}{2}}\lambda(V)^{\frac{p-2}{2}}\left(\frac{p-2}{2p\alpha-p-2}\right)^{\frac{p-2}{2}}(\psi(u)-\psi(0))^{\frac{2p\alpha-p-2}{2}}$$

$$\cdot u\mathbb{E}\left(\frac{1}{u}\int_0^u\int_V|g(s,y_\epsilon(s),y_\epsilon(s-\tau),v)-\bar{g}(y_\epsilon(s),y_\epsilon(s-\tau),v)|^p\lambda(dv)ds\right)$$

$$\leq K^{\frac{p+2}{2}}\lambda(V)^{\frac{p-2}{2}}\left(\frac{p-2}{2p\alpha-p-2}\right)^{\frac{p-2}{2}}\beta(u)u(\psi(u)-\psi(0))^{\frac{2p\alpha-p-2}{2}}$$
$$\cdot\left[1+\mathbb{E}\left(\sup_{0\leq t\leq u}|y_\epsilon(t)|^p\right)+\mathbb{E}\left(\sup_{0\leq t\leq u}|y_\epsilon(t-\tau)|^p\right)\right]. \tag{28}$$

Substituting (27) and (28) into (26), we obtain

$$I_{32}\leq A_{321}\epsilon^{\frac{p}{2}}(\psi(u)-\psi(0))^{p(\alpha-1)+1}+A_{322}\epsilon^{\frac{p}{2}}\beta(u)u(\psi(u)-\psi(0))^{\frac{2p\alpha-p-2}{2}}, \tag{29}$$

where

$$A_{321}=\frac{12^{p-1}D_p}{(1-C_1)^p\Gamma(\alpha)^p}\frac{(C_3^p+C_4)K^{p-1}}{p(\alpha-1)+1}\left[1+\mathbb{E}\left(\sup_{0\leq t\leq u}|y_\epsilon(t)|^p\right)+\mathbb{E}\left(\sup_{0\leq t\leq u}|y_\epsilon(t-\tau)|^p\right)\right],$$

$$A_{322}=\frac{6^{p-1}}{(1-C_1)^p\Gamma(\alpha)^p}D_pK^{\frac{p+2}{2}}\lambda(V)^{\frac{p-2}{2}}\left(\frac{p-2}{2p\alpha-p-2}\right)^{\frac{p-2}{2}}$$
$$\cdot\left[1+\mathbb{E}\left(\sup_{0\leq t\leq u}|y_\epsilon(t)|^p\right)+\mathbb{E}\left(\sup_{0\leq t\leq u}|y_\epsilon(t-\tau)|^p\right)\right].$$

Combining (13), (14), (17)–(21), (25), with (29), for $u\in(0,T]$ we obtain

$$\mathbb{E}\left(\sup_{0\leq t\leq u}|x_\epsilon(t)-y_\epsilon(t)|^p\right)$$
$$\leq A(u)+B(u)\int_0^u(\psi(u)-\psi(s))^{p(\alpha-1)}\psi'(s)$$
$$\cdot\left[\mathbb{E}\left(\sup_{0\leq\theta\leq s}|x_\epsilon(\theta)-y_\epsilon(\theta)|^p\right)+\mathbb{E}\left(\sup_{0\leq\theta\leq s}|x_\epsilon(\theta-\tau)-y_\epsilon(\theta-\tau)|^p]\right)\right]ds, \tag{30}$$

where

$$A(u)=A_{12}\epsilon^p(\psi(u)-\psi(0))^{\alpha p-1}u+A_{22}\epsilon^{\frac{p}{2}}u^{\frac{p}{2}-1}(\psi(u)-\psi(0))^{(\alpha-1)p+1}$$
$$+A_{321}\epsilon^{\frac{p}{2}}(\psi(u)-\psi(0))^{p(\alpha-1)+1}+A_{322}\epsilon^{\frac{p}{2}}\beta(u)u(\psi(u)-\psi(0))^{\frac{2p\alpha-p-2}{2}},$$

and

$$B(u)=A_{11}\epsilon^pu^{p-1}+A_{21}\epsilon^{\frac{p}{2}}u^{\frac{p}{2}-1}+A_{31}\epsilon^{\frac{p}{2}}\left(1+\lambda(V)^{\frac{p-2}{2}}u^{\frac{p-2}{2}}\right).$$

Set

$$\Sigma(u):=\mathbb{E}\left(\sup_{0\leq\theta\leq u}|x_\epsilon(\theta)-y_\epsilon(\theta)|^p\right).$$

Noting that $\mathbb{E}\left(\sup_{-\tau\leq\theta<0}|x_\epsilon(\theta)-y_\epsilon(\theta)|^p\right)=0$, then

$$\mathbb{E}\left(\sup_{0\leq\theta\leq s}|x_\epsilon(\theta-\tau)-y_\epsilon(\theta-\tau)|^p\right)=\Sigma(s-\tau).$$

Hence, it follows from (30) that

$$\Sigma(u) \leq A(u) + B(u)\int_0^u (\psi(u) - \psi(s))^{p(\alpha-1)}\psi'(s)(\Sigma(s) + \Sigma(s-\tau))ds.$$

For each $u \in [0, T]$, denote $\Phi(u) = \sup_{-\tau \leq t \leq u} \Sigma(t)$. Then,

$$\Sigma(s) \leq \Phi(s) \quad \text{and} \quad \Sigma(s-\tau) \leq \Phi(s).$$

Thus, one has

$$\Phi(u) = \sup_{-\tau \leq t \leq u} \Sigma(u) \leq A(u) + 2B(u)\int_0^u (\psi(u) - \psi(s))^{p(\alpha-1)}\psi'(s)\Phi(s)ds.$$

By using Lemma 5, we obtain

$$\Phi(u) \leq A(u)E_{p(\alpha-1)+1}\Big(2B(u)\Gamma(p(\alpha-1)+1)(\psi(u) - \psi(0))^{p(\alpha-1)+1}\Big).$$

Moreover, we have

$$\mathbb{E}\left(\sup_{0 \leq t \leq u} |x_\epsilon(t) - y_\epsilon(t)|^p\right) \leq A(u)E_{p(\alpha-1)+1}\Big(2B(u)\Gamma(p(\alpha-1)+1)(\psi(u) - \psi(0))^{p(\alpha-1)+1}\Big).$$

Choose $M > 0$ and $\beta \in (0, 1)$ such that for all $t \in (0, M\epsilon^{-\beta}] \subset (0, T]$

$$\mathbb{E}\left(\sup_{0 < t \leq M\epsilon^{-\beta}} |x_\epsilon(t) - y_\epsilon(t)|^p\right) \leq \bar{A}E_{p(\alpha-1)+1}\Big(2\bar{B}\Gamma(p(\alpha-1)+1)(\psi(T) - \psi(0))^{p(\alpha-1)+1}\Big)\epsilon^{1-\beta},$$

where

$$\bar{A} = A_{12}M\epsilon^{p-1}(\psi(T) - \psi(0))^{\alpha p-1} + A_{22}M^{\frac{p}{2}-1}\epsilon^{(\frac{p}{2}-1)(1-\beta)+\beta}(\psi(T) - \psi(0))^{(\alpha-1)p+1}$$
$$+ A_{321}\epsilon^{\frac{p}{2}-(1-\beta)}(\psi(T) - \psi(0))^{p(\alpha-1)+1} + A_{322}Mm\epsilon^{\frac{p}{2}-1}(\psi(T) - \psi(0))^{\frac{2p\alpha-p-2}{2}},$$

here, m is a positive bounded of function $\beta(\cdot)$, and

$$\bar{B} = A_{11}M^{p-1}\epsilon^{p-(p-1)\beta} + A_{21}M^{\frac{p}{2}-1}\epsilon^{\frac{p}{2}(1-\beta)+\beta} + A_{31}\epsilon^{\frac{p}{2}} + A_{31}\lambda(V)^{\frac{p-2}{2}}M^{\frac{p-2}{2}}\epsilon^{\frac{p}{2}(1-\beta)+\beta},$$

are two constants. Thus, for any given number $\delta > 0$, $\epsilon_1 \in (0, \epsilon_0]$ exists such that for each $\epsilon \in (0, \epsilon_1]$ and $t \in [-\tau, M\epsilon^{-\beta}]$,

$$\mathbb{E}\left(\sup_{t \in [-\tau, M\epsilon^{-\beta}]} |x_\epsilon(t) - y_\epsilon(t)|^p\right) \leq \delta.$$

□

Remark 1. *If $\psi(t) \equiv t$, $g \equiv 0$, and $\tau = 0$, then FSDDEs (3) reduces to FSDEs (1) in [18]. Therefore, Theorem 1 generalizes the main result of [18].*

Example 1. *Consider the following ψ-Caputo fractional stochastic delay differential equation (FSDDEs) with Poisson jumps :*

$$\begin{cases} {}^{C}D_0^{0.9,\sqrt{t}}\Big[x_\varepsilon(t) - \Big(\frac{1}{8}t^{\frac{1}{5}} + \frac{1}{9}\sin(x_\varepsilon(t))\Big)\Big] = \frac{1}{2}\varepsilon x_\varepsilon(t-\tau) + \frac{3\pi}{4}\sqrt{\varepsilon}\sin^3 t x_\varepsilon(t)\frac{dB_t}{dt} \\ \quad + \sqrt{\varepsilon}\int_V 3\tilde{N}(dt, dv), \quad t \in [0, 25], \\ x_\varepsilon(t) = 0.5, \quad -0.25 \leq t \leq 0, \end{cases} \tag{31}$$

where $\alpha = 0.9$, $\psi(t) = \sqrt{t}$, $T = 25$, $\tau = 0.25$, and

$$h(t, x_\varepsilon(t)) = \frac{1}{8}t^{\frac{1}{5}} + \frac{1}{9}\sin(x_\varepsilon(t)), \quad f(t, x_\varepsilon(t), x_\varepsilon(t-\tau)) = \frac{1}{2}x_\varepsilon(t-\tau),$$

$$\sigma(t, x_\varepsilon(t), x_\varepsilon(t-\tau)) = \frac{3\pi}{4}\sin^3 t \cdot x_\varepsilon(t), \quad g(t, x_\varepsilon(t), x_\varepsilon(t-\tau), v) = 3.$$

Then,

$$\bar{f}(y_\varepsilon(t), y_\varepsilon(t-\tau)) = \frac{1}{\pi}\int_0^\pi f(t, y_\varepsilon(t), y_\varepsilon(t-\tau))dt = \frac{1}{2}y_\varepsilon(t-\tau),$$

$$\bar{\sigma}(y_\varepsilon(t), y_\varepsilon(t-\tau)) = \frac{1}{\pi}\int_0^\pi \sigma(t, y_\varepsilon(t), y_\varepsilon(t-\tau))dt = y_\varepsilon(t),$$

$$\bar{g}(y_\varepsilon(t), y_\varepsilon(t-\tau), v) = \frac{1}{\pi}\int_0^\pi g(t, y_\varepsilon(t), y_\varepsilon(t-\tau), v)dt = 3.$$

Thus, we have the corresponding averaged FSDDEs with Poisson jumps :

$$\begin{cases} {}^C D_0^{0.9,\sqrt{t}}\left[y_\varepsilon(t) - \left(\frac{1}{8}t^{\frac{1}{5}} + \frac{1}{9}\sin(y_\varepsilon(t))\right)\right] = \frac{1}{2}\varepsilon y_\varepsilon(t-\tau) + \sqrt{\varepsilon}y_\varepsilon(t)\frac{dB_t}{dt} \\ \qquad +\sqrt{\varepsilon}\int_V 3\tilde{N}(dt, dv), \quad t \in [0, 25], \\ y_\varepsilon(t) = 0.5, \quad -0.25 \le t \le 0. \end{cases} \tag{32}$$

It is easy to check that the conditions of Theorem 1 are satisfied. So, as $\varepsilon \to 0$, the original solution x_ε and the average solution y_ε are equivalent in the sense of L^p ($p = 2$ or $p > 2$ with $\max\left\{\frac{p-1}{p}, \frac{p+2}{2p}\right\} < 0.9$). To test this, Equations (31) and (32) are calculated numerically and error $Err = |x_\varepsilon(t) - y_\varepsilon(t)|^3$ are given in Figures 1 and 2. So, the averaging principle for the ψ-Capuo FSDDE with Poisson jumps is successfully established.

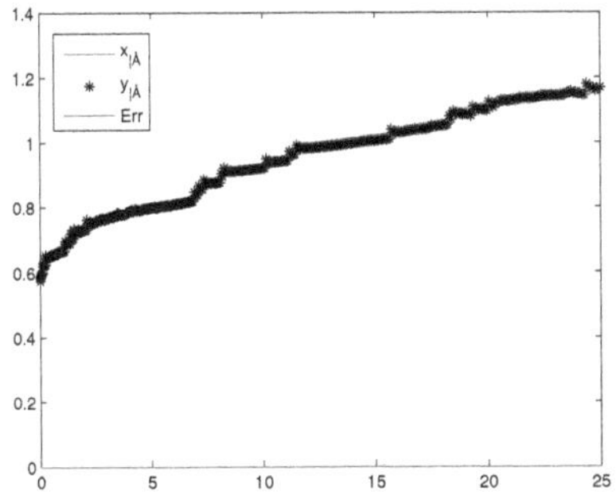

Figure 1. Comparison of x_ε and y_ε for Equations (31) and (32) with $\alpha = 0.9$ and $\varepsilon = 0.1$.

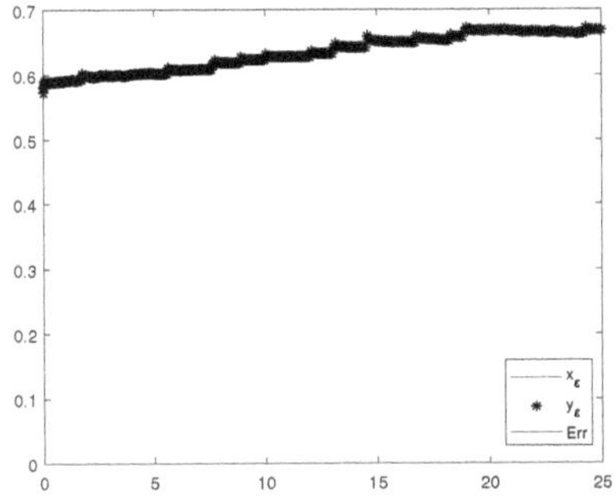

Figure 2. Comparison of x_ε and y_ε for Equations (31) and (32) with $\alpha = 0.9$ and $\varepsilon = 0.01$.

4. Conclusions

In this article, the averaging principle for FSDDEs in the sense of L^p has been proved. Hölders inequality, Jensen's inequality, Burkholder-Davis-Gundys inequality, Doobs martingale inequality, and fractional Gronwall's inequality are applied in the estimation. To the best of our knowledge, this is the first work dealing with the averaging principle for ψ-Capuo fractional stochastic delay differential equations with Poisson jumps. The obtained results generalize the two cases of $p = 2$ and the classical Caputo fractional derivative. For future research, the averaging principle for fractional stochastic neutral functional differential equations driven by the Rosenblatt process with delay and Poisson jumps is both interesting and important. It is worth further investigation in the future.

Author Contributions: Conceptualization, D.Y. and C.B.; formal analysis, D.Y. and J.W.; investigation, D.Y., J.W. and C.B.; and writing—review and editing, D.Y. and C.B. All authors have read and agreed to the published version of the manuscript.

Funding: This work was supported by Natural Science Foundation of China (11571136).

Data Availability Statement: No data were used to support this study.

Acknowledgments: The authors thanks reviewers for their careful reading of our manuscript and their many insightful comments and suggestions that have improved the quality of our manuscript.

Conflicts of Interest: The authors declare no conflict of interest.

References

1. Gaeta, G. Symmetry of stochastic non-variational differential equations. *Phys. Rep.* **2017**, *686*, 1–62. [CrossRef]
2. De Vecchi, F.C.D.; Morando, P.; Ugolini, S. Symmetries of stochastic differential equations: A geometric approach. *J. Math. Phys.* **2016**, *57*, 063504. [CrossRef]
3. Gaeta, G. Symmetry analysis of the stochastic logistic equation. *Symmetry* **2020**, *12*, 973. [CrossRef]
4. Hussain, S.; Elissa Nadia Madi, H.K.; Abdo, M.S. A numerical and analytical study of a stochastic epidemic SIR model in the light of white noise. *Adv. Math. Phys.* **2022**, *2022*, 1638571. [CrossRef]
5. Hussain, S.; Tunç, O.; ur Rahman, G.; Khan, H.; Nadia, E. Mathematical analysis of stochastic epidemic model of MERS-corona & application of ergodic theory. *Math. Comput. Simul.* **2023**, *207*, 130–150. [PubMed]
6. Hussain, S.; Elissa Nadia Madi, E.N.; Khan, H.; Gulzar, H.; Etemad, S.; Rezapour, S.; Kaabar, M.K.A. On the stochastic modeling of COVID-19 under the environmental white noise. *J. Funct. Spaces* **2022**, *2022*, 4320865. [CrossRef]
7. Alzabut, J.; Alobaidi, G.; Hussain, S.; Madi, E.N.; Khan, H. Stochastic dynamics of influenza infection: Qualitative analysis and numerical results. *Math. Biosci. Eng.* **2022**, *19*, 10316–10331. [CrossRef]
8. Khasminskii, R.Z. On the principle of averaging the Itö stochastic differential equations. *Kibernetika* **1968**, *4*, 260–279.
9. Golec, J.; Ladde, G. Averaging principle and systems of singularly perturbed stochastic differential equations. *J. Math. Phys.* **1990**, *31*, 1116–1123. [CrossRef]
10. Xu, Y.; Duan, J.Q.; Xu, W. An averaging principle for stochastic dynamical systems with Lvy noise. *Physica D* **2011**, *240*, 1395–1401. [CrossRef]
11. Mao, W.; You, S.; Wu, X.; Mao, X. On the averaging principle for stochastic delay differential equations with jumps. *Adv. Differ. Equ.* **2015**, *2015*, 70. [CrossRef]
12. Shah, N.A.; Alyousef, H.A.; El-Tantawy, S.A.; Shah, R.; Chung, J.D. Analytical investigation of fractional-order Korteweg-De-Vries-type equations under Atangana-Baleanu-Caputo operator: Modeling nonlinear waves in a plasma and fluid. *Symmetry* **2022**, *14*, 739. [CrossRef]
13. Shah, N.A.; Hamed, Y.S.; Abualnaja, K.M.; Chung, J.D.; Shah, R.; Khan, A. A Comparative analysis of fractional-order Kaup-Kupershmidt equation within different operators. *Symmetry* **2022**, *14*, 986. [CrossRef]
14. Alshammari, S.; Al-Sawalha, M.M.; Shah, R. Approximate analytical methods for a fractional-order nonlinear system of Jaulent-Miodek equation with energy-dependent Schrödinger potential. *Fractal Fract.* **2023**, *7*, 140. [CrossRef]
15. Xu, W.J.; Duan, J.Q.; Xu, W. An averaging principle for fractional stochastic differential equations with Levy noise. *Chaos* **2020**, *30*, 083126. [CrossRef] [PubMed]
16. Cui, J.; Bi, N.N. Averaging principle for neutral stochastic functional differential equations with impulses and non-Lipschitz coefficients. *Statist. Probab. Lett.* **2020**, *163*, 108775. [CrossRef]
17. Li, S.; Xie, Y. Averaging principle for stochastic 3D fractional Leray- model with a fast oscillation. *Stoch. Anal. Appl.* **2019**, *38*, 248–276. [CrossRef]
18. Wang, Z.; Lin, P. Averaging principle for fractional stochastic differential equations with L^p convergence. *Appl. Math. Lett.* **2022**, *130*, 108024. [CrossRef]

19. Xu, W.; Xu, W.; Zhang, S. The averaging principle for stochastic differential equations with Caputo fractional derivative. *Appl. Math. Lett.* **2019**, *93*, 79–84. [CrossRef]
20. Luo, D.; Zhu, Q.; Luo, Z. An averaging principle for stochastic fractional differential equations with time-delays. *Appl. Math. Lett.* **2020**, *105*, 106290. [CrossRef]
21. Xu, W.; Xu, W.; Lu, K. An averaging principle for stochastic differential equations of fractional order $0 < \alpha < 1$. *Fract. Calc. Appl. Anal.* **2020**, *23*, 908–919.
22. Ahmed, H.M. Impulsive conformable fractional stochastic differential equations with Poisson jumps. *Evol. Equations Control Theory* **2022**, *11*, 2073–2080. [CrossRef]
23. Ahmed, H.M.; El-Borai, M.M.; El-Owaidy, H.M.; Ghanem, A.S. Impulsive Hilfer fractional differential equations. *Adv. Differ. Equations* **2018**, *2018*, 226. [CrossRef]
24. Ahmed, H.M.; El-Borai, M.M.; Ramadan, M.E. Boundary controllability of nonlocal Hilfer fractional stochastic differential systems with fractional Brownian motion and Poisson jumps. *Adv. Differ. Equations* **2019**, *2019*, 82. [CrossRef]
25. Ahmed, H.M.; El-Borai, M.M.; Bab, A.S.O.E.; Ramadan, M.E. Approximate controllability of noninstantaneous impulsive Hilfer fractional integrodifferential equations with fractional Brownian motion. *Bound. Value Probl.* **2020**, *2020*, 120. [CrossRef]
26. Wang, J.; Ahmed, H.M. Null controllability of nonlocal Hilfer fractional stochastic differential equations. *Miskolc Math. Notes* **2017**, *18*, 1073–1083.
27. Ahmed, H.M.; Zhu, Q. The averaging principle of Hilfer fractional stochastic delay differential equations with Poisson jumps. *Appl. Math. Lett.* **2021**, *112*, 106755. [CrossRef]
28. Almeida, R. A Caputo fractional derivative of a function with respect to another function. *Commun. Nonlinear Sci.* **2017**, *44*, 460–481. [CrossRef]
29. Suechoei, A.; Ngiamsunthorn, P.S. Existence uniqueness and stability of mild solutions for semilinear ψ-Caputo fractional evolution equations. *Adv. Differ. Equ.* **2020**, *2020*, 114. [CrossRef]
30. Jiang, D.; Bai, C. On coupled Gronwall inequalities involving a ψ-fractional integral operator with its applications. *AIMS Math.* **2022**, *7*, 7728–7741. [CrossRef]
31. Jiang, D.; Bai, C. Existence Results for Coupled Implicit ψ-Riemann-Liouville Fractional Differential Equations with Nonlocal Conditions. *Axioms* **2022**, *2022*, 103. [CrossRef]
32. Yang, Q.; Bai, C.; Yang, D. Controllability of a class of impulsive ψ-Caputo fractional evolution equations of Sobolev type. *Axioms* **2022**, *2022*, 283. [CrossRef]
33. Xu, W.; Xu, W. An effective averaging theory for fractional neutral stochastic equations of order $0 < \alpha < 1$ with Poisson jumps. *Appl. Math. Lett.* **2020**, *106*, 106344.
34. Wang, P.; Wang, X.; Su, H. Input-to-state stability of impulsive stochastic infinite dimensional systems with Poisson jumps. *Automatica* **2021**, *128*, 109553. [CrossRef]
35. Deng, S.; Fei, W.; Liu, W.; Mao, X. The truncated EM method for stochastic differential equations with Poisson jumps. *J. Comput. Appl. Math.* **2019**, *355*, 232–257. [CrossRef]
36. Jarad, F.; Abdeljawad, T. Generalized fractional derivatives and Laplace transform. *Discrete Contin. Dyn. Syst. Ser. S.* **2019**, *13*, 709–722. [CrossRef]
37. Ahmadova, A.; Mahmudov, N.I. Existence and uniqueness results for a class of fractional stochastic neutral differential equations. *Chaos Solitons Fractals* **2020**, *139*, 110253. [CrossRef]
38. Mao, X. *Stochastic Differential Equations and Applications*; Ellis Horwood: Chichester, UK, 2008.
39. Applebaum, D. *Lévy Process and Stochastic Calculus*; Cambridge University Press: Cambridge, UK, 2009.
40. Kunita, H. Stochastic differential equations based on Lévy processes and stochastic flows of diffeomorphisms. In *Real and Stochastic Analysis. New Perspectives*; Birkhäuser: Basel, Switzerland, 2004; pp. 305–373.
41. Vanterler, J.; Sousa, J.V.V.; Oliveira, E.C. A Gronwall inequality and the Cauchy-type problem by means of ψ-Hilfer operator. *Differ. Equ. Appl.* **2019**, *11*, 87–106.

symmetry

Article

Weakly Coupled System of Semi-Linear Fractional θ-Evolution Equations with Special Cauchy Conditions

Abdelhamid Mohammed Djaouti

Preparatory Year Deanship, King Faisal University, Hofuf 31982, Saudi Arabia; adjaout@kfu.edu.sa

Abstract: In this paper, we consider a weakly system of fractional θ-evolution equations. Using the fixed-point theorem, a global-in-time existence of small data solutions to the Cauchy problem is proved for one single equation. Using these results, we prove the global existence for the system under some mixed symmetrical conditions that describe the interaction between the equations of the system.

Keywords: fractional derivatives; θ-evolution equation; weakly coupled system of equations; global existence

Citation: Mohammed Djaouti, A. Weakly Coupled System of Semi-Linear Fractional θ-Evolution Equations with Special Cauchy Conditions. *Symmetry* **2023**, *15*, 1341. https://doi.org/10.3390/sym15071341

Academic Editors: Sergei D. Odintsov, Francisco Martínez González and Mohammed K. A. Kaabar

Received: 10 May 2023
Revised: 15 June 2023
Accepted: 25 June 2023
Published: 30 June 2023

1. Introduction

In this paper, we show the existence of the global (in time) solutions with small data to the weakly coupled system of fractional wave equations

$$\begin{aligned} &D^{1+\lambda_1}\mu + (-\Delta)^{\frac{\theta_1}{2}} u = |v|^p, \qquad J^{1-\lambda_1}u(0,x) = u_{\lambda_1}(x), D^{\lambda_1}u(0,x) = 0,\\ &D^{1+\lambda_2}v + (-\Delta)^{\frac{\theta_2}{2}} v = |u|^q, \qquad J^{1-\lambda_2}u(0,x) = u_{\lambda_2}(x), D^{\lambda_2}v(0,x) = 0, \end{aligned} \tag{1}$$

where $\lambda_1, \lambda_2 \in (0,1), \theta_1, \theta_2$ are real positive numbers and $D^{1+\lambda}$ is the Riemann–Liouville fractional derivative defined by

$$D^{1+\lambda}f(t) := \partial_t^2 (J^{1-\lambda}f)(t) \tag{2}$$

with the Riemann–Liouville fractional integral operator

$$D^a f(t) := \frac{1}{\Gamma(a)} \int_0^t (t-s)^{a-1} f(s)ds, t > 0 \tag{3}$$

for $\Re(a) > 0$, and Γ is the Euler Gamma function.

Such mathematical models have promising applications in engineering and in other physical sciences, as well as in numerical simulations of some fractional nonlinear viscoelastic flow problems, and they impact the bioconvection on the free stream flow of a pseudoplastic nanofluid past a rotating cone.

At the outset, since the fractional equation interpolates the heat equation for $\lambda \to 0$ and the wave equation for $\lambda \to 1$ we will provide briefly some previous results of the wave equations and heat equation.

On the one hand, we consider the Cauchy problem for the semi-linear heat equation

$$u_t - \Delta u = |u|^p, \qquad u(0,x) = u_0(x).$$

Fujita in [1] proved that the exponent $p_{Fuj} := 1 + \frac{2}{n}$ is critical for the classical heat model, which means that we have the global (in time) existence of small data solutions for $p > p_{crit}$, and the blow up if we have the inverse $1 < p < p_{Fuj}$. In [2,3], the authors proved the blow-up for the critical case $p = p_{Fuj}$.

On the other hand, let us consider the Cauchy problem for the semi-linear wave equation

$$u_{tt} - \Delta u = |u|^p, \qquad u(0,x) = u_0(x), u_t(0,x) = u_1(x),$$

where the authors in [4] proved for $n = 3$ that the critical exponent is defined as a positive root of the quadratic equation

$$(n-1)p^2 - (n+1)p - 2 = 0.$$

The defined exponent by the last equation is called the Strauss exponent and denoted by p_S for further considerations, which means that we have the global (in time) existence of small data weak solutions for the above p_S , whereas the local (in time) existence for $p > 1$ and large data can be only expected. In [5,6], the author proved in $\mathbb{R}^2$ that the Strauss exponent p_S is critical. After that, the global existence for $n = 2, 3$ was treated in [7] and for $n \geq 4$ in [8,9]. The nonexistence of solutions for data compactly supported was studied in [10] for $1 < p < \frac{n+1}{n-1}$. For $n = 3$, the authors proved some optimal results in [11] for $p = 1 + \sqrt{2}$. For $n > 3$, a nonexistence result with small data proved in [12] for $1 < p < p_S$.

In 2017, D'Abbicco et al. [13] considered the semi-linear fractional wave equation

$$\partial_t^{1+\lambda} u - \Delta u = |u|^p, \qquad u(0,x) = u_0(x), u_t(0,x) = 0, \tag{4}$$

where $\lambda \in (0,1)$ with the fractional Riemann–Liouville fractional derivative. They proved the critical exponent for the global existence of a small data solution in a low space dimension. The Caputo fractional order and the existence of non-null Cauchy data was studied in [14].

In [15], the authors proved the global (in time) existence of small data solutions to semi-linear fraction θ-evolution equations with mass or power nonlinearity. A similar problem was treated in [16] by considering a memory term instead of the power nonlinearity.

In the first part of our main results, we show the global existence of a small data solution to the fractional Riemann–Liouville order to the semi-linear θ-evolution problem (7).

For the systems, let us first consider the weakly coupled system of damped wave equations semi-linear heat equations

$$\begin{aligned} u_t - \Delta u &= |v|^p, \ u(0,x) = u_0(x), \ u_t(0,x) = u_1(x), \\ v_t - \Delta v &= |u|^q, \ v(0,x) = v_0(x), \ v_t(0,x) = v_1(x), \end{aligned}$$

where $t \in [0, \infty), x \in \mathbb{R}^n, p, q > 1$ and $pq > 1$. The authors of [17] showed that the exponents p and q satisfying

$$\frac{n}{2} = \frac{\max\{p,q\} + 1}{pq - 1}$$

are critical, which means that the solutions exist globally for $\frac{n}{2} > \frac{\max\{p,q\}+1}{pq-1}$ and blow-up for the inverse case. For more details about the system of damped wave equations semi-linear heat equations, the reader can also see [18–21].

Some papers are considered for the weakly coupled systems of semilinear classical damped wave equations with power non-linearities. The problem we have in mind is

$$\begin{aligned} u_{tt} - \Delta u + u_t &= |v|^p, \ u(0,x) = u_0(x), \ u_t(0,x) = u_1(x), \\ v_{tt} - \Delta v + v_t &= |u|^q, \ v(0,x) = v_0(x), \ v_t(0,x) = v_1(x), \end{aligned} \tag{5}$$

where $t \in [0, \infty), x \in \mathbb{R}^n$. In 2007, Sun and Wang proved in [22] that if

$$\lambda := \frac{\max\{p;q\} + 1}{pq - 1} < \frac{n}{2}. \tag{6}$$

for $n = 1$ or 3, then the solution exists globally in time for small initial data, while, if $\lambda \geq \frac{n}{2}$, then every solution having positive average value does not exist globally. In [23], the authors generalized the previous results to the case where $n = 1,2,3$ and improved the time decay estimates for $n = 2$. In 2014, using the weighted energy method, Nishihara and Wakasugi proved, in [24], the critical exponent for any space dimensions. Considering the time-dependent dissipation terms, the authors of [25–27] proved the global (in time) existence of small data solutions under a plan condition , which presents the interplay between the exponents of power nonlinearities.

In our paper, we consider first the single equation from system (1) where we proved the global existence for some range of the exponent p under conditions related to the regularity of the data and the dimension. After that, we apply the results of the single equation to study the weakly coupled systems (1). We proved the global existence for the system with a loss of decay if one of the exponents of power nonlinearities did not satisfy the condition of the single equation.

The paper is organized as follows. In Section 2, we will show our main results of global (in time) existence with examples. Moreover, we mention some remarks of the interpolated cases of wave and heat equations. Next, in Section 3, we prove the existence of solution by applying Banach's fixed point. Appendix A concludes the paper.

2. Main Results

2.1. Single Equation of Fractional Integral Equation

In this section, we will show our main results where we start with the global (in time) existence of solutions to the single equation of the Cauchy problem. Using the formal representation of the solution to our equation, we obtain the estimates of the solutions, and finally we prove the existence using fixed-point theorem explained in the Appendix A.

$$D^{\lambda+1}u + (-\Delta)^{\frac{\theta}{2}}u = |u|^p, \qquad J^{1-\lambda}u(0,x) = u_\lambda(x), D^\lambda u(0,x) = 0 \tag{7}$$

where $\lambda \in (0,1), \theta > 0$.

Theorem 1. *Let $n \geq 1$, and the data u_λ are supposed to belong to $L^1 \cap L^p$. The following conditions are satisfied for the exponent p :*

$$p > 1 + \frac{1+\lambda}{\frac{n}{\theta}(1+\lambda) - \lambda}, \tag{8}$$

and

$$p < 1 + \frac{\theta}{n-\theta} \quad \text{if} \quad n > \theta. \tag{9}$$

Then, a small constant ϵ exists such that, if $\|u_\lambda\|_{L^1 \cap L^p} \leq \epsilon$, then there is a uniquely determined globally (in time) energy solution to (7) in $C([0,\infty), L^1 \cap L^p)$.
Furthermore, the solution satisfies the estimates:

$$\|u\|_{L^q} \lesssim (1+t)^{\lambda - \frac{n}{\theta}(1+\lambda)\left(1-\frac{1}{q}\right)} \|u_\lambda\|_{L^1 \cap L^p},$$

where $q \in [1,p]$.

The new type of date has a strong influence in the representation of the solution of (1) after [28], which leads to a quite different admissible range of the exponent p compared with the classical equations presented in [14].

Remark 1. *If $\lambda \to 0$, then the admissible range for the global (in time) existence corresponds with a Fujita like exponent $1 + \frac{\theta}{n}$. On the contrary for $\lambda \to 1$, we obtain a gap of continuity with respect to the Strauss exponent, which appeared in previous results as a critical exponent for the classical wave equation.*

Remark 2. *One can obtain the optimal for the exponent p in (8) using the scaling argument similarity to prove of the critical exponent to (4) illustrated in [14].*

Example 1. *We consider a concrete example by giving values to the parameters appearing in the theorem. Let us consider in $\mathbb{R}^3$ the following model:*

$$D^{\frac{3}{2}}u + (-\Delta)^{\frac{3}{4}}u = |u|^p, \qquad J^{\frac{1}{2}}u(0,x) = u_\lambda(x), D^{\frac{1}{2}}u(0,x) = 0.$$

Then, using Theorem 1, the admissible range for the global existence is $\frac{8}{5} < p < 2$.

2.2. Weakly Coupled System of Fractional Integral Equations

In this section, we apply the results of the previous theorem to study systems of weakly coupled fractional θ-evolution equations.

Theorem 2. *Let $n \geq 1$, and the data $u_{\lambda_1}, u_{\lambda_2}$ is supposed to belong to $(L^1 \cap L^p) \times (L^1 \cap L^q)$. The following conditions are satisfied for the exponent p and q:*

$$p < 1 + \frac{1+\lambda_2}{\frac{n}{\theta_2}(1+\lambda_2) - \lambda_2}, \quad q > 1 + \frac{1+\lambda_1}{\frac{n}{\theta_1}(1+\lambda_1) - \lambda_1}, \tag{10}$$

$$p < 1 + \frac{\theta}{n-\theta_1}, q < 1 + \frac{\theta}{n-\theta_2} \quad \text{if} \quad n > \min\{\theta_1; \theta_2\} \tag{11}$$

and

$$Q(\lambda_1, \lambda_2, \theta_1, \theta_2, q) > o. \tag{12}$$

Then, a small constant ϵ exists such that, if $\|u_{\lambda_1}\|_{L^1 \cap L^p} + \|v_{\lambda_2}\|_{L^1 \cap L^q} \leq \epsilon$, then there is a uniquely determined globally (in time) energy solution to (1) in $C([0,\infty), L^1 \cap L^p) \times C([0,\infty), L^1 \cap L^q)$. Furthermore, the solution satisfies the estimates:

$$\|u\|_{L^{r_1}} \lesssim (1+t)^{\lambda + L(p) - \frac{n}{\theta_1}(1+\lambda_1)\left(1-\frac{1}{r_1}\right)} \|u_\lambda\|_{L^1 \cap L^p},$$

$$\|v\|_{L^{r_2}} \lesssim (1+t)^{\lambda_2 - \frac{n}{\theta_2}(1+\lambda_2)\left(1-\frac{1}{r_2}\right)} \|v_\lambda\|_{L^1 \cap L^q},$$

where $L(p) = -\frac{n}{\theta_2}(1+\lambda_2)(p-1) + p\lambda_2, Q(\lambda_1, \lambda_2, \theta_1, \theta_2, q) = \left(\frac{n}{\theta_2}(1+\lambda_2) - \lambda_2\right)q^2 - \left(\frac{n}{\theta_1}(1+\lambda_1) - \frac{n}{\theta_2}(1+\lambda_2) - \lambda_1\right)q - \frac{n}{\theta_1}(1+\lambda_1)$ *and* $r_2 \in [1,p], r_2 \in [1,q]$.

Remark 3. *If we take in Theorem 2 the condition $p > 1 + \frac{1+\lambda_2}{\frac{n}{\theta_2}(1+\lambda_2)-\lambda_2}$, then we cannot feel any interplay between the equations of the system since it will behave as a single equation.*

Remark 4. *If we consider $p = 1 + \frac{1+\lambda_2}{\frac{n}{\theta_2}(1+\lambda_2)-\lambda_2}$ then, after using Proposition A1, we obtain a new decay generated by the* log *term appearing in the estimate of u, exactly, $(1+t)^{-1}\log(1+t) \approx (1+t)^{-1+\epsilon}$.*

Example 2. *Let us consider $\theta_1 = \theta_2 = 2$ in $\mathbb{R}^2$ and the parameter of the fractional derivative of the first equation $\lambda_1 \to 0$ and the second $\lambda_2 \to 1$. Then, with the Cauchy condition the model, we obtain*

$$\partial_t u + -\Delta u = |v|^p,$$
$$\partial_{tt} v + -\Delta v = |u|^q.$$

Applying Theorem 2, we obtain the global (in time) existence of the solution for $p < 3$ and $q > 2$.

Remark 5. *The reader can apply the last theorem for several examples. Giving values to some parameters such as the dimension or the order of the fractional derivative, we obtain the mixed condition that leads to the global existence .*

3. Philosophy of Our Approach

In this section, we will prove results for the Cauchy problems (1) and (7). Our main interest is to prove the global (in time) existence of small data solutions, which means the global existence after the perturbation of the null Cauchy condition $\|u_\lambda\|_{L^1\cap L^p} \leq \epsilon$. Such results imply immediate stability results for the zero solution .

3.1. Proof of Theorem 1

In this section, we deal with the following single equation:

$$\partial_t^{\lambda+1}u + (-\Delta)^{\frac{\theta}{2}}u = |u|^p, \qquad J^{1-\lambda}u(0,x) = u_\lambda(x), D^\lambda u(0,x) = 0. \tag{13}$$

We define the norm of the solution space $X(t)$, which we will propose in all of the proofs of the above theorems by

$$\|u\|_{X(t)} = \sup_{\tau\in[0,t]} (1+t)^{-\lambda}\{\|u(\tau,\cdot)\|_{L^1} + (1+t)^{\frac{n}{\theta}(1+\lambda)\left(1-\frac{1}{p}\right)}\|u(\tau,\cdot)\|_{L^p}\}, \tag{14}$$

We introduce the operator N by

$$N : u \in X(t) \to Nu = Nu(t,x) := u^{ln}(t,x) + u^{nl}(t,x),$$

where u^{ln} is a Sobolev solution to the Cauchy problem

$$\partial_t^{\lambda+1}u + (-\Delta)^{\frac{\theta}{2}}u = 0, \qquad J^{1-\lambda}u(0,x) = u_\lambda(x), D^\lambda u(0,x) = 0,$$

and u^{nl} is a Sobolev solution to the Cauchy problem

$$\partial_t^{\lambda+1}u + (-\Delta)^{\frac{\theta}{2}}u = |u|^p, \qquad J^{1-\lambda}u(0,x) = u_\lambda(x), D^\lambda u(0,x) = 0.$$

Using Fourier analysis together with Theorem A1 from Appendix A, we can show that the solutions of the previous problems can be presented by $u(t,x) = u^{ln}(t,x) + u^{nl}(t,x)$ as follows:

$$u^{ln}(t,x) = t^{\lambda-1}\mathcal{F}^{-1}\left(E_{1+\lambda,\lambda}\left(-t^{1+\lambda}|\xi|^\theta\right)\right)(t,x) *_{(x)} u_\lambda(x), \tag{15}$$

and

$$u^{nl}(t,x) = \int_0^t (t-s)^\lambda \mathcal{F}^{-1}\left(E_{1+\lambda,1+\lambda}\left(-t^{1+\lambda}|\xi|^\theta\right)\right)(t-s,x) *_{(x)} |u(s,x)|^p ds. \tag{16}$$

Following Proposition A2, our aim is to prove the following inequalities:

$$\|Nu\|_{X(t)} \lesssim \|u_\lambda\|_{L^1\cap L^p} + \|u\|^p_{X(t)}, \tag{17}$$

$$\|Nu - Nv\|_{X(t)} \lesssim \|u-v\|_{X(t)}\left(\|u\|^{p-1}_{X(t)} + \|v\|^{p-1}_{X(t)}\right). \tag{18}$$

After proving these both inequalities, we apply Banach's fixed-point theorem. In this way, we obtain the local (in time) existence of large data Sobolev solutions and the global (in time) existence of small data Sobolev solutions as well.

We split the prove of the first inequality (17) into the following inequalities:

$$\|u^{ln}\|_{X(t)} \lesssim \|u_\lambda\|_{L^1\cap L^p}, \tag{19}$$

and

$$\|u^{nl}\|_{X(t)} \lesssim \|u\|^p_{X(t)}. \tag{20}$$

To prove inequality (19) we have to derive the estimate of $\|\mathcal{F}^{-1}(E_{1+\lambda,\lambda}(-t^{1+\lambda}|\xi|^\theta))\|_{L^p}$ in order to use Young's inequality. Using the scaling property, we obtain

$$\left\|\mathcal{F}^{-1}\left(E_{1+\lambda,\lambda}\left(-t^{1+\lambda}|\xi|^\theta\right)\right)\right\|_{L^p} = t^{-\frac{n}{\theta}(1+\lambda)\left(1-\frac{1}{p}\right)}\left\|\mathcal{F}^{-1}\left(E_{1+\lambda,\lambda}\left(-|\xi|^\theta\right)\right)\right\|_{L^p}. \tag{21}$$

Indeed, after change of variable $\xi_1 = t^{1+\lambda}|\xi|$ we obtain

$$\begin{aligned} \mathcal{F}^{-1}\left(G\left(t^{1+\lambda}|\xi|^\theta\right)\right) &= t^{-\frac{n}{\theta}(1+\lambda)}\int_{\mathbb{R}^n} e^{it^{-\frac{1+\lambda}{\theta}}x\xi_1}G(|\xi_1|^\theta)d\xi_1 \\ &= t^{-\frac{n}{\theta}(1+\lambda)}\mathcal{F}^{-1}\left(G(|\xi|^\theta)\right)(t^{-\frac{1+\lambda}{\theta}}x). \end{aligned}$$

Using the last equality, we obtain

$$\begin{aligned} \left\|\mathcal{F}^{-1}\left(G\left(t^{1+\lambda}|\cdot|^\theta\right)\right)\right\|^p_{L^p} &= t^{-\frac{n}{\theta}(1+\lambda)p}\left\|\mathcal{F}^{-1}\left(G(|\cdot|^\theta)\right)(t^{-\frac{1+\lambda}{\theta}}x)\right\|_{L^p} \\ &= t^{-\frac{n}{\theta}(1+\lambda)p}\int_{\mathbb{R}^n}\left|\mathcal{F}^{-1}\left(G(|\cdot|^\theta)\right)(t^{-\frac{1+\lambda}{\theta}}x)\right|^p dx. \end{aligned}$$

The change of variable $y = t^{-\frac{1+\lambda}{\theta}}x$ leads to

$$\left\|\mathcal{F}^{-1}\left(G\left(t^{1+\lambda}|\cdot|^\theta\right)\right)\right\|^p_{L^p} = t^{-\frac{n}{\theta}(1+\lambda)p+\frac{n}{\theta}(1+\lambda)}\left\|\mathcal{F}^{-1}\left(G\left(|\cdot|^\theta\right)\right)\right\|^p_{L^p},$$

which completes the proof of 21.

Then, we restrict ourselves to the estimates of $\|\mathcal{F}^{-1}(E_{1+\lambda,\lambda}(-|\xi|^\theta))\|_{L^p}$. After applying Theorem A2 from the Appendix A, we obtain

$$\begin{aligned} E_{1+\lambda,\lambda}\left(-|\xi|^\theta\right) &= \frac{2}{1+\lambda}|\xi|^{-\theta\left(1-\frac{2}{1+\lambda}\right)}e^{\frac{\theta}{1+\lambda}\cos\left(\frac{\pi}{1+\lambda}\right)}cos\left(|\xi|^{\frac{\theta}{1+\lambda}}sin\frac{\pi}{1+\lambda}\right) \\ &\quad +\pi^{-1}|\xi|^{-\theta\left(1-\frac{2}{1+\lambda}\right)}\int_0^\infty \frac{s^{2+\lambda}}{s^{2(1+\lambda)}+2cos(\pi(1+\lambda))+1}e^{-s|\xi|^{\frac{\theta}{1+\lambda}}}ds\,\sin(\lambda\pi), \end{aligned}$$

which leads to

$$\mathcal{F}^{-1}\left(E_{1+\lambda,\lambda}\left(-|\xi|^\theta\right)\right) = \frac{2}{1+\lambda}A(s,x) + \pi^{-1}\sin(\lambda\pi)\int_0^\infty \frac{s^{2+\lambda}}{s^{2(1+\lambda)}+2cos(\pi(1+\lambda))+1}B(s,x)ds,$$

where

$$\begin{aligned} A(s,x) &= \mathcal{F}^{-1}\left(|\xi|^{-\theta\left(1-\frac{2}{1+\lambda}\right)}e^{\frac{\theta}{1+\lambda}\cos\left(\frac{\pi}{1+\lambda}\right)}cos\left(|\xi|^{\frac{\theta}{1+\lambda}}sin\frac{\pi}{1+\lambda}\right)\right), \\ B(s,x) &= \mathcal{F}^{-1}\left(|\xi|^{-\theta\left(1-\frac{2}{1+\lambda}\right)}e^{-s|\xi|^{\frac{\theta}{1+\lambda}}}\right). \end{aligned}$$

First, we consider $B(s,x)$. Similarly to (21), we have

$$\begin{aligned} \|B(s,\cdot)\|_{L^p} &= \left\| \mathcal{F}^{-1}\left(|\xi|^{-\theta\left(1-\frac{2}{1+\lambda}\right)} e^{-s|\xi|^{\frac{\theta}{1+\lambda}}} \right) \right\|_{L^p} \\ &= \left\| \mathcal{F}^{-1}\left(s^{(1+\lambda)\left(1-\frac{2}{1+\lambda}\right)} (s^{1+\lambda}|\xi|^\theta)^{-\left(1-\frac{2}{1+\lambda}\right)} e^{-(s^{1+\lambda}|\xi|^\theta)^{\frac{1}{1+\lambda}}} \right) \right\|_{L^p} \\ &= s^{\lambda-1-\frac{n}{\theta}(1+\lambda)\left(1-\frac{1}{p}\right)} \left\| \mathcal{F}^{-1}\left(|\xi|^{-\theta\left(1-\frac{2}{1+\lambda}\right)} e^{-|\xi|^{\frac{\theta}{1+\lambda}}} \right) \right\|_{L^p} \\ &= s^{\lambda-1-\frac{n}{\theta}(1+\lambda)\left(1-\frac{1}{p}\right)} \|B(1,\cdot)\|_{L^p}. \end{aligned}$$

Then,

$$\|B(s,\cdot)\|_{L^p} = s^{\lambda-1-\frac{n}{\theta}(1+\lambda)\left(1-\frac{1}{p}\right)} \|B(1,\cdot)\|_{L^p}. \tag{22}$$

Then,

$$\left\| \mathcal{F}^{-1}\left(E_{1+\lambda,\lambda}\left(-|\cdot|^\theta\right) \right) \right\|_{L^p} = \frac{2}{1+\lambda}\|A(s,\cdot)\|_{L^p} + \pi^{-1}\sin(\lambda\pi) \int_0^\infty \frac{s^{1+2\lambda-\frac{n}{\theta}(1+\lambda)\left(1-\frac{1}{p}\right)}}{s^{2(1+\lambda)} + 2\cos(\pi(1+\lambda)) + 1} \|B(1,\cdot)\|_{L^p} ds,$$

Using the last estimate together with (A5) from Remark A1, one can obtain the following estimate from Lemma 2.1 in [14] for $d = -\theta\left(1 - \frac{2}{1+\lambda}\right)$:

$$\mathcal{F}^{-1}\left(E_{1+\lambda,\lambda}\left(-|\xi|^\theta\right) \right) \in L^p \quad \text{if} \quad \frac{n}{\theta}\left(1 - \frac{1}{p}\right) < 2, \tag{23}$$

which satisfied (9) .
From (15) with (21), and after using Young's inequality, we obtain

$$\left\| u^{ln}(t,x) \right\|_{L^1} \lesssim (1+t)^{\lambda-1}\|u_\lambda\|_{L^1}, \tag{24}$$

$$\left\| u^{ln}(t,x) \right\|_{L^p} \lesssim (1+t)^{\lambda-1-\frac{n}{\theta}(1+\lambda)\left(1-\frac{1}{p}\right)} (\|u_\lambda\|_{L^1} + \|u_\lambda\|_{L^p}). \tag{25}$$

Replacing last estimates in the definition of the norm of solution space (14) leads to the desired estimate (19).
For the second estimate (20), under the same conditions requested for (23) we have

$$\left\| \mathcal{F}^{-1}\left(E_{1+\lambda,1+\lambda}\left(-t^{1+\lambda}|\xi|^\theta\right) \right) \right\|_{L^p} \lesssim (1+t)^{-\frac{n}{\theta}(1+\lambda)\left(1-\frac{1}{p}\right)}.$$

From (16), we obtain

$$\left\| u^{nl}(t,x) \right\|_{L^1} \lesssim \int_0^t (t-s)^\lambda \| |u(s,x)|^p \|_{L^1} ds, \tag{26}$$

and

$$\left\| u^{nl}(t,x) \right\|_{L^p} \lesssim \int_0^t (t-s)^{\lambda-\frac{n}{\theta}(1+\lambda)\left(1-\frac{1}{p}\right)} \| |u(s,x)|^p \|_{L^1} ds. \tag{27}$$

Using the definition of solution space from (14), we obtain

$$\begin{aligned} \left\| u^{nl}(t,x) \right\|_{L^1} &\lesssim \|u\|_{X(t)}^p \int_0^t (t-s)^\lambda (1+s)^{-\frac{n}{\theta}(1+\lambda)(p-1)+p\lambda} ds, \\ \left\| u^{nl}(t,x) \right\|_{L^p} &\lesssim \|u\|_{X(t)}^p \int_0^t (t-s)^{\lambda-\frac{n}{\theta}(1+\lambda)\left(1-\frac{1}{p}\right)} (1+s)^{-\frac{n}{\theta}(1+\lambda)(p-1)+p\lambda} ds. \end{aligned}$$

Using Proposition A1, we obtain

$$\left\|u^{nl}(t,x)\right\|_{L^1} \lesssim (1+t)^{\lambda}\|u\|^p_{X(t)}, \quad (28)$$

$$\left\|u^{nl}(t,x)\right\|_{L^p} \lesssim (1+t)^{\lambda-\frac{n}{\theta}(1+\lambda)\left(1-\frac{1}{p}\right)}\|u\|^p_{X(t)}, \quad (29)$$

provided that $\frac{n}{\theta}(1+\lambda)\left(1-\frac{1}{p}\right)-\lambda<1$ and $\frac{n}{\theta}(1+\lambda)(p-1)-p\lambda>1$, which are equivalent to (8) and (9), respectively.

Replacing the last estimates in the norm of solution space, we obtain (20), which complete, together with (19), the proof of the first inequality (17).

For the second condition (18), we assume that u and v belong to $X(t)$. Then,

$$Nu-Nv=\int_0^t (t-s)^{\lambda}\mathcal{F}^{-1}\left(E_{1+\lambda,1+\lambda}\left(-t^{1+\lambda}|\xi|^{\theta}\right)\right)(t-s,x)*_{(x)}\left(|u(s,x)|^p-|v(s,x)|^p\right)ds.$$

We control all norms appearing in $\|Nu-Nv\|_{X(t)}$. These are the norms $\|Nu-Nv\|_{L^1}$ and $\||D|^s(Nu-Nv)\|_{L^p}$.

Similarly to (26), we have

$$\|Nu-Nv\|_{L^1}\lesssim\int_0^t (t-s)^{\lambda}\left\|\left(|u(s,x)|^p-|v(s,x)|^p\right)\right\|_{L^1}ds.$$

Hölder's inequality implies

$$\left\||u(s,\cdot)|^p-|v(s,\cdot)|^p\right\|_{L^1}\lesssim\left\|u(s,\cdot)-v(s,\cdot)\right\|_{L^p}\left(\|u(s,\cdot)\|^{p-1}_{L^p}+\|v(s,\cdot)\|^{p-1}_{L^p}\right), \quad (30)$$

Using the norm of the solution space $X(t)$, we obtain

$$\left\|u(s,\cdot)-v(s,\cdot)\right\|_{L^p} \lesssim (1+s)^{-\frac{n}{\theta}(1+\lambda)(1-\frac{1}{p})+\lambda}\left\|u(s,\cdot)-v(s,\cdot)\right\|_{X(t)},$$

$$\|u(s,\cdot)\|^{p-1}_{L^p} \lesssim (1+s)^{\left(-\frac{n}{\theta}(1+\lambda)(1-\frac{1}{p})+\lambda\right)(p-1)}\left\|v(s,\cdot)\right\|^{(p-1)}_{X(t)},$$

$$\|u(s,\cdot)\|^{p-1}_{L^p} \lesssim (1+s)^{\left(-\frac{n}{\theta}(1+\lambda)(1-\frac{1}{p})+\lambda\right)(p-1)}\left\|v(s,\cdot)\right\|^{(p-1)}_{X(t)}.$$

Using the last estimates, we can obtain similarly to (28) and (29)

$$\|Nu-Nv\|_{L^1}\lesssim(1+t)^{\lambda}\|u-v\|_{X(t)}\left(\|u\|^{p-1}_{X(t)}+\|v\|^{p-1}_{X(t)}\right),$$

and

$$\|Nu-Nv\|_{L^p}\lesssim(1+t)^{\lambda-\frac{n}{\theta}(1+\lambda)\left(1-\frac{1}{p}\right)}\|u-v\|_{X(t)}\left(\|u\|^{p-1}_{X(t)}+\|v\|^{p-1}_{X(t)}\right).$$

Then, the proof of the second condition and the theorem is completed.

3.2. *Proof of Theorem 2*

We define the norm of the solution space $X(t)$ by

$$\|(u,v)\|_{X(t)}=\sup_{\tau\in[0,t]}\{M(\tau,u)+M(\tau,v)\} \quad (31)$$

where

$$M(\tau,u) = (1+t)^{-\lambda_1-L(p)}\left[\|u(\tau,\cdot)\|_{L^1}+(1+t)^{\frac{n}{\theta_1}(1+\lambda_1)\left(1-\frac{1}{p}\right)}\|u(\tau,\cdot)\|_{L^p}\right], \quad (32)$$

$$M(\tau,v) = (1+t)^{-\lambda_2}\left[\|v(\tau,\cdot)\|_{L^1}+(1+t)^{\frac{n}{\theta_2}(1+\lambda_2)\left(1-\frac{1}{q}\right)}\|v(\tau,\cdot)\|_{L^q}\right]. \quad (33)$$

Then, we introduce the operator N by

$$N : (u,v) \in X(t) \to N(u,v) = (u^{ln} + u^{nl}, v^{ln} + v^{nl}),$$

where

$$\begin{aligned} u^{ln}(t,x) &:= t^{\lambda_1 - 1}\mathcal{F}^{-1}\Big(E_{1+\lambda_1,\lambda_1}\Big(-t^{1+\lambda_1}|\xi|^{\theta_1}\Big)\Big)(t,x) *_{(x)} u_{\lambda_1}(x), \\ u^{nl}(t,x) &:= \int_0^t (t-s)^{\lambda_1}\mathcal{F}^{-1}\Big(E_{1+\lambda_1,1+\lambda_1}\Big(-t^{1+\lambda_1}|\xi|^{\theta_1}\Big)\Big)(t-s,x) *_{(x)} |v(s,x)|^p ds, \\ v^{ln}(t,x) &:= t^{\lambda_2 - 1}\mathcal{F}^{-1}\Big(E_{1+\lambda_2,\lambda_1}\Big(-t^{1+\lambda_2}|\xi|^{\theta_2}\Big)\Big)(t,x) *_{(x)} v_{\lambda_2}(x), \\ v^{nl}(t,x) &:= \int_0^t (t-s)^{\lambda_2}\mathcal{F}^{-1}\Big(E_{1+\lambda_2,1+\lambda_2}\Big(-t^{1+\lambda_2}|\xi|^{\theta_2}\Big)\Big)(t-s,x) *_{(x)} |u(s,x)|^q ds. \end{aligned}$$

If we consider the results Proposition A3, then our aim is to prove the following inequalities, which imply, among other things, the global existence of small data solutions:

$$\|N(u,v)\|_{X(t)} \lesssim \|u_{\lambda_1}\|_{L^1\cap L^p} + \|v_{\lambda_2}\|_{L^1\cap L^q} + \|(u,v)\|^p_{X(t)} + \|(u,v)\|^q_{X(t)}, \tag{34}$$

$$\begin{aligned} \|N(u,v) - N(\tilde{u},\tilde{v})\|_{X(t)} \lesssim\; & \|(u,v) - (\tilde{u},\tilde{v})\|_{X(t)} \big(\|(u,v)\|^{p-1}_{X(t)} + \|(\tilde{u},\tilde{v})\|^{p-1}_{X(t)} \\ & + \|(u,v)\|^{q-1}_{X(t)} + \|(\tilde{u},\tilde{v})\|^{q-1}_{X(t)}\big). \end{aligned} \tag{35}$$

Let us start by the first condition. Similarly to (24) and (25), we obtain

$$\begin{aligned} \Big\|u^{ln}(t,x)\Big\|_{L^1} &\lesssim (1+t)^{\lambda_1 - 1}\|u_{\lambda_1}\|_{L^1}, \\ \Big\|u^{ln}(t,x)\Big\|_{L^p} &\lesssim (1+t)^{\lambda_1 - 1 - \frac{n}{\theta_1}(1+\lambda_1)\left(1-\frac{1}{p}\right)}\big(\|u_{\lambda_1}\|_{L^1} + \|u_{\lambda_1}\|_{L^p}\big), \\ \Big\|v^{ln}(t,x)\Big\|_{L^1} &\lesssim (1+t)^{\lambda_2 - 1}\|v_{\lambda_2}\|_{L^1}, \\ \Big\|v^{ln}(t,x)\Big\|_{L^q} &\lesssim (1+t)^{\lambda_2 - 1 - \frac{n}{\theta_2}(1+\lambda_2)\left(1-\frac{1}{q}\right)}\big(\|v_{\lambda_2}\|_{L^1} + \|u_{\lambda_2}\|_{L^q}\big). \end{aligned}$$

The last estimates, together with the definition of the norm in (31), lead to

$$\|(u^{ln}, v^{ln})\|_{X(t)} \lesssim \|u_{\lambda_1}\|_{L^1\cap L^p} + \|v_{\lambda_2}\|_{L^1\cap L^q}. \tag{36}$$

Then, we complete the proof by showing the inequality

$$\|(u^{nl}, v^{nl})\|_{X(t)} \lesssim \|(u,v)\|^p_{X(t)} + \|(u,v)\|^q_{X(t)}. \tag{37}$$

For u^{nl}, we have

$$\Big\|u^{nl}(t,x)\Big\|_{L^1} \lesssim \int_0^t (t-s)^{\lambda_1}\||v(s,x)|^p\|_{L^1} ds,$$

and

$$\Big\|u^{nl}(t,x)\Big\|_{L^p} \lesssim \int_0^t (t-s)^{\lambda_1 - \frac{n}{\theta_1}(1+\lambda_1)\left(1-\frac{1}{p}\right)}\||v(s,x)|^p\|_{L^1} ds.$$

Using the definition of solution space from (31), we obtain

$$\begin{aligned} \Big\|u^{nl}(t,x)\Big\|_{L^1} &\lesssim \|(u,v)\|^p_{X(t)} \int_0^t (t-s)^{\lambda_1}(1+s)^{-\frac{n}{\theta_2}(1+\lambda_2)(p-1)+p\lambda_2} ds, \\ \Big\|u^{nl}(t,x)\Big\|_{L^p} &\lesssim \|(u,v)\|^p_{X(t)} \int_0^t (t-s)^{\lambda_1 - \frac{n}{\theta_1}(1+\lambda_1)\left(1-\frac{1}{p}\right)}(1+s)^{-\frac{n}{\theta_2}(1+\lambda_2)(p-1)+p\lambda_2} ds. \end{aligned}$$

From Proposition A1, one can obtain

$$\left\|u^{nl}(t,x)\right\|_{L^1} \lesssim \|(u,v)\|^p_{X(t)}(1+t)^{\lambda_1-\frac{n}{\theta_2}(1+\lambda_2)(p-1)+p\lambda_2} = \|(u,v)\|^p_{X(t)}(1+t)^{\lambda_1+L(p)}, \tag{38}$$

$$\left\|u^{nl}(t,x)\right\|_{L^p} \lesssim \|(u,v)\|^p_{X(t)}(1+t)^{\lambda_1-\frac{n}{\theta_1}(1+\lambda_1)\left(1-\frac{1}{p}\right)-\frac{n}{\theta_2}(1+\lambda_2)(p-1)+p\lambda_2} = \|(u,v)\|^p_{X(t)}(1+t)^{\lambda_1-\frac{n}{\theta_1}(1+\lambda_1)\left(1-\frac{1}{p}\right)+L(p)}, \tag{39}$$

provided that $\frac{n}{\theta}(1+\lambda_1)\left(1-\frac{1}{p}\right)-\lambda_1<1$, which is equivalent to (11).
For u^{nl}, we have

$$\left\|v^{nl}(t,x)\right\|_{L^1} \lesssim \int_0^t (t-s)^{\lambda_2}\||u(s,x)|^q\|_{L^1}ds,$$

and

$$\left\|v^{nl}(t,x)\right\|_{L^q} \lesssim \int_0^t (t-s)^{\lambda_2-\frac{n}{\theta_2}(1+\lambda_2)\left(1-\frac{1}{q}\right)}\||u(s,x)|^q\|_{L^1}ds.$$

Using the norm of the solution space, we obtain

$$\left\|v^{nl}(t,x)\right\|_{L^1} \lesssim \|(u,v)\|^p_{X(t)}\int_0^t (t-s)^{\lambda_2}(1+t)^{-Q(\lambda_1,\lambda_2,\theta_1,\theta_2,q)}ds,$$

and

$$\left\|v^{nl}(t,x)\right\|_{L^q} \lesssim \|(u,v)\|^p_{X(t)}\int_0^t (t-s)^{\lambda_2-\frac{n}{\theta_2}(1+\lambda_2)\left(1-\frac{1}{q}\right)}(1+t)^{-Q(\lambda_1,\lambda_2,\theta_1,\theta_2,q)}ds.$$

Proposition A1, together with (12), leads to

$$\left\|v^{nl}(t,x)\right\|_{L^1} \lesssim \|(u,v)\|^p_{X(t)}(1+t)^{\lambda_2}, \tag{40}$$

and

$$\left\|v^{nl}(t,x)\right\|_{L^q} \lesssim \|(u,v)\|^p_{X(t)}(1+t)^{\lambda_2-\frac{n}{\theta_2}(1+\lambda_2)\left(1-\frac{1}{q}\right)}ds, \tag{41}$$

provided that $\frac{n}{\theta}(1+\lambda_2)\left(1-\frac{1}{q}\right)-\lambda_2<1$, which is equivalent to (11).
From (38) to (41), we obtain (37), which implies, together with (36), the first condition (34). To prove (35), we assume that (u,v) and $(\tilde{u},\tilde{v})$ are two elements from the function space $X(t)$. Then, we have

$$\begin{aligned} N(u,v)-N(\widetilde{u},\widetilde{v}) &= \left(u^{nl}(t,x)-\widetilde{u}^{nl}(t,x), v^{nl}(t,x)-\widetilde{v}^{nl}(t,x)\right)\\ &= \Big(\int_0^t \mathcal{F}^{-1}\Big(E_{1+\lambda_1,1+\lambda_1}\Big(-t^{1+\lambda_1}|\xi|^{\theta_1}\Big)\Big)(t-s,x)*_{(x)}\left(|v(s,x)|^p-|\widetilde{v}(s,x)|^p\right)ds, \qquad (42)\\ &\quad \int_0^t \mathcal{F}^{-1}\Big(E_{1+\lambda_2,1+\lambda_2}\Big(-t^{1+\lambda_2}|\xi|^{\theta_2}\Big)\Big)(t-s,x)*_{(x)}\left(|u(s,x)|^q-|\widetilde{u}(s,x)|^q\right)ds\Big). \qquad (43) \end{aligned}$$

Similarly to the proof of the estimates (30), we can derive the following estimates for $0\le\tau\le t$:

$$\||v(\tau,\cdot)|^p-|\widetilde{v}(\tau,\cdot)|^p\|_{L^1} \lesssim (1+t)^{-\frac{n}{\theta_2}(1+\lambda_2)(p-1)+p\lambda_2}\|v-\widetilde{v}\|_{X(t)}\left(\|v\|^{p-1}_{X(t)}+\|\widetilde{v}\|^{p-1}_{X(t)}\right), \tag{44}$$

$$\||u(\tau,\cdot)|^q-|\widetilde{u}(\tau,\cdot)|^q\|_{L^1} \lesssim (1+t)^{-Q(\lambda_1,\lambda_2,\theta_1,\theta_2,q)}\|u-\widetilde{u}\|_{X(t)}\left(\|u\|^{q-1}_{X(t)}+\|\widetilde{u}\|^{q-1}_{X(t)}\right). \tag{45}$$

Using the last estimates, one may finally conclude, similarly to (38) to (41), the following estimates:

$$\left\| \int_0^t \mathcal{F}^{-1}\Big(E_{1+\lambda_1,1+\lambda_1}\Big(-t^{1+\lambda_1}|\xi|^{\theta_1}\Big)\Big)(t-s,x) *_{(x)} \big(|v(s,x)|^p - |\widetilde{v}(s,x)|^p\big)\, ds \right\|_{L^1}$$
$$\lesssim (1+t)^{\lambda_1+L(p)} \|v-\widetilde{v}\|_{X(t)} \Big(\|v\|^{p-1}_{X(t)} + \|\widetilde{v}\|^{p-1}_{X(t)}\Big),$$
$$\left\| \int_0^t \mathcal{F}^{-1}\Big(E_{1+\lambda_1,1+\lambda_1}\Big(-t^{1+\lambda_1}|\xi|^{\theta_1}\Big)\Big)(t-s,x) *_{(x)} \big(|v(s,x)|^p - |\widetilde{v}(s,x)|^p\big)\, ds \right\|_{L^p}$$
$$\lesssim (1+t)^{\lambda_1-\frac{n}{\theta_1}(1+\lambda_1)\left(1-\frac{1}{p}\right)+L(p)} \|v-\widetilde{v}\|_{X(t)} \Big(\|v\|^{p-1}_{X(t)} + \|\widetilde{v}\|^{p-1}_{X(t)}\Big),$$
$$\left\| \int_0^t \mathcal{F}^{-1}\Big(E_{1+\lambda_2,1+\lambda_2}\Big(-t^{1+\lambda_2}|\xi|^{\theta_2}\Big)\Big)(t-s,x) *_{(x)} \big(|u(s,x)|^q - |\widetilde{u}(s,x)|^q\big)\, ds\Big) \right\|_{L^1}$$
$$\lesssim (1+t)^{\lambda_2} \|v-\widetilde{v}\|_{X(t)} \Big(\|v\|^{p-1}_{X(t)} + \|\widetilde{v}\|^{p-1}_{X(t)}\Big),$$
$$\left\| \int_0^t \mathcal{F}^{-1}\Big(E_{1+\lambda_2,1+\lambda_2}\Big(-t^{1+\lambda_2}|\xi|^{\theta_2}\Big)\Big)(t-s,x) *_{(x)} \big(|u(s,x)|^q - |\widetilde{u}(s,x)|^q\big)\, ds\Big) \right\|_{L^q}$$
$$\lesssim (1+t)^{\lambda_2-\frac{n}{\theta_2}(1+\lambda_2)\left(1-\frac{1}{q}\right)} \|v-\widetilde{v}\|_{X(t)} \Big(\|v\|^{p-1}_{X(t)} + \|\widetilde{v}\|^{p-1}_{X(t)}\Big),$$

In this way, we can conclude the proof of the last condition (35) and the theorem.

4. Concluding Remarks

- We need to prove the blow-up for the system an interaction between the exponents of both equations. However, the method of scaling is not suitable to prove the blow-up result for the system since we have no interactions between the exponents. Moreover, the influence of each equation to the other one generated a condition presented by several parameters, fractional derivatives, dimensions, and others. For this reason, we will devote the blow-up problem in a forthcoming project using another approach.
- The applications of our results in real world problems and phenomena can be investigated after mathematical modeling by choosing the suitable parameters involved in our problem, such as dimension, and by taking the experimental values into consideration.

Funding: This work was supported by the Deanship of Scientific Research, Vice Presidency for Graduate Studies and Scientific Research, King Faisal University, Saudi Arabia (Project No. GRANT3371).

Data Availability Statement: Not applicable.

Acknowledgments: This work was supported by the Deanship of Scientific Research, Vice Presidency for Graduate Studies and Scientific Research, King Faisal University, Saudi Arabia (Project No. GRANT3371).

Conflicts of Interest: The author declares that there is no competing interest.

Appendix A

Theorem A1. *Let $\lambda \in (0,1), a_\lambda \in \mathbb{R}$. Then, the unique solution solution to*

$$\partial_t^{\lambda+1} f + |\xi|^\theta f = g(t), \qquad J^{1-\lambda} f(0) = a_\lambda, D^\lambda g(0) = 0. \tag{A1}$$

is given by

$$f(t) = t^{\lambda-1} E_{1+\lambda,\lambda}\Big(-t^{1+\lambda}|\xi|^\theta\Big) a_\lambda + \int_0^t (t-s)^\lambda E_{1+\lambda,1+\lambda}\Big(-t^{1+\lambda}|\xi|^\theta\Big)(t-s,\cdot) g(t) ds, \tag{A2}$$

where $E_{1+\lambda,\mu}$ are the Mittag–Leffler functions defined by

$$E_{1+\lambda,\mu}(z) = \sum_{k=0}^{\infty} \frac{z^k}{\Gamma(k+\lambda k+\mu)}.$$

For the proof, see [28].

Theorem A2. *Let $0 < \lambda < 2, \mu \in \mathbb{R}$, and $m \in \mathbb{N}$, with $m \geq \frac{\mu}{1+\lambda} - 1$. Then, for the real number $z > 0$, the following holds :*

$$E_{1+\lambda,\mu}(z^{1+\lambda}) = \frac{2}{1+\lambda} z^{1-\mu} e^{z\cos(\frac{\pi}{1+\lambda})} \cos(z \sin(\frac{\pi}{1+\lambda}) - \frac{\pi}{1+\lambda}(\mu - 1)) \tag{A3}$$

$$+ \sum_{k=1}^{m} \frac{(-1)^{k-1}}{\Gamma(\mu - k(1+\lambda))} z^{k(1+\lambda)} + \Omega_m(z), \tag{A4}$$

where

$$\Omega_m(z) = \frac{(-1)^m z^{1-\mu}}{\pi} (I_{1,m}(z) \sin(\pi(\mu - (m+1)(1+\lambda)) + I_{2,m}(z) \sin(\pi(\mu - m(1+\lambda)))),$$

and

$$I_{j,m}(z) = \int_0^\infty \frac{s^{(m+j)(1+\lambda)-\mu}}{s^{2(1+\lambda)} 2\cos(\pi(1+\lambda)) s^{1+\lambda} + 1} e^{sz} ds.$$

Remark A1. *The integral $I_{j,m}(z)$ is uniformly bounded if*

$$-1 < m + j - 1 + \frac{1-\mu}{1+\lambda} < 1. \tag{A5}$$

For the proof, see [29].

Proposition A1. *Let $a \in \mathbb{R} < 1$ and $b \in \mathbb{R}$. Then,*

$$\int_0^t (t-s)^{-a}(1+s)^{-b} ds \lesssim \begin{cases} (1+t)^{-a} & \text{if } a < 1 < b; \\ (1+t)^{-1}\log(1+t) & \text{if } a < 1 = b; \\ (1+t)^{1-a-b} & \text{if } a, b < 1. \end{cases} \tag{A6}$$

The reader can find the proof of Proposition A1 in [14].

Proposition A2. *The operator N maps $X(t)$ into itself and has one and only one fixed point $u \in X(t)$ if the following inequalities hold:*

$$\|Nu\|_{X(t)} \leq C_0(t)\|(u_0, u_1)\|_{\mathcal{A}_{m,s}} + C_1(t)\|u\|^p_{X(t)}, \tag{A7}$$

$$\|Nu - Nv\|_{X(t)} \leq C_2(t)\|u - v\|_{X(t)}(\|u\|^{p-1}_{X(t)} + \|v\|^{p-1}_{X(t)}), \tag{A8}$$

where $C_1(t), C_2(t) \longrightarrow 0$ for $t \longrightarrow +0$ and $C_0(t), C_1(t), C_2(t) \leq C$ for all $t \in [0, \infty)$.

For the proof, see [30].

Proposition A3. *Let us suppose that for any $(u_0, u_1), (v_0, v_1) \in \mathcal{A}_{m,s_1} \times \mathcal{A}_{m,s_2}$, the mapping N satisfies the following estimates:*

$$\|N(u,v)\|_{X(t)} \leq C_0(t)\Big(\|(u_0,u_1)\|_{\mathcal{A}_{m_1,s_1}} + \|(v_0,v_1)\|_{\mathcal{A}_{m_2,s_2}}\Big) + C_1(t)\Big(\|(u,v)\|^p_{X(t)} + \|(u,v)\|^q_{X(t)}\Big), \tag{A9}$$

$$\|N(u,v) - N(\tilde{u},\tilde{v})\|_{X(t)} \leq C_2(t)\|(u,v) - (\tilde{u},\tilde{v})\|_{X(t)} \times (\|(u,v)\|^{p-1}_{X(t)} + \|(\tilde{u},\tilde{v})\|^{p-1}_{X(t)} + \|(u,v)\|^{q-1}_{X(t)} + \|(\tilde{u},\tilde{v})\|^{q-1}_{X(t)}), \tag{A10}$$

where $C_1(t), C_2(t) \longrightarrow 0$ for $t \longrightarrow +0$ and $C_0(t), C_1(t), C_2(t) \leq C$ for all $t \in [0, \infty)$. Then, N maps $X(t)$ into itself and has one and only one fixed point $(u, v) \in X(t)$.

For the proof, see [26].

References

1. Fujita, H.T. On the blowing-up of solutions of the Cauchy problem for $\partial_t u = \Delta u + u^{1+\lambda}$. *J. Fac. Sci. Univ. Tokyo Sect.* **1966**, *13*, 109–124
2. Hayakawa, K. On the growing up problem for semi-linear heat equations. *Proc. Jpn. Acad.* **1973**, *49*, 503–505
3. Kobayashi, K.; Sirao, T.; Tanaka, H. The Critical Exponent(s) for the Semilinear Fractional Diffusive Equation. *J. Math. Soc. Jpn.* **1977**, *29*, 407–424.
4. Strauss, W.A. Nonlinear scattering theory at low energy. *J. Funct. Anal.* **1981**, *41*, 209–409. [CrossRef]
5. Glassey, R.T. Existence in the large for $\Box u = F(u)$ in two space dimensions. *Math. Z.* **1981**, *178*, 233–261. [CrossRef]
6. Glassey, R.T. Finite-time blow-up for solutions of nonlinear wave equations. *Math. Z.* **1981**, *177*, 323–340. [CrossRef]
7. Schaeffer, J. The equation $\partial_{tt} u + -\Delta u = |u|^p$ for the critical value of p. *Proc. R. Soc. Edinb. Sect. A* **1985**, *101*, 31–44. [CrossRef]
8. Yordanov, B.T.; Zhang, Q.S. Finite time blow up for critical wave equations in high dimensions. *J. Funct. Anal.* **2006**, *231*, 361–374. [CrossRef]
9. Zhou, Y. Blow up of solutions to semilinear wave equations with critical exponent in high dimensions. *Chin. Ann. Math. Ser. B* **2007**, *28*, 205–212. [CrossRef]
10. Kato, T. Blow-up of solutions of some nonlinear hyperbolic equations. *Commun. Pure Appl. Math.* **1980**, *33*, 501–505. [CrossRef]
11. John, F. Blow-up of solutions of nonlinear wave equations in three space dimensions. *Manuscr. Math.* **1979**, *28*, 235–268. [CrossRef]
12. Sideris, T.C. Nonexistence of global solutions to semilinear wave equations in high dimensions. *J. Differ. Equ.* **1984**, *52*, 378–406. [CrossRef]
13. D'Abbicco, M.; Ebert, M.R.; Picon, T. Global Existence of Small Data Solutions to the Semilinear Fractional Wave Equation. In *Trends in Mathematics*, 1st ed.; Springer: Boston, MA, USA, 2017; pp. 465–471.
14. D'Abbicco, M.; Ebert, M.R.; Picon, T.H. The Critical Exponent(s) for the Semilinear Fractional Diffusive Equation. *J. Fourier Anal. Appl.* **2019**, *25*, 696–731. [CrossRef]
15. Kainane, M.A.; Reissig, M. Semi-linear fractional $\sigma-$ evolution equations with mass or power non-linearity. *Nonlinear Differ. Equ. Appl.* **2018**, *25*, 42. [CrossRef]
16. Kainane, M.A. Global Existence of Small Data Solutions to Semi-linear Fractional $\sigma-$ Evolution Equations with Mass and Nonlinear Memory. *Mediterr. J. Math.* **2020**, *17*, 159. [CrossRef]
17. Escobedo, M.; Herrero, A. Boundedness and blow up for a semilinear reaction-diffusion system. *J. Differ. Equ.* **1991**, *89*, 176–202. [CrossRef]
18. Andreucci, D.; Herrero, M.A.; Velázquez, J.L. Liouville theorems and blow up behaviour in semilinear reaction diffusion systems. *Ann. Inst. Poincaré Anal. Non Linéaire* **1997**, *14*, 1–53. [CrossRef]
19. Rencławowicz, J. Blow up, global existence and growth rate estimates in nonlinear parabolic systems. *Colloq. Math.* **2000**, *86*, 43–66. [CrossRef]
20. Escobedo, M.; Levine, H.A. Critical blowup and global existence numbers for a weakly coupled system of reaction-diffusion equations. *Arch. Rational Mech. Anal.* **1995**, *129*, 47–100. [CrossRef]
21. Snoussi, S.; Tayachi, S. Global existence, asymptotic behavior and self-similar solutions for a class of semilinear parabolic systems. *Nonlinear Anal.* **2002**, *48*, 13–35. [CrossRef]
22. Sun. F.; Wang, M. Existence and nonexistence of global solutions for a non-linear hyperbolic system with damping. *Nonlinear Anal.* **2007**, *66*, 2889–2910. [CrossRef]
23. Narazaki, T. Global solutions to the Cauchy problem for the weakly coupled of damped wave equations. *Discrete Contin. Dyn. Syst.* **2009**, *2009*, 592–601.
24. Nishihara, K.; Wakasugi, Y. Critical exponant for the Cauchy problem to the weakly coupled wave system. *Nonlinear Anal.* **2014**, *108*, 249–259. [CrossRef]
25. Mohammed Djaouti, A.; Reissig, M. *Weakly Coupled Systems of Semilinear Effectively Damped Waves with Different Time-Dependent Coefficients in the Dissipation Terms and Different Power Nonlinearities*; D'Abbicco, M., Ebert, M., Georgiev, V., Ozawa, T., Eds.; New Tools for Nonlinear PDEs and Application; Trends in Mathematics; Birkhäuser: Basel, Switzerland, 2019; pp. 209–409.
26. Mohammed Djaouti, A. Semilinear Systems of Weakly Coupled Damped Waves. Ph.D. Thesis, TU Bergakademie Freiberg, Freiberg, Germany, 2018.
27. Mohammed Djaouti, A. Modified different nonlinearities for weakly coupled systems of semilinear effectively damped waves with different time-dependent coefficients in the dissipation terms. *Adv. Differ. Equ.* **2021**, 66. [CrossRef]
28. Kilbas, A.A.; Srivastava, H.M.; Trujillo, J.J. *Theory and Applications of Fractional Differential Equations*; Elsevier: Amsterdam, The Netherlands, 2006; p. 204.

29. Popov, A.Y.; Sedletskii, A.M. Distribution of roots of Mittag-Leffler functions. *J. Math. Sci.* **2013**, *190*, 209–409. [CrossRef]
30. Mohammed Djaouti, A.; Reissig, M. Weakly coupled systems of semilinear effectively damped waves with time-dependent coefficient, different power nonlinearities and different regularity of the data. *Nonlinear Anal.* **2018**, *175*, 28–55. [CrossRef]

symmetry

MDPI

Article

Numerical Simulation for a Hybrid Variable-Order Multi-Vaccination COVID-19 Mathematical Model

Nasser Sweilam [1,*,†], Seham M. Al-Mekhlafi [2,3,†], Reem G. Salama [4] and Tagreed A. Assiri [5]

1 Department of Mathematics, Faculty of Science, Cairo University, Giza 12613, Egypt
2 Department of Mathematics, Faculty of Education, Sana'a University, Sana'a 1247, Yemen
3 Department of Engineering Mathematics and Physics, Future University in Egypt, New Cairo 11835, Egypt
4 Department of Mathematics, Faculty of Science, Beni-Suef University, Beni-Suef 62521, Egypt
5 Department of Mathematics, Faculty of Science, Umm Al-Qura University, Makkah 21961, Saudi Arabia
* Correspondence: nsweilam@sci.cu.edu.eg
† These authors contributed equally to this work.

Abstract: In this paper, a hybrid variable-order mathematical model for multi-vaccination COVID-19 is analyzed. The hybrid variable-order derivative is defined as a linear combination of the variable-order integral of Riemann–Liouville and the variable-order Caputo derivative. A symmetry parameter σ is presented in order to be consistent with the physical model problem. The existence, uniqueness, boundedness and positivity of the proposed model are given. Moreover, the stability of the proposed model is discussed. The theta finite difference method with the discretization of the hybrid variable-order operator is developed for solving numerically the model problem. This method can be explicit or fully implicit with a large stability region depending on values of the factor Θ. The convergence and stability analysis of the proposed method are proved. Moreover, the fourth order generalized Runge–Kutta method is also used to study the proposed model. Comparative studies and numerical examples are presented. We found that the proposed model is also more general than the model in the previous study; the results obtained by the proposed method are more stable than previous research in this area.

Keywords: variable-order hybrid operator; Pfizer vaccine; Moderna vaccine; Janssen vaccine; theta finite difference method; generalized fourth order Runge–Kutta method

MSC: 65L05; 37N30; 65M06

Citation: Sweilam, N.; Al-Mekhlafi, S.M.; Salama, R.G.; Assiri, T.A. Numerical Simulation for a Hybrid Variable-Order Multi-Vaccination COVID-19 Mathematical Model. *Symmetry* **2023**, *15*, 869. https://doi.org/10.3390/sym15040869

Academic Editors: Francisco Martínez González, Mohammed K. A. Kaabar and Juan Luis García Guirao

Received: 4 February 2023
Revised: 26 March 2023
Accepted: 30 March 2023
Published: 6 April 2023

1. Introduction

Coronaviruses are a large family of viruses known to cause illnesses ranging from the common cold to more serious illnesses such as severe acute respiratory syndrome [1]. The World Health Organization has designated this variant as a variant of serious concern. The United States Centers for Disease Control and Prevention has granted Emergency Use Authorization to the following vaccines: Pfizer-BioNTech with 95% efficacy against symptomatic COVID-19, Moderna vaccine with 94.5% efficacy and Janssen vaccine manufactured by Johnson & Johnson, which has an efficacy rating of 67%, as well as many others [1,2]. SARS-CoV-2 vaccinations have been shown to be effective against infections, including both silent and symptomatic cases, of severe COVID-19 illness and deaths [2]. Mathematical modeling is a valuable tool to study disease spread and control very effectively. Several mathematical models have been proposed in the literature to study and understand the novel complex transmission pattern of the COVID-19 pandemic; see, for example, [3–8].

In the meantime, there are now extensive articles explaining the advantage of fractional order models for studying real mathematical models in various fields [9]. The variable-order fractional derivatives (VOFDs) can describe the effects of the long variable memory of a time-dependent system. In [10], Samko et al. proposed this interesting extension of the classical calculation of fractions. In the concept of fractional derivative with variable order, the order may vary either as a function of the independent differentiation variable (t) or as a function of another (possibly spatial) variable (x), or both. Therefore, the derivative models described using variable-order fractional derivatives are useful and appropriate for the epidemic models. We can obtain the results of fractional order and integer order as a special case from variable-order mathematical models [11–18].

In this article, we will present the theta finite different method with the discretization of new hybrid fractional operator. This operator is called the constant proportional Caputo variable-order fractional derivative (CPC-Θ FDM) and is used to study the proposed model numerically. In the literature, the theta finite differences method (ΘFDM) method, also called the weighted average finite differences method (WAFDM), is one of the finite difference methods [19,20]. This method could be an explicit method or an implicit method (more stable and efficient), depending on the weight factor $\Theta \in [0,1]$. Using Caputo and Riesz–Feller derivatives, this method was developed for a nonstandard finite difference method [21,22].

The goal of this work is to present and analyze a hybrid variable-order fractional model of multi-vaccination for COVID-19. The new variable-order hybrid derivatives are defined as the linear combination of the variable-order Riemann–Liouville integral and the variable-order derivative of Caputo. This is one of the most effective and reliable of these operators; it is more general than the Caputo fractional operator. Positivity, boundedness and stability will be proved in the current model.

Moreover, one of the aims of this article is developing CPC-Θ FDM for solving the variable-order fractional differential equations numerically and we will compare the obtained results with the results obtained with the fourth order generalized Runge–Kutta method (GRK4M) [23] and the method in [24]. Moreover, we extended the method in [24] to variable order. The analysis of stability and convergence of the proposed method will be studied. Numerical simulations will be given to confirm the efficiency and wide applicability of the proposed method.

To our knowledge, no numerical investigations of a hybrid variable-order fractional for multi-vaccination for a COVID-19 mathematical model utilizing CPC-Θ FDM have been conducted.

This paper is organized as follows: Some notations and definitions of variable-order fractional derivatives are introduced in Section 2. In Section 3, the model with a hybrid variable order is presented; moreover, the positivity, boundedness, existence and uniqueness of the solutions and the stability of the present model are discussed. In Section 4, the numerical methods GRK4M and CPC-Θ SFDM are studied; moreover, stability analyses for these methods are proved. In Section 5, numerical simulations are presented. The conclusions are ultimately outlined in Section 6.

2. Notations and Preliminaries

In this section, we review several key definitions of variable-order calculus that will be utilized throughout the remainder of this article.

Definition 1. *Caputo's derivatives (right–left side variable-order fractional $\alpha(t)$) are defined, respectively, as follows [25]:*

$$({}^{C}D_{b-}^{\alpha(t)}f)(x) = ({}_{t}^{C}D_{b}^{\alpha(t)}f)(t) = \frac{(-1)^n}{\Gamma(n-\alpha(t))}\int_t^b \frac{f^{(n)}(s)}{(s-t)^{-n+\alpha(t)+1}}ds, \quad b > t, \tag{1}$$

$$({}^{C}D_{a+}^{\alpha(t)}f)(t) = ({}_{a}^{C}D_{t}^{\alpha(t)}f)(t) = \frac{1}{\Gamma(n-\alpha(t))}\int_a^t \frac{f^{(n)}(s)}{(t-\xi)^{-n+\alpha(t)+1}}ds, \quad t > a, \tag{2}$$

$f(t) \in AC^n[a,b],\ n = 1 + [\Re(\alpha(t))],\ \Re(\alpha(t)) \notin \mathbb{N}_0.$

Definition 2. *Let* $1 > \alpha(t) > 0$, $-\infty < a < b < +\infty$; *the right–left side variable-order fractional Riemann–Liouville's integral and* $f(t) \in AC^n[a,b]$ *are given as follows* [25]:

$$ {}_tI_b^{\alpha(t)} f(t) = \left[\int_t^b f(s)(t-s)^{\alpha(t)-1} ds \right] \frac{1}{\Gamma(\alpha(t))}, \ t < b, \tag{3} $$

$$ {}_aI_t^{\alpha(t)} f(t) = \left[\int_a^t f(s)(t-s)^{\alpha(t)-1} ds \right] \frac{1}{\Gamma(\alpha(t))}, \ t > a. \tag{4} $$

$\alpha(t) \in \mathbb{C}$.

Definition 3 ([26]). *The variable-order fractional Caputo proportional operator (CP) is given as follows:*

$$ \begin{aligned} {}_0^{CP}D_t^{\alpha(t)} y(t) &= \int_0^t (\Gamma(1-\alpha(t)))^{-1}(t-s)^{-\alpha(t)}(y'(s)K_0(s,\alpha(t)) + y(s)K_1(s,\alpha(t)))ds, \\ &= \left(\frac{\Gamma(1-\alpha(t))^{-1}}{t^{\alpha(t)}} \right)(y'(t)K_0(t,\alpha(t)) + y(t)K_1(t,\alpha(t))). \end{aligned} \tag{5} $$

$K_1(\alpha(t),t) = (-\alpha(t)+1)t^{\alpha(t)}, \quad K_0(\alpha(t),t) = t^{(1-\alpha(t))}\alpha(t), \quad 1 > \alpha(t) > 0.$

Alternatively, the constant proportional Caputo (CPC) variable-order fractional hybrid operator can be formulated as follows [26]:

$$ \begin{aligned} {}_0^{CPC}D_t^{\alpha(t)} y(t) &= \left(\int_0^t (t-s)^{-\alpha(t)} \frac{1}{\Gamma(1-\alpha(t))} (K_1(\alpha(t))y(s) + y'(s)K_0(\alpha(t)))ds \right) \\ &= K_1(\alpha(t)) {}_0^{RL}I_t^{1-\alpha(t)} y(t) + K_0(\alpha(t)) {}_0^{C}D_t^{\alpha(t)} y(t), \end{aligned} \tag{6} $$

$K_0(\alpha(t)) = Q^{(-\alpha(t)+1)}\alpha(t), \ \ K_1(\alpha(t)) = Q^{\alpha(t)}(-\alpha(t)+1)$, where Q is a constant.

Definition 4. *Moreover, its inverse operator is* [26]:

$$ {}_0^{CPC}I_t^{\alpha(t)} y(t) = \left(\int_0^t exp\left[\frac{K_1(\alpha(t))}{K_0(\alpha(t))}(t-s) \right] {}_0^{RL}D_t^{1-\alpha(t)} y(s) ds \right) \frac{1}{K_0(\alpha(t))}. \tag{7} $$

3. A Hybrid Variable-Order Mathematical Model

A variable-order multiple vaccination model for COVID-19 is presented below; it is an extension of the model given in [24]. To satisfy the dimensional fit between the two sides of the resulting variable-order fraction equations, the variable-order operator is modified by an auxiliary parameter σ. As a result, the dimension of the left side is $(\text{day})^{-1}$ [27]. The following is the updated variable-order nonlinear fractional mathematical model:

$$ \frac{1}{\sigma^{1-\alpha(t)}} {}_0^{CPC}D_t^{\alpha(t)} S = \Lambda - \nu_1 S - \nu_2 S - \nu_3 S - \lambda S - \mu S, $$

$$ \begin{aligned} \frac{1}{\sigma^{1-\alpha(t)}} {}_0^{CPC}D_t^{\alpha(t)} V_1 &= \nu_1 S - (1-\xi_1)\lambda V_1 - \mu V_1, \\ \frac{1}{\sigma^{1-\alpha(t)}} {}_0^{CPC}D_t^{\alpha(t)} V_2 &= \nu_2 S - (1-\xi_2)\lambda V_2 - \mu V_2, \\ \frac{1}{\sigma^{1-\alpha(t)}} {}_0^{CPC}D_t^{\alpha(t)} V_3 &= \nu_3 S - (1-\xi_3)\lambda V_3 - \mu V_3, \\ \frac{1}{\sigma^{1-\alpha(t)}} {}_0^{CPC}D_t^{\alpha(t)} A &= f_3(1-\xi_3)\lambda V_3 + f_2(1-\xi_2)\lambda V_2 + f_1(1-\xi_1)\lambda V_1 - (\gamma_A + \mu)A + p\lambda S, \end{aligned} $$

$$
\begin{aligned}
\frac{1}{\sigma^{1-\alpha(t)}}{}_{0}^{CPC}D_t^{\alpha(t)} I_U =&(1-p)\lambda S-(\gamma_{IU}+d_{IU}+\alpha_1\mu)I_U,\\
\frac{1}{\sigma^{1-\alpha(t)}}{}_{0}^{CPC}D_t^{\alpha(t)} I_V =&(1-f_2)(1-\xi_2)\lambda V_2+(1-f_3)(1-\xi_3)+(1-f_1)(1-\xi_1)\lambda V_1\lambda V_3\\
&-(\gamma_{IV}+(1-\phi)\alpha\mu+d_{IV})I_V,
\end{aligned}
$$

$$
\begin{aligned}
\frac{1}{\sigma^{1-\alpha(t)}}{}_{0}^{CPC}D_t^{\alpha(t)} I_S =&\alpha_1(1-\phi)I_V-(d_{IS}+\mu+\gamma_{IS})I_S+\alpha_1 I_U,\\
\frac{1}{\sigma^{1-\alpha(t)}}{}_{0}^{CPC}D_t^{\alpha(t)} R =&\gamma_A A+\gamma_{IU}I_U+\gamma_{IV}I_V+\gamma_{IS}I_S-\mu R.
\end{aligned}
\tag{8}
$$

$$
\lambda=\beta N_H^{-1}\Bigg(I_U+\theta A+\eta_v I_v\Bigg),
$$

$$
S+V_1+V_2+V_3+A+I_U+I_V+I_S+R=N_H(t),
$$

with the initial conditions

$$
\begin{aligned}
S(0)=s_0\geq 0,\ V_1(0)=v_{10}\geq 0,\ V_2=v_{20}\geq 0,\ V_3=v_{30}\geq 0,\ A=a_0\geq 0,\ I_U=i_{u0}\geq 0,\\
I_V=i_{v0}\geq 0,\ I_S=i_{s0},\ R(0)=r_0\geq 0.
\end{aligned}
\tag{9}
$$

Figure 1 shows the flowchart of the model (8). Table 1 shows the definitions of variables for system (8). The hypotheses of the model for the rate of each type of vaccination are the same as in [24], as follows:

1 $N_H(t)=S+V_1+V_2+V_3+A+I_u+I_v+I_s+R$.
2 Vaccination simulations of the proposed model in the strategy implementing only the Pfizer vaccine ($f_1\neq 0,\ \xi_1\neq 0, \phi_1\neq 0,\ v_1\neq 0$), where these parameters are defined as in Table 2.
3 Vaccination simulations of the proposed model in the strategy implementing only Moderna vaccine ($\xi_2\neq 0, f_2\neq 0,\ v_2\neq 0, \phi_2\neq 0$).
4 Vaccination simulations of the proposed model in the strategy implementing only Janssen vaccine ($\xi_3\neq 0, f_3\neq 0,\ v_3\neq 0, \phi_3\neq 0$).

We can verify the boundedness of the solution for the suggested model (8) as follows:

$$
\begin{aligned}
&\frac{1}{\sigma^{1-\alpha(t)}}({}_{0}^{CPC}D_t^{\alpha(t)}S+{}_{0}^{CPC}D_t^{\alpha(t)}R+{}_{0}^{CPC}D_t^{\alpha(t)}V_3+{}_{0}^{CPC}D_t^{\alpha(t)}V_2+{}_{1}^{CtV}+\\
&{}_{0}^{CPC}D_t^{\alpha(t)}A+{}_{0}^{CPC}D_t^{\alpha(t)}S+{}_{0}^{CPC}D_t^{\alpha(t)}I_s+{}_{0}^{CPC}D_t^{\alpha(t)}I_V+{}_{0}^{CPC}D_t^{\alpha(t)}I_U)=\sigma^{-1+\alpha(t)}{}_{0}^{CPC}D_t^{\alpha(t)}N_H(t),\\
&\frac{1}{\sigma^{1-\alpha(t)}}{}_{0}^{CPC}D_t^{\alpha(t)}N_H(t)=\Lambda-\mu N_H(t)-[d_{IV}I_V+d_{IU}I_U+d_{IS}I_S],\quad N_H(0)=A\geq 0,
\end{aligned}
\tag{10}
$$

$$
\Lambda-(\mu+3\delta)N_H\leq\frac{dN_H}{dt}<\Lambda-\mu N_H,\quad \delta=min\{d_{IV},d_{IU},d_{IS}\}.
$$

Therefore, we have $N_H(t)\leq\Lambda\mu^{-1}$, at $t\longrightarrow\infty$. The feasible region

$$
\Omega=\{S,A,I_U,I_V,I_S,R,V_3,V_1,V_2\in\mathbb{R}^9,\ N_H(t)\leq\Lambda\mu^{-1}\}.
$$

System (8) has a solution in Ω. This verifies the boundedness of the solution.

Table 1. Variables of system (8).

Variable	Interpretation
R	Humans who have recovered
S	Unvaccinated susceptible individuals
V_3	Vaccinated using vaccination number three (Oxford Johnson & Johnson)
V_2	Vaccinated using vaccination number two (Moderna)
V_1	Vaccinated using vaccination number one (Pfizer)
I_S	Individuals with severe sickness and hospitalization who are symptomatic (vaccinated and unvaccinated) (under complete isolation)
I_V	Symptomatic people who have been vaccinated
I_U	Symptomatic people who have not been immunized
A	Asymptomatic individuals (vaccinated and unvaccinated)

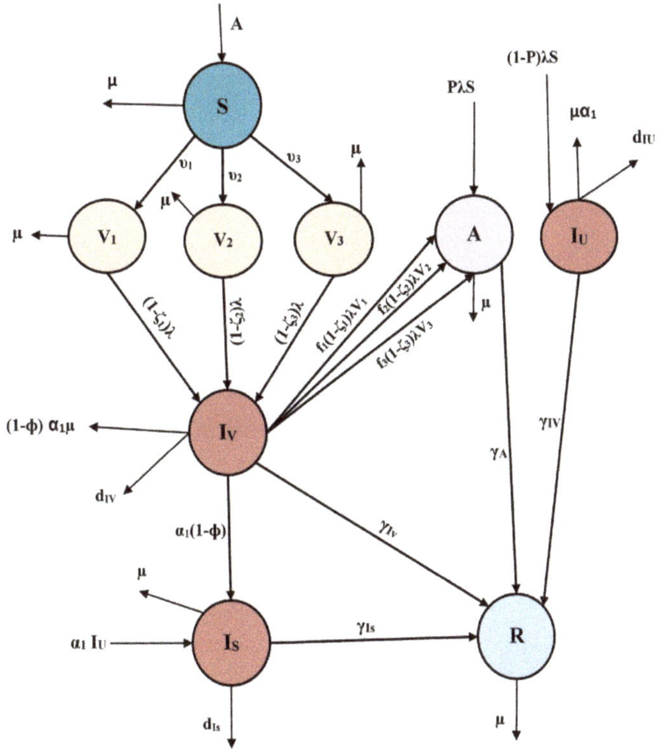

Figure 1. Flowchart for system (8).

Theorem 1. *Using (9), for $t \geq 0$ solutions of (8) are still nonnegative.*

Proof. Using (9), we obtain [28]:

$$
\begin{aligned}
&\frac{1}{\sigma^{1-\alpha(t)}}{}_0^{CPC}D_t^{\alpha(t)}S\ |_{S=0} = \Lambda \geq 0,\\
&\frac{1}{\sigma^{1-\alpha(t)}}{}_0^{CPC}D_t^{\alpha(t)}V_1\ |_{V_1=0} = V_1 S \geq 0,\\
&\frac{1}{\sigma^{1-\alpha(t)}}{}_0^{CPC}D_t^{\alpha(t)}V_2\ |_{V_2=0} = V_2 S \geq 0,\\
&\frac{1}{\sigma^{1-\alpha(t)}}{}_0^{CPC}D_t^{\alpha(t)}V_3\ |_{V_3=0} = V_3 S \geq 0,\\
&\frac{1}{\sigma^{1-\alpha(t)}}{}_0^{CPC}D_t^{\alpha(t)}A\ |_{A=0} = (1-\xi_2)f_2\lambda V_2 + (1-\xi_3)f_3\lambda V_3 + p\lambda S + (1-\xi_1)f_1\lambda V_1 \geq 0,\\
&\frac{1}{\sigma^{1-\alpha(t)}}{}_0^{CPC}D_t^{\alpha(t)}I_U\ |_{I_U=0} = (1-p)\lambda S \geq 0,\\
&\frac{1}{\sigma^{1-\alpha(t)}}{}_0^{CPC}D_t^{\alpha(t)}I_V\ |_{I_V=0} = (1-\xi_2)\lambda(1-f_2)V_2 + (1-\xi_3)\lambda V_3(1-f_3) + (1-\xi_1)\lambda V_1(1-f_1) \geq 0,\\
&\frac{1}{\sigma^{1-\alpha(t)}}{}_0^{CPC}D_t^{\alpha(t)}I_S\ |_{I_S=0} = \alpha_1 I_U + (1-\phi)\alpha_1 I_V \geq 0,\\
&\frac{1}{\sigma^{-\alpha(t)+1}}{}_0^{CPC}D_t^{\alpha(t)}R\ |_{R=0} = \gamma_A A + \gamma_{IU} I_U + \gamma_{IV} I_V + \gamma_{IS} I_S \geq 0.
\end{aligned}
\tag{11}
$$

□

3.1. Uniqueness and Existence

The existence and uniqueness of the solutions of the proposed model will be established using Banach fixed point theorem. Let system (8) be written as follows [4]:

$$ {}_0^{CPC}D_t^{\alpha(t)}\varepsilon(t) = \varpi(\varepsilon(t), t), \quad \varepsilon(0) = \varepsilon_0 \geq 0, \tag{12} $$

$\varepsilon(t) = \begin{pmatrix} S, A, I_U, I_V, I_S, R, V_3, V_1, V_2 \end{pmatrix}^T$ represents the variables of the proposed system (8) and ϖ is a vector that represents the equations in the right of the system (8).

$$
\begin{pmatrix} \varpi_1\\ \varpi_2\\ \varpi_3\\ \varpi_4\\ \varpi_5\\ \varpi_6\\ \varpi_7\\ \varpi_8\\ \varpi_9 \end{pmatrix}
=
\begin{pmatrix}
\sigma^{1-\alpha(t)}(\Lambda - v_1 S - v_2 S - v_3 S - \lambda S - \mu S)\\
\sigma^{1-\alpha(t)}(v_1 S - (1-\xi_1)\lambda V_1 - \mu V_1)\\
\sigma^{1-\alpha(t)}(v_1 S - (1-\xi_2)\lambda V_2 - \mu V_2)\\
\sigma^{1-\alpha(t)}(v_1 S - (1-\xi_3)\lambda V_3 - \mu V_3)\\
\sigma^{1-\alpha(t)}((1-\xi_2)\lambda f_2 V_2 + (1-\xi_1)\lambda f_1 V_1 + (1-\xi_3)\lambda f_3 V_3 - (\gamma_A+\mu)A) + p\lambda S\\
\sigma^{1-\alpha(t)}((1-p)\lambda S - (\gamma_{IU} + d_{IU} + \alpha_1\mu)I_U)\\
\sigma^{1-\alpha(t)}((1-\xi_2)\lambda(1-f_2)V_2 + (1-\xi_3)\lambda(1-f_3)V_3 + (1-\xi_1)\lambda(1-f_1)V_1 - (d_{IV} + \alpha(1-\phi)\mu)I_V + \gamma_{IV})\\
\sigma^{1-\alpha(t)}(\alpha_1 I_U - (d_{IS} + \mu + \gamma_{IS})I_S) + \alpha_1(1-\phi)I_V\\
\sigma^{1-\alpha(t)}(\gamma_{IU} I_U + \gamma_{IV} I_V + \gamma_{IS} I_S - \mu R + \gamma_A A)
\end{pmatrix},
$$

with an initial condition ε_0. Furthermore, Lipschitz requirements as in [4] are satisfied:

$$ \|\varpi(\varepsilon_1(t), t) - \varpi(\varepsilon_2(t), t)\| \leq W^0 \|\varepsilon_1(t) - \varepsilon_2(t)\|, \quad W^0 \in \mathbb{R}. \tag{13} $$

Theorem 2. *If the following conditions are met:*

$$ \frac{W^0 F_{max}^{\alpha(t)} X_{max}^{\alpha(t)}}{\Gamma(\alpha(t)-1)K_0(\alpha(t))} < 1, \tag{14} $$

the hybrid variable-order fractional model (8) has a unique solution.

Proof. Applying (6) in (12), we have:

$$ \varepsilon(t) = \varepsilon(t_0) + \frac{1}{K_0(\alpha(t))}\int_0^t exp(-\frac{K_1(\alpha(t))}{K_0(\alpha(t))}(t-s))_0^{RL}D_t^{1-\alpha(t)}\varpi(\varepsilon(s), s)ds. \tag{15} $$

Let $B: C(K, \mathbb{R}^9) \longrightarrow C(K, \mathbb{R}^9)$ and $K = (0, T)$; then:

$$B[\varepsilon(t)] = \varepsilon(t_0) + \frac{1}{K_0(\alpha(t))} \int_0^t exp(-\frac{K_1(\alpha(t))}{K_0(\alpha(t))}(t-s))_0^{RL} D_t^{1-\alpha(t)} \varpi(\varepsilon(s), s)ds. \quad (16)$$

We have:

$$B[\varepsilon(t)] = \varepsilon(t).$$

The supremum norm on K is represented by $\|.\|_K$. Thus

$$\|\varepsilon(t)\|_K = \sup_{t \in K} \|\varepsilon(t)\|, \quad \varepsilon(t) \in C(K, \mathbb{R}^9).$$

So, $\|.\|_K$ with $C(K, \mathbb{R}^9)$ is a Banach space. Then, the following relation holds:

$$\Lambda \|\varphi(s,t)\|_K \|\varepsilon(s)\|_K \geq \| \int_0^t \varphi(s,t)\varepsilon(s)ds\|, \quad 0 < t < \Lambda < \infty$$

with $\varphi(s,t) \in C(K^2, \mathbb{R}^9)$ $\varepsilon(t) \in C(K, \mathbb{R}^9)$,
then $\sup_{t,s \in K} |\varphi(s,t)| = \|\varphi(s,t)\|_K$.
Relation (16) can be written as:

$$\begin{aligned}
\|B[\varepsilon_1(t)] - B[\varepsilon_2(t)]\|_K \leq & \|\frac{1}{K_0(\alpha(t))} \int_0^t exp(-\frac{K_1(\alpha(t))}{K_0(\alpha(t))}(t-s))({}_0^{RL} D_t^{1-\alpha(t)} \varpi(\varepsilon_1(s), s) \\
& - {}_0^{RL} D_t^{1-\alpha(t)} \varpi(\varepsilon_2(s), s))ds\|_K. \\
\leq & \frac{F_{max}^{\alpha(t)}}{K_0(\alpha(t))\Gamma(\alpha(t)-1)} \| \int_0^t (t-s)^{\alpha(t)-2} (\varpi(\varepsilon_1(s), s) - \varpi(\varepsilon_2(s), s))ds\|_K, \\
\leq & \frac{F_{max}^{\alpha(t)} X_{max}^{\alpha(t)}}{K_0(\alpha(t))\Gamma(\alpha(t)-1)} \|\varpi(\varepsilon_1(t), t) - \varpi(\varepsilon_2(t), t)\|_K, \\
\leq & \frac{W^0 F_{max}^{\alpha(t)} X_{max}^{\alpha(t)}}{K_0(\alpha(t))\Gamma(\alpha(t)-1)} \|\varepsilon_1(t) - \varepsilon_2(t)\|_K. \quad (17)
\end{aligned}$$

Then

$$\|B[\varepsilon_1(t)] - B[\varepsilon_2(t)]\|_K \leq L\|\varepsilon_1(t) - \varepsilon_2(t)\|_K, \quad (18)$$

where

$$L = \frac{W^0 F_{max}^{\alpha(t)} X_{max}^{\alpha(t)}}{K_0(\alpha(t))\Gamma(\alpha(t)-1)}.$$

B is a contraction operator if $1 > L$. So (8) has a unique solution. □

Table 2. The definition of all parameters of system (8).

Parameter	Interpretation	Baseline Value (per day^{-1})	Reference
Λ	Recruitment rate	$\frac{29,200,000}{75 \times 365}$ day^{-1}	[29]
β	Rate of effective transmission	0.00016708	[24]
μ	Natural death rate	$\frac{1}{75 \times 365}$ day^{-1}	[29]
ξ_3	Efficacy of the Janssen vaccine	0.67	[1]
ξ_2	Efficacy of the Moderena vaccine	0.945	[30]
ξ_1	Efficacy of the Pfizer vaccine	0.95	[31]
v_3	Rate of Janssen vaccination	0.00053 day^{-1}	[24]
v_2	Rate of Moderena vaccination	0.0042 day^{-1}	[24]
v_1	Rate of Pfizer vaccination	0.0059 day^{-1}	[24]
p	Unvaccinated susceptibles who move to the asymptomatic stage are a small percentage of the total	0.5	[24]
θ	A parameter was changed to limit the transmissibility of asymptomatic people	0.7	[32]
ϕ	Vaccine effectiveness against severe COVID-19 sickness	0.8	[2]
f_i	The percentage of susceptibles who received the vaccine and went on to develop subclinical disease	0.5	[24]
$\gamma_A, \gamma_{IU}, \gamma_{IV}, \gamma_{IS}$	Individuals in A, I_U, I_V and I_S classes, respectively; the programme has a high rate of recovery	0.13978 day^{-1}	[24]
d_{IU}, d_{IV}, d_{IS}	Death rates from disease for people in the I_U, I_V and I_S groups, respectively	0.015	[32]
α_1	The rate at which severe COVID-19 sickness develops	0.3	[32]

3.2. Local Stability

The basic reproduction number is calculated in this section. The next generation operator method is used to investigate the local stability of the disease-free equilibrium (DFE), which is given by solving $\frac{1}{\sigma^{1-\alpha(t)}}{}_{0}^{CPC}D_t^{\alpha(t)}(.) = 0$ of model (8) and considering $I_U = I_V = I_S = 0$. Then, we obtained D_0, where D_0 is the DFE and is given by [33]:

$$D_0 = (\tilde{S}, \tilde{V}_3, \tilde{V}_2, \tilde{V}_1, \tilde{A}, \tilde{I}_V, \tilde{I}_S, \tilde{I}_U, \tilde{R}) =$$
$$\Big(\frac{\Lambda}{(\nu_3 + \nu_2 + \nu_1 + \mu)}, \frac{\nu_3\Lambda}{(\nu_3 + \nu_2 + \nu_1 + \mu)}, \frac{\nu_2\Lambda}{(\nu_3 + \nu_2 + \nu_1 + \mu)}, \frac{\nu_1\Lambda}{(\nu_1 + \nu_2 + \nu_3 + \mu)},$$
$$\frac{\Lambda}{(\nu_1 + \nu_\lambda 2 + \nu_3 + \mu)}, 0, 0, 0, 0\Big).$$

As a result, the matrix V of the transfer of individuals between compartments and the matrix F of new infection terms are provided by

$$F = \sigma^{1-\alpha(t)} \begin{pmatrix} \frac{\beta\theta\tilde{Q}}{\tilde{N}_H} & \frac{\beta\tilde{Q}}{\tilde{N}_H} & \frac{\beta\eta_V\tilde{Q}}{\tilde{N}_H} & 0 \\ (1-p)\frac{\beta\theta\tilde{S}}{\tilde{N}_H} & (1-p)\frac{\beta\tilde{S}}{\tilde{N}_H} & (1-p)\frac{\beta\eta_V\tilde{S}}{\tilde{N}_H} & 0 \\ \frac{\beta\theta\tilde{v}}{\tilde{N}_H} & \frac{\beta\tilde{v}}{\tilde{N}_H} & \frac{\beta\eta_V\tilde{v}}{\tilde{N}_H} & 0 \\ 0 & 0 & 0 & 0 \end{pmatrix},$$

with $\tilde{v} = (1-\xi_3)(1-f_3)\tilde{V}_3 + (1-\xi_2)(1-f_2)\tilde{V}_2 + (1-\xi_1)(1-f_1)\tilde{V}_1$,
$\tilde{Q} = (1-\xi_3)f_3\tilde{V}_3 + (1-\xi_1)f_1\tilde{V}_1 + (1-\xi_2)f_2\tilde{V}_2 + p\tilde{S}$.

$$V = \sigma^{1-\alpha(t)} \begin{pmatrix} \mu + \gamma_A & 0 & 0 & 0 \\ 0 & \gamma_{IU} + d_{IU} + \alpha_1 + \mu & 0 & 0 \\ 0 & 0 & \gamma_{IV} + d_{IV} + (1-\phi)\alpha_1 + \mu & 0 \\ 0 & -\alpha_1 & -(1-\phi)\alpha_1 & \gamma_{IU} + d_{IU} + \mu \end{pmatrix}.$$

The model's basic reproduction number, denoted by R_0, is given by [34,35]:

$$\rho(FV^{-1}) = R_0 = \sigma^{1-\alpha(t)}\beta\Big(\frac{(1-p)E_1E_3\mu + E_1E_2\eta_V Y_1 + E_2E_3\eta_A\theta Y_2}{\mu(\nu_1 + \nu_2 + \nu_3)E_1E_2E_3}\Big). \quad (19)$$

with $E_1 = (\gamma_A + \mu)$,
$E_2 = (\gamma_{IU} + d_{IU} + \alpha_1 + \mu)$,
$E_3 = (\alpha_1(1-\phi) + d_{IV} + \mu + \gamma_{IV})$,
$Y_1 = (1-\xi_3)(1-f_3)\nu_3 + (1-\xi_1)(1-f_1)\nu_1 + (1-\xi_2)(1-f_2)\nu_2$,
$Y_2 = (1-\xi_3)f_3\nu_3 + (1-\xi_2)f_2\nu_2 + \mu p(1-\xi_1)f_1\nu_1$.

Theorem 3. *The disease-free equilibrium point D_0 of model (8) is locally asymptotically stable (LAS) if $R_0 < 1$ and unstable if $R_0 > 1$.*

Proof. The Jacobian matrix of the system (8) at the DFE is used to investigate the local stability of model (8) [33,36].

$$J(D_0) = \sigma^{1-\alpha(t)} \begin{pmatrix} X & 0 & 0 & 0 & A_1 & A_2 & A_3 & 0 & 0 \\ \nu_1 & -\mu & 0 & 0 & B_1 & B_2 & B_3 & 0 & 0 \\ \nu_2 & 0 & -\mu & 0 & F_1 & F_2 & F_3 & 0 & 0 \\ \nu_3 & 0 & 0 & -\mu & G_1 & G_2 & G_3 & 0 & 0 \\ 0 & 0 & 0 & 0 & M_1 & M_2 & M_3 & 0 & 0 \\ 0 & 0 & 0 & 0 & N_1 & N_2 & N_3 & 0 & 0 \\ 0 & 0 & 0 & 0 & Z_1 & Z_2 & Z_3 & 0 & 0 \\ 0 & 0 & 0 & 0 & 0 & \alpha_1 & (1-\phi)\alpha_1 & -E_4 & 0 \\ 0 & 0 & 0 & 0 & \gamma_A & \gamma_{IU} & \gamma_{IV} & \gamma_{IS} & -\mu \end{pmatrix},$$

where $X = -(\nu_1 + \nu_2 + \nu_3 + \mu)$, $A_1 = -\frac{\beta\theta\tilde{S}_H}{\tilde{N}_H}$, $A_2 = -\frac{\beta\tilde{S}_H}{\tilde{N}_H}$, $A_3 = -\frac{\beta\eta_V\tilde{S}_H}{\tilde{N}_H}$,
$B_1 = -(1-\xi_1)\frac{\beta\theta\tilde{V}_1}{\tilde{N}_H}$, $B_2 = -(1-\xi_1)\frac{\beta\tilde{V}_1}{\tilde{N}_H}$, $B_3 = -(1-\xi_1)\frac{\beta\eta_V\tilde{V}_1}{\tilde{N}_H}$,
$F_1 = -(1-\xi_2)\frac{\beta\theta\tilde{V}_2}{\tilde{N}_H}$, $F_2 = -(1-\xi_2)\frac{\beta\tilde{V}_2}{\tilde{N}_H}$, $F_3 = -(1-\xi_2)\frac{\beta\eta_V\tilde{V}_2}{\tilde{N}_H}$,
$G_1 = -(1-\xi_3)\frac{\beta\theta\tilde{V}_3}{\tilde{N}_H}$, $G_2 = -(1-\xi_3)\frac{\beta\tilde{V}_3}{\tilde{N}_H}$, $G_3 = -(1-\xi_3)\frac{\beta\eta_V\tilde{V}_3}{\tilde{N}_H}$,
$M_1 = \frac{\beta\theta\tilde{Q}}{\tilde{N}_H} - E_1$, $M_2 = \frac{\beta\tilde{Q}}{\tilde{N}_H}$, $M_3 = \frac{\beta\eta_V\tilde{Q}}{\tilde{N}_H}$,
$N_1 = \frac{\beta\theta(1-p)\tilde{S}_H}{\tilde{N}_H}$, $N_2 = \frac{\beta(1-p)\tilde{S}_H}{\tilde{N}_H} - E_2$, $N_3 = \frac{\beta(1-p)\eta_V\tilde{S}_H}{\tilde{N}_H}$,
$Z_1 = \frac{\beta\theta\tilde{v}}{\tilde{N}_H}$, $Z_2 = \frac{\beta\tilde{v}}{\tilde{N}_H}$, $Z_3 = \frac{\beta\eta_V\tilde{v}}{\tilde{N}_H} - E_3$,
$E_4 = (\gamma_{IS} + d_{IS} + \mu)$.
The characteristic equation:
$(\nu_3 + \nu_1 + \nu_2 + \mu + \lambda)(\lambda^3 + (E_1 + E_2 + E_3 - \frac{(1-p)\tilde{S} + \eta_V\tilde{v} + \eta_V\theta\tilde{Q}}{\tilde{N}_H}\beta)\lambda^2$
$+ (E_1E_2 + E_1E_3 + E_2E_3 - \beta[(E_1+E_3)(1-p)\tilde{S} + (E_1+E_2)\eta_V\tilde{v}$
$+ (E_2+E_3)\eta_A\theta\tilde{Q}])\lambda + E_1E_2E_3(1-R_0))(\mu+\lambda)^4(\lambda+E_4) = 0.$
Then, we have
$(\lambda+\mu) = 0, (\lambda+E_4) = 0, (\lambda + \nu_1 + \nu_2 + \nu_3 + \mu) = 0;$
the arguments are $arg(\lambda_k) > \frac{\pi}{a} > k\frac{2\pi}{a} > \frac{\pi}{M} > \frac{\pi}{2M}$, where $k = 0, 1, 2, 3, \ldots, a-1$.

$$(\lambda^3 + (E_1 + E_2 + E_3 - \beta\frac{(1-p)\tilde{S} + \eta_V\tilde{v} + \eta_V\theta\tilde{Q}}{\tilde{N}})\lambda^2 + (E_1E_2 + E_1E_3 + E_2E_3 - \beta[(E_1+E_3)(1-p)\tilde{S} + (E_1+E_2)\eta_V\tilde{v} + (E_2+E_3)\eta_A\theta\tilde{Q}])\lambda + E_1E_2E_3(1-R_0)) = 0.$$

We can rewrite the above equation as:

$$\lambda^3 + a\lambda^2 + b\lambda + c = 0, \tag{20}$$

where

$$a = (E_1 + E_2 + E_3 - \beta\frac{(1-p)\tilde{S} + \eta_V\tilde{v} + \eta_V\theta\tilde{Q}}{\tilde{N}}),$$
$$b = (E_1E_2 + E_1E_3 + E_2E_3 - \beta[(E_1+E_3)(1-p)\tilde{S} + (E_1+E_2)\eta_V\tilde{v} + (E_2+E_3)\eta_A\theta\tilde{Q}]),$$
$$c = E_1E_2E_3(1-R_0).$$

$$\lambda^3 + a\lambda^2 + b\lambda + c = 0, \tag{21}$$

We obtain

$$\lambda^3 + a\lambda^2 + b\lambda + c = (\lambda - \zeta_{11})(\lambda^2 - \tau\lambda + \zeta_{11}), \tag{22}$$

$$\tau = -(a + \zeta_{11}), \tag{23}$$

$$\zeta_{11} = b + \zeta_{11}(a + \zeta_{11}), \tag{24}$$

$$c = -\zeta_{11}\delta_{11}, \tag{25}$$

Hence, the other two roots are given by

$$\zeta_{11,2,3} = \frac{1}{2}(\tau \pm \sqrt{\triangle}), \tag{26}$$

$$\triangle = \tau^2 - 4\delta_{11} = a^2 - 2a\zeta_{11} - (3\zeta_{11}^2 + 4b). \tag{27}$$

These two roots are complex conjugate when $\triangle < 0$, real and distinct when $\triangle > 0$, and real and conincident when $\triangle = 0$.
Considering that $\triangle = 0$ occurs $a = \zeta_{11} \pm 2\sqrt{\zeta_{11}^2 + b}$, we have that if $\zeta_{11}^2 + b < 0$, then $\triangle > 0$ and two distinct real roots given by

$$\zeta_{11,2,3} = \frac{1}{2}(\tau \pm \sqrt{\triangle}).$$

If $\zeta_{11}^2 + b = 0$ then $\triangle = (a - \zeta_{11})^2$ and two distinct real roots exist given by

$$\zeta_{11,2,3} = \frac{1}{2}(\tau \pm |a - \zeta_{11}|).$$

So that $\zeta_{11,2} = -\zeta_{11,1}$ and $\zeta_{11,3} = -a$, if $\zeta_{11,1}^2 + b > 0$ and $(\zeta_{11} - 2\sqrt{\zeta_{11}^2 + b}) < a < (\zeta_{11} + 2\sqrt{\zeta_{11}^2 + b})$, then $\triangle < 0$ and two complex conjugate roots exist, given by $\zeta_{11,2,3} = \alpha_{11} \pm iB_{11}$ where $\alpha_{22} = \frac{\tau}{2}$, $B_{11} = \frac{\sqrt{4\delta_{11} - \tau_2}}{2} = \sqrt{\delta_{11} - \alpha_{11}^2}$. $a = (\zeta_{11} - 2\sqrt{(\zeta_{11}^2) + b})$ or $a = (\zeta_{11} - 2\sqrt{(\zeta_{11}^2) + a_2})$, then $(\triangle = 0)$ and two concident real roots exist given by $\zeta_{11,2} = \zeta_{11,3} = \frac{\tau}{2} = \frac{a + \zeta_{11}}{2}$ $a < (\zeta_{11} - 2\sqrt{(\zeta_{11}^2) + a_2})$ or $a_1 > (\zeta_{11} - 2\sqrt{(\zeta_{11}^2) + b})$. Then, $\triangle = 0$ and two distinct real roots exists given by

$$\zeta_{11,2,3} = \frac{1}{2}(\tau \pm \sqrt{\triangle}).$$

Applying the Routh–Hurwitz criterion [37], Equation (27) has roots with negative real parts if and only if $R_0 < 1$. Thus, the DFE is locally asymptotically stable. □

4. Numerical Methods for Solving the Proposed Model

4.1. GRK4M

Consider the fractional derivatives with variable order given by the following equation:

$$ {}_0^C D_t^{\alpha(t)} \varepsilon(t) = f(t, \varepsilon(t)), \;\; T_f \geq t > 0, \;\; 1 \geq \alpha(t) > 0, \tag{28}$$

$$\varepsilon(0) = \varepsilon_o.$$

Using GRK4M [23], the approximate solution of (28) is:

$$\varepsilon_{n+1} = \varepsilon_n + \frac{1}{6}(K_1 + 2K_2 + 2K_3 + K_4), \tag{29}$$

$$K_1 = Yf(t_n, \varepsilon_n),$$

$$K_2 = Yf(t_n + \frac{1}{2}Y, \varepsilon_n + \frac{1}{2}K_1),$$

$$K_3 = Yf(t_n + \frac{1}{2}Y, \varepsilon_n + \frac{1}{2}K_2),$$

$$K_4 = Yf(t_n + Y, \varepsilon_n + K_3),$$

where $Y = \dfrac{\tau^{\alpha(t_n)}}{\Gamma(\alpha(t_n) + 1)}$.

4.2. Stability of GRK4M

To investigate the stability of GRK4M, we shall utilize the following test problem of variable-order linear differential equation for simplicity:

$$ {}_0^C D_t^{\alpha(t)} \varepsilon(t) = \varepsilon(t) v, \;\; T_f \geq t > 0, \;\; v < 0, \;\; 1 \geq \alpha(t) > 0, \tag{30}$$

$$\varepsilon(0) = \varepsilon_o.$$

As in [23], Equation (30) is written as follows:

$$\varepsilon(t_{i+1}) = \varepsilon(t_i) + \frac{1}{6}\frac{\upsilon\tau^{\alpha(t_i)}}{\Gamma(1+\alpha(t_i))}\varepsilon(t_i), \; i = 0,1,\dots,n-1. \tag{31}$$

Then, we have the following equation [38]:

$$\varepsilon(t_{i+1}) = (1 + \frac{1}{6}\frac{\tau^{\alpha(t_i)}\upsilon}{\Gamma(1+\alpha(t_i))})^i\varepsilon_0. \tag{32}$$

The condition of stability [38]:

$$-1 < (\frac{1}{6}\frac{\tau^{\alpha(t_i)}\upsilon}{\Gamma(1+\alpha(t_i))} + 1) < 1.$$

4.3. *CPC-ΘFDM*

Consider:

$${}_0^{CPC}D_t^{\alpha(t)}\varepsilon(t) = \xi(t,\varepsilon(t)), \;\; \varepsilon(0) = \varepsilon_0, \;\; 1 \geq \alpha(t) > 0. \tag{33}$$

Relationship (6) can be expressed as follows:

$$\begin{aligned}{}_0^{CPC}D_t^{\alpha(t)}\varepsilon(t) &= \frac{1}{\Gamma(1-\alpha(t))}\int_0^t (t-s)^{-\alpha(t)}(K_1(\alpha(t))\varepsilon(s) + K_0(\alpha(t))\varepsilon'(s))ds, \\ &= K_1(\alpha){}_0^{RL}I_t^{1-\alpha(t)}\varepsilon(t) + K_0(\alpha(t)){}_0^{C}D_t^{\alpha(t)}\varepsilon(t), \\ &= K_1(\alpha){}_0^{RL}D_t^{\alpha(t)-1}\varepsilon(t) + K_0(\alpha(t)){}_0^{C}D_t^{\alpha(t)}\varepsilon(t),\end{aligned} \tag{34}$$

Using ΘFDM and GL-approximation, we can discretize (34) as shown below:

$$\begin{aligned}{}_0^{CPC}D_t^{\alpha(t)}\varepsilon(t)|_{t=t^n} = {}& \frac{K_1(\alpha(t_n))}{\tau^{\alpha(t_n)-1}}\left(\varepsilon_{n+1} + \sum_{i=1}^{n+1}\omega_i\varepsilon_{n+1-i}\right) \\ &+ \frac{K_0(\alpha(t_n))}{\tau^{\alpha_n}}\left(\varepsilon_{n+1} - \sum_{i=1}^{n+1}\varrho_i\varepsilon_{n+1-i} - \varsigma_{n+1}\varepsilon_0\right),\end{aligned} \tag{35}$$

$$\begin{aligned}&\frac{K_1(\alpha(t_n))}{\tau^{\alpha(t_n)-1}}\left(\varepsilon_{n+1} + \sum_{i=1}^{n+1}\omega_i\varepsilon_{n+1-i}\right) + \frac{K_0(\alpha(t_n))}{\tau^{\alpha(t_n)}}\left(\varepsilon_{n+1} - \sum_{i=1}^{n+1}\varrho_i\varepsilon_{n+1-i} - \varsigma_{n+1}\varepsilon_0\right) \\ &= (\Theta)\xi(\varepsilon(t_n),t_n) + (1-\Theta)\xi(\varepsilon(t_{n+1}),t_{n+1}),\end{aligned} \tag{36}$$

where, $\omega_0 = 1$, $\omega_i = (1 - \frac{\alpha(t_n)}{i})\omega_{i-1}$, $t^n = n\tau, \tau = \frac{T_f}{N}$, N is a natural number, $\varrho_i = (-1)^{i-1}\binom{\alpha(t_n)}{i}$, $\varrho_1 = \alpha(t_n)$, $\varsigma_i = \frac{i^{\alpha(t_n)}}{\Gamma(1-\alpha(t_n))}$. Moreover, consider that [39]:

$$0 < \varrho_{i+1} < \varrho_i < \dots < \varrho_1 = \alpha(t_n) < 1,$$

$$0 < \varsigma_{i+1} < \varsigma_i < \dots < \varsigma_1 = \frac{1}{\Gamma(-\alpha(t_n)+1)}, \qquad i = 1,2,\dots,n+1.$$

Remark 1. *If* $K_1(\alpha(t)) = 0$ *and* $K_0(\alpha(t)) = 1$ *in (36), we can obtain the discretization of Caputo operator with theta finite difference technique (C-Θ FDM).*

4.4. *CPC-ΘFDM Stability Analysis*

The stability of method (36) will be considered here. We shall utilize the test problem of variable-order linear differential equation, for simplicity:

$$({}_0^{CPC}D_t^{\alpha(t)})\varepsilon(t) = A\varepsilon(t), \;\; t > 0, \;\; A < 0, \;\; 0 < \alpha(t) \leq 1. \tag{37}$$

By (34) and GL-approximation, we can discretize (37) as shown below:

$$\frac{K_1(\alpha(t_n))}{\tau^{\alpha(t_n)-1}}\left(\varepsilon_{n+1}+\sum_{i=1}^{n+1}\omega_i\varepsilon_{n+1-i}\right)+\frac{K_0(\alpha(t_n))}{\tau^{\alpha(t_n)}}\left(\varepsilon_{n+1}-\sum_{i=1}^{n+1}\varrho_i\varepsilon_{n+1-i}-\varsigma_{n+1}\varepsilon_0\right)$$
$$=\Theta A\varepsilon_n+(1-\Theta)A\varepsilon_{n+1}; \tag{38}$$

put $C=\frac{K_1(\alpha(t_n))}{\tau^{\alpha(t_n)-1}}$, $B=\frac{K_0(\alpha(t_n))}{\tau^{\alpha(t_n)}}$. Then, from boundness theorem [40], we have:

$$\varepsilon_{n+1}=\frac{1}{C+B}\left(A\varepsilon_n-C\sum_{i=1}^{n+1}\omega_i\varepsilon_{n+1-i}+B\left(\sum_{i=1}^{n+1}\varrho_i\varepsilon_{n+1-i}+\varsigma_{n+1}\varepsilon_0\right)\right)\le\varepsilon_n, \tag{39}$$

This means $\varepsilon_0\ge\varepsilon_1\ge\ldots\ge\varepsilon_{n-1}\ge\varepsilon_n\ge\varepsilon_{n+1}$. Then, method (36) is stable.

4.5. Convergence of the Method

Equation (34) can be discretized as shown below:

$$ {}^{CPC}_{\ 0}D_t^{\alpha(t)}\varepsilon(t)|_{t=t^n}=\frac{K_1(\alpha(t_n))}{\tau^{\alpha(t_n)-1}}\left(\varepsilon_{n+1}+\sum_{i=1}^{n+1}\omega_i\varepsilon_{n+1-i}\right)$$
$$+\frac{K_0(\alpha(t_n))}{\tau^{\alpha_n}}\left(\varepsilon_{n+1}-\sum_{i=1}^{n+1}\varrho_i\varepsilon_{n+1-i}-\varsigma_{n+1}\varepsilon_0\right), \tag{40}$$

$$\frac{K_1(\alpha(t_n))}{\tau^{\alpha(t_n)-1}}\left(\varepsilon_{n+1}+\sum_{i=1}^{n+1}\omega_i\varepsilon_{n+1-i}\right)+\frac{K_0(\alpha(t_n))}{\tau^{\alpha(t_n)}}\left(\varepsilon_{n+1}-\sum_{i=1}^{n+1}\varrho_i\varepsilon_{n+1-i}-\varsigma_{n+1}\varepsilon_0\right)$$
$$-\Theta\xi(\varepsilon(t_n),t_n)-(1-\Theta)\xi(\varepsilon(t_{n+1}),t_{n+1})=T_{Rn}, \tag{41}$$

where

$$\|T_{Rn}\|_\infty<W,\quad W=C\max_{0\le i\le n+1}|\varepsilon_{i+1}|,$$

$$C=(\tau^{\alpha(t_i)-1}+\tau^{\alpha(t_i)}).$$

The proposed method is convergent because it is stable and consistent [41], then (41) is convergent.

5. Numerical Results

In the following, we solved (8) numerically using GRK4M (29) and CPC-ΘFDM (36). Using CPC-Θ FDM for solving (8), we obtained $(9N+9)$ of the nonlinear algebraic system with $(9N+9)$ unknown.

$$\left(S,V_1,V_2,V_3,A,I_U,I_V,I_S,R\right)$$

can be solved using an appropriate iterative method based on the assumed beginning conditions. For the real data, we use [24]; the authors in this reference used the literature to obtain some parameter values and the remaining values were fitted to the data for the state of Texas, USA. They fitted the data of (8) solutions with the data for the state of Texas from 13 March to 29 June 2021 [29,42]. The model was fitted with three datasets, Moderna, Janssen, and Pfizer, with immunization data for Texas state. The three vaccination rates v_1, v_2 and v_3 corresponding to each vaccine as well as the effective contact rate for COVID-19 transmission, β, are estimated. According to publicly available data, the total population of the state of Texas, USA, for the year 2021 was 29,200,000 [1]. Let $R(0)=5000$, $V_2(0)=4{,}016{,}005$, $A(0)=50{,}000$, $V_3(0)=129{,}859$, $S(0)=24{,}000{,}000$, $I_U(0)=17{,}000$, $I_V(0)=15{,}000$, $V_1(0)=4{,}115{,}127$ and $I_S(0)=10{,}000$. The parameter values are given in Table 2. To show that the proposed scheme is efficient, we compare the results that we obtained in this paper with the results that were found in reference [24], which are given in Figure 2 in constant fractional order. Figure 3 shows the behavior of the approximate solution of (8) (using the method in [24]) with different values of $\alpha(t)$. As can be seen from this figure, when the value of the fractional derivative changes over time, the results are different and this can dramatically affect the behavior of the model. This confirms the generality of the variable-order derivatives. Unfortunately, this method gives us unstable solutions, as in Figure 4a, when the value of the step size equals one. Moreover, we obtained the stable solutions using the proposed method CPC-ΘFDM and $\Theta = 0$, in the fully implicit case given in this paper. This confirms that the method in [24] is stable only when the step size is very small, while our used method is stable regardless of the value of the step size. Figure 5 shows the behavior of the approximate solution of (8) (using CPC-ΘFDM and $\Theta = 0.5, Q = 0.00025$) with different values of $\alpha(t)$. The approximate solution behavior of (8) is shown in Figure 6 ($\Theta = 1$ and using CPC-ΘFDM) with different values of $\alpha(t)$, $Q = 0.00025$. The approximate solution behavior of (8) (using GRK4M with different values of $\alpha(t)$) is shown in Figure 7. Figure 8 shows the behavior of the approximate solution of (8) (using CPC-Θ FDM when $K_0(\alpha(t)) = 1$, $K_1(\alpha(t)) = 0$ and $\Theta = 0$) with different values of $\alpha(t)$. We noted that by comparing our results with different variable orders and constant orders as given in [24] and Figure 5, the result in the case of constant order is agreement. Moreover, by compering the results given in Figures 7 and 8, the result given using CPC-ΘFDM (fully implicit case) is convergent, better than the results given using GRK4 when we use nonlinear $\alpha(t)$. Figure 9 shows the relation between R and I_v, I_u, I_s, A using CPC-ΘFDM (fully implicit case) and nonlinear $\alpha(t)$. Furthermore, we found that the variable-order derivative order model is a more general model than the fractional order model given in [24] and integer order; a new behavior of the solution appears by using different values of $\alpha(t)$. Moreover, we can obtain the fractional Caputo operator as a special case from the CPC operator when $K_0(\alpha(t)) = 1$, $K_1(\alpha(t)) = 0$. Moreover, we can obtain the fractional Caputo operator as a special case from the CPC operator if $K_0(\alpha(t)) = 1$, $K_1(\alpha(t)) = 0$. The solutions obtained using the new method CPC-ΘFDM can be explicit ($\Theta = 1$) or implicit ($0 \leq \Theta \leq 1$,) and fully implicit with accurate solution when ($\Theta = 0$).

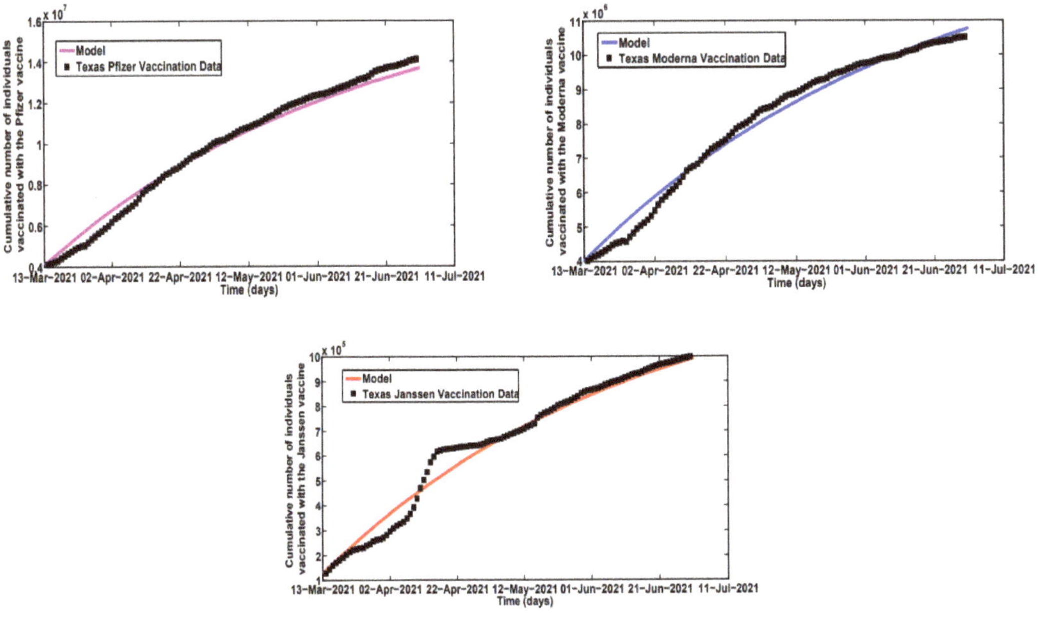

Figure 2. Real data [24] versus fitting model (8).

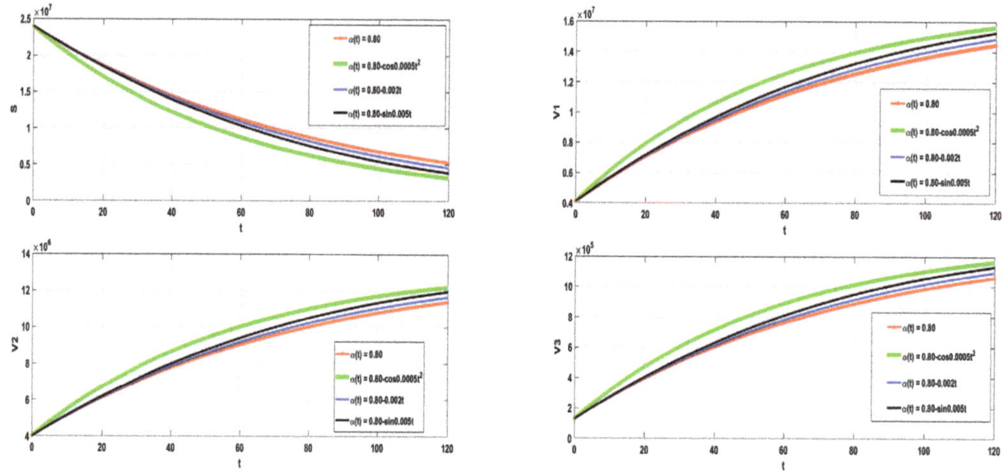

Figure 3. The solution behavior using the method in [24] with different values of $\alpha(t)$.

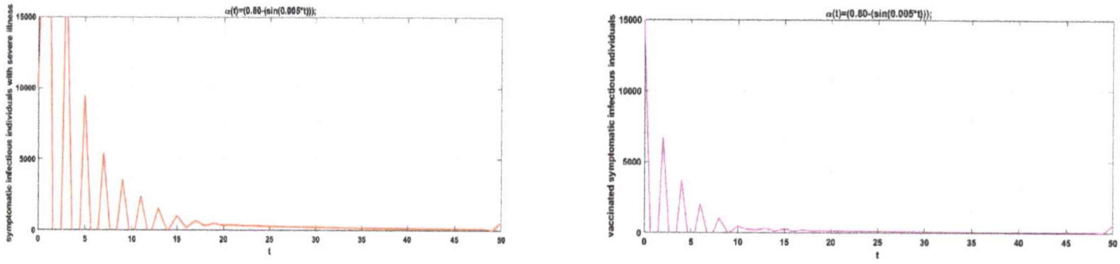

(a)

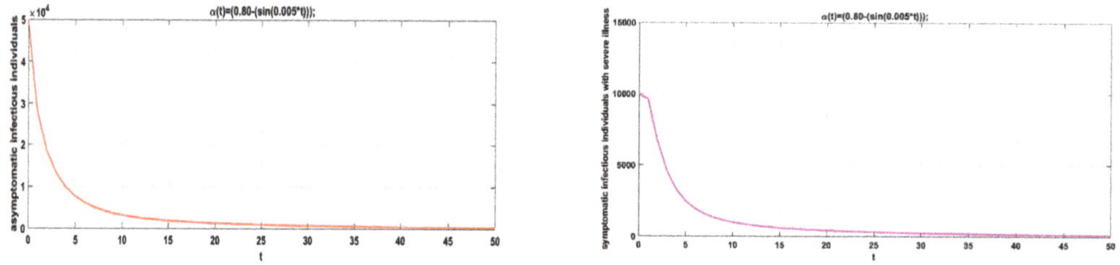

(b)

Figure 4. The solution behavior using the method [24] in (a) and using CPC-ΘFDM and $\Theta = 0$, in (b).

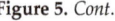

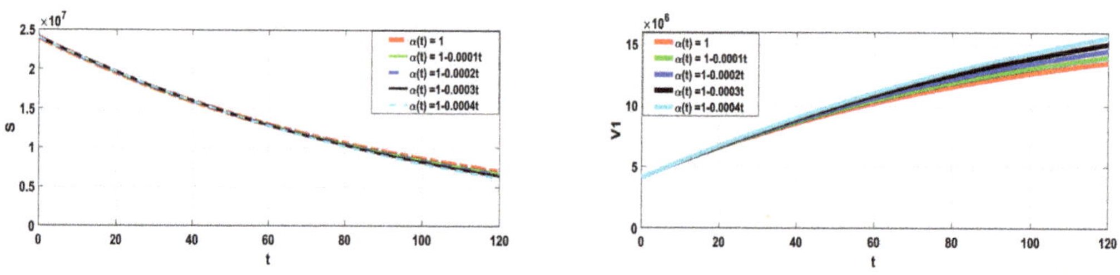

Figure 5. Cont.

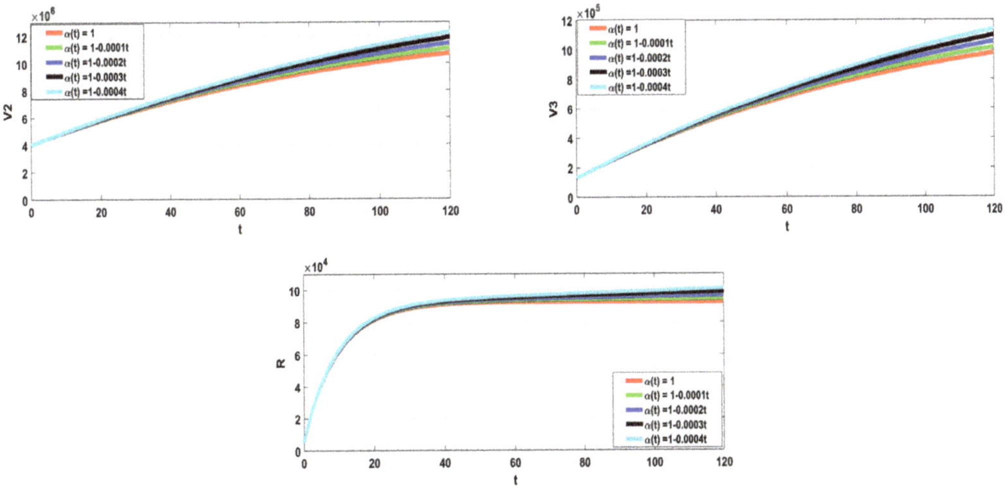

Figure 5. The solution behavior acquired via CPC-ΘFDM and $\Theta = 0.5$, of (8).

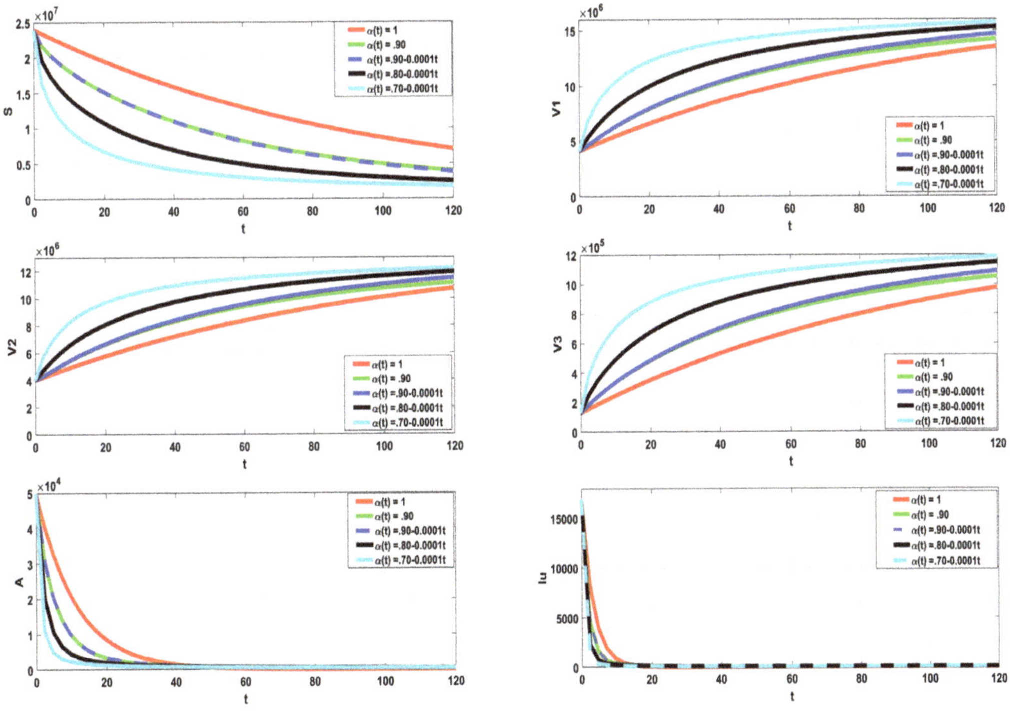

Figure 6. *Cont.*

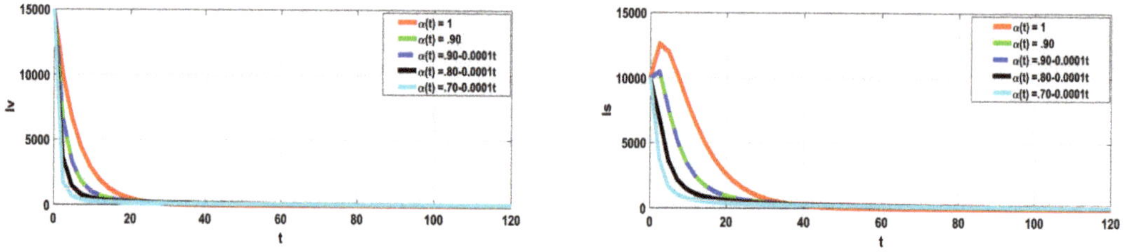

Figure 6. The solution behavior acquired via CPC-ΘFDM and $\Theta = 1$ of (8).

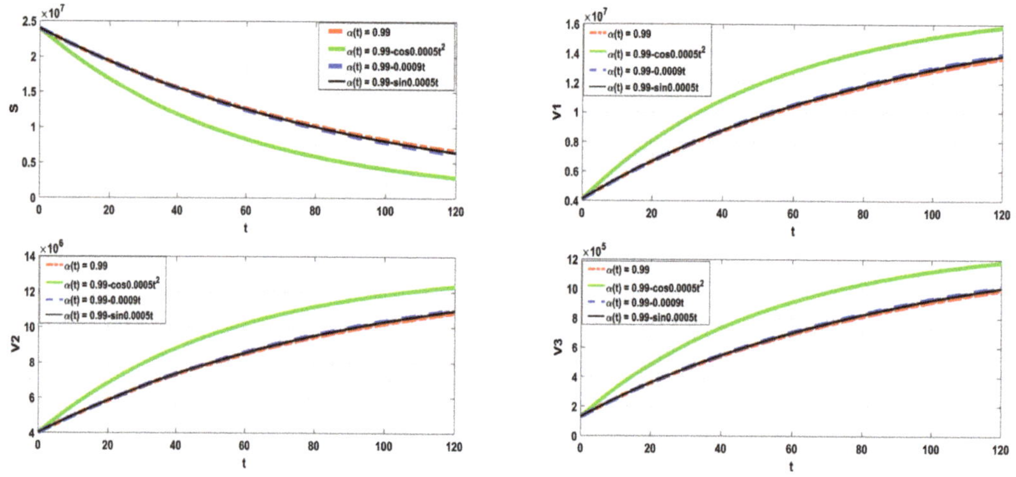

Figure 7. The solution behavior acquired via GRK4M of (8).

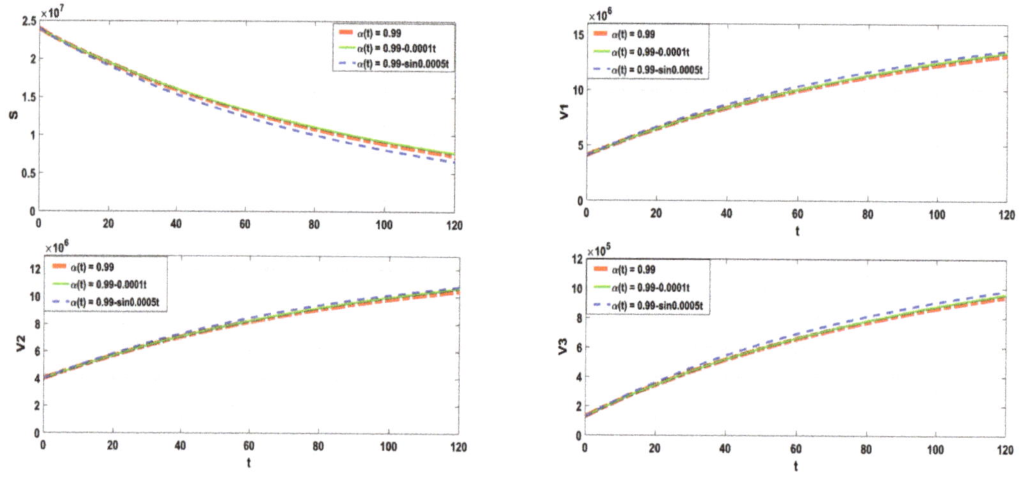

Figure 8. The solution behavior acquired via CPC-ΘFDM and $\Theta = 0$ of (8).

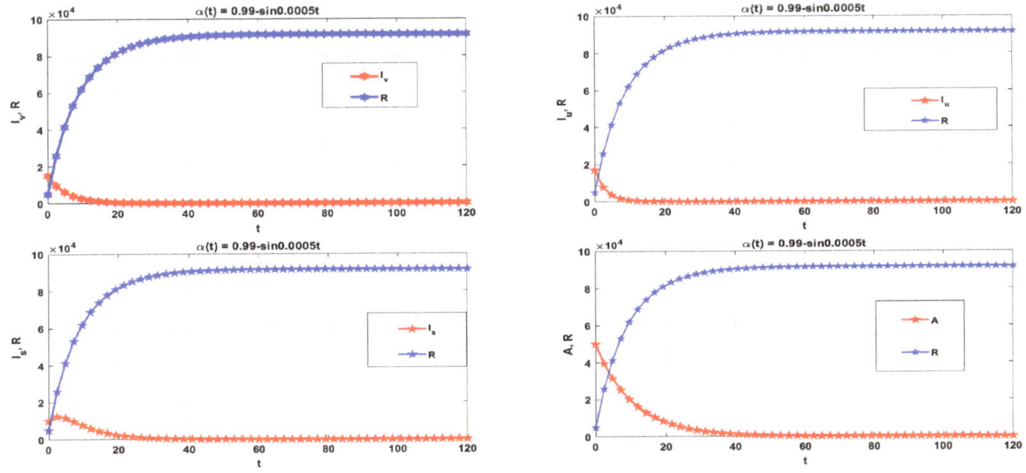

Figure 9. The relation between the variables concerning nonlinear $\alpha(t)$ using CPC-ΘFDM and $\Theta = 0$.

6. Conclusions

A novel hybrid variable-order fractional multi-vaccination model for COVID-19 is presented in this paper in order to further explore the spread of COVID-19. The main advantage of the hybrid variable-order fractional operator is that it can be defined as a linear combination of the variable-order integral of Riemann–Liouville and the variable-order Caputo derivative; it is one of the most effective and reliable operators and it is more general than the Caputo fractional operator. The proposed model's dynamics are improved and its complexity is increased by employing variable-order fractional derivatives. Furthermore, the variable-order fractional Caputo operator can be derived as a special case from the CPC operator. Existence, boundedness, uniqueness, positivity and stability of the proposed model are established for the model. To be compatible with the physical model, a new parameter σ is added. The proposed model is numerically studied using CPC-ΘFDM and GRK4M. CPC-ΘFDM depends on the values of the factor Θ. It can be explicit ($\Theta = 1$) or fully implicit ($\Theta = 0$) with a large stability region. We compared our results with the real data from the state of Texas in the United States. Moreover, the results obtained from the CPC-ΘFDM are more stable than the results obtained from the proposed method in [24]. As a result, some graphs are provided for various linear and non-linear variable-order derivatives. In the future, the presented study can be extended to optimal control and to examine the impact of multiple vaccination strategies on the dynamics of COVID-19 in a population.

Author Contributions: Conceptualization, T.A.A.; Methodology, N.S. and S.M.A.-M.; Software, S.M.A.-M. and R.G.S.; Formal analysis, N.S.; Investigation, R.G.S. and T.A.A.; Resources, S.M.A.-M.; Writing—original draft, T.A.A.; Writing —review and editing, N.S. and R.G.S. All authors have read and agreed to the published version of the manuscript.

Funding: This research received no external funding.

Data Availability Statement: Not applicable.

Conflicts of Interest: The authors declare no conflict of interest.

References

1. United States Food and Drug Administration. FDA Takes Key Action in Fight Against COVID-19 By Issuing Emergency Use Authorization for First COVID-19 Vaccine. 2020. Available online: https://www.fda.gov/news-events/press-announcements/fda-takes-key-action-fight-against-COVID-19-issuing-emergency-use-authorization-first-COVID-19 (accessed on 17 June 2021).
2. Interim Clinical Considerations for Use of COVID-19 Vaccines Currently Authorized in the United States. Available online: https://www.cdc.gov/vaccines/COVID-19/clinical-considerations/COVID-19-vaccines-us.html (accessed on 14 July 2021).
3. Machado, J.A.T.; Lope, A.M. Rare and extreme events: The case of COVID-19 pandemic. *Nonlinear Dyn.* **2020**, *100*, 2953–2972. [CrossRef] [PubMed]
4. Bonyah, E.; Sagoe, A.K.; Kumar, D.; Deniz, S. Fractional optimal control dynamics of Coronavirus model with Mittag-Leffler law. *Ecol. Complex.* **2020**, *45*, 100880. [CrossRef]
5. Ali, A.; Alshammari, F.S.; Islam, S.; Khan, M.A.; Ullah, S. Modeling and analysis of the dynamics of novel coronavirus (COVID-19) with Caputo fractional derivative. *Results Phys.* **2020**, *20*, 103669. [CrossRef] [PubMed]
6. Danane, J.; Hammouch, Z.; Allali, K.; Rashid, S.; Singh, J. A fractional-order model ofcoronavirus disease 2019 (COVID-19) with governmental action and individual reaction. *Math. Meth. Appl. Sci.* **2021**, 1–14. [CrossRef]
7. Yadav, S.; Kumar, D.; Singh, J.; Baleanu, D. Analysis and dynamics of fractional order COVID-19 model with memory effect. *Results Phys.* **2021**, *24*, 104017. [CrossRef]
8. Sinan, M.; Ali, A.; Shah, K.; Assiri, T.; Nofal, T.A. Stability analysis and optimal control of COVID-19 pandemic SEIQR fractional mathematical model with harmonic mean type incidence rate and treatment. *Results Phys.* **2021**, *22*, 103873. [CrossRef]
9. Conejero, J.A.; Franceschi, J.; Picó-Marco, E. Fractional vs. Ordinary Control Systems: What Does the Fractional Derivative Provide? *Mathematics* **2022**, *10*, 2719. [CrossRef]
10. Samko, S.G.; Ross, B. Integration and differentiation to a variable fractional order. *Integral Transform. Spec. Funct.* **1993**, *1*, 277–300. [CrossRef]
11. Solís-Pérez, J.E.; Gómez-Aguilar, J.F. Novel numerical method for solving variable-order fractional differential equations with power, exponential and Mittag-Leffler laws. *Chaos Solitons Fractals* **2018**, *14*, 175–185. [CrossRef]
12. Sun, H.; Chang, A.; Zhang, Y.; Chen, W. A review on variable-order fractional differential equations: Mathematical foundations, physical models and its applications. *Fract. Calc. Appl. Anal.* **2019**, *22*, 27–59. [CrossRef]
13. Sweilam, N.H.; Al-Mekhlafi, S.M. Numerical study for multi-strain tuberculosis(TB) model of variable-order fractional derivatives. *J. Adv. Res.* **2016**, *7*, 271–283. [CrossRef]
14. Sweilam, N.H.; L-Mekhlafi, S.M.A.; Shatta, S.A.; Baleanu, D. Numerical study for two types variable-order Burgers' equations with proportional delay. *Appl. Numer. Math.* **2020**, *156*, 364–376. [CrossRef]
15. Sweilam, N.H.; Assiri, T.; Hasan, M.A. Numerical solutions of nonlinear fractional Schrödinger equations using nonstandard discretizations, Numerical Solutions of Nonlinear Fractional Schrödinger Equations. *Numer. Methods Partial. Differ. Equ.* **2017**, *33*, 1399–1419. [CrossRef]
16. Sweilam, N.H.; Assiri, T. Numerical scheme for solving the space-time variable order nonlinear fractional wave equation. *Prog. Fract. Differ. Appl.* **2015**, *1*, 269–280. [CrossRef]
17. Bha, I.A.; Mishra, L.N. Numerical solutions of Volterra integral equations of third kind and its convergence analysis. *Symmetry* **2022**, *14*, 2600.
18. Pathak, V.K.; Mishra, L.N. Application of Fixed point theorem to solvability for non-linear fractional Hadamard functional integral e quations. *Mathematics* **2022**, *10*, 2400. [CrossRef]
19. Smith, G.D. Numerical solution of partial differential equations: Finite difference methods. In *Oxford Applied Mathematics and Computing Science Series*; Oxford University Press: Oxford, UK, 1985.
20. Yuste, S.B. Weighted average finite difference methods for fractional diffusion equations. *J. Comput. Phys.* **2006**, *216*, 264–274. [CrossRef]
21. Sweilam, N.H.; Hasan, M.M.A. An Improved method for nonlinear variable-order Lévy-Feller advection-dispersion equation. *Bull. Malays. Math. Sci. Soc.* **2019**, *42*, 3021–3046. [CrossRef]
22. Sweilam, N.H.; Hasan, M.M.A.; Al-Mekhlafi, S.M.; Al khatib, S. Time fractional of nonlinear heat-wave propagation in a rigid thermal conductor: Numerical treatment. *AEJ—Alex. Eng. J.* **2022**, *61*, 10153–10159. [CrossRef]
23. Milici, C.; Machado, J.T.; Draganescu, G. Application of the Euler and Runge–Kutta generalized methods for FDE and symbolic packages in the analysis of some fractional attractors. *Int. J. Nonlinear Sci. Numer. Simul.* **2020**, *21*, 159–170. [CrossRef]
24. Omame, A.; Okuonghae, D.; Nwajeri, U.K.; Onyenegecha, C.P. A fractional-order multi-vaccination model for COVID-19 with non-singular kernel. *Alex. Eng. J.* **2022**, *16*, 6089–6104. [CrossRef]
25. Sun, H.G.; Chen, W.; Wei, H.; Chen, Y.Q. A comparative study of constant-order and variable-order fractional models in characterizing memory property of systems. *Eur. Phys. J. Spec. Top.* **2011**, *193*, 185–192. [CrossRef]
26. Baleanu, D.; Fernandez, A.; Akgül, A. On a fractional operator combining proportional and classical differintegrals. *Mathematics* **2020**, *8*, 360. [CrossRef]
27. Ullah, M.Z.; Baleanu, D. A new fractional SICA model and numerical method for the transmission of HIV/AIDS. *Math. Meth. Appl. Sci.* **2021**, *44*, 4648–4659. [CrossRef]
28. Lin, W. Global existence theory and chaos control of fractional differential equations. *J. Math. Anal. Appl.* **2007**, *332*, 709–726. [CrossRef]

29. Texas Population, Census Reporter. Available online: https://censusreporter.org/profiles/04000US48-texas/ (accessed on 26 June 2021).
30. United States Food and Drug Administration. FDA Briefing Document Moderna COVID-19 Vaccine. 2020. Available online: https://www.fda.gov/media/144434/download (accessed on 17 June 2021).
31. United States Food and Drug Administration. FDA Briefing Document Pfizer-BioNTech COVID-19 Vaccine. 2020. Available online: https://www.fda.gov/media/144245/download (accessed on 17 June 2021).
32. Okuonghae, D.; Omame, A. Analysis of a mathematical model for COVID-19 population dynamics in Lagos, Nigeria. *Chaos Solitons Fractals* **2020**, *139*, 110032. [CrossRef]
33. Chu, Y.-M.; Yassen, M.F.; Ahmad, I.; Sunthrayuth, P.; Khan, M.A. A fractional SARS-COV-2 model with Atangana-Baleanu derivative: Application to fourth wave. *Fractals* **2022**, *30*, 2240210. . [CrossRef]
34. Driessche, P.; Watmough, P. Reproduction numbers and sub-threshold endemic equilibria for compartmental models of disease transmission. *Math. Biosci.* **2002**, *180*, 29–48. [CrossRef]
35. Fosu, G.O.; Akweittey, E. Albert-adu-sackey, Next-generation matrices and basic reproductive numbers for all phases of the Coronavirus disease. *Open J. Math. Sci.* **2020**, *4*, 261–272. [CrossRef]
36. Sekerci, Y. Climate change forces plankton species to move to get rid of extinction: Mathematical modeling approach. *Eur. Phys. J. Plus* **2020**, *135*, 794. [CrossRef]
37. Al-Mekhlafi, S.M.; Sweilam, N.H. *Numerical Studies for Some Tuberculosis Models*; LAP LAMBERT Academic Publishing: London, UK, 2016; 156p.
38. Sweilam, N.H.; L-Mekhlafi, S.M.A.; Alshomrani, A.S.; Baleanu, D. Comparative study for optimal control nonlinear variable-order fractional tumor model. *Chaos Solitons Fractals* **2020**, *136*, 109810. [CrossRef]
39. Scherer, R.; Kalla, S.; Tang, Y.; Huang, J. The Grünwald-Letnikov method for fractional differential equations. *Comput. Math. Appl.* **2011**, *62*, 902–917. [CrossRef]
40. Arenas, A.J.; Gonzàlez-Parra, G.; Chen-Charpentierc, B.M. Construction of nonstandard finite difference schemes for the SI and SIR epidemic models of fractional order. *Math. Comput. Simul.* **2016**, *121*, 48–63. [CrossRef]
41. Yuste, S.B.; Quintana-Murillo, J. A finite difference method with non-uniform time steps for fractional diffusion equations. *Comput. Phys. Commun.* **2012**, *183*, 2594–2600. [CrossRef]
42. COVID-19 Vaccinations in the US. Available online: https://data.cdc.gov/Vaccinations/COVID-19-Vaccinations-in-the-United-States-Jurisdi/uns (accessed on 13 March 2021).

symmetry

MDPI

Article

Studies on Special Polynomials Involving Degenerate Appell Polynomials and Fractional Derivative

Shahid Ahmad Wani [1,†], Kinda Abuasbeh [2,*,†], Georgia Irina Oros [3,*,†] and Salma Trabelsi [2,†]

1 Department of Applied Sciences, Symbiosis Institute of Technology, Symbiosis International (Deemed University) (SIU), Lavale, Pune 412115, Maharashtra, India; shahid.wani@sitpune.edu.in
2 Department of Mathematics and Statistics, College of Science, King Faisal University, Hofuf 31982, Al Ahsa, Saudi Arabia; satrabelsi@kfu.edu.sa
3 Department of Mathematics and Computer Science, Faculty of Informatics and Sciences, University of Oradea, 410087 Oradea, Romania
* Correspondence: kabuasbeh@kfu.edu.sa (K.A.); georgia_oros_ro@yahoo.co.uk (G.I.O.)
† These authors contributed equally to this work.

Abstract: The focus of the research presented in this paper is on a new generalized family of degenerate three-variable Hermite–Appell polynomials defined here using a fractional derivative. The research was motivated by the investigations on the degenerate three-variable Hermite-based Appell polynomials introduced by R. Alyosuf. We show in the paper that, for certain values, the well-known degenerate Hermite–Appell polynomials, three-variable Hermite–Appell polynomials and Appell polynomials are seen as particular cases for this new family. As new results of the investigation, the operational rule for this new generalized family is introduced and the explicit summation formula is established. Furthermore, using the determinant formulation of the Appell polynomials, the determinant form for the new generalized family is obtained and the recurrence relations are also determined considering the generating expression of the polynomials contained in the new generalized family. Certain applications of the generalized three-variable Hermite–Appell polynomials are also presented showing the connection with the equivalent results for the degenerate Hermite–Bernoulli and Hermite–Euler polynomials with three variables.

Keywords: Hermite polynomials; Appell polynomials; three-variable Hermite-based Appell polynomials; fractional derivative; integral transforms; operational rule

MSC: 26A33; 33B10; 33C45

Citation: Wani, S.A.; Abuasbeh, K.; Oros, G.I.; Trabelsi, S. Studies on Special Polynomials Involving Degenerate Appell Polynomials and Fractional Derivative. *Symmetry* **2023**, *15*, 840. https://doi.org/10.3390/sym15040840

Academic Editors: Francisco Martínez González and Mohammed K. A. Kaabar

Received: 15 March 2023
Revised: 29 March 2023
Accepted: 29 March 2023
Published: 31 March 2023

1. Introduction and Preliminaries

Fractional calculus, a branch of mathematical analysis, examines the possibility of using the differentiation operators of real or complex number powers. Theoretical studies successfully employ fractional calculus operators, which are also applicable in a variety of science and engineering domains. A comprehensive overview of the theory and applications of the fractional-calculus operators can be seen in recent review papers [1,2].

A powerful method for dealing with fractional derivatives is the combination of integral transforms and special polynomials; see, for instance, [3].

For $\min\{\mathrm{Re}(\nu), \mathrm{Re}(b)\} > 0$, the integral of the form [4] (p. 218),

$$\int_0^\infty e^{-bt} t^{\nu-1} dt = \Gamma(\nu)\, b^{-\nu}, \tag{1}$$

is called Euler's integral of the second kind. Consequently, the following consequences are obtained in [3]:

$$\Gamma(\nu)\left(\alpha-\frac{\partial}{\partial u}\right)^{-\nu}f(u)= \int_0^\infty e^{-\alpha t}t^{\nu-1}\,e^{t\frac{\partial}{\partial u}}f(u)dt=\int_0^\infty e^{-\alpha t}t^{\nu-1}\,f(u+t)dt, \tag{2}$$

and

$$\Gamma(\nu)\left(\alpha-\frac{\partial^2}{\partial u^2}\right)^{-\nu}f(u)=\int_0^\infty e^{-\alpha t}t^{\nu-1}\,e^{t\frac{\partial^2}{\partial u^2}}f(u)dt. \tag{3}$$

Particularly in recent years, a number of generalizations of special functions in mathematical physics have seen a significant evolution. Many mathematical physics issues can be solved analytically thanks to the recent developments in special functions theory, which have various wide-range applications. Multi-variable and multi-index special functions represent a substantial improvement in the theory of generalized special functions. Both in the realm of pure mathematics and in real-world applications, special functions have been recognized for their importance. To address the problems appearing in the theory of abstract algebra and partial differential equations, the necessity for multi-variable and multi-index special functions is acknowledged. In physics, the Hermite polynomials are used to produce the quantum harmonic oscillator's eigenstates and to solve the Schrodinger equation for the harmonic oscillator. They are also employed as Gaussian quadrature in numerical analysis and the notion of multiple-index, multiple-variate Hermite polynomials were given by Hermite in [5]. Degenerate q-Hermite polynomials are defined by means of generating function in [6], and significant properties have been determined.

Recently, additional extensions of special polynomials have been built on the foundation of Euler's integral. When establishing operational definitions and generating relations for the generalized and innovative forms of special polynomials in [3], Dattoli em et al. employed Euler's integral. Thus, using (1), a generalization of a number of special polynomials including hybrid special polynomials was introduced by several authors. Extended Laguerre–Appell polynomials are considered for research in [7]. A new class of q-Sheffer–Appell polynomials was introduced and studied in [8] and certain positive linear operators together with the Sheffer–Appell polynomial sequences were investigated in [9]. Fractional calculus aspects were connected to special polynomials involving Appell sequences in the study presented in [10]. Complex Appell polynomials and their degenerate-type polynomials were studied in [11] and it iwas shown that the results can be applied to complex Bernoulli polynomials and complex Euler polynomials. Further studies involve Gould–Hopper-based Frobenius–Genocchi polynomials, Lagrange–Hermite polynomials [12] and generalized Legendre–Laguerre–Appell polynomials are investigated through fractional calculus.

In a recent study, R. Alyosuf [13] introduced degenerate three-variable Hermite-based Appell polynomials (D-3VHAP) listed by the generating relation:

$$\sum_{m=0}^{\infty} {}_{\mathcal{J}}R_m(u,v,w;\chi)\frac{t^m}{m!}=\mathcal{Y}(t,u,v,w;\chi)=R(t)(1+\chi)^{\frac{ut}{\chi}}(1+\chi)^{\frac{vt^2}{\chi}}(1+\chi)^{\frac{wt^3}{\chi}}, \tag{4}$$

which possess the series definition:

$$\sum_{k=0}^{m}\binom{m}{k}\,\mathcal{J}_{m-k}(u,v,w;\chi)\,R_k={}_{\mathcal{J}}R_m(u,v,w;\chi), \tag{5}$$

and are represented by operational rule:

$$\exp\left(\frac{v\,\chi}{log(1+\chi)}\,D_u^2+w\Big(\frac{\chi}{log(1+\chi)}\Big)^2D_u^3\right)\{R_m(u)\}={}_{\mathcal{J}}R_m(u,v,w;\chi), \tag{6}$$

where, $R_m(u)$ are Appell polynomials [14]given by generating relation:

$$\sum_{k=0}^{\infty} R_k(u)\frac{t^k}{k!} = R(t)\exp(ut), \tag{7}$$

with $R(t)$ being the convergent power series given by:

$$\sum_{k=0}^{\infty} R_k\frac{t^k}{k!} = R(t), \qquad R_0 \neq 0. \tag{8}$$

Fractional operators offer a more accurate representation of complex systems that cannot be modeled using integer-order derivatives. Hence, they have significant applications in various fields, including numerous branches of mathematics, physics [15], engineering [16], and finance [17]. For example, the behavior of viscoelastic materials, biological systems and electrical networks can be described using fractional operators [18]. Additionally, fractional operators have applications in electromagnetics, where those operators are used to describe the behavior of electromagnetic waves in media with fractional-order dielectric and magnetic properties [19]. Other applications of fractional calculus can be seen in [20].

The work of Datolli and colleagues [3] and that of R. Alyusof [13] served as a source of inspiration and motivation for the investigation reported in this paper due to the tremendous relevance of fractional operators. The generalized form of a convoluted degenerate hybrid special polynomial family is constructed here by using the fractional operator called Eulers' integral given by (1). A generalized degenerate Hermite-based Appell polynomial family denoted by ${}_{\mathcal{J}}R_{m,\nu}(u,v,w;\chi,\beta)$ is introduced using the generating expression:

$$\frac{R(z)(1+\chi)^{\frac{uz}{\chi}}}{\left[\beta - \frac{(vz^2+wz^3)}{\chi}\, log(1+\chi)\right]^{\nu}} = \sum_{m=0}^{\infty} {}_{\mathcal{J}}R_{m,\nu}(u,v,w;\chi,\beta)\frac{z^m}{m!}. \tag{9}$$

These hybrid special polynomials could be useful in image processing and computer vision to enhance image quality and extract features. Further, they have applications in financial mathematics, where they model the behavior of stock prices, interest rates, and other financial variables.

The focus of the present article is to present the study on the features of the generalized forms of the hybrid degenerate special polynomials connected to the Hermite polynomials through the extensive use of integral transforms and operational principles. The main contributions of the paper are contained in Sections 2 and 3, after a comprehensive introduction where all the necessary previously known results are listed. The novelty starts in Section 2, where fractional derivatives are used to introduce a generalized version of degenerate three-variable Hermite–Appell polynomials. These polynomials are further investigated and for them, summation formula, determinant form and recurrence relations are also deduced. Section 3 includes several applications of the new results involving generalized degenerate three-variable Hermite-Appell polynomials as well as equivalent results for the degenerate Hermite–Bernoulli and Hermite–Euler polynomials with three variables.

2. Generalized Forms of Mixed Special Polynomials

We first establish the following result before introducing the generalized version of the degenerate three-variable Hermite–Appell polynomials:

Theorem 1. *For the generalized degenerate three-variable Hermite–Appell polynomials ${}_{\mathcal{J}}R_{m,\nu}(u,v,w;\chi,\beta)$, the following operational rule holds true:*

$$\left(\beta - \left(v\frac{\chi}{log(1+\chi)}\,D_u^2 + w\left(\frac{\chi}{log(1+\chi)}\right)^2 D_u^3\right)\right)^{-\nu} R_m(u) = {}_{\mathcal{J}}R_{m,\nu}(u,v,w;\chi,\beta). \tag{10}$$

Proof. Substituting b with $\alpha - \left(v\frac{\chi}{log(1+\chi)}\ D_u^2 + w\left(\frac{\chi}{log(1+\chi)}\right)^2\ D_u^3\right)$ in integral (1) and the resulting equation on $R_m(u)$, we discover

$$\left(\alpha - \left(v\frac{\chi}{log(1+\chi)}\ D_u^2 + w\left(\frac{\chi}{log(1+\chi)}\right)^2\ D_u^3\right)\right)^{-\nu} R_m(u)$$
$$= \frac{1}{\Gamma(\nu)}\int_0^\infty e^{-\alpha t}t^{\nu-1}\exp\left(vt\frac{\chi}{log(1+\chi)}\ D_u^2 + wt\left(\frac{\chi}{log(1+\chi)}\right)^2\ D_u^3\right) R_m(u)dt, \tag{11}$$

which in view of Equation (6) gives

$$\left(\alpha - \left(v\frac{\chi}{log(1+\chi)}\ D_u^2 + w\left(\frac{\chi}{log(1+\chi)}\right)^2\ D_u^3\right)\right)^{-\nu} R_m(u) =$$
$$\frac{1}{\Gamma(\nu)}\int_0^\infty e^{-\alpha t}t^{\nu-1}{}_{\mathcal{J}}R_m(u, vt, wt; \chi)dt. \tag{12}$$

A new family of polynomials is defined by the transform on the right-hand side of Equation (12). Using the symbol ${}_{\mathcal{J}}R_{m,\nu}(u, v, w; \chi, \beta)$ to identify this unique family of polynomials, we may create the generalized degenerate three-variable Hermite Appell polynomials (D3VHAP) given by expression

$${}_{\mathcal{J}}R_{m,\nu}(u, v, w; \chi, \beta) = \frac{1}{\Gamma(\nu)}\int_0^\infty e^{-\alpha t}t^{\nu-1}{}_{\mathcal{J}}R_m(u, vt, wt; \chi)dt. \tag{13}$$

In view of Equations (12) and (13), assertion (10) follows. □

Next, we prove the following result, which will be applied to construct the generating function of the generalized D3VHAP ${}_{\mathcal{J}}R_{m,\nu}(u, v, w; \chi, \beta)$:

Theorem 2. *For the generalized D3VHAP ${}_{\mathcal{J}}R_{m,\nu}(u, v, w; \chi, \beta)$, the following generating expression holds true:*

$$\frac{R(z)(1+\chi)^{\frac{uz}{\chi}}}{\left[\beta - \frac{(vz^2+wz^3)}{\chi}\ log(1+\chi)\right]^\nu} = \sum_{m=0}^{\infty} {}_{\mathcal{J}}R_{m,\nu}(u, v, w; \chi, \beta)\frac{z^m}{m!} \tag{14}$$

Proof. When we multiply both sides of expression (13) by $\frac{z^m}{m!}$ and summing over m adding the results, we obtain

$$\sum_{m=0}^{\infty} {}_{\mathcal{J}}R_{m,\nu}(u, v, w; \chi, \beta)\frac{z^m}{m!} = \sum_{m=0}^{\infty} \frac{1}{\Gamma(\nu)}\int_0^\infty e^{-\beta t}t^{\nu-1}{}_{\mathcal{J}}R_m(u, vt, wt; \chi)\frac{z^m}{m!}dt. \tag{15}$$

Using Equation (4) with t replaced by z in the right-hand side of Equation (15), it follows that

$$\sum_{m=0}^{\infty} {}_{\mathcal{J}}R_{m,\nu}(u, v, w; \chi, \beta)\frac{z^m}{m!} = \sum_{m=0}^{\infty} \frac{1}{\Gamma(\nu)}\int_0^\infty e^{-\beta t}t^{\nu-1}R(z)(1+\chi)^{\frac{uz+vz^2t+wz^3t}{\chi}}dt, \tag{16}$$

which in view of expression (1) yields assertion (14). □

Corollary 1. *For $R(z) = 1$, the generalized D3VHAP ${}_{\mathcal{J}}R_{m,\nu}(u, v, w; \chi, \beta)$ reduces to the degenerate three-variable Hermite polynomials $\mathcal{J}_{m,\nu}(u, v, w; \chi, \beta)$, therefore the corresponding operational rule and generating function for these polynomials are given by the expressions:*

$$\left(\beta - \left(v\frac{\chi}{log(1+\chi)}\ D_u^2 + w\left(\frac{\chi}{log(1+\chi)}\right)^2\ D_u^3\right)\right)^{-\nu} u^m = \mathcal{J}_{m,\nu}(u, v, w; \chi, \beta) \tag{17}$$

and

$$\frac{(1+\chi)^{\frac{uz}{\chi}}}{\left[\beta - \frac{(vz^2+wz^3)}{\chi}\, log(1+\chi)\right]^{\nu}} = \sum_{m=0}^{\infty} \mathcal{J}_{m,\nu}(u,v,w;\chi,\beta)\frac{z^m}{m!}, \tag{18}$$

respectively.

Remark 1. *For $\beta = \nu = 1$, the generalized D3VHAP ${}_{\mathcal{J}}R_{m,\nu}(u,v,w;\chi,\beta)$ reduces to the degenerate Hermite–Appell polynomials ${}_{\mathcal{J}}R_m(u,v,w;\chi)$ [13].*

Remark 2. *For $\alpha = \nu = 1$ and $\chi \to 0$, the generalized D3VHAP ${}_{\mathcal{J}}R_m(u,v,w;\chi)$ becomes the 3VHAP [21].*

Remark 3. *For $\alpha = \nu = 1$, $v = w = 0$ and $\chi \to 0$, the generalized D3VHAP ${}_{\mathcal{J}}R_m(u,v,w;\chi)$ reduces to the Appell polynomials [14].*

The next step is to prove the explicit summation formula for the generalized D3VHAP ${}_{\mathcal{J}}R_{m,\nu}(u,v,w;\chi,\beta)$:

Theorem 3. *For, ${}_{\mathcal{J}}R_{m,\nu}(u,v,w;\chi,\beta)$ i-e, the generalized D3VHAP, the below listed explicit summation formula in terms of the generalized D3VHP $\mathcal{J}_{m,\nu}(u,v,w;\chi,\beta)$ holds true:*

$${}_{\mathcal{J}}R_{m,\nu}(u,v,w;\chi,\beta) = \sum_{r=0}^{m} \binom{m}{r} R_r\, \mathcal{J}_{m-r,\nu}(u,v,w;\chi,\beta) \tag{19}$$

Proof. By inserting Equations (18) and (8) into the left-hand side of the expression (14), assertion (19) is obtained. □

Corollary 2. *The determinant formulation listed in [22] (p. 1533) of the Appell polynomials is used to obtain the determinant form of the generalized D3VHAP:*

$$R_0(u) = \frac{1}{\gamma_0}, \quad \gamma_0 = \frac{1}{R_0}, \tag{20}$$

$$R_m(u) = \frac{(-1)^m}{(\gamma_0)^{n+1}} \begin{vmatrix} 1 & u & u^2 & \cdots & u^{m-1} & u^m \\ \gamma_0 & \gamma_1 & \gamma_2 & \cdots & \gamma_{m-1} & \gamma_m \\ 0 & \gamma_0 & \binom{2}{1}\gamma_1 & \cdots & \binom{m-1}{1}\gamma_{m-2} & \binom{m}{1}\gamma_{m-1} \\ 0 & 0 & \gamma_0 & \cdots & \binom{m-1}{2}\gamma_{m-3} & \binom{m}{2}\gamma_{m-2} \\ . & . & . & \cdots & . & . \\ . & . & . & \cdots & . & . \\ 0 & 0 & 0 & \cdots & \gamma_0 & \binom{m}{m-1}\gamma_1 \end{vmatrix}, \tag{21}$$

$$\gamma_m = -\frac{1}{R_0}\left(\sum_{k=1}^{m} \binom{m}{k} R_k\, \gamma_{m-k}\right), m = 1,2,3,\cdots,$$

where $\gamma_0, \gamma_1, \cdots, \gamma_m \in \mathbb{R}$, $\gamma_0 \neq 0$.

Theorem 4. *For the generalized D3VHAP ${}_{\mathcal{J}}R_{m,\nu}(u,v,w;\chi,\beta)$, the following determinant form holds true:*

$${}_{\mathcal{J}}R_{0,\nu}(u,v,w;\chi,\beta) = \frac{1}{\gamma_0}\, \mathcal{J}_{m,\nu}(u,v,w;\chi,\beta), \qquad \gamma_0 = \frac{1}{R_0}, \tag{22}$$

$$
{}_{\mathcal{J}}R_{m,\nu}(u,v,w;\chi,\beta)
$$

$$
= \frac{(-1)^m}{(\gamma_0)^{m+1}} \begin{vmatrix} \mathcal{J}_{0,\nu}(u,v,w;\chi,\beta) & \mathcal{J}_{1,\nu}(u,v,w;\chi,\beta) & \cdots & \mathcal{J}_{m-1,\nu}(u,v,w;\chi,\beta) & \mathcal{J}_{m,\nu}(u,v,w;\chi,\beta) \\ \gamma_0 & \gamma_1 & \cdots & \gamma_{m-1} & \beta_n \\ 0 & \gamma_0 & \cdots & \binom{m-1}{1}\gamma_{m-2} & \binom{m}{1}\gamma_{m-1} \\ 0 & 0 & \cdots & \binom{m-1}{2}\gamma_{m-3} & \binom{m}{2}\gamma_{m-2} \\ . & . & \cdots & . & . \\ . & . & \cdots & . & . \\ 0 & 0 & \cdots & \gamma_0 & \binom{m}{m-1}\gamma_1 \end{vmatrix}, \tag{23}
$$

$$
\gamma_m = -\frac{1}{R_0}\left(\sum_{k=1}^{m}\binom{m}{k}R_k\gamma_{m-k}\right), \qquad m = 1,2,3,\cdots,
$$

where $\gamma_0, \gamma_1, \cdots, \gamma_m \in \mathbb{R}$, $\gamma_0 \neq 0$ *and* $\mathcal{J}_{m,\nu}(u,v,w;\chi,\beta)$ $(m = 0,1,\cdots)$ *are the generalized D3VHP defined by Equation (18).*

Proof. Taking $m = 0$ in Equation (19) and then using Equation (17) in the resultant equation, it follows that:

$$
{}_{\mathcal{J}}R_{m,\nu}(u,v,w;\chi,\beta) = \frac{1}{\gamma_0}\,\mathcal{J}_{0,\nu}(u,v,w;\chi,\beta), \quad \gamma_0 = \frac{1}{R_0}. \tag{24}
$$

Expansion of the determinant in Equation (20) with respect to the first row gives

$$
\begin{aligned}
R_m(u) = \frac{(-1)^m}{(\gamma_0)^{m+1}} &\begin{vmatrix} \gamma_1 & \gamma_2 & \cdots & \gamma_{m-1} & \gamma_m \\ \gamma_0 & \binom{2}{1}\gamma_1 & \cdots & \binom{m-1}{1}\gamma_{m-2} & \binom{m}{1}\gamma_{m-1} \\ 0 & \gamma_0 & \cdots & \binom{m-1}{2}\gamma_{m-3} & \binom{m}{2}\gamma_{m-2} \\ . & . & \cdots & . & . \\ . & . & \cdots & . & . \\ 0 & 0 & \cdots & \gamma_0 & \binom{m}{m-1}\gamma_1 \end{vmatrix} \\
-\frac{(-1)^m u}{(\gamma_0)^{m+1}} &\begin{vmatrix} \gamma_0 & \gamma_2 & \cdots & \gamma_{m-1} & \gamma_m \\ 0 & \binom{2}{1}\gamma_1 & \cdots & \binom{m-1}{1}\gamma_{m-2} & \binom{m}{1}\gamma_{m-1} \\ 0 & \gamma_0 & \cdots & \binom{m-1}{2}\gamma_{m-3} & \binom{m}{2}\gamma_{m-2} \\ . & . & \cdots & . & . \\ . & . & \cdots & . & . \\ 0 & 0 & \cdots & \gamma_0 & \binom{m}{m-1}\gamma_1 \end{vmatrix} \\
+\frac{(-1)^m u^2}{(\gamma_0)^{m+1}} &\begin{vmatrix} \gamma_0 & \gamma_1 & \cdots & \gamma_{m-1} & \gamma_m \\ 0 & \gamma_0 & \cdots & \binom{m-1}{1}\gamma_{m-2} & \binom{m}{1}\gamma_{m-1} \\ 0 & 0 & \cdots & \binom{m-1}{2}\gamma_{m-3} & \binom{m}{2}\gamma_{m-2} \\ . & . & \cdots & . & . \\ . & . & \cdots & . & . \\ 0 & 0 & \cdots & \gamma_0 & \binom{m}{m-1}\gamma_1 \end{vmatrix}
\end{aligned} \tag{25}
$$

$$+\cdots+\frac{(-1)^{2m-1}u^{m-1}}{(\gamma_0)^{m+1}}\begin{vmatrix} \gamma_0 & \gamma_1 & \gamma_2 & \cdots & \gamma_m \\ 0 & \gamma_0 & \binom{2}{1}\gamma_1 & \cdots & \binom{m}{1}\gamma_{m-1} \\ 0 & 0 & \gamma_0 & \cdots & \binom{m}{2}\gamma_{m-2} \\ . & . & . & \cdots & . \\ . & . & . & \cdots & . \\ 0 & 0 & 0 & \cdots & \binom{m}{m-1}\gamma_1 \end{vmatrix}$$

$$+\frac{u^m}{(\gamma_0)^{m+1}}\begin{vmatrix} \gamma_0 & \gamma_1 & \gamma_2 & \cdots & \gamma_{m-1} \\ 0 & \gamma_0 & \binom{2}{1}\gamma_1 & \cdots & \binom{m-1}{1}\gamma_{m-2} \\ 0 & 0 & \gamma_0 & \cdots & \binom{m-1}{2}\gamma_{m-3} \\ . & . & . & \cdots & . \\ . & . & . & \cdots & . \\ 0 & 0 & 0 & \cdots & \gamma_0 \end{vmatrix}. \tag{26}$$

Since each minor in Equation (26) is independent of u, operating $\left(\beta-\left(v\frac{\chi}{log(1+\chi)}D_u^2+w\left(\frac{\chi}{log(1+\chi)}\right)^2D_u^3\right)\right)^{-\nu}$ on both sides of Equation (26) and then using Equations (10) and (17), we find

$$_{\mathcal{J}}R_{m,\nu}(u,v,w;\chi,\beta)=\frac{(-1)^m\mathcal{J}_{0,\nu}(u,v,w;\chi,\beta)}{(\gamma_0)^{m+1}}\begin{vmatrix} \gamma_1 & \gamma_2 & \cdots & \gamma_{m-1} & \gamma_m \\ \gamma_0 & \binom{2}{1}\gamma_1 & \cdots & \binom{m-1}{1}\gamma_{m-2} & \binom{m}{1}\gamma_{m-1} \\ 0 & \gamma_0 & \cdots & \binom{m-1}{2}\gamma_{m-3} & \binom{m}{2}\gamma_{m-2} \\ . & . & \cdots & . & . \\ . & . & \cdots & . & . \\ 0 & 0 & \cdots & \gamma_0 & \binom{m}{m-1}\gamma_1 \end{vmatrix}$$

$$-\frac{(-1)^m\mathcal{J}_{1,\nu}(u,v,w;\chi,\beta)}{(\gamma_0)^{m+1}}\begin{vmatrix} \gamma_0 & \gamma_2 & \cdots & \gamma_{m-1} & \gamma_m \\ 0 & \binom{2}{1}\gamma_1 & \cdots & \binom{m-1}{1}\gamma_{m-2} & \binom{m}{1}\gamma_{m-1} \\ 0 & \gamma_0 & \cdots & \binom{m-1}{2}\gamma_{m-3} & \binom{m}{2}\gamma_{m-2} \\ . & . & \cdots & . & . \\ . & . & \cdots & . & . \\ 0 & 0 & \cdots & \gamma_0 & \binom{m}{m-1}\gamma_1 \end{vmatrix}$$

$$+\frac{(-1)^m\mathcal{J}_{2,\nu}(u,v,w;\chi,\beta)}{(\gamma_0)^{m+1}}\begin{vmatrix} \gamma_0 & \gamma_1 & \cdots & \gamma_{m-1} & \gamma_m \\ 0 & \gamma_0 & \cdots & \binom{m-1}{1}\gamma_{m-2} & \binom{m}{1}\gamma_{m-1} \\ 0 & 0 & \cdots & \binom{m-1}{2}\gamma_{m-3} & \binom{m}{2}\gamma_{m-2} \\ . & . & \cdots & . & . \\ . & . & \cdots & . & . \\ 0 & 0 & \cdots & \gamma_0 & \binom{m}{m-1}\gamma_1 \end{vmatrix}+\cdots$$

$$
+\frac{(-1)^{2m-1}\mathcal{J}_{m-1,v}(u,v,w;\chi,\beta)}{(\gamma_0)^{m+1}}\begin{vmatrix} \beta_0 & \gamma_1 & \gamma_2 & \cdots & \gamma_m \\ 0 & \gamma_0 & \binom{2}{1}\gamma_1 & \cdots & \binom{m}{1}\gamma_{m-1} \\ 0 & 0 & \gamma_0 & \cdots & \binom{m}{2}\gamma_{m-2} \\ . & . & . & \cdots & . \\ . & . & . & \cdots & . \\ 0 & 0 & 0 & \cdots & \binom{m}{m-1}\gamma_1 \end{vmatrix}
$$

$$
+\frac{\mathcal{J}_{m,v}(u,v,w;\chi,\beta)}{(\gamma_0)^{m+1}}\begin{vmatrix} \gamma_0 & \gamma_1 & \gamma_2 & \cdots & \gamma_{m-1} \\ 0 & \gamma_0 & \binom{2}{1}\gamma_1 & \cdots & \binom{m-1}{1}\gamma_{m-2} \\ 0 & 0 & \gamma_0 & \cdots & \binom{m-1}{2}\gamma_{m-3} \\ . & . & . & \cdots & . \\ . & . & . & \cdots & . \\ 0 & 0 & 0 & \cdots & \binom{m}{m-1}\gamma_1 \end{vmatrix}. \tag{27}
$$

Combining the components in Equation (27), the right-hand side leads to the theorem's proof (12). □

Next, we derive the recurrence relations of the generalized D3VHAP ${}_{\mathcal{J}}R_{m,v}(u,v,w;\chi,\beta)$ by considering their generating expression. A recurrence relation is an equation that iteratively creates a sequence or multidimensional array of values after one or more initial terms are given. The definition of each subsequent term in the series or array depends on the preceding terms. The listed recurrence relations of the generalized D3VHAP ${}_{\mathcal{J}}R_{m,v}(u,v,w;\chi,\beta)$ are discovered by differentiating generating function (14) with respect to u, v, w, and β:

$$
\begin{aligned}
\frac{\chi}{log(1+\chi)}\frac{\partial}{\partial u}\Big({}_{\mathcal{J}}R_{m,v}(u,v,w;\chi,\beta)\Big) &= m\ {}_{\mathcal{J}}R_{m-1,v}(u,v,w;\chi,\beta),\\
\frac{\chi}{log(1+\chi)}\frac{\partial}{\partial v}\Big({}_{\mathcal{J}}R_{m,v}(u,v,w;\chi,\beta)\Big) &= v\ m(m-1)\ {}_{\mathcal{J}}R_{m-2,v+1}(u,v,w;\chi,\beta),\\
\frac{\chi}{log(1+\chi)}\frac{\partial}{\partial w}\Big({}_{\mathcal{J}}R_{m,v}(u,v,w;\chi,\beta)\Big) &= v\ m(m-1)(m-2)\ {}_{\mathcal{J}}R_{m-3,v+1}(u,v,w;\chi,\beta),\\
\frac{\partial}{\partial \beta}\Big({}_{\mathcal{J}}R_{m,v}(u,v,w;\chi,\beta)\Big) &= -v\ {}_{\mathcal{J}}R_{m,v+1}(u,v,w;\chi,\beta).
\end{aligned}
$$

Given the aforementioned relationships, we have

$$
\frac{\chi}{log(1+\chi)}\frac{\partial}{\partial v}\Big({}_{\mathcal{J}}R_{m,v}(u,v,w;\chi,\beta)\Big) = -\frac{\partial^3}{\partial u^2\partial\beta}\,{}_{\mathcal{J}}R_{m,v}(u,v,w;\chi,\beta),
$$

$$
\left(\frac{n}{log(1+\chi)}\right)^2\frac{\partial}{\partial w}\Big({}_{\mathcal{J}}R_{m,v}(u,v,w;\chi,\beta)\Big) = -\frac{\partial^4}{\partial u^3\partial\beta}\,{}_{\mathcal{J}}R_{m,v}(u,v,w;\chi,\beta).
$$

3. Applications

A variety of members of the Appell polynomial family can be obtained depending on the proper choice for the function $\mathcal{R}(t)$. Several applications in number theory, combinatorics, numerical analysis, and other areas of practical mathematics make use of these polynomials and numbers of Bernoulli, Euler, and Genocchi. The Taylor expansion, the trigonometric and hyperbolic tangent and cotangent functions, and the sums of powers of natural numbers are only a few examples of mathematical formulas where the Bernoulli numbers can be found. In close proximity to the trigonometric and hyperbolic secant function origins, the Euler numbers enter the Taylor expansion. In graph theory, automata theory, and calculating the number of up–down ascending sequences, the Genocchi numbers are useful.

Thus, for suitable selection of $R(z)$ in (14), the following generating expressions for degenerate 3VH-Bernoulli, Euler and Genocchi polynomials hold:

$$\frac{\frac{z}{e^z-1}(1+\chi)^{\frac{uz}{\chi}}}{\left[\beta-\frac{(vz^2+wz^3)}{\chi}\,log(1+\chi)\right]^{\nu}}=\sum_{m=0}^{\infty}{}_{\mathcal{J}}\mathfrak{B}_{m,\nu}(u,v,w;\chi,\beta)\frac{z^m}{m!},$$

$$\frac{\frac{z}{e^z+1}(1+\chi)^{\frac{uz}{\chi}}}{\left[\beta-\frac{(vz^2+wz^3)}{\chi}\,log(1+\chi)\right]^{\nu}}=\sum_{m=0}^{\infty}{}_{\mathcal{J}}\mathfrak{E}_{m,\nu}(u,v,w;\chi,\beta)\frac{z^m}{m!}$$

and

$$\frac{\frac{2z}{e^z+1}(1+\chi)^{\frac{uz}{\chi}}}{\left[\beta-\frac{(vz^2+wz^3)}{\chi}\,log(1+\chi)\right]^{\nu}}=\sum_{m=0}^{\infty}{}_{\mathcal{J}}G_{m,\nu}(u,v,w;\chi,\beta)\frac{z^m}{m!}.$$

The generalized D3VH-Bernoulli polynomials ${}_{\mathcal{J}}\mathfrak{B}_{m,\nu}(u,v,w;\chi,\beta)$ and the generalized D3VH-Euler polynomials ${}_{\mathcal{J}}\mathfrak{E}_{m,\nu}(u,v,w;\chi,\beta)$ in view of (10) are defined using the following operational rules:

$$\left(\beta-\left(v\frac{\chi}{log(1+\chi)}\,D_u^2+w\left(\frac{\chi}{log(1+\chi)}\right)^2 D_u^3\right)\right)^{-\nu}\mathfrak{B}_m(u)={}_{\mathcal{J}}R_{m,\nu}(u,v,w;\chi,\beta)$$

and

$$\left(\beta-\left(v\frac{\chi}{log(1+\chi)}\,D_u^2+w\left(\frac{\chi}{log(1+\chi)}\right)^2 D_u^3\right)\right)^{-\nu}\mathfrak{E}_m(u)={}_{\mathcal{J}}R_{m,\nu}(u,v,w;\chi,\beta),$$

respectively.

Appell polynomials are involved in various identities. The operational formalism outlined in the preceding section can be used to acquire the appropriate identification using the generalized Hermite–Appell polynomials. To do this, we take the following course of action:

The operator $\left(\beta-\left(v\frac{\chi}{log(1+\chi)}\,D_u^2+w\left(\frac{\chi}{log(1+\chi)}\right)^2 D_u^3\right)\right)^{-\nu}$, referred to as $(\mathcal{O})$, is applied on both sides of a given relation.

We have the four applications listed below.

1. Consider first the following connections involving Bernoulli polynomials [23] (pp. 29–30):

$$\mathfrak{B}_m(u+1)-\mathfrak{B}_m(u)=m\,u^{m-1},\quad m=0,1,2,...$$

$$\sum_{k=0}^{m-1}\binom{m}{k}\mathfrak{B}_k(u)=mu^{m-1},\quad m=2,3,4,...$$

$$\mathfrak{B}_m(ku)=k^{m-1}\sum_{k=0}^{m-1}\mathfrak{B}_m\left(u+\frac{k}{m}\right),\quad m=0,1.2,...;\quad k=1,2,3,...$$

The identities that contain the generalized D3VH-Bernoulli polynomials ${}_{\mathcal{J}}\mathfrak{B}_{m,\nu}(u,v,w;\chi,\beta)$ are obtained by applying the operator $(\mathcal{O})$ to earlier expressions and taking into account operational rules (14) and (17) on the resulting expressions. They are listed as follows:

$$_{\mathcal{J}}\mathfrak{B}_{m,v}(u+1,v,w;\chi,\beta) - {_{\mathcal{J}}\mathfrak{B}_{m,v}}(u,v,w;\chi,\beta) = m\ {_{\mathcal{J}}\mathfrak{B}_{m-1,v}}(u,v,w;\chi,\beta), \quad m = 0,1,2...,$$

$$\sum_{k=0}^{m-1}\binom{m}{k}{_{\mathcal{J}}\mathfrak{B}_{m,v}}(u,v,w;\chi,\beta) = m\ {_{\mathcal{J}}\mathfrak{B}_{m-1,v}}(u,v,w;\chi,\beta), \quad m = 2,3,4...,$$

$$_{\mathcal{J}}\mathfrak{B}_{m,v}(ku,k^2v,k^3w;\chi,\beta) = k^{m-1}\sum_{k=0}^{m-1}{_{\mathcal{J}}\mathfrak{B}_{m-1,v}}(u+k/m,v,w;\chi,\beta), \quad m = 0,1,2,...;\ k = 1,2,\cdots.$$

2. We now use the the following relationships involving Euler polynomials [23] (pp. 29–30):

$$\mathfrak{E}_m(u+1) + \mathfrak{E}_m(u) = 2u^m.$$

$$\mathfrak{E}_m(kx) = k^m\sum_{k=0}^{m-1}(-1)^k\mathfrak{E}_m\left(u+\frac{k}{m}\right) \quad m = 0,1,2...;\ k\ odd,$$

The following identities involving the generalized D3VH-Euler polynomials ${_{\mathcal{J}}\mathfrak{E}_{m,v}}(u,v,w;\chi,\beta)$ are obtained:

$$_{\mathcal{J}}\mathfrak{E}_{m,v}(u+1,v,w;\chi,\beta) + {_{\mathcal{J}}\mathfrak{E}_{m,v}}(u,v,w;\chi,\beta) = 2\ \mathcal{J}_{m,v}(u,v,w;\chi,\beta).$$

$$_{\mathcal{J}}\mathfrak{E}_{m,v}(ku,k^2v,k^3w;\chi,\beta) = k^m\sum_{k=0}^{m-1}(-1)^k{_{\mathcal{J}}\mathfrak{E}_{m,v}}(u+k/m,v,w;\chi,\beta), \quad m = 0,1.2,...;\ k\ odd.$$

3. Next, we review the relationships between Bernoulli and Euler polynomials [23] (pp. 29–30), which are listed below:

$$\mathfrak{B}_m(u) = 2^{-m}\sum_{k=0}^{m}\binom{m}{k}\mathfrak{B}_{m-k}\mathfrak{E}_k(2u), \quad m = 0,1,2...,$$

$$\mathfrak{E}_m(u) = \frac{2^{m+1}}{m+1}\left[\mathfrak{B}_{m+1}\left(\frac{u+1}{2}\right) - \mathfrak{B}_{m+1}\left(\frac{u}{2}\right)\right], \quad m = 0,1,2...,$$

$$\mathfrak{E}_m(ku) = -\frac{2k^m}{m+1}\sum_{k=0}^{m-1}(-1)^k\mathfrak{B}_{m+1}\left(\frac{u+k}{m}\right), \quad m = 0,1,2...;k\ even.$$

When we apply the operator $(\mathcal{O})$ to the prior listed equations, we obtain:

$$_{\mathcal{J}}\mathfrak{B}_{m,v}(u,v,w;\chi,\beta) = 2^{-m}\sum_{k=0}^{m}\binom{m}{k}\mathfrak{B}_{m-k}{_{\mathcal{J}}\mathfrak{E}_{m,v}}(2u,4v,8w;\chi,\beta), \quad m = 0,1,2...,$$

$$_{\mathcal{J}}\mathfrak{E}_{m,v}(u,v,w;\chi,\beta) = \frac{2^{m+1}}{m+1}\Big[{_{\mathcal{J}}R_{m+1,v}}(\frac{u+1}{2},\frac{v}{4},\frac{w}{8};\chi,\beta) - {_{\mathcal{J}}\mathfrak{B}_{m+1,v}}(\frac{u}{2},\frac{v}{4},\frac{w}{8};\chi,\beta)\Big], \quad m = 0,1,2,...$$

$$_{\mathcal{J}}\mathfrak{E}_{m,v}(ku,k^2v,k^3w;\chi,\beta) = -\frac{2k^m}{m+1}\sum_{k=0}^{m-1}(-1)^k{_{\mathcal{J}}\mathfrak{B}_{m+1,v}}(\frac{u+k}{m},v,w;\chi,\beta), \quad m = 0,1.2,\cdots;\ k\ even.$$

4. Further, the determinant definition of the generalized D3VH-Bernoulli polynomials ${_{\mathcal{J}}\mathfrak{B}_{m,v}}(u,v,w;\chi,\beta)$ is derived by assuming $\gamma_0 = 1$ and $\gamma_i = \frac{1}{i+1}$ $(i = 1,2,\cdots,n)$ in (22) and (23) and the determinant formulation of the generalized D3VH-Euler polynomials $\mathcal{J}Em,v(u,v,w;\chi,\beta)$ is derived by taking $\gamma 0 = 1$ and $\gamma i = \frac{1}{2}$ $(i = 1,2,\cdots,n)$ in expressions (22) and (23).

The examples above show how the operational connection between the Appell and generalized D3VHAP polynomials may be used to find solutions for the generalized D3VHAP polynomials.

4. Conclusions

Inspired by the study conducted in [13], where three-variable degenerate Hermite-based Appell polynomials have been introduced and studied, the new generalized family of degenerate three-variable Hermite–Appell polynomials ${}_{\mathcal{J}}R_{m,\nu}(u,v,w;\chi,\beta)$ is introduced in Section 2 of this paper. For these polynomials ${}_{\mathcal{J}}R_{m,\nu}(u,v,w;\chi,\beta)$, Theorem 1 provides the operational rule. Theorem 2 gives the generating expression for the function ${}_{\mathcal{J}}R_{m,\nu}(u,v,w;\chi,\beta)$ and the connection of this family to the degenerate Hermite–Appell polynomials, three-variable Hermite–Appell polynomials and Appell polynomials. The explicit summation formula for polynomials ${}_{\mathcal{J}}R_{m,\nu}(u,v,w;\chi,\beta)$ is proved in Theorem 3 and the determinant form for the generalized family is obtained in Theorem 4. The recurrence relations of the generalized three-variable degenerate Hermite-based Appell polynomials are also derived. In Section 3, certain applications of the results obtained in Section 2 are presented giving the equivalent results for the degenerate Hermite–Bernoulli and Hermite–Euler polynomials with three variables. These generalized degenerate hybrid special polynomials associated with Hermite polynomials have a wide range of applications in mathematics and physics. These polynomials may arise naturally in the study of quantum mechanics, in probability theory, where these polynomials may be related to the normal distribution, which is one of the most important distributions in probability theory. In approximation theory, these polynomials can be used as a basis for approximating functions and serve as a powerful tool for numerical analysis. Further, in statistical mechanics, Hermite polynomials are used to calculate the partition function and thermodynamic properties of ideal gases and can be used in Fourier analysis to decompose functions into a sum of orthogonal functions.

By using operational approaches, the development of new functional families is facilitated as well as the derivation of the characteristics of those functional families linked to regular and generalized special functions. Dattoli and his colleagues recognized the significance of the use of operational techniques in the study of special functions that are intended to provide explicit solutions for families of partial differential equations, including those of the Heat and D'Alembert type, and their applications; see, for example [3,24,25] when applied to multi-variable generalized special functions in conjunction with the monomiality principle. This article's method can be utilized as a helpful tool in novel analytical techniques for the solutions of a large class of partial differential equations that are regularly encountered in physical issues.

Further, future research can be conducted in order to find the symmetric identities and determinant forms for these polynomials. Additionally, implicit summation formulae can be taken as future observations.

Author Contributions: Conceptualization, S.A.W., K.A., G.I.O. and S.T.; methodology, S.A.W., K.A., G.I.O. and S.T.; software, S.A.W. and G.I.O.; validation, S.A.W., K.A., G.I.O. and S.T.; formal analysis, S.A.W., K.A., G.I.O. and S.T.; investigation, S.A.W., K.A., G.I.O. and S.T.; resources, S.A.W., K.A., G.I.O. and S.T.; data curation, S.A.W., K.A., G.I.O. and S.T.; writing—original draft preparation, S.A.W., K.A. and S.T.; writing—review and editing, S.A.W., K.A., G.I.O. and S.T.; visualization, S.A.W., K.A., G.I.O. and S.T.; supervision, S.A.W.; project administration, S.A.W.; funding acquisition, K.A. and S.T. All authors have read and agreed to the published version of the manuscript.

Funding: This work was supported by the Deanship of Scientific Research, Vice Presidency for Graduate Studies and Scientific Research, King Faisal University, Saudi Arabia (Grant No. 2930).

Institutional Review Board Statement: Not applicable.

Informed Consent Statement: Not applicable.

Data Availability Statement: Not applicable.

Conflicts of Interest: The authors declare no conflict of interest.

References

1. Baleanu, D.; Agarwal, R.P. Fractional calculus in the sky. *Adv. Differ. Equ.* **2021**, *2021*, 117. [CrossRef]
2. Srivastava, H.M. An Introductory Overview of Fractional-Calculus Operators Based Upon the Fox-Wright and Related Higher Transcendental Functions. *J. Adv. Eng. Comput.* **2021**, *5*, 135–166. [CrossRef]
3. Dattoli, G.; Ricci, P.E.; Cesarano, C.; Vázquez, L. Special polynomials and fractional calculus. *Math. Comput. Model.* **2003**, *37*, 729–733. [CrossRef]
4. Srivastava, H.M.; Manocha, H.L. *A Treatise on Generating Functions*; Halsted Press: New York, NY, USA, 1984.
5. Hermite, C. Sur un nouveau dévelopment en séries de functions. *C. R. Acad. Sci. Paris* **1864**, *58*, 93–100.
6. Ryoo, C.-S.; Kang, J.-Y. Some Identities Involving Degenerate q-Hermite Polynomials Arising from Differential Equations and Distribution of Their Zeros. *Symmetry* **2022**, *14*, 706. [CrossRef]
7. Khan, S.; Wani, S.A. Extended Laguerre-Appell polynomials via fractional operators and their determinant forms. *Turk. J. Math.* **2018**, *42*, 1686–1697. [CrossRef]
8. Yasmin, G.; Muhyi, A.; Araci, S. Certain Results of q -Sheffer–Appell Polynomials. *Symmetry* **2019**, *11*, 159. [CrossRef]
9. Jeelani, M.B.; Alnahdi, A.S. Approximation by Operators for the Sheffer–Appell Polynomials. *Symmetry* **2022**, *14*, 2672. [CrossRef]
10. Khan, S.; Wani, S.A. Fractional calculus and generalized forms of special polynomials associated with Appell sequences. *Georgian Math. J.* **2019**, *28*, 261–270. [CrossRef]
11. Kim, D. A Note on the Degenerate Type of Complex Appell Polynomials. *Symmetry* **2019**, *11*, 1339. [CrossRef]
12. Muhiuddin, G.; Khan, W.A.; Duran, U.; Al-Kadi, D. A New Class of Higher-Order Hypergeometric Bernoulli Polynomials Associated with Lagrange–Hermite Polynomials. *Symmetry* **2021**, *13*, 648. [CrossRef]
13. Alyosuf, R. Quasi-monomiality principle and certain properties of degenerate hybrid special polynomials. *Symmetry* **2023**, *15*, 407. [CrossRef]
14. Appell, P. Sur une classe de polynômes. *Annales Scientifiques de l'École Normale Supérieure* **1880**, *9*, 119–144. [CrossRef]
15. Kumar, D.; Baleanu, D. Fractional calculus and its applications in physics. *Front. Phys.* **2019**, *7*, 81. [CrossRef]
16. Sun, H.; Zhang, Y.; Baleanu, D.; Chen, W.; Chen, Y. A new collection of real world applications of fractional calculus in science and engineering. *Commun. Nonlinear Sci. Numer. Simul.* **2018**, *64*, 213–231. [CrossRef]
17. Farman, M.; Akgül, A.; Baleanu, D.; Imtiaz, S.; Ahmad, A. Analysis of Fractional Order Chaotic Financial Model with Minimum Interest Rate Impact. *Fractal Fract.* **2020**, *4*, 43. [CrossRef]
18. Patnaik, S.; Hollkamp, J.P.; Semperlotti, F. Applications of variable-order fractional operators: A review. *Proc. R. Soc. A* **2020**, *476*, 20190498. [CrossRef]
19. Kachhia, K.B.; Atangana, A. Electromagnetic waves described by a fractional derivative of variable and constant order with non singular kernel. *Discrete Contin. Dyn. Syst. Ser. S* **2021**, *14*, 2357–2371. [CrossRef]
20. Qiao, L.; Xu, D.; Qiu, W. The formally second-order BDF ADI difference/compact difference scheme for the nonlocal evolution problem in three-dimensional space. *Appl. Numer. Math.* **2022**, *172*, 359–381. [CrossRef]
21. Khan, S.; Yasmin, G.; Khan, R.; Hassan, N.A.M. Hermite-based Appell polynomials: Properties and applications. *J. Math. Anal. Appl.* **2009**, *351*, 756–764. [CrossRef]
22. Costabile, F.A.; Longo, E. A determinantal approach to Appell polynomials. *J. Comput. Appl. Math.* **2010**, *234*, 1528–1542. [CrossRef]
23. Magnus, W.; Oberhettinger, F.; Soni, R.P. *Formulas and Theorems for Special Functions of Mathematical Physics*; Springer: New York, NY, USA, 1956.
24. Dattoli, G. Hermite-Bessel and Laguerre-Bessel functions: A by-product of the monomiality principle. *Adv. Spec. Funct. Appl.* **1999**, *1*, 147–164.
25. Dattoli, G. Generalized polynomials operational identities and their applications. *J. Comput. Appl. Math.* **2000**, *118*, 111–123. [CrossRef]

symmetry

MDPI

Article

Numerical Method for Solving Fractional Order Optimal Control Problems with Free and Non-Free Terminal Time

Oday I. Al-Shaher [1,2], M. Mahmoudi [1] and Mohammed S. Mechee [3,*]

1 Department of Mathematics, University of Qom , Qom 3716146611, Iran
2 The General Directorate of Education in Najaf, Najaf 540011, Iraq
3 Information Technology Research and Development Center (ITRDC), University of Kufa, Najaf 540011, Iraq
* Correspondence: mohammeds.abed@uokufa.edu.iq

Abstract: The optimal control theory in mathematics aims to study the finding of control for a dynamic system over time, where an objective function is optimized. It has a broad range of applications in engineering, operations research, and science. The main purpose of this study is to provide numerical algorithms for two cases of optimal control problems of fractional order that involve fractional order derivatives with free and non-free terminal time. In addition to comparing the numerical results for three test problems with exact solutions of these problems, various computer simulations are also introduced.

Keywords: optimal control; fractional differential equations (FDEs); fractional optimal control problems (FOCPs); free terminal time

Citation: Al-Shaher, O.I.; Mahmoudi, M.; Mechee, M.S. Numerical Method for Solving Fractional Order Optimal Control Problems with Free and Non-Free Terminal Time. *Symmetry* **2023**, *15*, 624. https://doi.org/10.3390/sym15030624

Academic Editor: Francisco Martínez González

Received: 24 January 2023
Revised: 14 February 2023
Accepted: 23 February 2023
Published: 2 March 2023

1. Introduction

Optimal control is the study of finding a dynamic control system over time in order to optimize an objective function. It has many uses in operations research, engineering, and science. For instance, the dynamic system could be a spacecraft with controls that correspond to rocket thrusters, and the goal could be to reach the moon using the least amount of fuel. In terms of result, the dynamic system could be a country's economy with the goal to minimize unemployment; in this scenario, the controls could be fiscal and monetary policy. It is also possible to integrate operations research issues into the framework of optimal control theory by using a dynamic system. Additionally, a branch of mathematics and physics known as the fractional dynamics examines how objects and systems behave by differentiating fractional orders. Research on fractional dynamical systems has produced novel findings that have attracted the interest of a significant audience of professionals, including mathematicians, physicists, applied researchers, and practitioners. This is due to the topic's wide applications in science and technology. In contrast to integer-order models, however, fractional-order models offer the potential to express non-local relations in time and space using power law memory kernels [1]. Consequently, this indicates that they offer more accurate and more realistic results. Moreover, the standard integral and differential calculus are generalized to any order in fractional calculus. If the order of the fractional derivative operator is an integer m, we obtain an m-fold integral when m is negative and the traditional derivative of order m when m is positive. Furthermore, for the review of the literature on numerical studies of fractional optimal control problems (FOCPs), Agrawal [2] preformed a formulation and numerical scheme for FOCPs, the work in [3] introduced the numerical solution of some types of FOCPs, while Bhrawy et al. [4] introduced an accurate numerical technique for solving FOCPs. Furthermore, a new method for the numerical solution of FOCPs was introduced in [5]. Furthermore, to solve multidimensional FOCPs with a quadratic performance index, the authors of [6] developed a practical numerical method for the purpose of solving FOCPs, and Doha et al. [7] investigated an effective numerical method based on the shifted orthonormal Jacobi polynomials. However, the generalized differential transform approach was used in [8] to introduce the numerical solutions of the coupled space-and-time Burgers equations. Lotfi et al. [9] introduced a numerical technique for solving FOCPs, Pooseh et al. [10] introduced a numerical scheme to solve FOCPs, Zhao and Li [11] solved the time–space fractional telegraph equation using the fractional difference-finite element, and Mechee and Senu [12] studied the numerical solution of fractional differential equations of Lane–Emden type by the method of collocation. For the space fractional diffusion equations, Zhou et al. [13] studied the quasi-compact finite difference schemes, and Bhrawy et al. [14] investigated a new Jacobi spectral collocation approach for fractional coupled Schrödinger systems and 1 + 1 fractional Schrödinger equations. At the same time, for the review of the literature on Legendre polynomials, using a Chebyshev–Legendre operational technique, Bhrawy et al. [15] solved the fractional optimal control for dynamical systems problems (FOCDSs). In fact, Yousefi et al. [16] employed a Legendre multiwavelet collocation approach in order to solve the FOCPs. In contrast, Bhrawy and Ezz-Eldien [17] used a new Legendre operational technique for solving delay FOCPs, in similar to Dirichlet boundary conditions, Heydari et al. [18] solved fractional partial differential equations (FPDEs) using the Legendre wavelets method. On the other hand, for the solution of fractional sub-diffusion and reaction sub-diffusion equations, Doha et al. [7] utilized an effective Legendre spectral tau matrix formulation, Khan and Khalil [19] provided a new approach that is based on Legendre polynomials. In parallel to these researchers, Sweilam and Al-Ajami [20] solved some types of FOCPs using the Legendre spectral-collocation method; additionally, some authors studied different cases of fractional differential equations. To solve the space fractional order diffusion equation, Sweilam et al. [21] utilized the second sort of shifted Chebyshev polynomials, but a discrete method for solving FOCPs was introduced in [22], while ref. [23] established a fractional adaptation strategy for lateral control of an AGV; whereas, Pinto and Tenreiro Machado [24] introduced the fractional dynamics of computer virus propagation, Pooseh et al. [25] studied the FOCPs with free terminal time by using operational matrices of Bernstein polynomials, Jafari and Tajadodi [26] solved the FOCPs, and Jesus and Tenreiro Machado [27] investigated the fractional control of heat diffusion systems. Thereafter, for a review of more literature on the applications, Ahmad and El-Khazali [28] introduced the fractional-order dynamical models of love and David et al. [29] studied fractional-order calculus, meanwhile, analog fractional-order controllers for temperature and motor control applications were studied in [30,31] introduced a 2D dynamic analysis of the model of disturbances in the calcium neuronal model and its implications in neurodegenerative disease; the work in [32] introduced the fractional sub-equation method and its applications to nonlinear fractional PDEs, whereas Kreyszig [33] studied historical apologia, fundamental ideas, as well as certain applications. Lastly, a fractional-order iterative learning control (FOILC) design challenge for linear time-varying systems with nonuniform trial durations was addressed in [34]. Additionally, a closed-loop FOILC updating legislation has been provided for activities with variable trial lengths. A central idea that unifies the coordination, prioritization, and execution of digital transformations within a firm was investigated in [35] in organizations that needed to build management procedures to oversee initiatives to investigate new digital technologies. For the purpose of tracking control of fractional-order linear systems, Zhao et al. [36] developed a revolutionary FOILC approach. In the meantime, the same beginning condition assumption is relaxed with the introduction of an initial state learning mechanism. For the FOCPs exposed to fractional systems with equality and inequality constraints, Sabermahani and Ordokhani [37] investigated fractional optimal control issues using the Fibonacci wavelets and Galerkin approach.

The free and non-free terminal time optimal control for dynamical systems (OCDSs) is introduced in this study. Additionally, the direct search approach to the unconstrained optimization problem is investigated. The proposed numerical methods for solving the optimal control problems of fractional orders with free and non-free terminal time are then constructed. The algorithm of the known procedure as Hooke and Jeeves's method is used in the computation.

2. Main Problem

A dynamic system's optimal control problem is the task of determining the control law that minimizes a performance index in terms of the state and control variables. Many authors have recently studied a wide range of optimization issues related to the integer optimal control of differential systems. In this research, we propose a novel numerical method for approximating the solutions of the fractional optimal control systems in both cases with free- and non-free terminal time.

Case I: Non-Free Terminal Time

Consider

$$\min_{x(\tau),u(\tau)} J(\tau, x(\tau), u(\tau)) = \min_{x(\tau),u(\tau)} \int_{\tau_0}^{\tau_1} P(\tau, x(\tau), u(\tau))d\tau, \tag{1}$$

subjected to the constricted dynamical system

$$\alpha\dot{x}(\tau) + \beta D^{\gamma}x(\tau) = \gamma(\tau)x(\tau) + f(\tau)u(\tau) + g(\tau), \quad \tau_0 \le \tau \le \tau_1, \quad 0 \le \gamma \le 1. \tag{2}$$

The constricted boundary conditions are as follows:

$$x(\tau_0) = \zeta, \quad x(\tau_1) = \eta, \tag{3}$$

where $\alpha, \beta \neq 0$,

Case II: Free Terminal Time

Consider the FOCP in the equations

$$\min_{x(\tau),u(\tau),T} J(\tau, T, x(\tau), u(\tau)) = \min_{x(\tau),u(\tau),T} \int_{\tau_0}^{T} P(\tau, x(\tau), u(\tau))d\tau, \tag{4}$$

subjected to the constricted dynamical system in Equation (2) with the free terminal time:

$$x(t_0) = c, \quad x(T) = d, \tag{5}$$

where T is a free parameter.

Firstly, for using the proposed numerical approach, we use the basic polynomials to approximate the state variable $x(\tau)$ with the control variable $u(\tau)$, and the known functions $e(\tau), f(\tau)$, and $g(\tau)$ are given. The second stage of the numerical method involves using a search method such as the Hooke and Jeeves method to optimize the parameters of the approximation in case of the problem of fractional order of optimal control systems with free terminal time together to the parameter of T in case of non-free terminal time. The manuscript is organized as follows. Section 3 introduces the basic definitions and background related to the problem of this study, while Section 4 presents the numerical methods and studies the proposed numerical method for solving FOCPs with free and non-free terminal time. Furthermore, Section 5 introduces the implementations of test examples for solving two types of fractional optimal control dynamical systems with free and non-free terminal time. Lastly, this paper ends with a discussion and conclusions in Section 6.

3. Preliminary

We have introduced the basic definitions and background related to the problem of this study.

3.1. Basic Definitions of the Fractional Derivatives and (FOCDS) with Free and Non-Free Terminal Time

The fundamental definitions of fractional derivatives as well as the free and non-free terminal times of fractional-order optimal control problems are introduced in this subsection.

Definition 1. *Non-Free Terminal Time (FOC) Problem*

The (FOC) problem in Equations (1)–(3) is said to be a non-free terminal time if we have a constraint in t_1 that means it is fixed else free terminal time (FOC) if $t_1 = T$ is not a fixed parameter. The following are two famous fractional derivatives since a large number of scholars have worked to establish a fractional derivative. In the literature, the fractional derivative often was presented in integral form. Two famous fractional derivatives are known as follows:

(i) *Let $f : [a, \infty) \rightarrow \Re$. and $a > 0$. The fractional definition of f using the Riemann–Liouville derivative for $\alpha \in [n-1, n)$ is defined by:*

Definition 2. *Riemann–Liouville Fractional Integral The left and right Riemann–Liouville (RL) fractional integral operators of order $\alpha > 0$*

$$ {}^{a}D_{\tau}^{\alpha} y(\tau) = \frac{1}{\Gamma(n-\alpha)} \frac{d^n}{d\tau^n} \int_a^{\tau} \frac{y(\tau)}{(\tau - x)^{n-\alpha-1}} dx, \tag{6} $$

and

$$ {}^{a}D_{\tau}^{\alpha} y(\tau) = \frac{1}{\Gamma(n-1)} \frac{d^n}{d\tau^n} \int_a^{\tau} (\tau - s)^{n-\alpha-1} y(s) ds, \tag{7} $$

respectively, such that n is an integer and $n - 1 < \alpha < n, n \in N$. Additionally, (ii) the Caputo derivative definition of f, for $\alpha \in [n-1, n)$, is defined as follows:

Definition 3. *The Fractional Caputo Derivative*

$$ {}^{a}D_{\tau}^{\alpha} f(\tau) = \frac{1}{\Gamma(n-\alpha)} \int_a^{\tau} \frac{f^{(n)}(\tau)}{(\tau - x)^{\alpha-n-1}} dx, \tag{8} $$

where n is an integer and $n - 1 < \alpha < n$, $n \in N$ The fractional integral and derivative in the Definitions 2 and 3 satisfy the linearity properties for the fractional integrals and derivatives for $\alpha > 0, n - 1 \leq \alpha < n$.

3.2. *Hooke and Jeeves Direct Search Method Analysis*

In this subsection, the direct search method for solving the unconstrained optimization problem

$$ \min_{X} \quad f(X), \tag{9} $$

where the objective function f maps $\Re^n$ into $\Re \bigcup \{+\infty\}$ and $X = (x_1, x_2, x_3, \ldots, x_n)$, is introduced.

3.2.1. Algorithm of Hooke and Jeeves Method

1. Set k = 0;
2. Choose an initial point $X(k)$ and indicate the variable increments with $\triangle_i$ for $i = 1, 2, \ldots, N$, where the factor of step reduction is a > 1, and the termination parameter is ϵ;
3. Use $X(k)$ as the base point for an experimental move. Consider the result of the exploratory maneuver to be X. Set $Z(k+1) = X$ and proceed to Step 4 if the exploratory move is successful; otherwise, proceed to Step 3;
4. Is $||\triangle|| < \epsilon$? If so, terminate; otherwise, set A= A/a for $i = 1, 2, \ldots, N$ and go to Step 3;
5. Perform the pattern move after setting k = k+1: $Xp(k+1) = Xp(k) + X(k) - X(k+1)$;
6. Perform another exploratory move using Xp as the base point. Let the result be $X(k+1)$;
7. Is $f(X(k+1)) < f(X(k))$? If so, go to Step 5; otherwise, go to Step 4.

3.2.2. The Convergence of Hooke and Jeeves Method

The set of points produced by the direct algorithm is consistently dense in the search region for all box selection methods. When $N_{max} = 1$ and $H_{min} = 0$, the proposed algorithm's properties of convergence are examined. The sequence of the solutions of the problem in Equation (9) is $\{X(0), X(1), \ldots, X(k), X(k+1), \ldots\}$, which is obtained using the Hooke and Jeeves method. This sequence satisfies the convergence conditions according to the condition in step three.

Consider $\xi \in E$ to be arbitrary, where

$$E = z_m + h_{meso}[-1;1]^n.$$

For each valid box selection approach, let $\{\delta_r\}_{r=1}^{\infty}$ represent the points produced by strategy Γ. Let

$$\Delta(r) = \max_{\Gamma} \max_{\xi \in E} \min_{\mathbf{i=1,2,\ldots,r}} ||\xi_i - \delta_i||.$$

Then, $\Delta(r) \longrightarrow 0$ as $r \longrightarrow \infty$.

4. The Numerical Method

In this section, the proposed numerical method for solving (FOC) problems with free- and non-free terminal time is introduced.

4.1. Proposed Algorithm

In this subsection, we write the algorithm of the proposed numerical approach for approximating the solutions of (FOC) in two cases: (FOC) problems with free- and non-free terminal time. The steps of this algorithm are written as follows:

- Algorithm of non-free terminal time (FOC) problem:
  1. Choose a suitable approximated base.

  $$\Omega = \{\Omega_0(t), \Omega_1(t), \Omega_2(t), \ldots, \Omega_n(t)\},$$

  2. Construct an approximated solution of (FOC),

  $$x(t) = \sum_{i=0}^{n} c_i\Omega_i(t) = c_0\Omega_0(t) + c_1\Omega_1(t) + \cdots + c_n\Omega_n(t). \quad (10)$$

     In Equations (1) and (2) which satisfy the boundary conditions in Equation (3) using the approximated base.
  3. In case the differential equation in Equation (2) is given as explicit formula in the control function $u(t)$, then we have to evaluate the function $u(t)$;
  4. Substitute the approximated formulas of the functions $x(t)$ and $u(t)$ in Equation (1);
  5. Use a suitable minimizing search methods such as the Hooke and Jeeves method to find the minimal parameter(s) in Equation (1).
- Algorithm of free terminal time (FOC) problem:
  1. Perform steps 1–4 in the previous algorithm;
  2. Use suitable minimizing search methods such as the Hooke and Jeeves method to find the best parameters (minimal) including the parameter T in Equation (1).

  where T is the free parameter.

4.2. Dual Discreet Problem

- Algorithm of non-free terminal time (FOC) problem:

1. From Equation (10), consider

$$x(t) = \sum_{i=0}^{n} c_i \Omega_i(t), \tag{11}$$

with the boundary conditions leading

$$c_1 = \gamma_0 \frac{\eta - \frac{\zeta \Omega_0(t_1)}{\Omega_0(t_0)} - \sum_{i=2}^{n} c_i \left(\frac{\Omega_i(t_0)\Omega_0(t_1)}{\Omega_0(t_0)} - \Omega_i(t_1) \right)}{\Omega_1(t_1)}, \tag{12}$$

where

$$\gamma_0 = \frac{\Omega_0(t_0)\Omega_1(t_1) - \Omega_1(t_0)\Omega_0(t_1)}{\Omega_0(t_0)\Omega_1(t_1)},$$

and

$$c_0 = \frac{\zeta - c_1\Omega_1(t_0) - \sum_{i=2}^{n} c_i \Omega_i(t_0)}{\Omega_0(t_0)}, \tag{13}$$

hence,

$$\begin{aligned} x(t) &= \frac{\zeta - \sum_{i=2}^{n} c_i \Omega_i(t_0)}{\Omega_0(t_0)} \Omega_0(t) + \gamma_0 \frac{\eta - \frac{\zeta \Omega_0(t_1)}{\Omega_0(t_0)} - \sum_{i=2}^{n} c_i \left(\frac{\Omega_i(t_0)\Omega_0(t_1)}{\Omega_0(t_0)} - \Omega_i(t_1) \right)}{\Omega_1(t_1)} \\ &+ \left(\Omega_1(t) - \frac{\Omega_1(t_0)}{\Omega_0(t_0)} \Omega_0(t) \right) + \sum_{i=2}^{n} c_i \Omega_i(t). \end{aligned} \tag{14}$$

2. From the differential equation in Equation (2), we obtain the control function $u(t)$ as a function $u(t) = f$, then, we have to evaluate the function $u(t) = \psi(c_2, c_3, \dots, c_n, \Omega_0(t), \Omega_1(t), \dots, \Omega_n(t))$;
3. From Equation (1), we obtain the optimal problem Minimum $\phi(c_2, c_3, \dots, c_n)$, in case of the free terminal time (FOC) problem, and Minimum $\phi(c_2, c_3, \dots, c_n, T)$, in case of the non-free terminal time (FOC) problem.

where T is free parameter.

5. Implementations (Numerical Examples)

In this section, we introduce two types of dynamical problems. The numerical method introduced in Section 4 has been used for solving the optimal control problems of integer and fractional order with free and non-free terminal time.

Example 1. *Let us take into consideration the following (FOC) problem with non-free terminal time introduced by* [10,25].

$$\min_{x(\tau),u(\tau)} J(\tau, x(\tau), u(\tau)) = \min_{x(\tau),u(\tau)} \int_0^1 (\tau u(\tau) - (\gamma+2)x(\tau))^2 d\tau, \tag{15}$$

subjected to the dynamic system

$$D^{\gamma} x(\tau) + \dot{x}(\tau) = \tau^2 + u(\tau), \tag{16}$$

with the boundary conditions (BCs)

$$x(0) = 0, \quad x(1) = \frac{2}{3+\gamma}, \tag{17}$$

where the exact solution is given by

$$(x(\tau), u(\tau)) = (\frac{2\tau^{2+\gamma}}{\Gamma(3+\gamma)}, \frac{2\tau^{1+\gamma}}{\Gamma(2+\gamma)}).$$

Using the approximation base $\Omega(t) = \{\tau^2, \tau, 1\}$, *we have the approximation of* $x(\tau)$ *as*

$$x(\tau) = c_0 + c_1\tau + c_2\tau^2. \tag{18}$$

If we use the BCs in Equation (17), we obtain $c_0 = 0$ *and* $c_1 = \frac{2}{3+\gamma} - c_2$. *Then, the following approximation is obtained*

$$x(t) = t(c_2 + \frac{2}{3+\gamma} - c_2 t). \tag{19}$$

Then, we have

$$u(\tau) = x(\tau) = (1 + \tau^{1-\gamma})(c_2 - \frac{2}{3+\gamma} + 2c_2\tau) - \tau^2. \tag{20}$$

Substitute Equations (19) and (20) in the problem of minimizing in the Equation (15) to obtain the optimal values of c_2, *and the non-free terminal parameter* T. *Hence, using the Hooke and Jeeves method for the problem in parameter* c_2, *the approximation of the problem is plotted in Figure 1a.*

Example 2. *Consider the following integer-order optimal control problem with non-free terminal time:*

$$\min_{x(\tau),u(\tau),T} J(\tau, x(\tau), u(\tau), T) = \min_{x(\tau),u(\tau),T} \int_0^T (\tau u(\tau) - 2x(\tau))^2 d\tau, \tag{21}$$

subjected to the dynamic system

$$\dot{x}(\tau) + \dot{x}(\tau) = \tau^2 + u(\tau), \tag{22}$$

with the boundary conditions

$$x(0) = 0, \quad x(1) = 1, \tag{23}$$

where the exact solution is given by

$$(x(\tau), u(\tau)) = (\tau(2-\tau), -\tau^2 + 2\tau + 2).$$

Consider the solution of the optimization problem in Equations (21)–(23) written as follows

$$x(\tau) = \tau(1 + c_2 - c_2\tau). \tag{24}$$

This satisfies the boundary conditions in Equation (23). Then, we have

$$u(\tau) = -\tau^2 + 2c_2 t + 3c_2 - 1. \tag{25}$$

Substitute Equations (24) and (25) in minimizing the problem in Equation (21) to obtain the optimal values of $c_2 = 0.989$ *and* $T = 0.997$. *Hence, the approximation of the problem is written as* $x(\tau) = \tau(2-\tau)$ *and then, it is plotted in Figure 1b.*

Example 3. *Let us consider the following optimal control problem of fractional order with non-free terminal time which was introduced by [25]*

$$\min_{x(\tau),u(\tau),T} J(\tau, x(\tau), u(\tau), T) = \min_{x(\tau),u(\tau),T} \int_0^T (\tau u(\tau) - (\gamma+2)x(\tau))^2 d\tau, \tag{26}$$

subject to the control system

$$D_\tau^\gamma x(\tau) + \dot{x}(\tau) = \tau^2 + u(\tau), \tag{27}$$

and the boundary conditions

$$x(0) = 0, \quad x(T) = 1. \tag{28}$$

Consider the solution of the optimization problem in Equations (26)–(28) written as follows:

$$x(\tau) = c_2\tau^2 + (\frac{1}{T} - c_2 T)\tau, \tag{29}$$

which satisfied the boundary conditions in Equation (28). Then, we have

$$u(\tau) = -\tau^2 + 2c_2\tau + 3c_2 - 1. \tag{30}$$

Substitute Equations (29) and (30) in minimizing the problem in Equation (26) to obtain the optimal solutions using the Hooke and Jeeves method. Hence, the approximation of the problem is plotted in Figure 1c.

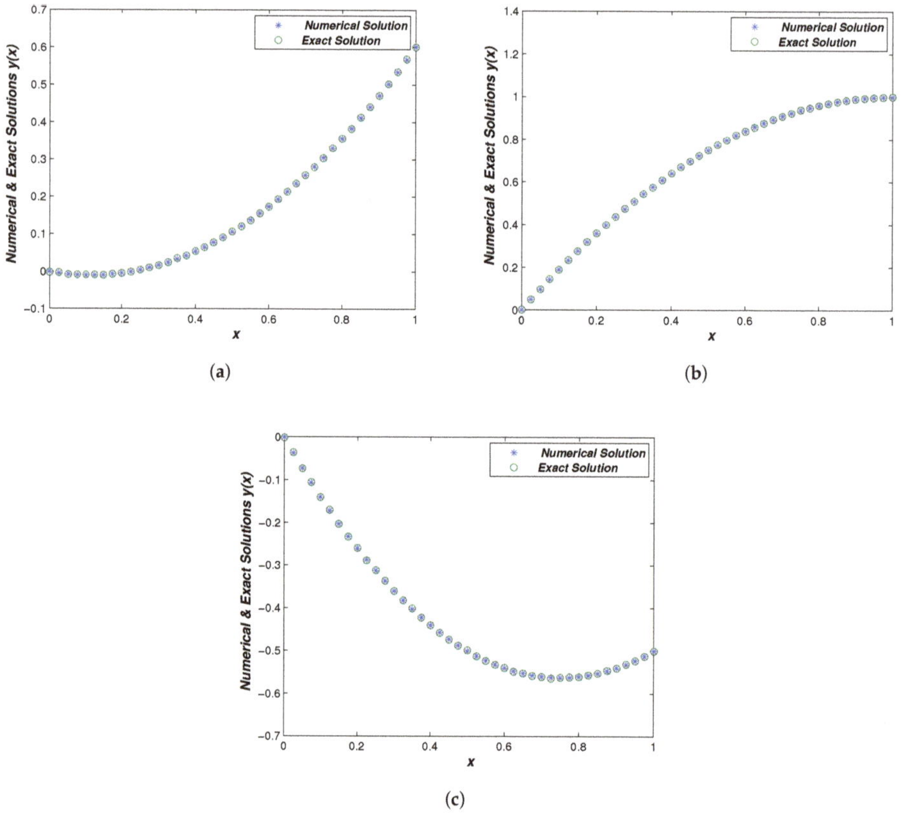

Figure 1. A Comparison of Numerical Solutions of (FrOCDS) Evaluated by the Hooke and Jeeves Direct Search Method for the Implementations in (**a**) Example 1, (**b**) Example 2, and (**c**) Example 3.

6. Discussion and Conclusions

The main purpose of this study is to introduce numerical methods for solving two cases of fractional-order optimal dynamical control systems with free and non-free terminal time. The study also offers a comparison of the numerical results obtained by using the proposed method with the exact solutions for three test problems. From the numerical results of the proposed method, we observe that the method is applicable to a class of (FOC) problems with free or non-free terminal time. Moreover, the proposed method achieves good agreement with exact solutions. As a result, the new method is efficient and provides encouraging results. This direction of this research can be extended in the future to new directions such as improving the numerical studies of stochastic optimal control problems to the continuity of the research in this domain.

Author Contributions: Literature Review, O.I.A.-S.; Investigation, O.I.A.-S.; Methodology, O.I.A.-S., and M. M.; Project administration, M.M.S.; Resources, O.I.A.-S.; Software, M.S.M.; Supervision, M.S.M. and M.M.; Validation, M.S.M., and M.M.; Writing—original draft, O.I.A.-S.; Writing—review and editing, M.S.M. and M.M. All authors have read and agreed to the published version of the manuscript.

Funding: This research received no external funding.

Institutional Review Board Statement: Not applicable.

Informed Consent Statement: Not applicable.

Data Availability Statement: Not applicable.

Conflicts of Interest: The authors declare no conflict of interest.

References

1. Longuski, J.M.; Guzmán, J.J.; Prussing, J.E. *Optimal Control with Aerospace Applications*; Springer:Cham, Switzerland, 2014.
2. Agrawal, O.P. A formulation and numerical scheme for fractional optimal control problems. *J. Vib. Control.* **2008**, *14*, 1291–1299. [CrossRef]
3. Sweilam, N.; Hassan, A.-A.; Tamer, M.; Hoppe, R.H.W. Numerical solution of some types of fractional optimal control problems. *Sci. World J.* **2013**, *2013*, 306237. [CrossRef] [PubMed]
4. Bhrawy, A.H.; Doha, E.H.; Baleanu, D.; Ezz-Eldien, S.S.; Abdelkawy, M.A. An accurate numerical technique for solving fractional optimal control problems. *Differ. Equ.* **2015**, *15*, 23.
5. Akbarian, T.; Keyanpour, M. A new approach to the numerical solution of fractional order optimal control problems. *Appl. Appl. Math.* **2013**, *8*, 523–534.
6. Bhrawy, A.H.; Doha, E.H.; Tenreiro Machado, J.A.; Ezz-Eldien, S.S. An efficient numerical scheme for solving multi-dimensional fractional optimal control problems with a quadratic performance index. *Asian J. Control* **2015**, *17*, 2389–2402. [CrossRef]
7. Doha, E.H.; Bhrawy, A.H.; Ezz-Eldien, S.S. An efficient Legendre spectral tau matrix formulation for solving fractional subdiffusion and reaction subdiffusion equations. *J. Comput. Nonlinear Dyn.* **2015**, *10*, 021019. [CrossRef]
8. Liu, J.; Hou, G. Numerical solutions of the space-and time-fractional coupled Burgers equations by generalized differential transform method. *Appl. Math. Comput.* **2011**, *217*, 7001–7008. [CrossRef]
9. Lotfi, A.; Dehghan, M.; Yousefi, S.A. A numerical technique for solving fractional optimal control problems. *Comput. Math. Appl.* **2011**, *62*, 1055–1067. [CrossRef]
10. Pooseh, S.; Almeida, R.; Torres, D.F.M. A numerical scheme to solve fractional optimal control problems. *Conf. Pap. Sci.* **2013**, *2013*, 165298. [CrossRef]
11. Zhao, Z.; Li, C. Fractional difference/finite element approximations for the time–space fractional telegraph equation. *Appl. Math. Comput.* **2012**, *219*, 2975–2988. [CrossRef]
12. Mechee, M.S.; Senu, N. *Numerical Study of Fractional Differential Equations of Lane-Emden Type by Method of Collocation*; Scientific Research Publishing: Wuhan, China, 2012.
13. Zhou, H.; Tian, W.Y.; Deng, W. Quasi-compact finite difference schemes for space fractional diffusion equations. *J. Sci. Comput.* **2013**, *56*, 45–66. [CrossRef]
14. Bhrawy, A.H.; Doha, E.H.; Ezz-Eldien, S.S.; Van Gorder, R.A. A new Jacobi spectral collocation method for solving 1 + 1 fractional Schrödinger equations and fractional coupled Schrödinger systems. *Eur. Phys. J. Plus* **2014**, *129*, 1–21. [CrossRef]
15. Bhrawy, A.H.; Ezz-Eldien, S.S.; Doha, E.H.; Abdelkawy, M.A.; Baleanu, D. Solving fractional optimal control problems within a Chebyshev–Legendre operational technique. *Int. J. Control* **2017**, *90*, 1230–1244. [CrossRef]
16. Yousefi, S.A.; Lotfi, A.; Dehghan, M. The use of a Legendre multiwavelet collocation method for solving the fractional optimal control problems. *J. Vib. Control.* **2011**, *17*, 2059–2065. [CrossRef]

17. Bhrawy, A.H.; Ezz-Eldien, S.S. A new Legendre operational technique for delay fractional optimal control problems. *Calcolo* **2016**, *53*, 521–543. [CrossRef]
18. Heydari, M.H.; Hooshmandasl, M.R.; Mohammadi, F. Legendre wavelets method for solving fractional partial differential equations with Dirichlet boundary conditions. *Appl. Math. Comput.* **2014**, *234*, 267–276. [CrossRef]
19. Khan, R.A.; Khalil, H. A new method based on legendre polynomials for solution of system of fractional order partial differential equations. *Int. J. Comput. Math.* **2014**, *91*, 2554–2567. [CrossRef]
20. Sweilam, N.H.; Al-Ajami, T.M. Legendre spectral-collocation method for solving some types of fractional optimal control problems. *J. Adv. Res.* **2015**, *6*, 393–403. [CrossRef]
21. Sweilam, N.H.; Nagy, A.M.; El-Sayed, A.A. Second kind shifted Chebyshev polynomials for solving space fractional order diffusion equation. *Chaos Solitons Fractals* **2015**, *73*, 141–147. [CrossRef]
22. Almeida, R.; Torres, D.F.M. A discrete method to solve fractional optimal control problems. *Nonlinear Dyn.* **2015**, *80*, 1811–1816. [CrossRef]
23. Suárez, J.I.; Vinagre, B.M.; Chen, Y. A fractional adaptation scheme for lateral control of an AGV. *J. Vib. Control* **2008**, *14*, 1499–1511. [CrossRef]
24. Pinto, C.; Tenreiro Machado, J.A. Fractional dynamics of computer virus propagation. *Math. Probl. Eng.* **2014**, *2014*, 476502. [CrossRef]
25. Pooseh, S.; Almeida, R.; Torres, D.F.M. Fractional order optimal control problems with free terminal time. *arXiv* **2013**, arXiv:1302.1717.
26. Jafari, H.; Tajadodi, H. Fractional order optimal control problems via the operational matrices of Bernstein polynomials. *UPB Sci. Bull.* **2014**, *76*, 115–128.
27. Jesus, I.S.; Tenreiro, M.J.A. Fractional control of heat diffusion systems. *Nonlinear Dyn.* **2008**, *54*, 263–282. [CrossRef]
28. Ahmad, W.M.; El-Khazali, R. Fractional-order dynamical models of love. *Chaos Solitons Fractals* **2007**, *33*, 1367–1375. [CrossRef]
29. David, S.A.; Linares, J.L.; Pallone, E.M.J.A. Fractional order calculus: Historical apologia, basic concepts and some applications. *Rev. Bras. Ensino Física* **2011**, *33*, 4302. [CrossRef]
30. Bohannan, G.W. Analog fractional order controller in temperature and motor control applications. *J. Vib. Control.* **2008**, *14*, 1487–1498. [CrossRef]
31. Joshi, H.J.; Brajesh, K. 2D dynamic analysis of the disturbances in the calcium neuronal model and its implications in neurodegenerative disease. *Cogn. Neurodynam.* **2022**, 1–12. [CrossRef]
32. Zhang, S.; Zhang, H.-Q. Fractional sub-equation method and its applications to nonlinear fractional PDEs. *Phys. Lett. A* **2011**, *375*, 1069–1073. [CrossRef]
33. Kreyszig, E. *Introductory Functional Analysis with Applications*; John Wiley & Sons: Hoboken, NJ, USA, 1991; Volume 17.
34. Zhao, Y.; Li, Y.; Liu, H. Fractional-order Iterative Learning Control with Nonuniform Trial Lengths. *Int. J. Control. Autom. Syst.* 2022, *20*, 3167–3176. [CrossRef]
35. Matt, C.; Hess, T.; Benlian, A. Digital transformation strategies. *Bus. Inf. Syst. Eng.* **2015**, *57*, 339–343. [CrossRef]
36. Zhao, Y.; Li, Y.; Zhang, F.; Liu, H. Iterative learning control of fractional-order linear systems with nonuniform pass lengths. *Trans. Inst. Meas. Control.* **2022**, *44*, 3071–3080. [CrossRef]
37. Sabermahani, S.; Ordokhani, Y. Fibonacci wavelets and Galerkin method to investigate fractional optimal control problems with bibliometric analysis. *J. Vib. Control.* **2021**, *27*, 1778–1792. [CrossRef]

symmetry

MDPI

Article

Fractional Weighted Midpoint-Type Inequalities for *s*-Convex Functions

Nassima Nasri [1], Fatima Aissaoui [2], Keltoum Bouhali [1,3], Assia Frioui [2], Badreddine Meftah [2], Khaled Zennir [3] and Taha Radwan [3,4,*]

[1] Departement de Mathématiques, Université 20 Août 1955, Skikda Bp 26 Route El-Hadaiek, Skikda 21000, Algeria
[2] Département de Mathématiques, Faculté des Mathématiques, de l'informatique et des Sciences de la Matière, Université 8 Mai 1945 Guelma, Guelma 24000, Algeria
[3] Department of Mathematics, College of Science and Arts, Qassim University, Ar-Rass 51452, Saudi Arabia
[4] Department of Mathematics and Statistics, Faculty of Management Technology and Information Systems, Port Said University, Port Said 42511, Egypt
* Correspondence: t.radwan@qu.edu.sa

Abstract: In the present paper, we first prove a new integral identity. Using that identity, we establish some fractional weighted midpoint-type inequalities for functions whose first derivatives are extended s-convex. Some special cases are discussed. Finally, to prove the effectiveness of our main results, we provide some applications to numerical integration as well as special means.

Keywords: fractional derivatives; weighted integral; midpoint formula; integral inequalities; s-convex functions

Citation: Nasri, N.; Aissaoui, F.; Bouhali, K.; Frioui, A.; Meftah, B.; Zennir, K.; Radwan, T. Fractional Weighted Midpoint-Type Inequalities for *s*-Convex Functions. *Symmetry* **2023**, *15*, 612. https://doi.org/10.3390/sym15030612

Academic Editors: Sergei D. Odintsov, Junesang Choi, Francisco Martínez González and Mohammed K. A. Kaabar

Received: 1 January 2023
Revised: 7 February 2023
Accepted: 23 February 2023
Published: 28 February 2023

1. Introduction

It is well known that convexity is one of the most fundamental principles of analysis that is widely used in several fields of pure and applied sciences. Especially, in the classical theory of optimization where convexity causes it to be possible to obtain necessary and sufficient global optimality conditions; in consumer theory in economics, information theory as well as in the field of inequalities where the relationship is closely linked. For papers related to convexity and integral inequalities we refer readers to [1–5].

A real function defined on E is called convex; if for all $x, z \in E$ and all $a \in [0,1]$, we have

$$g(ax + (1-a)z) \leq ag(x) + (1-a)g(z).$$

We note that all convex function on a finite interval, and $[\varrho, \varpi]$ must satisfy the so called Hermite–Hadamard inequality (see [6]).

$$g\left(\frac{\varrho+\varpi}{2}\right) \leq \frac{1}{\varpi-\varrho}\int_{\varrho}^{\varpi} g(x)dx \leq \frac{g(\varrho)+g(\varpi)}{2}. \tag{1}$$

Inequality (1) can be seen as a second definition of convex functions equivalent to the first one for continuous function (see [7]); it is a character of which all convex functions must satisfy at least the left- or right-hand side.

Pearce and Pečarić [8] introduced the following inequality connected with (1)

$$\left|g\left(\frac{\varrho+\varpi}{2}\right) - \frac{1}{\varpi-\varrho}\int_{\varrho}^{\varpi} g(x)dx\right| \leq \frac{\varpi-\varrho}{4}\left(\frac{|g'(\varrho)|^q + |g'(\varpi)|^q}{2}\right)^{\frac{1}{q}},$$

where $q \geq 1$.

Kirmaci [9] proved that, for all function f such that $|g'|$ or $|g'|^q$ are convex, the following inequalities hold:

$$\left|g\left(\frac{\varrho+\varpi}{2}\right)-\frac{1}{\varpi-\varrho}\int_{\varrho}^{\varpi}g(x)dx\right|\leq\frac{\varpi-\varrho}{8}(|g'(\varrho)|+|g'(\varpi)|),$$

where $q \geq 1$. Furthermore, they proved the following result

$$\left|g\left(\frac{\varrho+\varpi}{2}\right)-\frac{1}{\varpi-\varrho}\int_{\varrho}^{\varpi}g(x)dx\right|$$
$$\leq\frac{\varpi-\varrho}{16}\left(\frac{4}{p+1}\right)^{\frac{1}{p}}\left(\left(3|g'(\varrho)|^q+|g'(\varpi)|^q\right)^{\frac{1}{q}}+\left(|g'(\varrho)|^q+3|g'(\varpi)|^q\right)^{\frac{1}{q}}\right),$$

where $q,p>1$ with $\frac{1}{p}+\frac{1}{q}=1$.

İşcan et al. [10] showed the following midpoint inequalities for P-functions (see (3) below):

$$\left|g\left(\frac{\varrho+\varpi}{2}\right)-\frac{1}{\varpi-\varrho}\int_{\varrho}^{\varpi}g(x)dx\right|\leq\frac{\varpi-\varrho}{4}\left(|g'(\varrho)|^c+|g'(\varpi)|^c\right)^{\frac{1}{c}},$$

where $c \geq 1$. Furthermore, they proved the following result:

$$\left|g\left(\frac{\varrho+\varpi}{2}\right)-\frac{1}{\varpi-\varrho}\int_{\varrho}^{\varpi}g(x)dx\right|$$
$$\leq\frac{\varpi-\varrho}{4}\left(\frac{1}{b+1}\right)^{\frac{1}{b}}\left(\left(|g'(\varrho)|^c+\left|g'\left(\frac{\varrho+\varpi}{2}\right)\right|^c\right)^{\frac{1}{c}}+\left(\left|g'\left(\frac{\varrho+\varpi}{2}\right)\right|^c+|g'(\varpi)|^c\right)^{\frac{1}{c}}\right)$$
$$\leq\frac{\varpi-\varrho}{4}\left(\frac{1}{b+1}\right)^{\frac{1}{b}}\left(\left(2|g'(\varrho)|^c+|g'(\varpi)|^c\right)^{\frac{1}{c}}+\left(|g'(\varrho)|^c+2|g'(\varpi)|^c\right)^{\frac{1}{c}}\right),$$

where $c,b>1$ with $\frac{1}{c}+\frac{1}{b}=1$.

Nowadays, fractional calculus has become a popular implement for scientists. It has been successfully used in various fields of science and engineering see [11–18]. Its main strength in the description of memory and genetic properties of different materials and processes has aroused great interest for researchers in different domains. This innovative idea of fractional calculus has attracted many researchers in recent years, several generalizations, extensions, refinements, and finding a counterpart have appeared (see [19–26]).

In [6], Sarikaya and Yildirim established the analogue fractional of inequality (1) as follows:

$$g\left(\frac{\varrho+\varpi}{2}\right)\leq\frac{2^{\alpha-1}\Gamma(\alpha+1)}{(\varpi-\varrho)^{\alpha}}\left(J^{\alpha}_{(\frac{\varrho+\varpi}{2})+}g(\varpi)+J^{\alpha}_{(\frac{\varrho+\varpi}{2})-}g(\varrho)\right)\leq\frac{g(\varrho)+g(\varpi)}{2}.$$

Furthermore, the authors investigate the following fractional midpoint inequalities for convex-first derivatives

$$\left|\frac{2^{\alpha-1}\Gamma(\alpha+1)}{(\varpi-\varrho)^{\alpha}}\left(J^{\alpha}_{(\frac{\varrho+\varpi}{2})+}g(\varpi)+J^{\alpha}_{(\frac{\varrho+\varpi}{2})-}g(\varrho)\right)-g\left(\frac{\varrho+\varpi}{2}\right)\right|$$
$$\leq\frac{\varpi-\varrho}{4(\alpha+1)}\left(\left(\frac{(\alpha+1)|g'(\varrho)|^q+(\alpha+3)|g'(\varpi)|^q}{2(\alpha+2)}\right)^{\frac{1}{q}}\right.$$

$$+\left(\frac{(\alpha+3)|g'(\varrho)|^q+(\alpha+1)|g'(\varpi)|^q}{2(\alpha+2)}\right)^{\frac{1}{q}}\Bigg)$$

and

$$\left|\frac{2^{\alpha-1}\Gamma(\alpha+1)}{(\varpi-\varrho)^\alpha}\left(J^\alpha_{(\frac{\varrho+\varpi}{2})+}g(\varpi)+J^\alpha_{(\frac{\varrho+\varpi}{2})-}g(\varrho)\right)-g\left(\frac{\varrho+\varpi}{2}\right)\right|$$
$$\leq\frac{\varpi-\varrho}{4}\left(\frac{1}{\alpha p+1}\right)^{\frac{1}{p}}\left(\left(\frac{|g'(\varrho)|^q+3|g'(\varpi)|^q}{4}\right)^{\frac{1}{q}}+\left(\frac{3|g'(\varrho)|^q+|g'(\varpi)|^q}{4}\right)^{\frac{1}{q}}\right)$$
$$\leq\frac{\varpi-\varrho}{4}\left(\frac{4}{\alpha p+1}\right)^{\frac{1}{p}}(|g'(\varrho)|+|g'(\varpi)|),$$

where $\alpha>0, p,q>1$ with $\frac{1}{p}+\frac{1}{q}=1, \Gamma$ is the gamma function and $J^\alpha_{(\frac{\varrho+\varpi}{2})+}$ and $J^\alpha_{(\frac{\varrho+\varpi}{2})-}$ are the Riemann–Liouville integrals (see Definition 1 below).

Motivated by the above results, here, we first prove a new integral identity and, then, by using this identity, we establish some fractional weighted midpoint-type inequalities for functions that the first derivatives are extended s-convex functions. We also derive some known results and, state applications in numerical integration and in special means are presented to prove the effectiveness of our main results.

The paper is organized as follows. In the next section, we provide some auxiliary results as a preliminaries. In Section 3, we provide the main results and proofs. In Section 4, we will provide an applications of our analysis to illustrate our main results. In Section 5, we conclude our work.

2. Preliminaries

In this section, we recall certain notions concerning special functions, some classes of convex functions, and the Riemann–Liouville integral operator.

A non-negative function $g : E \subset [0,\infty) \to \mathbb{R}$ is said to be s-convex in the second sense for some fixed $s \in (0,1]$, if

$$g(ax+(1-a)z)\ \leq a^s g(x)+(1-a)^s g(z), \tag{2}$$

holds for all $x,z \in E$ and $a \in [0,1]$.

Whereas, a non-negative function $g : E \to \mathbb{R}$ is said to be P-convex; if for all $x,z \in E$ and all $a \in (0,1)$, we have

$$g(ax+(1-a)z) \leq g(x)+g(z). \tag{3}$$

A non-negative function $g : E \to \mathbb{R}$ is said to be s-Godunova–Levin function, where $s \in [0,1]$; if for all $x,z \in E$, and all $a \in (0,1)$, we have

$$g(ax+(1-a)z) \leq \frac{g(x)}{a^s}+\frac{g(z)}{(1-a)^s}. \tag{4}$$

A non-negative function $g : E \subset [0,\infty) \to \mathbb{R}$ is said to be an extended s-convex for some fixed $s \in [-1,1]$; if for all $x,z \in E$ and all $a \in (0,1)$, we have

$$g(ax+(1-a)z)\ \leq a^s g(x)+(1-a)^s g(z). \tag{5}$$

Definition 1 ([12]). *Let $\Omega \in L^1[\varrho,\varpi]$. The Riemann–Liouville integrals $J^\alpha_{\varrho+}\Omega$ and $J^\alpha_{\varpi-}\Omega$ of order $\alpha>0$ with $\varpi>\varrho\geq 0$ are defined by*

$$J^\alpha_{\varrho+}\Omega(d) = \frac{1}{\Gamma(\alpha)}\int_\varrho^d (d-a)^{\alpha-1}\Omega(a)da, \quad d>\varrho,$$

$$J^{\alpha}_{\varpi^-}\Omega(d) = \frac{1}{\Gamma(\alpha)}\int_{d}^{\varpi}(a-d)^{\alpha-1}\Omega(a)da, \quad \varpi > d,$$

respectively, where

$$\Gamma(\alpha) = \int_{0}^{\infty} e^{-a}a^{\alpha-1}da,$$

and $J^{0}_{\varrho^+}\Omega(d) = J^{0}_{\varpi^-}\Omega(d) = \Omega(d)$.

For any complex numbers k, l such that $Re(k) > 0$ and $Re(l) > 0$. The beta function is provided by

$$B(k,l) = \int_{0}^{1} a^{k-1}(1-a)^{l-1}da = \frac{\Gamma(k)\Gamma(l)}{\Gamma(k+l)}.$$

3. Main Results and Proofs

To prepare the proofs of our main results, we will need the following Lemma.

Lemma 1. *Let $g : E = [\varrho, \varpi] \to \mathbb{R}$ be a differentiable map on I° (I° is the interior of I), with $\varrho < \varpi$, and let $w : [\varrho, \varpi] \to \mathbb{R}$ be symmetric as regards $\frac{\varrho+\varpi}{2}$. If $g, w \in L[\varrho, \varpi]$, then*

$$L^{\alpha}[w]g\left(\frac{\varrho+\varpi}{2}\right) - L^{\alpha}[wg]$$
$$= \frac{(\varpi-\varrho)^2}{4}\left(\int_{0}^{1} p_1(a)g'\left(a\varrho + (1-a)\frac{\varrho+\varpi}{2}\right)da - \int_{0}^{1} p_2(a)g'\left((1-a)\frac{\varrho+\varpi}{2} + a\varpi\right)da\right).$$

where

$$p_1(a) = \int_{a}^{1}(1-b)^{\alpha-1}w\left(b\varrho + (1-b)\frac{\varrho+\varpi}{2}\right)db, \tag{6}$$

$$p_2(a) = \int_{a}^{1}(1-b)^{\alpha-1}w\left(b\varpi + (1-b)\frac{\varrho+\varpi}{2}\right)db, \tag{7}$$

and

$$L^{\alpha}[g] = \left(\frac{2}{\varpi-\varrho}\right)^{\alpha-1}\Gamma(\alpha)\left(J^{\alpha}_{\left(\frac{\varrho+\varpi}{2}\right)^-}g(\varrho) + J^{\alpha}_{\left(\frac{\varrho+\varpi}{2}\right)^+}g(\varpi)\right). \tag{8}$$

Proof. Let

$$I = I_1 - I_2, \tag{9}$$

where

$$I_1 = \int_{0}^{1} p_1(a)g'\left(a\varrho + (1-a)\frac{\varrho+\varpi}{2}\right)da,$$

and

$$I_2 = \int_{0}^{1} p_2(a)g'\left(a\varpi + (1-a)\frac{\varrho+\varpi}{2}\right)da.$$

Integrating by parts I_1, we obtain

$$\int_{0}^{1} p_1(a)g'\left(a\varrho + (1-a)\frac{\varrho+\varpi}{2}\right)da$$

$$
\begin{aligned}
&= \int_0^1 \left[\int_a^1 (1-b)^{\alpha-1} w\left(b\varrho + (1-b)\frac{\varrho+\varpi}{2}\right) db \right] g'\left(a\varrho + (1-a)\frac{\varrho+\varpi}{2}\right) da \\
&= -\frac{2}{\varpi-\varrho}\left[\int_a^1 (1-b)^{\alpha-1} w\left(b\varrho + (1-b)\frac{\varrho+\varpi}{2}\right) db \right] g\left(a\varrho + (1-a)\frac{\varrho+\varpi}{2}\right) \Bigg|_{a=0}^{a=1} \\
&\quad - \frac{2}{\varpi-\varrho}\int_0^1 (1-a)^{\alpha-1} w\left(a\varrho + (1-a)\frac{\varrho+\varpi}{2}\right) g\left(a\varrho + (1-a)\frac{\varrho+\varpi}{2}\right) da \\
&= \frac{2}{\varpi-\varrho}\left[\int_0^1 (1-b)^{\alpha-1} w\left(b\varrho + (1-b)\frac{\varrho+\varpi}{2}\right) db \right] g\left(\frac{\varrho+\varpi}{2}\right) \\
&\quad - \frac{2}{\varpi-\varrho}\int_0^1 (1-a)^{\alpha-1} w\left(a\varrho + (1-a)\frac{\varrho+\varpi}{2}\right) g\left(a\varrho + (1-a)\frac{\varrho+\varpi}{2}\right) da \\
&= \left(\frac{2}{\varpi-\varrho}\right)^{\alpha+1} \left[\int_\varrho^{\frac{\varrho+\varpi}{2}} (u-\varrho)^{\alpha-1} w(u) du \right] g\left(\frac{\varrho+\varpi}{2}\right) \\
&\quad - \left(\frac{2}{\varpi-\varrho}\right)^{\alpha+1} \int_\varrho^{\frac{\varrho+\varpi}{2}} (u-\varrho)^{\alpha-1} w(u) g(u) du \qquad (10) \\
&= \left(\frac{2}{\varpi-\varrho}\right)^{\alpha+1} \Gamma(\alpha)\left(J^\alpha_{\left(\frac{\varrho+\varpi}{2}\right)^-} w(\varrho)\right) g\left(\frac{\varrho+\varpi}{2}\right) - \left(\frac{2}{\varpi-\varrho}\right)^{\alpha+1} \Gamma(\alpha) J^\alpha_{\left(\frac{\varrho+\varpi}{2}\right)^-}(wg)(\varrho).
\end{aligned}
$$

Similarly, we have

$$
\begin{aligned}
&\int_0^1 p_2(a) g'\left(a\varpi + (1-a)\frac{\varrho+\varpi}{2}\right) da \\
&= \int_0^1 \left(\left[\int_a^1 (1-b)^{\alpha-1} w\left(b\varpi + (1-b)\frac{\varrho+\varpi}{2}\right) db \right] g'\left(a\varpi + (1-a)\frac{\varrho+\varpi}{2}\right) da \right. \\
&= \frac{2}{\varpi-\varrho}\left[\int_a^1 (1-b)^{\alpha-1} w\left(b\varpi + (1-b)\frac{\varrho+\varpi}{2}\right) db \right] g\left(a\varpi + (1-a)\frac{\varrho+\varpi}{2}\right) \Bigg|_{a=0}^{a=1} \\
&\quad + \frac{2}{\varpi-\varrho}\int_0^1 (1-a)^{\alpha-1} w\left(a\varpi + (1-a)\frac{\varrho+\varpi}{2}\right) g\left(a\varpi + (1-a)\frac{\varrho+\varpi}{2}\right) da \\
&= -\frac{2}{\varpi-\varrho}\left[\int_0^1 (1-b)^{\alpha-1} w\left(b\varpi + (1-b)\frac{\varrho+\varpi}{2}\right) db \right] g\left(\frac{\varrho+\varpi}{2}\right) \\
&\quad + \frac{2}{\varpi-\varrho}\int_0^1 (1-a)^{\alpha-1} w\left(a\varpi + (1-a)\frac{\varrho+\varpi}{2}\right) g\left(a\varpi + (1-a)\frac{\varrho+\varpi}{2}\right) da \\
&= -\left(\frac{2}{\varpi-\varrho}\right)^{\alpha+1} \left[\int_{\frac{\varrho+\varpi}{2}}^{\varpi} (\varpi-u)^{\alpha-1} w(u) du \right] g\left(\frac{\varrho+\varpi}{2}\right) \\
&\quad + \left(\frac{2}{\varpi-\varrho}\right)^{\alpha+1} \int_{\frac{\varrho+\varpi}{2}}^{\varpi} (\varpi-u)^{\alpha-1} w(u) g(u) du \qquad (11)
\end{aligned}
$$

$$= -\left(\frac{2}{\varpi-\varrho}\right)^{\alpha+1}\Gamma(\alpha)\left(J^{\alpha}_{\left(\frac{\varrho+\varpi}{2}\right)^{+}}w(\varpi)\right)g\left(\frac{\varrho+\varpi}{2}\right)+\left(\frac{2}{\varpi-\varrho}\right)^{\alpha+1}\Gamma(\alpha)J^{\alpha}_{\left(\frac{\varrho+\varpi}{2}\right)^{+}}(wg)(\varpi).$$

Substituting (10) and (11) into (9), then multiplying the resulting equality by $\frac{(\varpi-\varrho)^2}{4}$ and using (8), we obtain the desired result. □

Theorem 1. *Let $g : [\varrho,\varpi] \to \mathbb{R}$ be a differentiable function on (ϱ,ϖ) such that $g' \in L([\varrho,\varpi])$ with $0 \le \varrho < \varpi$, and let $w : [\varrho,\varpi] \to \mathbb{R}$ be a continuous and symmetric function as regards $\frac{\varrho+\varpi}{2}$. If $|g'|$ is an extended s-convex for some fixed $s \in (-1,1]$, then we have*

$$\begin{aligned}&\left|L^{\alpha}[w]g\left(\frac{\varrho+\varpi}{2}\right)-L^{\alpha}[wg]\right|\\ \le&\frac{(\varpi-\varrho)^2}{4\alpha}\|w\|_{[\varrho,\varpi],\infty}\\ &\times\left(\frac{\Gamma(s+1)\Gamma(\alpha+1)|g'(\varrho)|+2\Gamma(s+\alpha+1)\left|g'\left(\frac{\varrho+\varpi}{2}\right)\right|+\Gamma(s+1)\Gamma(\alpha+1)|g'(\varpi)|}{\Gamma(s+\alpha+2)}\right),\end{aligned}$$

where Γ is the gamma function.

Proof. Using Lemma 1, the absolute value and s-convexity of $|g'|$ provide

$$\begin{aligned}&\left|L^{\alpha}[w]g\left(\frac{\varrho+\varpi}{2}\right)-L^{\alpha}[wg]\right|\\ \le&\frac{(\varpi-\varrho)^2}{4}\left(\int_0^1|p_1(a)|\left|g'\left(a\varrho+(1-a)\frac{\varrho+\varpi}{2}\right)\right|da\right.\\ &\left.+\int_0^1|p_2(a)|\left|g'\left(a\varpi+(1-a)\frac{\varrho+\varpi}{2}\right)\right|da\right)\\ \le&\frac{(\varpi-\varrho)^2}{4}\|w\|_{[\varrho,\varpi],\infty}\left(\int_0^1\left(\int_a^1(1-b)^{\alpha-1}db\right)\left|g'\left(a\varrho+(1-a)\frac{\varrho+\varpi}{2}\right)\right|da\right.\\ &\left.+\int_0^1\left(\int_a^1(1-b)^{\alpha-1}db\right)\left|g'\left(a\varpi+(1-a)\frac{\varrho+\varpi}{2}\right)\right|da\right)\\ \le&\frac{(\varpi-\varrho)^2}{4\alpha}\|w\|_{[\varrho,\varpi],\infty}\left(\int_0^1(1-a)^{\alpha}\left(a^s|g'(\varrho)|+(1-a)^s\left|g'\left(\frac{\varrho+\varpi}{2}\right)\right|\right)da\right.\\ &\left.+\int_0^1(1-a)^{\alpha}\left(a^s|g'(\varpi)|+(1-a)^s\left|g'\left(\frac{\varrho+\varpi}{2}\right)\right|\right)da\right)\\ =&\frac{(\varpi-\varrho)^2}{4\alpha}\|w\|_{[\varrho,\varpi],\infty}\\ &\times\left(\frac{\Gamma(s+1)\Gamma(\alpha+1)|g'(\varrho)|+2\Gamma(s+\alpha+1)\left|g'\left(\frac{\varrho+\varpi}{2}\right)\right|+\Gamma(s+1)\Gamma(\alpha+1)|g'(\varpi)|}{\Gamma(s+\alpha+2)}\right).\end{aligned}$$

Then, the proof is now completed. □

Corollary 1. *In Theorem 1, if we use:*

1. $s = 0$, *we obtain*

$$\left|L^{\alpha}[w]g\left(\frac{\varrho+\varpi}{2}\right)-L^{\alpha}[wg]\right|$$
$$\leq\frac{(\varpi-\varrho)^2}{4\alpha(\alpha+1)}\|w\|_{[\varrho,\varpi],\infty}\left(|g'(\varrho)|+2\left|g'\left(\frac{\varrho+\varpi}{2}\right)\right|+|g'(\varpi)|\right).$$

2. $s = 1$, *we obtain*

$$\left|L^{\alpha}[w]g\left(\frac{\varrho+\varpi}{2}\right)-L^{\alpha}[wg]\right|$$
$$\leq\frac{(\varpi-\varrho)^2}{2\alpha(\alpha+1)}\|w\|_{[\varrho,\varpi],\infty}\left(\frac{|g'(\varrho)|+2(\alpha+1)\left|g'\left(\frac{\varrho+\varpi}{2}\right)\right|+|g'(\varpi)|}{2(\alpha+2)}\right).$$

Corollary 2. *In Theorem 1, if we use $\alpha = 1$, we obtain*

$$\left|g\left(\frac{\varrho+\varpi}{2}\right)\int_{\varrho}^{\varpi}w(\mathrm{N})d\mathrm{N}-\int_{\varrho}^{\varpi}w(\mathrm{N})g(\mathrm{N})d\mathrm{N}\right|$$
$$\leq\frac{(\varpi-\varrho)^2}{4(s+1)(s+2)}\|w\|_{[\varrho,\varpi],\infty}\left(|g'(\varrho)|+2(s+1)\left|g'\left(\frac{\varrho+\varpi}{2}\right)\right|+|g'(\varpi)|\right).$$

Remark 1. *In Corollary 2, if we use $s \in (0,1]$, we obtain the first inequality of Corollary 2.2.1 in [27]. Moreover, if we use $s = 0$ and $s = 1$, we obtain Corollary 2 and Corollary 3 in [28] respectively.*

Corollary 3. *In Theorem 1, if we choose:*

1. $w(u) = \frac{1}{\varpi-\varrho}$, *we obtain*

$$\left|g\left(\frac{\varrho+\varpi}{2}\right)-\frac{2^{\alpha-1}}{(\varpi-\varrho)^{\alpha}}\Gamma(\alpha+1)\left(J^{\alpha}_{\left(\frac{\varrho+\varpi}{2}\right)^-}g(\varrho)+J^{\alpha}_{\left(\frac{\varrho+\varpi}{2}\right)^+}g(\varpi)\right)\right|$$
$$\leq\frac{\varpi-\varrho}{4\Gamma(s+\alpha+2)}\left(\Gamma(s+1)\Gamma(\alpha+1)|g'(\varrho)|+2\Gamma(s+\alpha+1)\left|g'\left(\frac{\varrho+\varpi}{2}\right)\right|\right.$$
$$\left.+\Gamma(s+1)\Gamma(\alpha+1)|g'(\varpi)|\right).$$

2. $w(u) = \frac{1}{\varpi-\varrho}$ *and $\alpha = 1$, we obtain*

$$\left|g\left(\frac{\varrho+\varpi}{2}\right)-\frac{1}{\varpi-\varrho}\int_{\varrho}^{\varpi}g(u)du\right|$$
$$\leq\frac{\varpi-\varrho}{4(s+2)(s+1)}\left(|g'(\varrho)|+2(s+1)\left|g'\left(\frac{\varrho+\varpi}{2}\right)\right|+|g'(\varpi)|\right).$$

Corollary 4. *In Theorem 1, using the s-convexity of $|g'|$, i.e.,*

$$\left|g'\left(\frac{\varrho+\varpi}{2}\right)\right|\leq\frac{|g'(\varrho)|+|g'(\varpi)|}{2^{s-1}(1+s)},$$

we obtain

$$\left|L^{\alpha}[w]g\left(\frac{\varrho+\varpi}{2}\right)-L^{\alpha}[wg]\right|$$

$$\leq \frac{(\varpi-\varrho)^2}{4\alpha(1+s)}\|w\|_{[\varrho,\varpi],\infty}\left(\frac{2^{2-s}\Gamma(s+\alpha+1)+\Gamma(s+2)\Gamma(\alpha+1)}{\Gamma(s+\alpha+2)}\right)(|g'(\varrho)|+|g'(\varpi)|).$$

Corollary 5. *In Corollary 4, if we use:*

1. $\alpha = 1$, *we obtain*

$$\left|g\left(\frac{\varrho+\varpi}{2}\right)\int_{\varrho}^{\varpi} w(\mathrm{N})d\mathrm{N} - \int_{\varrho}^{\varpi} w(\mathrm{N})g(\mathrm{N})d\mathrm{N}\right|$$
$$\leq \frac{(2^{2-s}+1)(\varpi-\varrho)^2}{4(1+s)(s+2)}\|w\|_{[\varrho,\varpi],\infty}(|g'(\varrho)|+|g'(\varpi)|).$$

2. $w(u) = \frac{1}{\varpi-\varrho}$, *we obtain*

$$\left|g\left(\frac{\varrho+\varpi}{2}\right) - \frac{2^{\alpha-1}}{(\varpi-\varrho)^{\alpha}}\Gamma(\alpha+1)\left(J^{\alpha}_{\left(\frac{\varrho+\varpi}{2}\right)^-}g(\varrho) + J^{\alpha}_{\left(\frac{\varrho+\varpi}{2}\right)^+}g(\varpi)\right)\right|$$
$$\leq \frac{\varpi-\varrho}{4(1+s)}\left(\frac{2^{2-s}\Gamma(s+\alpha+1)+\Gamma(s+2)\Gamma(\alpha+1)}{\Gamma(s+\alpha+2)}\right)(|g'(\varrho)|+|g'(\varpi)|).$$

3. $w(u) = \frac{1}{\varpi-\varrho}$ *and* $\alpha = 1$, *we obtain*

$$\left|g\left(\frac{\varrho+\varpi}{2}\right) - \frac{1}{\varpi-\varrho}\int_{\varrho}^{\varpi} g(u)du\right| \leq \frac{(2^{2-s}+1)(\varpi-\varrho)}{4(1+s)(s+2)}(|g'(\varrho)|+|g'(\varpi)|).$$

Remark 2. *Corollary 5, the third point will be reduced to Theorem 2.2 in [9] when* $s = 1$.

Theorem 2. *Let* $g : [\varrho,\varpi] \to \mathbb{R}$ *be a differentiable function on* (ϱ,ϖ) *such that* $g' \in L([\varrho,\varpi])$ *with* $0 \leq \varrho < \varpi$, *and let* $w : [\varrho,\varpi] \to \mathbb{R}$ *be a continuous and symmetric function with respect to* $\frac{\varrho+\varpi}{2}$. *If* $|g'|^q$ *is an extended s-convex for some fixed* $s \in (-1,1]$ *and* $q > 1$ *with* $\frac{1}{p}+\frac{1}{q} = 1$, *then we have*

$$\left|L^{\alpha}[w]g\left(\frac{\varrho+\varpi}{2}\right) - L^{\alpha}[wg]\right|$$
$$\leq \frac{(\varpi-\varrho)^2}{4\alpha(p\alpha+1)^{\frac{1}{p}}}\|w\|_{[\varrho,\varpi],\infty}\left(\left(\frac{|g'(\varrho)|^q + \left|g'\left(\frac{\varrho+\varpi}{2}\right)\right|^q}{s+1}\right)^{\frac{1}{q}} + \left(\frac{|g'(\varpi)|^q + \left|g'\left(\frac{\varrho+\varpi}{2}\right)\right|^q}{s+1}\right)^{\frac{1}{q}}\right),$$

where $B(.,.)$ *is the beta function.*

Proof. Using Lemma 1, the absolute value, Hölder's inequality, and s-convexity of $|g'|$, we obtain

$$\left|L^{\alpha}[w]g\left(\frac{\varrho+\varpi}{2}\right) - L^{\alpha}[wg]\right|$$
$$\leq \frac{(\varpi-\varrho)^2}{4}\left(\int_0^1 |p_1(a)|\left|g'\left(a\varrho+(1-a)\frac{\varrho+\varpi}{2}\right)\right|da\right.$$
$$\left.+\int_0^1 |p_2(a)|\left|g'\left(a\varpi+(1-a)\frac{\varrho+\varpi}{2}\right)\right|da\right)$$

$$\leq \frac{(\varpi-\varrho)^2}{4}\left(\left(\int_0^1 |p_1(a)|^p da\right)^{\frac{1}{p}}\left(\int_0^1 \left|g'\left(a\varrho+(1-a)\frac{\varrho+\varpi}{2}\right)\right|^q da\right)^{\frac{1}{q}}\right.$$

$$\left.+\left(\int_0^1 |p_2(a)|^p da\right)^{\frac{1}{p}}\left(\int_0^1 \left|g'\left(a\varpi+(1-a)\frac{\varrho+\varpi}{2}\right)\right|^q da\right)^{\frac{1}{q}}\right)$$

$$\leq \frac{(\varpi-\varrho)^2}{4}\|w\|_{[\varrho,\varpi],\infty}\left(\int_0^1\left(\int_a^1 (1-b)^{\alpha-1}db\right)^p da\right)^{\frac{1}{p}}$$

$$\times\left(\left(\int_0^1\left(a^s|g'(\varrho)|^q+(1-a)^s\left|g'\left(\frac{\varrho+\varpi}{2}\right)\right|^q\right)da\right)^{\frac{1}{q}}\right.$$

$$\left.+\left(\int_0^1\left(a^s|g'(\varpi)|^q+(1-a)^s\left|g'\left(\frac{\varrho+\varpi}{2}\right)\right|^q\right)da\right)^{\frac{1}{q}}\right)$$

$$\leq \frac{(\varpi-\varrho)^2}{4\alpha}\|w\|_{[\varrho,\varpi],\infty}\left(\int_0^1 (1-a)^{p\alpha}da\right)^{\frac{1}{p}}$$

$$\times\left(\left(\frac{|g'(\varrho)|^q+\left|g'\left(\frac{\varrho+\varpi}{2}\right)\right|^q}{s+1}\right)^{\frac{1}{q}}+\left(\frac{|g'(\varpi)|^q+\left|g'\left(\frac{\varrho+\varpi}{2}\right)\right|^q}{s+1}\right)^{\frac{1}{q}}\right)$$

$$=\frac{(\varpi-\varrho)^2}{4\alpha(p\alpha+1)^{\frac{1}{p}}}\|w\|_{[\varrho,\varpi],\infty}\left(\left(\frac{|g'(\varrho)|^q+\left|g'\left(\frac{\varrho+\varpi}{2}\right)\right|^q}{s+1}\right)^{\frac{1}{q}}+\left(\frac{|g'(\varpi)|^q+\left|g'\left(\frac{\varrho+\varpi}{2}\right)\right|^q}{s+1}\right)^{\frac{1}{q}}\right).$$

The proof is now finished. □

Corollary 6. *In Theorem 2, if we use:*

1. $s=0$, *we obtain*

$$\left|L^\alpha[w]g\left(\frac{\varrho+\varpi}{2}\right)-L^\alpha[wg]\right|$$
$$\leq\frac{(\varpi-\varrho)^2}{4\alpha(p\alpha+1)^{\frac{1}{p}}}\|w\|_{[\varrho,\varpi],\infty}$$
$$\times\left(\left(|g'(\varrho)|^q+\left|g'\left(\frac{\varrho+\varpi}{2}\right)\right|^q\right)^{\frac{1}{q}}+\left(|g'(\varpi)|^q+\left|g'\left(\frac{\varrho+\varpi}{2}\right)\right|^q\right)^{\frac{1}{q}}\right).$$

2. $s=1$, *we obtain*

$$\left|L^\alpha[w]g\left(\frac{\varrho+\varpi}{2}\right)-L^\alpha[wg]\right|$$
$$\leq\frac{(\varpi-\varrho)^2}{4\alpha(p\alpha+1)^{\frac{1}{p}}}\|w\|_{[\varrho,\varpi],\infty}$$

$$\times\left(\left(\frac{|g'(\varrho)|^q+\left|g'\left(\frac{\varrho+\varpi}{2}\right)\right|^q}{2}\right)^{\frac{1}{q}}+\left(\frac{|g'(\varpi)|^q+\left|g'\left(\frac{\varrho+\varpi}{2}\right)\right|^q}{2}\right)^{\frac{1}{q}}\right).$$

Corollary 7. *In Theorem 2, if we use $\alpha=1$, we obtain*

$$\left|g\left(\frac{\varrho+\varpi}{2}\right)\int_{\varrho}^{\varpi}w(\mathrm{N})d\mathrm{N}-\int_{\varrho}^{\varpi}w(\mathrm{N})g(\mathrm{N})d\mathrm{N}\right|$$

$$\leq\frac{(\varpi-\varrho)^2}{4(p+1)^{\frac{1}{p}}}\|w\|_{[\varrho,\varpi],\infty}\left(\left(\frac{|g'(\varrho)|^q+\left|g'\left(\frac{\varrho+\varpi}{2}\right)\right|^q}{s+1}\right)^{\frac{1}{q}}+\left(\frac{|g'(\varpi)|^q+\left|g'\left(\frac{\varrho+\varpi}{2}\right)\right|^q}{s+1}\right)^{\frac{1}{q}}\right).$$

Remark 3. *In Corollary 7, if we assume that $s\in(0,1]$, we obtain Theorem 2.4 in [27]. Moreover, if we use $s=1$, we obtain Corollary 7 in [28], respectively.*

Corollary 8. *In Theorem 2, if we choose*

1. $w(u)=\frac{1}{\varpi-\varrho}$, *we obtain*

$$\left|g\left(\frac{\varrho+\varpi}{2}\right)-\frac{2^{\alpha-1}}{(\varpi-\varrho)^{\alpha}}\Gamma(\alpha+1)\left(J^{\alpha}_{\left(\frac{\varrho+\varpi}{2}\right)^-}g(\varrho)+J^{\alpha}_{\left(\frac{\varrho+\varpi}{2}\right)^+}g(\varpi)\right)\right|$$

$$\leq\frac{\varpi-\varrho}{4(p\alpha+1)^{\frac{1}{p}}}\left(\left(\frac{|g'(\varrho)|^q+\left|g'\left(\frac{\varrho+\varpi}{2}\right)\right|^q}{s+1}\right)^{\frac{1}{q}}+\left(\frac{|g'(\varpi)|^q+\left|g'\left(\frac{\varrho+\varpi}{2}\right)\right|^q}{s+1}\right)^{\frac{1}{q}}\right).$$

2. $w(u)=\frac{1}{\varpi-\varrho}$ *and $\alpha=1$, we obtain*

$$\left|g\left(\frac{\varrho+\varpi}{2}\right)-\frac{1}{\varpi-\varrho}\int_{\varrho}^{\varpi}g(u)du\right|$$

$$\leq\frac{\varpi-\varrho}{4(p+1)^{\frac{1}{p}}}\left(\left(\frac{|g'(\varrho)|^q+\left|g'\left(\frac{\varrho+\varpi}{2}\right)\right|^q}{s+1}\right)^{\frac{1}{q}}+\left(\frac{|g'(\varpi)|^q+\left|g'\left(\frac{\varrho+\varpi}{2}\right)\right|^q}{s+1}\right)^{\frac{1}{q}}\right).$$

Remark 4. *Corollary 8, the second point will be reduced to Corollary 6 in [10] when $s=0$.*

Corollary 9. *In Theorem 2, using the s-convexity of $|g'|^q$, i.e.,*

$$\left|g'\left(\frac{\varrho+\varpi}{2}\right)\right|^q\leq\frac{|g'(\varrho)|^q+|g'(\varpi)|^q}{2^{s-1}(1+s)},$$

we obtain

$$\left|L^{\alpha}[w]g\left(\frac{\varrho+\varpi}{2}\right)-L^{\alpha}[wg]\right|$$

$$\leq\frac{(\varpi-\varrho)^2}{4\alpha(p\alpha+1)^{\frac{1}{p}}}\|w\|_{[\varrho,\varpi],\infty}\left(\left(\frac{(1+s+2^{1-s})|g'(\varrho)|^q+2^{1-s}|g'(\varpi)|^q}{(1+s)^2}\right)^{\frac{1}{q}}\right.$$

$$+\left(\frac{2^{1-s}|g'(\varrho)|^q+(1+s+2^{1-s})|g'(\varpi)|^q}{(1+s)^2}\right)^{\frac{1}{q}}\Bigg).$$

Corollary 10. *In Corollary 9:*

1. *If we use $\alpha=1$, we obtain*

$$\left|g\left(\frac{\varrho+\varpi}{2}\right)\int_{\varrho}^{\varpi}w(N)dN-\int_{\varrho}^{\varpi}w(N)g(N)dN\right|$$
$$\leq\frac{(\varpi-\varrho)^2}{4(p+1)^{\frac{1}{p}}}\|w\|_{[\varrho,\varpi],\infty}\Bigg(\left(\frac{(1+s+2^{1-s})|g'(\varrho)|^q+2^{1-s}|g'(\varpi)|^q}{(1+s)^2}\right)^{\frac{1}{q}}$$
$$+\left(\frac{2^{1-s}|g'(\varrho)|^q+(1+s+2^{1-s})|g'(\varpi)|^q}{(1+s)^2}\right)^{\frac{1}{q}}\Bigg).$$

2. *If we choose $w(u)=\frac{1}{\varpi-\varrho}$, we obtain*

$$\left|g\left(\frac{\varrho+\varpi}{2}\right)-\frac{2^{\alpha-1}}{(\varpi-\varrho)^{\alpha}}\Gamma(\alpha+1)\left(J^{\alpha}_{\left(\frac{\varrho+\varpi}{2}\right)^-}g(\varrho)+J^{\alpha}_{\left(\frac{\varrho+\varpi}{2}\right)^+}g(\varpi)\right)\right|$$
$$\leq\frac{\varpi-\varrho}{4(p\alpha+1)^{\frac{1}{p}}}\Bigg(\left(\frac{(1+s+2^{1-s})|g'(\varrho)|^q+2^{1-s}|g'(\varpi)|^q}{(1+s)^2}\right)^{\frac{1}{q}}$$
$$+\left(\frac{2^{1-s}|g'(\varrho)|^q+(1+s+2^{1-s})|g'(\varpi)|^q}{(1+s)^2}\right)^{\frac{1}{q}}\Bigg).$$

3. *If we choose $w(u)=\frac{1}{\varpi-\varrho}$ and $\alpha=1$, we obtain*

$$\left|g\left(\frac{\varrho+\varpi}{2}\right)-\frac{1}{\varpi-\varrho}\int_{\varrho}^{\varpi}g(u)du\right|$$
$$\leq\frac{\varpi-\varrho}{4(p+1)^{\frac{1}{p}}}\Bigg(\left(\frac{(1+s+2^{1-s})|g'(\varrho)|^q+2^{1-s}|g'(\varpi)|^q}{(1+s)^2}\right)^{\frac{1}{q}}$$
$$+\left(\frac{2^{1-s}|g'(\varrho)|^q+(1+s+2^{1-s})|g'(\varpi)|^q}{(1+s)^2}\right)^{\frac{1}{q}}\Bigg).$$

Remark 5.

1. *Corollary 10, the first point will be reduced to Corollary 2.3 in [9] when $s=1$.*
2. *The second point of Corollary 10 will be reduced to Theorem 6 in [6] when $s=1$.*
3. *Corollary 10, the third point will be reduced to Theorem 2.3 in [9] when $s=1$.*

Corollary 11. *In Corollary 9, if we use the discrete power mean inequality, we obtain*

$$\left|L^{\alpha}[w]g\left(\frac{\varrho+\varpi}{2}\right)-L^{\alpha}[wg]\right|$$
$$\leq\frac{(\varpi-\varrho)^2}{2\alpha(p\alpha+1)^{\frac{1}{p}}}\|w\|_{[\varrho,\varpi],\infty}\left(\frac{1+s+2^{2-s}}{(1+s)^2}\right)^{\frac{1}{q}}\left(\frac{|g'(\varrho)|^q+|g'(\varpi)|^q}{2}\right)^{\frac{1}{q}}.$$

Corollary 12. *In Corollary 11:*

1. *If we use $\alpha = 1$, we obtain*

$$\left| g\left(\frac{\varrho+\varpi}{2}\right)\int_{\varrho}^{\varpi} w(\mathrm{N})d\mathrm{N} - \int_{\varrho}^{\varpi} w(\mathrm{N})g(\mathrm{N})d\mathrm{N} \right|$$

$$\leq \frac{(\varpi-\varrho)^2}{2(p+1)^{\frac{1}{p}}}\|w\|_{[\varrho,\varpi],\infty}\left(\frac{1+s+2^{2-s}}{(1+s)^2}\right)^{\frac{1}{q}}\left(\frac{|g'(\varrho)|^q+|g'(\varpi)|^q}{2}\right)^{\frac{1}{q}}.$$

2. *If we choose $w(u) = \frac{1}{\varpi-\varrho}$, we obtain*

$$\left| g\left(\frac{\varrho+\varpi}{2}\right) - \frac{2^{\alpha-1}}{(\varpi-\varrho)^{\alpha}}\Gamma(\alpha+1)\left(J^{\alpha}_{\left(\frac{\varrho+\varpi}{2}\right)^-}g(\varrho) + J^{\alpha}_{\left(\frac{\varrho+\varpi}{2}\right)^+}g(\varpi)\right)\right|$$

$$\leq \frac{\varpi-\varrho}{2(p\alpha+1)^{\frac{1}{p}}}\left(\frac{1+s+2^{2-s}}{(1+s)^2}\right)^{\frac{1}{q}}\left(\frac{|g'(\varrho)|^q+|g'(\varpi)|^q}{2}\right)^{\frac{1}{q}}.$$

3. *If we choose $w(u) = \frac{1}{\varpi-\varrho}$ and $\alpha = 1$, we obtain*

$$\left| g\left(\frac{\varrho+\varpi}{2}\right) - \frac{1}{\varpi-\varrho}\int_{\varrho}^{\varpi} g(u)du \right|$$

$$\leq \frac{\varpi-\varrho}{2(p+1)^{\frac{1}{p}}}\left(\frac{1+s+2^{2-s}}{(1+s)^2}\right)^{\frac{1}{q}}\left(\frac{|g'(\varrho)|^q+|g'(\varpi)|^q}{2}\right)^{\frac{1}{q}}.$$

Theorem 3. *Let $g : [\varrho,\varpi] \to \mathbb{R}$ be a differentiable function on (ϱ,ϖ) such that $g' \in L([\varrho,\varpi])$ with $0 \leq \varrho < \varpi$, and let $w : [\varrho,\varpi] \to \mathbb{R}$ be a continuous and symmetric function with respect to $\frac{\varrho+\varpi}{2}$. If $|g'|^q$ is an extended s-convex for some fixed $s \in (-1,1]$ and $q \geq 1$, then we have*

$$\left| L^{\alpha}[w]g\left(\frac{\varrho+\varpi}{2}\right) - L^{\alpha}[wg]\right|$$

$$\leq \frac{(\varpi-\varrho)^2}{4\alpha(\alpha+1)^{1-\frac{1}{q}}}\|w\|_{[\varrho,\varpi],\infty}\left(\left(B(s+1,\alpha+1)|g'(\varrho)|^q + \frac{1}{\alpha+s+1}\left|g'\left(\frac{\varrho+\varpi}{2}\right)\right|^q\right)^{\frac{1}{q}}\right.$$

$$\left. + \left(B(s+1,\alpha+1)|g'(\varpi)|^q + \frac{1}{\alpha+s+1}\left|g'\left(\frac{\varrho+\varpi}{2}\right)\right|^q\right)^{\frac{1}{q}}\right),$$

where $B(.,.)$ is the beta function.

Proof. Using Lemma 1, the absolute value, power mean inequality, and s-convexity of $|g'|$, we obtain

$$\left| L^{\alpha}[w]g\left(\frac{\varrho+\varpi}{2}\right) - L^{\alpha}[wg]\right|$$

$$\leq \frac{(\varpi-\varrho)^2}{4}\left(\int_0^1 |p_1(a)|\left|g'\left(a\varrho+(1-a)\frac{\varrho+\varpi}{2}\right)\right|da\right.$$

$$\left. + \int_0^1 |p_2(a)|\left|g'\left(a\varpi+(1-a)\frac{\varrho+\varpi}{2}\right)\right|da\right)$$

$$
\begin{aligned}
&\leq \frac{(\varpi-\varrho)^2}{4}\left(\left(\int_0^1 |p_1(a)|dt\right)^{1-\frac{1}{q}}\left(\int_0^1 |p_1(a)|\left|g'\left(a\varrho+(1-a)\frac{\varrho+\varpi}{2}\right)\right|^q da\right)^{\frac{1}{q}}\right.\\
&\quad\left.+\left(\int_0^1 |p_2(a)|da\right)^{1-\frac{1}{q}}\left(\int_0^1 |p_2(a)|\left|g'\left(a\varpi+(1-a)\frac{\varrho+\varpi}{2}\right)\right|^q da\right)^{\frac{1}{q}}\right)\\
&\leq \frac{(\varpi-\varrho)^2}{4}\|w\|_{[\varrho,\varpi],\infty}\left(\int_0^1\left(\int_a^1 (1-b)^{\alpha-1}db\right)da\right)^{1-\frac{1}{q}}\\
&\quad\times\left(\left(\int_0^1\left(\int_a^1 (1-b)^{\alpha-1}db\right)\left(a^s|g'(\varrho)|^q+(1-a)^s\left|g'\left(\frac{\varrho+\varpi}{2}\right)\right|^q\right)da\right)^{\frac{1}{q}}\right.\\
&\quad\left.+\left(\int_0^1\left(\int_a^1 (1-b)^{\alpha-1}db\right)\left(a^s|g'(\varpi)|^q+(1-a)^s\left|g'\left(\frac{\varrho+\varpi}{2}\right)\right|^q\right)da\right)^{\frac{1}{q}}\right)\\
&=\frac{(\varpi-\varrho)^2}{4\alpha}\|w\|_{[\varrho,\varpi],\infty}\left(\int_0^1 (1-a)^\alpha da\right)^{1-\frac{1}{q}}\\
&\quad\times\left(\left(\int_0^1 (1-a)^\alpha\left(a^s|g'(\varrho)|^q+(1-a)^s\left|g'\left(\frac{\varrho+\varpi}{2}\right)\right|^q\right)da\right)^{\frac{1}{q}}\right.\\
&\quad\left.+\left(\int_0^1 (1-a)^\alpha\left(a^s|g'(\varpi)|^q+(1-a)^s\left|g'\left(\frac{\varrho+\varpi}{2}\right)\right|^q\right)da\right)^{\frac{1}{q}}\right)\\
&=\frac{(\varpi-\varrho)^2}{4\alpha(\alpha+1)^{1-\frac{1}{q}}}\|w\|_{[\varrho,\varpi],\infty}\\
&\quad\times\left(\left(|g'(\varrho)|^q\int_0^1 (1-a)^\alpha a^s da+\left|g'\left(\frac{\varrho+\varpi}{2}\right)\right|^q\int_0^1 (1-a)^{\alpha+s}da\right)^{\frac{1}{q}}\right.\\
&\quad\left.+\left(|g'(\varpi)|^q\int_0^1 (1-a)^\alpha a^s da+\left|g'\left(\frac{\varrho+\varpi}{2}\right)\right|^q\int_0^1 (1-a)^{\alpha+s}da\right)^{\frac{1}{q}}\right)\\
&=\frac{(\varpi-\varrho)^2}{4\alpha(\alpha+1)^{1-\frac{1}{q}}}\|w\|_{[\varrho,\varpi],\infty}\left(\left(B(s+1,\alpha+1)|g'(\varrho)|^q+\frac{1}{\alpha+s+1}\left|g'\left(\frac{\varrho+\varpi}{2}\right)\right|^q\right)^{\frac{1}{q}}\right.\\
&\quad\left.+\left(B(s+1,\alpha+1)|g'(\varpi)|^q+\frac{1}{\alpha+s+1}\left|g'\left(\frac{\varrho+\varpi}{2}\right)\right|^q\right)^{\frac{1}{q}}\right).
\end{aligned}
$$

The proof is now completed. □

Corollary 13. *In Theorem 3, if we use:*

1. $s=0$, *we get*

$$
\left|L^\alpha[w]g\left(\frac{\varrho+\varpi}{2}\right)-L^\alpha[wg]\right|
$$

$$\leq \frac{(\varpi-\varrho)^2}{4\alpha(\alpha+1)}\|w\|_{[\varrho,\varpi],\infty}\left(\left(|g'(\varrho)|^q+\left|g'\left(\frac{\varrho+\varpi}{2}\right)\right|^q\right)^{\frac{1}{q}}\right.$$
$$\left.+\left(|g'(\varpi)|^q+\left|g'\left(\frac{\varrho+\varpi}{2}\right)\right|^q\right)^{\frac{1}{q}}\right).$$

2. *If we use $s=1$, we obtain*

$$\left|L^{\alpha}[w]g\left(\frac{\varrho+\varpi}{2}\right)-L^{\alpha}[wg]\right|$$
$$\leq \frac{(\varpi-\varrho)^2}{4\alpha(\alpha+1)}\|w\|_{[\varrho,\varpi],\infty}\left(\left(\frac{1}{\alpha+2}|g'(\varrho)|^q+\frac{\alpha+1}{\alpha+2}\left|g'\left(\frac{\varrho+\varpi}{2}\right)\right|^q\right)^{\frac{1}{q}}\right.$$
$$\left.+\left(\frac{1}{\alpha+2}|g'(\varpi)|^q+\frac{\alpha+1}{\alpha+2}\left|g'\left(\frac{\varrho+\varpi}{2}\right)\right|^q\right)^{\frac{1}{q}}\right).$$

3. *If we choose $\alpha=1$, we obtain*

$$\left|g\left(\frac{\varrho+\varpi}{2}\right)\int_{\varrho}^{\varpi}w(\mathrm{N})d\mathrm{N}-\int_{\varrho}^{\varpi}w(\mathrm{N})g(\mathrm{N})d\mathrm{N}\right|$$
$$\leq \frac{(\varpi-\varrho)^2}{8}\|w\|_{[\varrho,\varpi],\infty}\left(\frac{2}{(s+1)(s+2)}\right)^{\frac{1}{q}}\left(\left(|g'(\varrho)|^q+(s+1)\left|g'\left(\frac{\varrho+\varpi}{2}\right)\right|^q\right)^{\frac{1}{q}}\right.$$
$$\left.+\left(|g'(\varpi)|^q+(s+1)\left|g'\left(\frac{\varrho+\varpi}{2}\right)\right|^q\right)^{\frac{1}{q}}\right).$$

Remark 6. *In the third point of Corollary 13, if we assume that $s\in(0,1]$, we obtain Theorem 2.2 in [27]. Moreover, if we use $s=1$, we obtain Corollary 12 in [28].*

Corollary 14. *In Theorem 3, if we choose:*

1. $w(u)=\frac{1}{\varpi-\varrho}$, *we obtain*

$$\left|g\left(\frac{\varrho+\varpi}{2}\right)-\frac{2^{\alpha-1}}{(\varpi-\varrho)^{\alpha}}\Gamma(\alpha+1)\left(J^{\alpha}_{\left(\frac{\varrho+\varpi}{2}\right)^-}g(\varrho)+J^{\alpha}_{\left(\frac{\varrho+\varpi}{2}\right)^+}g(\varpi)\right)\right|$$
$$\leq \frac{\varpi-\varrho}{4(\alpha+1)^{1-\frac{1}{q}}}\left(\left(B(s+1,\alpha+1)|g'(\varrho)|^q+\frac{1}{\alpha+s+1}\left|g'\left(\frac{\varrho+\varpi}{2}\right)\right|^q\right)^{\frac{1}{q}}\right.$$
$$\left.+\left(B(s+1,\alpha+1)|g'(\varpi)|^q+\frac{1}{\alpha+s+1}\left|g'\left(\frac{\varrho+\varpi}{2}\right)\right|^q\right)^{\frac{1}{q}}\right).$$

2. *If we choose $w(u)=\frac{1}{\varpi-\varrho}$ and $\alpha=1$, we obtain*

$$\left|g\left(\frac{\varrho+\varpi}{2}\right)-\frac{1}{\varpi-\varrho}\int_{\varrho}^{\varpi}g(u)du\right|$$
$$\leq \frac{\varpi-\varrho}{8}\left(\frac{2}{(s+1)(s+2)}\right)^{\frac{1}{q}}\left(\left(|g'(\varrho)|^q+(s+1)\left|g'\left(\frac{\varrho+\varpi}{2}\right)\right|^q\right)^{\frac{1}{q}}\right.$$

$$+\left(\left|g'(\varpi)\right|^q+(s+1)\left|g'\left(\frac{\varrho+\varpi}{2}\right)\right|^q\right)^{\frac{1}{q}}\Bigg).$$

Corollary 15. *In Theorem 3, using the s-convexity of $|g'|$, we obtain*

$$\left|L^\alpha[w]g\left(\frac{\varrho+\varpi}{2}\right)-L^\alpha[wg]\right|\leq\frac{(\varpi-\varrho)^2}{4\alpha(\alpha+1)^{1-\frac{1}{q}}}\|w\|_{[\varrho,\varpi],\infty}$$
$$\times\left(\left(\frac{(1+s)(\alpha+s+1)B(s+1,\alpha+1)+2^{1-s}}{(1+s)(\alpha+s+1)}|g'(\varrho)|^q+\frac{2^{1-s}}{(1+s)(\alpha+s+1)}|g'(\varpi)|^q\right)^{\frac{1}{q}}\right.$$
$$\left.+\left(\frac{2^{1-s}}{(1+s)(\alpha+s+1)}|g'(\varrho)|^q+\frac{(1+s)(\alpha+s+1)B(s+1,\alpha+1)+2^{1-s}}{(1+s)(\alpha+s+1)}|g'(\varpi)|^q\right)^{\frac{1}{q}}\right).$$

Corollary 16. *In Corollary 9, if we use:*

1. $\alpha=1$, *we obtain*

$$\left|g\left(\frac{\varrho+\varpi}{2}\right)\int_\varrho^\varpi w(\mathrm{N})d\mathrm{N}-\int_\varrho^\varpi w(\mathrm{N})g(\mathrm{N})d\mathrm{N}\right|$$
$$\leq\frac{(\varpi-\varrho)^2}{8}\|w\|_{[\varrho,\varpi],\infty}\left(\frac{2}{(1+s)(s+2)}\right)^{\frac{1}{q}}\left(\left(\left(1+2^{1-s}\right)|g'(\varrho)|^q+2^{1-s}|g'(\varpi)|^q\right)^{\frac{1}{q}}\right.$$
$$\left.+\left(2^{1-s}|g'(\varrho)|^q+\left(1+2^{1-s}\right)|g'(\varpi)|^q\right)^{\frac{1}{q}}\right).$$

2. $w(u)=\frac{1}{\varpi-\varrho}$, *we obtain*

$$\left|g\left(\frac{\varrho+\varpi}{2}\right)-\frac{2^{\alpha-1}}{(\varpi-\varrho)^\alpha}\Gamma(\alpha+1)\left(J^\alpha_{\left(\frac{\varrho+\varpi}{2}\right)^-}g(\varrho)+J^\alpha_{\left(\frac{\varrho+\varpi}{2}\right)^+}g(\varpi)\right)\right|$$
$$\leq\frac{\varpi-\varrho}{4(\alpha+1)^{1-\frac{1}{q}}}\left(\left(\frac{(1+s)(\alpha+s+1)B(s+1,\alpha+1)+2^{1-s}}{(1+s)(\alpha+s+1)}|g'(\varrho)|^q\right.\right.$$
$$\left.+\frac{2^{1-s}}{(1+s)(\alpha+s+1)}|g'(\varpi)|^q\right)^{\frac{1}{q}}+\left(\frac{2^{1-s}}{(1+s)(\alpha+s+1)}|g'(\varrho)|^q\right.$$
$$\left.\left.+\frac{(1+s)(\alpha+s+1)B(s+1,\alpha+1)+2^{1-s}}{(1+s)(\alpha+s+1)}|g'(\varpi)|^q\right)^{\frac{1}{q}}\right).$$

3. *If we choose* $w(u)=\frac{1}{\varpi-\varrho}$ *and* $\alpha=1$, *we obtain*

$$\left|g\left(\frac{\varrho+\varpi}{2}\right)-\frac{1}{\varpi-\varrho}\int_\varrho^\varpi g(u)du\right|$$
$$\leq\frac{\varpi-\varrho}{8}\left(\frac{2}{(1+s)(s+2)}\right)^{\frac{1}{q}}\left(\left(\left(1+2^{1-s}\right)|g'(\varrho)|^q+2^{1-s}|g'(\varpi)|^q\right)^{\frac{1}{q}}\right.$$
$$\left.+\left(2^{1-s}|g'(\varrho)|^q+\left(1+2^{1-s}\right)|g'(\varpi)|^q\right)^{\frac{1}{q}}\right).$$

Remark 7. *Corollary 16, the second point will be reduced to Theorem 5 in [6] when $s=1$.*

Corollary 17. *In Corollary 15, if we use the discrete power mean inequality, we obtain*

$$\begin{aligned}&\left|L^{\alpha}[w]g\left(\tfrac{\varrho+\varpi}{2}\right)-L^{\alpha}[wg]\right|\\ \le\ &\frac{(\varpi-\varrho)^2}{2\alpha(\alpha+1)^{1-\frac{1}{q}}}\|w\|_{[\varrho,\varpi],\infty}\left(\frac{(1+s)(\alpha+s+1)B(s+1,\alpha+1)+2^{2-s}}{(1+s)(\alpha+s+1)}\right)^{\frac{1}{q}}\left(\frac{|g'(\varrho)|^q+|g'(\varpi)|^q}{2}\right)^{\frac{1}{q}}.\end{aligned}$$

Corollary 18. *In Corollary 17, if we use:*

1. $\alpha=1$, *we obtain*

$$\begin{aligned}&\left|g\left(\frac{\varrho+\varpi}{2}\right)\int_{\varrho}^{\varpi}w(\mathrm{N})d\mathrm{N}-\int_{\varrho}^{\varpi}w(\mathrm{N})g(\mathrm{N})d\mathrm{N}\right|\\ \le&\frac{(\varpi-\varrho)^2}{4}\|w\|_{[\varrho,\varpi],\infty}\left(\frac{1+2^{2-s}}{(1+s)(s+2)}\right)^{\frac{1}{q}}\left(|g'(a)|^q+|g'(\varpi)|^q\right)^{\frac{1}{q}}.\end{aligned}$$

2. $w(u)=\frac{1}{\varpi-\varrho}$, *we obtain*

$$\begin{aligned}&\left|g\left(\frac{\varrho+\varpi}{2}\right)-\frac{2^{\alpha-1}}{(\varpi-\varrho)^{\alpha}}\Gamma(\alpha+1)\left(J^{\alpha}_{\left(\frac{\varrho+\varpi}{2}\right)^-}g(\varrho)+J^{\alpha}_{\left(\frac{\varrho+\varpi}{2}\right)^+}g(\varpi)\right)\right|\\ \le&\frac{\varpi-\varrho}{2(\alpha+1)^{1-\frac{1}{q}}}\left(\frac{(1+s)(\alpha+s+1)B(s+1,\alpha+1)+2^{2-s}}{(1+s)(\alpha+s+1)}\right)^{\frac{1}{q}}\left(\frac{|g'(\varrho)|^q+|g'(\varpi)|^q}{2}\right)^{\frac{1}{q}}.\end{aligned}$$

3. $w(u)=\frac{1}{\varpi-\varrho}$ *and* $\alpha=1$, *we obtain*

$$\left|g\left(\frac{\varrho+\varpi}{2}\right)-\frac{1}{\varpi-\varrho}\int_{\varrho}^{\varpi}g(u)du\right|\le\frac{\varpi-\varrho}{4}\left(\frac{1+2^{2-s}}{(1+s)(s+2)}\right)^{\frac{1}{q}}\left(|g'(\varrho)|^q+|g'(\varpi)|^q\right)^{\frac{1}{q}}.$$

Remark 8. *Corollary 18, the first point will be reduced to Theorem 2 in [8] when* $s=1$.

4. Applications

4.1. Weighted Midpoint Quadrature

Let Υ be the partition of the points $\varrho=\wp_0<\wp_1<\ldots<\wp_n=\varpi$ of the interval $[\varrho,\varpi]$, and consider the quadrature formula

$$\int_{\varrho}^{\varpi}w(u)g(u)du=\lambda_w(g,\Upsilon)+R_w(g,\Upsilon),$$

where

$$\lambda_w(g,\Upsilon)=\sum_{i=0}^{n-1}g\left(\frac{\wp_i+\wp_{i+1}}{2}\right)\int_{\wp_i}^{\wp_{i+1}}w(u)du$$

and $R_w(g,\Upsilon)$ is the associated approximation error.

Proposition 1. *Let* $g:[\varrho,\varpi]\to\mathbb{R}$ *be a differentiable function on* (ϱ,ϖ) *with* $0\le\varrho<\varpi$ *and* $g'\in L^1[\varrho,\varpi]$, *and let* $w:[\varrho,\varpi]\to\mathbb{R}$ *be symmetric as regards* $\frac{\varrho+\varpi}{2}$. *If* $|g'|$ *is s-convex function, then for* $n\in\mathbb{N}$ *we have*

$$|R_w(g,\Upsilon)|\le\frac{(2^{2-s}+1)}{4(1+s)(s+2)}\|w\|_{[\varrho,\varpi],\infty}\sum_{i=0}^{n-1}(\wp_{i+1}-\wp_i)^2\left(|g'(\wp_i)|+|g'(\wp_{i+1})|\right).$$

Proof. Applying Corollary 5 on the subintervals $[\wp_i, \wp_{i+1}]$ $(i = 0, 1, ..., n-1)$ of the partition Y, we obtain

$$\left| g\left(\frac{\wp_i + \wp_{i+1}}{2}\right) \int_{\wp_i}^{\wp_{i+1}} w(u)du - \int_{\wp_i}^{\wp_{i+1}} w(u)g(u)du \right|$$

$$\leq \frac{(2^{2-s}+1)(\wp_{i+1}-\wp_i)^2}{4(1+s)(s+2)} \|w\|_{[\wp_i,\wp_{i+1}],\infty} \left(|g'(\wp_i)| + |g'(\wp_{i+1})|\right).$$

Add the above inequalities for all $i = 0, 1, ..., n-1$ and using the triangular inequality to obtain the desired result. □

Proposition 2. *Let $g : [\varrho, \varpi] \to \mathbb{R}$ be a differentiable function on (ϱ, ϖ) with $0 \leq \varrho < \varpi$ and $g' \in L^1[\varrho, \varpi]$, and let $w : [\varrho, \varpi] \to \mathbb{R}$ be symmetric as regards $\frac{\varrho+\varpi}{2}$. If $|g'|^q$ is a s-convex function, then for $n \in \mathbb{N}$ we have*

$$|R(g, Y)| \leq \frac{\|w\|_{[\varrho,\varpi],\infty}}{2(p+1)^{\frac{1}{p}}} \sum_{i=0}^{n-1} (\wp_{i+1}-\wp_i)^2 \left(\frac{1+s+2^{2-s}}{(1+s)^2}\right)^{\frac{1}{q}} \left(\frac{|g'(\wp_i)|^q + |g'(\wp_{i+1})|^q}{2}\right)^{\frac{1}{q}}.$$

Proof. Applying Corollary 12 on the subintervals $[\wp_i, \wp_{i+1}]$ $(i = 0, 1, ..., n-1)$ of the partition Y, we obtain

$$\left| g\left(\frac{\wp_i + \wp_{i+1}}{2}\right) \int_{\wp_i}^{\wp_{i+1}} w(u)du - \int_{\wp_i}^{\wp_{i+1}} w(u)g(u)du \right|$$

$$\leq \frac{(\wp_{i+1}-\wp_i)^2}{2(p+1)^{\frac{1}{p}}} \|w\|_{[\wp_i,\wp_{i+1}],\infty} \left(\frac{1+s+2^{2-s}}{(1+s)^2}\right)^{\frac{1}{q}} \left(\frac{|g'(\wp_i)|^q + |g'(\wp_{i+1})|^q}{2}\right)^{\frac{1}{q}}.$$

Add the above inequalities for all $i = 0, 1, ..., n-1$ and using the triangular inequality to obtain the desired result. □

Proposition 3. *Let $g : [\varrho, \varpi] \to \mathbb{R}$ be a differentiable function on (ϱ, ϖ) with $0 \leq \varrho < \varpi$ and $g' \in L^1[\varrho, \varpi]$, and let $w : [\varrho, \varpi] \to \mathbb{R}$ be symmetric as regards $\frac{\varrho+\varpi}{2}$. If $|g'|^q$ is a s-convex function, then, for $n \in \mathbb{N}$, we have*

$$|R(g, Y)| \leq \frac{\|w\|_{[a_i,b],\infty}}{4} \left(\frac{1+2^{2-s}}{(1+s)(s+2)}\right)^{\frac{1}{q}} \sum_{i=0}^{n-1} (\wp_{i+1}-\wp_i)^2 \left(|g'(\wp_i)|^q + |g'(\wp_{i+1})|^q\right)^{\frac{1}{q}}.$$

Proof. Applying Corollary 18 on the subintervals $[\wp_i, \wp_{i+1}]$ $(i = 0, 1, ..., n-1)$ of the partition Y, we obtain

$$\left| g\left(\frac{\wp_i + \wp_{i+1}}{2}\right) \int_{\wp_i}^{\wp_{i+1}} w(u)du - \int_{\wp_i}^{\wp_{i+1}} w(u)g(u)du \right|$$

$$\leq \frac{(\wp_{i+1}-\wp_i)^2}{4} \|w\|_{[\wp_i,\wp_{i+1}],\infty} \left(\frac{1+2^{2-s}}{(1+s)(s+2)}\right)^{\frac{1}{q}} \left(|g'(\wp_i)|^q + |g'(\wp_{i+1})|^q\right)^{\frac{1}{q}}.$$

Add the above inequalities for all $i = 0, 1, ..., n-1$ and using the triangular inequality to obtain the desired result. □

4.2. Application to Special Means

Let ϱ, ϖ be two arbitrary real numbers:

The Arithmetic mean:

$$A(\varrho, \varpi) = \frac{\varrho + \varpi}{2}.$$

The Logarithmic mean:

$$L(\varrho,\varpi)=\frac{\varpi-\varrho}{\ln\varpi-\ln\varrho},\ \varrho,\varpi>0,\ \varrho\neq\varpi.$$

The p-Logarithmic mean:

$$L_p(\varrho,\varpi)=\left(\frac{\varpi^{p+1}-\varrho^{p+1}}{(p+1)(\varpi-\varrho)}\right)^{\frac{1}{p}},\ \varrho,\varpi>0,\ \varrho\neq\varpi \text{ and } p\in\mathbb{R}\backslash\{-1,0\}.$$

Proposition 4. *Let $\varrho,\varpi\in\mathbb{R}$ with $0<\varrho<\varpi$, then we have*

$$\left|A^{\frac{3}{2}}(\varrho,\varpi)-L^{\frac{3}{2}}_{\frac{3}{2}}(\varrho,\varpi)\right|\leq\frac{\varpi-\varrho}{10}\left(\varrho^{\frac{1}{2}}+3\left(\frac{\varrho+\varpi}{2}\right)^{\frac{1}{2}}+\varpi^{\frac{1}{2}}\right).$$

Proof. Using Corollary 3 for function $g(k)=k^{\frac{3}{2}}$ whose derivative $g'(k)=\frac{3}{2}k^{\frac{1}{2}}$ is $\frac{1}{2}$-convex. □

Proposition 5. *Let $\varrho,\varpi\in\mathbb{R}$ with $0<\varrho<\varpi$, then we have*

$$\left|A^{-1}(\varrho,\varpi)-L^{-1}(\varrho,\varpi)\right|\leq\frac{(\varpi-\varrho)\sqrt{3}}{12}\left(\left(\frac{2\varrho+\varpi}{\varrho\varpi}\right)^{\frac{1}{2}}+\left(\frac{\varpi+2\varrho}{\varrho\varpi}\right)^{\frac{1}{2}}\right).$$

Proof. Applying Corollary 17 with $q=2$ to the function $g(k)=\frac{1}{k}$ whose derivative $|g'(k)|^2=\frac{1}{k}$ is P-function. □

5. Conclusions

In this study, we considered the weighted midpoint-type integral inequalities for s-convex first derivatives using Riemann–Liouville integrals operators, where the main novelties of the paper are provided by a new identity regarding the weighted midpoint-type inequalities being presented and some new fractional weighted midpoint-type inequalities for functions whose first derivatives are s-convex being established. Some special cases are derived and the applications of our results are provided.

Author Contributions: Writing—original draft preparation, N.N., F.A. and A.F.; writing—review and editing, K.B.; visualization, B.M.; supervision, K.Z.; project administration, T.R.; funding acquisition, T.R. All authors have read and agreed to the published version of the manuscript.

Funding: This research received no external funding.

Data Availability Statement: Not applicable.

Acknowledgments: Researchers would like to thank the Deanship of Scientific Research, Qassim University for funding the publication of this project.

Conflicts of Interest: The authors declare no conflict of interest.

References

1. Dragomir, S.S.; Pećarixcx, J.; Persson, L.E. Some inequalities of Hadamard type. *Soochow J. Math.* **1995**, *21*, 335–341.
2. Kaijser, S.; Nikolova, L.; Persson, L.-E.; Wedestig, A. Hardy-type inequalities via convexity. *Math. Inequal. Appl.* **2005**, *8*, 403–417. [CrossRef]
3. Kashuri, A.; Meftah, B.; Mohammed, P.O. Some weighted Simpson type inequalities for differentiable s-convex functions and their applications. *J. Fract. Calc. Nonlinear Syst.* **2020**, *1*, 75–94. [CrossRef]
4. Saker, S.H.; Abdou, D.M.; Kubiaczyk, I. Opial and P6 lya type inequalities via convexity. *Fasc. Math.* **2018**, *60*, 145–159. [CrossRef]
5. Vivas-Cortez, M.; Abdeljawad, T.; Mohammed, P.O.; Rangel-Oliveros, Y. Simpson's integral inequalities for twice differentiable convex functions. *Math. Probl. Eng.* **2020**, *2020*, 1936461. [CrossRef]

6. Sarikaya, M.Z.; Yildirim, H. On Hermite-Hadamard type inequalities for Riemann–Liouville fractional integrals. *Miskolc Math. Notes* **2016**, *17*, 1049–1059. [CrossRef]
7. Mitroi, F.-C.; Nikodem, K.; Wąsowicz, S. Hermite-Hadamard inequalities for convex set-valued functions. *Demonstratio Math.* **2013**, *46*, 655–662. [CrossRef]
8. Pearce, C.E.M.; Pećarixcx, J. Inequalities for differentiable mappings with application to special means and quadrature formulæ. *Appl. Math. Lett.* **2000**, *13*, 51–55. [CrossRef]
9. Kirmaci, U.S. Inequalities for differentiable mappings and applications to special means of real numbers and to midpoint formula. *Appl. Math. Comput.* **2004**, *147*, 137–146. [CrossRef]
10. İşcan, İ; Set, E.; Özdemir, M.E. Some new general integral inequalities for *P*-functions. *Malaya J. Mat.* **2014**, *2*, 510–516.
11. Gorenflo, R.; Mainardi, F. *Fractional Calculus: Integral and Differential Equations of Fractional Order*; Fractals and Fractional Calculus in Continuum Mechanics (Udine, 1996), 223–276. CISM Courses and Lect., 378; Springer: Vienna, Austria, 1997.
12. Kilbas, A.A.; Srivastava, H.M.; Trujillo, J.J. *Theory and Applications of Fractional Differential Equations*; North-Holland Mathematics Studies 204; Elsevier Science B.V.: Amsterdam, The Netherlands, 2006.
13. Eskandari, Z.; Avazzadeh, Z.; Ghaziani, R.K.; Li, B. Dynamics and bifurcations of a discrete-time Lotka-Volterra model using nonstandard finite difference discretization method. *Math. Meth. Appl. Sci.* **2022**. [CrossRef]
14. Li, B.; Zhang, Y.; Li, X.; Eskandari, Z.; He, Q. Bifurcation analysis and complex dynamics of a Kopel triopoly model. *J. Comp. Appl. Math.* **2023**, *426*, 115089. [CrossRef]
15. Li, B.; Liang, H.; He, Q. Multiple and generic bifurcation analysis of a discrete Hindmarsh-Rose model. *Chaos Solitons Fractals* **2021**, *146*, 110856. [CrossRef]
16. Li, B.; Liang, H.; Shi, L. Complex dynamics of Kopel model with nonsymmetric response between oligopolists. *Chaos Solitons Fractals* **2022**, *156*, 111860. [CrossRef]
17. Xu, C.; Rahman, M.; Baleanu, D. On fractional-order symmetric oscillator withoffset-boosting control. *Nonlinear Anal. Model. Control.* **2022**, *27*, 994–1008.
18. Zhang, L.; Rahman, M.; Ahmad, S.; Riaz, M.B. Dynamics of fractional order delay model of coronavirus disease. *AIMS Math.* **2022**, *7*, 4211–4232. [CrossRef]
19. Kalsoom, H.; Vivas-Cortez, M.; Latif, M.A.; Ahmad, H. Weighted midpoint Hermite-Hadamard-Fejér type inequalities in fractional calculus for harmonically convex functions. *Fractal Fract.* **2021**, *5*, 252.
20. Kamouche, N.; Ghomrani, S.; Meftah, B. Fractional Simpson like type inequalities for differentiable *s*-convex functions. *J. Appl. Math. Stat. Inform.* **2022**, *18*, 73–91. [CrossRef]
21. Kashuri, A.; Meftah, B.; Mohammed, P.O.; Lupa, A.A.; Abdalla, B.; Hamed, Y.S.; Abdeljawad, T. Fractional weighted Ostrowski type inequalities and their applications. *Symmetry* **2021**, *13*, 968. [CrossRef]
22. Khanna, N.; Zothansanga, A.; Kaushik, S.K.; Kumar, D. Some properties of fractional Boas transforms of wavelets. *J. Math.* **2021**, 2021, 6689779. [CrossRef]
23. Mohammed, P.O.; Abdeljawad, T. Modification of certain fractional integral inequalities for convex functions. *Adv. Differ. Equ.* **2020**, *2020*, 69 [CrossRef]
24. Mohammed, P.O.; Aydi, H.; Kashuri, A.; Hamed, Y.S.; Abualnaja, K.M. Midpoint inequalities in fractional calculus defined using positive weighted symmetry function kernels. *Symmetry* **2021**, *13*, 550. [CrossRef]
25. Rahman, G.; Abdeljawad, T.; Jarad, F.; Khan, A.; Nisar, K.S. Certain inequalities via generalized proportional Hadamard fractional integral operators. *Adv. Difference Equ.* **2019**, *2019*, 454. [CrossRef]
26. Set, E.; Gözpinar, A. A study on Hermite–Hadamard-type inequalities via new fractional conformable integrals. *Asian-Eur. J. Math.* **2021**, *14*, 2150016. [CrossRef]
27. Liu, Z. On inequalities of Hermite-Hadamard type involving an *s*-convex function with applications. *Issues Anal.* **2016**, *5*, 3–20. [CrossRef]
28. Azzouza, N.; Meftah, B. Some weighted integral inequalities for differentiable *beta*-convex functions. *J. Interdiscip. Math.* **2022**, *25*, 373–393. [CrossRef]

symmetry

MDPI

Article

Existence of Global and Local Mild Solution for the Fractional Navier–Stokes Equations

Muath Awadalla [1,*], Azhar Hussain [2,*], Farva Hafeez [3] and Kinda Abuasbeh [1]

[1] Department of Mathematics and Statistics, College of Science, King Faisal University, Hafuf, Al Ahsa 31982, Saudi Arabia
[2] Department Mathematics, University of Chakwal, Chakwal 48800, Pakistan
[3] Department of Mathematics and Statistics, University of Lahore, Sargodha 40100, Pakistan
* Correspondence: mawadalla@kfu.edu.sa (M.A.); azhar.hussain@uoc.edu.pk (A.H.)

Abstract: Navier–Stokes equations (NS-equations) are applied extensively for the study of various waves phenomena where the symmetries are involved. In this paper, we discuss the NS-equations with the time-fractional derivative of order $\beta \in (0,1)$. In fractional media, these equations can be utilized to recreate anomalous diffusion equations which can be used to construct symmetries. We examine the initial value problem involving the symmetric Stokes operator and gravitational force utilizing the Caputo fractional derivative. Additionally, we demonstrate the global and local mild solutions in $H^{\alpha,p}$. We also demonstrate the regularity of classical solutions in such circumstances. An example is presented to demonstrate the reliability of our findings.

Keywords: Navier–Stokes equations; Caputo fractional derivatives; mild solutions; regularity

MSC: 34A08; 34A12

Citation: Awadalla, M.; Hussain, A.; Hafeez, F.; Abuasbeh, K. Existence of Global and Local Mild Solution for the Fractional Navier–Stokes Equations. *Symmetry* **2023**, *15*, 343. https://doi.org/10.3390/sym15020343

Academic Editor: Mihai Postolache

Received: 21 December 2022
Revised: 18 January 2023
Accepted: 21 January 2023
Published: 26 January 2023

1. Introduction

Because of their importance in fluid mechanics, the Navier–Stokes equations have been extensively studied by various researchers. NS-equations are partial differential equations that describe the flow of incompressible fluid. These equations are generalization of the equations devised by Swiss mathematician Leonhard Euler in the eighteen century to describe the flow of incompressible and frictionless fluids. The NS-equations are useful because they describe the physics of many scientific and engineering phenomena. These can be used to simulate weather, ocean currents, water flow in a pipe, and airflow around a wing etc. The difference between the NS-equations and the Euler equations is that the NS-equations account for viscosity, whereas the Euler equations exclusively simulate inviscid flow.

As a result, the NS-equations are parabolic equations, which have exceptional analytic features. In a purely mathematical sense, the NS-equations are extremely interesting. Despite its extensive range of applications, it is still unknown if smooth solutions always exist in three dimensions, that is, whether these are infinite and differentiable at all points in the domain. The existence and smoothness problem is known as the Navier–Stokes problem.

Different scholars focus on mass and momentum conservation and describe useful phenomena concerning the motion of the incompressible fluid flow, ranging from large-scale atmospheric motions to the lubricant in ball bearings; see, Varnhorn [1], as well as Cannone [2]. Similarly, Rieusset [3] discussed the existence, uniqueness and regularity of NS-equations.

Jean Leray was a French mathematician who work on both PDEs and algebraic topology and explained a fascinating phenomenon. The Leray projection is a linear operator that is useful in the theory of partial differential equations, particularly in the subject of fluid dynamics. It can be considered as a projection on a vector field with no divergence. In the Stokes equations and NS-equations, it is applied to eliminate both the pressure term and the divergence-free term; see [4].

Aljandro Rangel-Huerts and Blanca Bermudez solved NS-equations using two unique formulations with moderate and high Reynolds numbers. They used two numerical solutions of lid-driven cavity and Taylor vortex problems. These problems can be solved by using stream function vorticity in two dimensions of NS-equations; see [5]. Moreover, Gallgher [6], Giga [7], Rejaiba [8], Kozono [9], Sell [10] and Choe [11] found unique results on the regularity of weak and strong solutions. Emilia Bazhlekova et al. [12] analyzed the Rayleigh Stokes' problems. Rayleigh problem is also known as Stokes' first problem which is a problem of determining the flow created by a sudden movement of an infinitely long plate from rest named after Lord Rayleigh and Sir George Stokes. The authors studied the Reyleigh problems involving RL-fractional derivative. They worked on smooth and non-smoothness initial data for Sobolev regularity of homogeneous problems.

On the contrary, fractional calculus has received a lot of attention in recent years. Many of the fundamental piece of calculus are related to fluid mechanics like total derivative, gradients, divergence and rotation. Fractional calculus proved that the topic indeed is very promising like in control theory of dynamical system, porous structure, viscoelasticity and among others; see, e.g., Hilfer [13], Herrmann [14], and Zhou [15–17]. Such models are important not just in Physics but also in pure mathematics. Recently, experimental data and theoretical analysis have shown that the diffusion equation fails to describes the diffusion phenomena in porous media. Basically, the diffusion equation is a parabolic PDE. In Physics, it describe the microscopic behavior of many microparticles in Brownain motion.

Do NS-equations describe all the motion of the fluid? Serkan Solmaz gave an interesting fact that the NS-equations encompass all types of fluid motion in case they are combined with a related mathematical model such as multi-phase flow, chemical reaction and turbulent etc. It is significant to specify the degree of error throughout the analysis in which the NS-equations enable a reasonable range of error. Thereby, these are the most famous equations that examine the motion of fluid reliably. Different authors talked about the time fractional NS-equations; see [18–20]. Moreover, to the best of our insight there are not many results on the existence, uniqueness and regularity of mild solution for time fractional NS-equations.

Keeping this in view, we discuss the time fractional NS-equations in an open set $\Omega \subset R^m (m \geq 3)$:

$$\begin{cases} \partial_t^\beta v - \mu\Delta v + (v\cdot\nabla)v = -\nabla p + \rho g + \mu\nabla^2\vec{v}, 0 < t, \\ \nabla\cdot v = 0, \\ \frac{v}{\partial\Omega} = 0, \\ v(0,y) = ax + b, \end{cases} \tag{1}$$

where $\rho\left(\frac{\partial v}{\partial t} + (v\cdot\nabla)v\right) = \rho\frac{Dv}{Dt}, g$ is a gravitational force or body force, $-\nabla p$ is a pressure gradient, $\mu\nabla^2\overrightarrow{v}$ is viscous term or diffusion term, $\rho\frac{Dv}{Dt}$ is local acceleration and ∂_t^β be the Caputo fractional derivative with order $\beta \in (0,1), y \in \Omega$ and the time $0 < t$. By applying a well-known Helmholtz projector P on (1) for getting rid of the pressure term, one has

$$\begin{cases} \partial_t^\beta v - \mu P\Delta v + P(v\cdot\nabla)v = Pg, 0 < t, \\ \nabla\cdot v = 0, \\ \frac{v}{\partial\Omega} = 0, \\ v(0,y) = b. \end{cases}$$

B is the Stokes operator under consideration, where b is the initial velocity and $-\mu P\Delta$ is the Dirichlet boundary condition. The abstract form of (1) is

$$\begin{cases} {}^{C}D_{t}^{\beta}v = -Bv + F(v,w) + Pg, 0 < t, \\ v(0) = b, \end{cases} \quad (2)$$

where $-P(v \cdot \nabla)w = F(v,w)$.

The arrangement of the paper is as: In Section 2, we review some helpful preliminaries. In Section 3, study of the global and local existence of mild solutions of problem (2) in $H^{\beta,p}$ is conducted. In Section 4, the regularity of classic solutions in Q_p will be discussed. At last, an example will be presented.

2. Preliminaries

In this section, we discuss some known definitions, notations and results.

Suppose that, $\omega = \{(y_1, ..., y_m) : y_m > 0\}$ be an open subset of R^m where $m \geq 3$ and $1 < p < \infty$. Then there exists a bounded projection

$$C_{\varrho}^{\infty}(\omega) = \{v \in (C^{\infty}(\omega))^m : \nabla \cdot v = 0, vhascompactin\omega\},$$

as well as the null space is the closure of

$$\{v \in (C^{\infty}(\omega))^m : v = \nabla\varphi, \varphi \in C^{\infty}(\omega)\}.$$

Suppose that, $Q_p = \overline{C_{\varrho}^{\infty}(\omega)}^{|\cdot|}$, be the closed subspace of $(L^p(\omega))^m$. $(M^{n,p}(\omega))^m$ be a Sobolev space along the norm $|\cdot|_{n,p}$.

$B = -\mu P\Delta$ is said to be the Stokes operator in Q_p whose domain is $D_p(B) = D_p(\Delta) \bigcap h_p$. Here

$$D_p(\Delta) = \{v \in (M^{2,p}(\omega))^m : \frac{v}{\partial\omega} = 0\}.$$

It is noted that $-B$ is a closed linear operator as well as generates the bounded analytic semi-group $\{e^{-tB}\}$ on Q_p.

We present new fractional power space definitions that are connected to $-B$. For $\alpha > 0$ as well as $v \in Q_p$, define

$$B^{-\alpha}v = \frac{1}{\Gamma(\alpha)} \int_0^{\infty} t^{\alpha-1} e^{-tB} v dt.$$

$B^{-\alpha}$ is bounded and one-to-one operator on Q_p. Suppose that B^{α} is the inverse of $B^{-\alpha}$. For $\alpha > 0$, indicate the space $H^{\alpha,p}$ according to the range $B^{-\alpha}$ along the norm

$$|v|_{H^{\alpha,p}} = |B^{\alpha}v|_p.$$

It is not difficult to see that e^{-tB} restrict to be a bounded analytic semi-group on $H^{\alpha,p}$, for further details; see [21].

Suppose that Y is a Banach space as well as Q is the interval of $\mathbb{R}$. All continuous Y valued functions are represented by $C(Q,Y)$. So for $0 < \zeta < 1, C^{\zeta}(Q,Y)$ indicates for the set of all functions is Holder continuous along the exponent ζ.

Assume that $\beta \in (0,1)$ as well as $w : [0,\infty) \to Y$, the fractional integral with the order β along the lower limit zero for the function w is defined as

$$I_t^{\beta}w(t) = \int_0^{\infty} h_{\beta}(t-s)w(s)ds, 0 < t,$$

the R.H.S is point-wise defined on the interval $[0,\infty)$, where h_β is said to be the Riemann-Liouville kernel

$$h_{\beta}(t) = \frac{t^{\beta-1}}{\Gamma(\beta)}, 0 < t.$$

${}^{C}D_t^{\beta}$ indicates the Caputo fractional derivative operator with order β. It can be describe as

$${}^{C}D_t^{\beta}w(t) = \frac{d}{dt}[I_t^{1-\beta}(w(t) - w(0))] = \frac{d}{dt}\left(\int_0^t h_{1-\beta}(t-s)(w(t) - w(0))ds\right), 0 < t.$$

Generally, for $w = [0,\infty) \times R^m \to R^m$, Caputo fractional derivative w.r.t time for the function w can be defined as

$$\partial_t^{\beta}v(t,y) = \partial_t\left(\int_0^t h_{1-\beta}(t-s)(v(t,y) - v(t,0))ds\right), 0 < t,$$

for further details; see [22]. Now, we define generalized Mittag-Leffler functions:

$$E_\beta(-t^\beta B) = \int_0^\infty \mathcal{M}_\beta(s)e^{-st^\beta B}ds, E_{\beta,\beta}(-t^\beta B) = \int_0^\infty \beta s\mathcal{M}_\beta(s)e^{-st^\beta B}ds,$$

where $\mathcal{M}(\theta)$ is Mainardi's Wright Type function defined as

$$\mathcal{M}_\beta(\theta) = \sum_{g=0}^{\infty} \frac{\theta^m}{m!\Gamma(1-\beta(1+m))}.$$

Lemma 1. *In uniform operator topology, $0 < t$, $E_\beta(-t^\beta B)$ and $E_{\beta,\beta}(-t^\beta B)$ are continuous. On the interval $[r,\infty]$, the continuity is uniform for every $0 < r$.*

Lemma 2. *Let $0 < \beta < 1$. At that point the following properties holds:*
(i) for every $v \in Y$, $\lim_{t\to 0^+} E_\beta(-t^\beta B)v = v$;
(ii) for every $v \in D(B)$ and $0 < t$, ${}^{C}D_\beta^t E_\beta(-t^\beta B)v = -BE_\beta(-t^\beta B)v$;
(iii) for every $v \in Y$, $E_\beta'(-t^\beta B)v = -t^{\beta-1}BE_{\beta,\beta}(-t^\beta B)v$;
(iv) for $0 < t$, $E_\beta(-t^\beta B)v = I_t^{1-\beta}(t^{\beta-1}E_{\beta,\beta}(-t^\beta B)v)$.

Definition 1. *A function $v : [0,\infty) \to H^{\alpha,p}$ is said to be the global mild solution of (2) in $H^{\alpha,p}$, if $v \in C([0,\infty), H^{\alpha,p})$ and for $t \in [0,\infty)$*

$$\begin{aligned} v(t) &= E_\beta(-t^\beta B)b + \int_0^t (t-s)^{\beta-1}E_{\beta,\beta}(-(t-s)^\beta B)F(v(s),w(s))ds \\ &+ \int_0^t (t-s)^{\beta-1}E_{\beta,\beta}(-(t-s)^\beta B)Pg(s)ds. \end{aligned} \quad (3)$$

Definition 2. *Suppose that $0 < \mathfrak{T} < \infty$. A local mild solution of problem (2) in $H^{\alpha,p}$ or in Q_p, is a function $v : [0,\mathfrak{T}] \to H^{\alpha,p}$ (Q_p), if $v \in C([0,\mathfrak{T}], H^{\alpha,p})$ as well as v fulfils (3) for interval $t \in [0,\mathfrak{T}]$.*

$$\begin{aligned} \varphi(t) &= \int_0^t (t-s)^{\beta-1}E_{\beta,\beta}(-(t-s)^\beta B)g(s)ds \\ \mathcal{U}(v,w) &= \int_0^t (t-s)^{\beta-1}E_{\beta,\beta}(-(t-s)^\beta B)F(v(s),w(s))ds. \end{aligned}$$

Lemma 3. *Suppose that $(Y, \|\cdot\|_Y)$ is a Banach space, $O : Y \times Y \to Y$ be a bi-linear operator as well as K be a non-negative real number in such a way that*

$$\|O(v,w)\|_Y \leq K\|v\|_Y\|w\|_Y, forall v, w \in Y.$$

Then, for some $v_0 \in Y$ with $\|v_0\|_Y < \frac{1}{4K}$, the relation $v = v_0 + O(v,w)$ must have a unique solution $v \in Y$.

The system (2) is equal to the following integral:

$$v(t) = b + \frac{1}{\Gamma(\beta)} \int_0^t (t-s)^{\beta-1} \Big(Bv(s) + F(v(s), w(s)) + Pg(s) \Big) ds, 0 \le t, \tag{4}$$

provided the integral (4) exist.

Theorem 1. *If (4) holds, then*

$$\begin{aligned} v(t) &= E_\beta(-t^\beta B)b + \int_0^t (t-s)^{\beta-1} E_{\beta,\beta}(-(t-s)^\beta B) F(v(s), w(s)) ds \\ &+ \int_0^t (t-s)^{\beta-1} E_{\beta,\beta}(-(t-s)^\beta B) Pg(s) ds, \end{aligned}$$

where

$$E_\beta(-t^\beta B) = \int_0^\infty M_\beta(\theta) T(t^\beta \theta) d\theta, E_{\beta,\beta}(-t^\beta B) = \int_0^\infty \beta\theta M_\beta(\theta) T(t^\beta\theta) d\theta.$$

Proof. Let $\lambda > 0$

$$\nu(\lambda) = \int_0^\infty e^{-\lambda s} v(s) ds, \mu(\lambda) = \int_0^\infty e^{-\lambda s} g(s) ds.$$

Apply Laplace Transformation on (4)

$$\nu(\lambda) = \lambda^{\beta-1}(\lambda^\beta I - B)^{-1} b + (\lambda^\beta I - B)^{-1} \mu(\lambda),$$

for $t \ge 0$

$$\nu(\lambda) = \lambda^{\beta-1} \int_0^\infty e^{-\lambda^\beta s} T(s) b ds + \int_0^\infty e^{-\lambda^\beta s} T(s) \mu(\lambda) ds.$$

Let

$$\phi_\beta(\theta) = \frac{\beta}{\theta^{\beta+1}} M_\beta(\theta^{-\beta}), \beta \in (0,1),$$

and its Laplace Transform is given by

$$\int_0^\infty e^{-\lambda\theta} \phi_\beta(\theta) d\theta = e^{-\lambda^\beta}, \tag{5}$$

using (4), so

$$\begin{aligned} \lambda^{\beta-1} \int_0^\infty e^{-\lambda^\beta s} T(s) b ds &= \int_0^\infty \beta(\lambda t)^{\beta-1} e^{-(\lambda t)^\beta} T(t^\beta) b dt \\ &= \int_0^\infty -\frac{1}{\lambda} \frac{d}{dt} \Big(\int_0^\infty e^{(-\lambda t)^\beta} \phi_\beta(\theta) d\theta \Big) T(t^\beta) b dt \\ &= \int_0^\infty \int_0^\infty \frac{-\lambda\theta}{-\lambda} e^{-\lambda t\theta} \phi_\beta(\theta) T(t^\beta) b dt \\ &= \int_0^\infty \int_0^\infty \theta \phi_\beta(\theta) e^{-\lambda t\theta} T(t^\beta) b dt d\theta \\ &= \int_0^\infty \int_0^\infty \phi_\beta(\theta) e^{-\lambda t} T(\frac{t^\beta}{\theta^\beta}) b d\theta dt \\ &= \int_0^\infty e^{-\lambda t} \Big[\int_0^\infty \phi_\beta(\theta) T(\frac{t^\beta}{\theta^\beta}) b \Big] d\theta dt \\ &= \mathcal{L}\Big[\int_0^\infty M_\beta(\theta) T(t^\beta\theta) b d\theta \Big](\lambda) \\ &= \mathcal{L}[E_\beta(-t^\beta B) b](\lambda). \end{aligned} \tag{6}$$

Similarly

$$\begin{aligned}\int_0^\infty e^{-\lambda^\beta s}T(s)\mu(\lambda)ds &= \int_0^\infty\int_0^\infty \beta t^{\beta-1}e^{(-\lambda t)^\beta}T(t^\beta)e^{-\lambda s}[F(v(s),w(s))+Pg(s)]dsdt \\ &= \int_0^\infty\int_0^\infty\int_0^\infty \beta t^{\beta-1}\phi_\beta(\theta)e^{-\lambda t\theta}T(t^\beta)e^{-\lambda s}[F(v(s),w(s))+Pg(s)]d\theta dsdt \\ &= \int_0^\infty\int_0^\infty\int_0^\infty \beta\frac{t^{\beta-1}}{\theta^\beta}\phi_\beta(\theta)T(\frac{t^\beta}{\theta^\beta})e^{-\lambda(t+s)}[F(v(s),w(s))+Pg(s)]d\theta dsdt \\ &= \int_0^\infty e^{-\lambda t}\Big[\beta\int_0^t\int_0^\infty \phi_\beta(\theta)T\Big(\frac{(t-s)^\beta}{\theta^\beta}\Big)\frac{(t-s)^{\beta-1}}{\theta^\beta} \\ &\quad [F(v(s),w(s))+Pg(s)]d\theta ds\Big]dt. \end{aligned} \tag{7}$$

Combining Equations (5)–(7), one has

$$\begin{aligned}\nu(\lambda) &= \int_0^\infty e^{-\lambda t}\Big[\int_0^\infty \phi_\beta(\theta)T(\frac{t^\beta}{\theta^\beta})bd\theta+\beta\int_0^t\int_0^\infty \phi_\beta(\theta)T\Big(\frac{(t-s)^\beta}{\theta^\beta}\Big)\frac{(t-s)^{\beta-1}}{\theta^\beta} \\ &\quad [F(v(s),w(s))+Pg(s)]d\theta ds\Big].\end{aligned}$$

By applying the Laplace Transform,

$$\begin{aligned}& v(t) \\ =& \int_0^\infty \phi_\beta(\theta)T(\frac{t^\beta}{\theta^\beta})bd\theta+\beta\int_0^t\int_0^\infty \phi_\beta(\theta)T\Big(\frac{(t-s)^\beta}{\theta^\beta}\Big)\frac{(t-s)^{\beta-1}}{\theta^\beta}[F(v(s),w(s))+Pg(s)]d\theta ds \\ =& \int_0^\infty M_\beta(\theta)T(t^\beta\theta)bd\theta+\beta\int_0^t\int_0^\infty \theta(t-s)^{\beta-1}M_\beta(\theta)T((t-s)^\beta\theta)[F(v(s),w(s))+Pg(s)]d\theta ds \\ =& E_\beta(-t^\beta B)b+\int_0^t(t-s)^{\beta-1}E_{\beta,\beta}(-t^\beta B)[F(v(s),w(s))+Pg(s)].\end{aligned}$$

We rewrite the above equation

$$v(t)=b+\frac{1}{\Gamma(\beta)}\int_0^t(t-s)^{\beta-1}\Big(Bv(s)+F(v(s),w(s))+Pg(s)\Big)ds.$$

Thus, the proof is complete. □

Proposition 1. *Prove that*

(i) $E_{\beta,\beta}(-t^\beta B)=\frac{1}{2\pi i}\int_{\Gamma\theta}E_{\beta,\beta}(-\nu t^\beta)(\nu I+B)^{-1}d\nu$;

(ii) $B^\gamma E_{\beta,\beta}(-t^\beta B)=\frac{1}{2\pi i}\int_{\Gamma\theta}\nu^\gamma E_{\beta,\beta}(-\nu t^\beta)(\nu I+B)^{-1}d\nu$

Proof. (i) Since $\int_0^\infty \beta sM_\beta(s)e^{-st^\beta B}ds=E_{\beta,\beta}(-t)$, by using Fabini's Theorem, we get

$$\begin{aligned}E_{\beta,\beta}(-t) &= \int_0^\infty \beta sM_\beta(s)e^{-st^\beta B}ds \\ &= \frac{1}{2\pi i}\int_0^\infty \beta sM_\beta(s)\int_{\Gamma\theta}e^{-\nu st^\beta}(\nu I+B)^{-1}d\nu ds \\ &= \frac{1}{2\pi i}\int_0^\infty \beta sM_\beta(s)e^{-\nu st^\beta}ds\int_{\Gamma\theta}(\nu I+B)^{-1}d\nu \\ &= \frac{1}{2\pi i}\int_{\Gamma\theta}E_{\beta,\beta}(-\nu t^\beta)(\nu I+B)^{-1}d\nu.\end{aligned}$$

(ii) We follow the same steps

$$
\begin{aligned}
B^{\gamma}E_{\beta,\beta}(-t^{\beta}B) &= \int_0^{\infty} \beta s M_{\beta}(s) B^{\gamma} e^{-st^{\beta}B} ds \\
&= \frac{1}{2\pi i}\int_0^{\infty} \beta s M_{\beta}(s) \int_{\Gamma\theta} \nu^{\gamma} e^{-\nu s t^{\beta}} (\nu I + B)^{-1} d\nu ds
\end{aligned}
$$

$$
\begin{aligned}
B^{\gamma}E_{\beta,\beta}(-t^{\beta}B) &= \frac{1}{2\pi i}\int_0^{\infty} \nu^{\gamma} \beta s M_{\beta}(s) e^{-\nu s t^{\beta}} ds \int_{\Gamma\theta} (\nu I + B)^{-1} d\nu \\
&= \frac{1}{2\pi i}\int_{\Gamma\theta} \nu^{\gamma} E_{\beta,\beta}(-\nu t^{\beta})(\nu I + B)^{-1} d\nu.
\end{aligned}
$$

□

3. Global and Local Existence in $H^{\alpha,p}$

In this section, our main purpose is to build up sufficient conditions for the existence and uniqueness of the mild solution of problem (2) in $H^{\alpha,p}$. We suppose that

Hypothesis 1 (H1). *Pg is said to be continuous for $0 < t$ and $|Pg(t)|_p = s(t^{-\beta(1-\alpha)})$ as $t \to 0$ for $1 > \alpha > 0$.*

Lemma 4. *See ([23]). Suppose that $1 < p < \infty$ and $\alpha_1 \le \alpha_2$. Then, at that point there exist a constant $\mathfrak{C} = \mathfrak{C}(\alpha_1, \alpha_2)$ in such a way that*

$$
|e^{-tB} w|_{H^{\alpha_2,p}} \le \mathfrak{C} t^{-(\alpha_2-\alpha_1)} |w|_{H^{\alpha_1,p}}, 0 < t,
$$

for $w \in H^{\alpha_1,p}$. Moreover, $\lim_{t\to 0} t^{(\alpha_2-\alpha_1)} |e^{-tB} w|_{H^{\alpha_2,p}} = 0$.

Lemma 5. *Suppose that $1 < p < \infty$ and $\alpha_1 \le \alpha_2$. For any $R > 0$ there is a constant $\mathfrak{C}_1 = \mathfrak{C}_1(\alpha_1, \alpha_2) > 0$ in such a way that*

$$
|E_{\beta}(-t^{\beta}B)w|_{H^{\alpha_2,p}} \le \mathfrak{C}_1 t^{-\beta(\alpha_2-\alpha_1)} |w|_{H^{\alpha_1,p}} and |E_{\beta,\beta}(-t^{\beta}B)w|_{H^{\alpha_2,p}} \le \mathfrak{C}_1 t^{-\beta(\alpha_2-\alpha_1)} |w|_{H^{\alpha_1,p}}
$$

for all $w \in H^{\alpha_1,p}$ as well as $t \in (0, R]$. Moreover,

$$
\lim_{t\to 0} t^{\beta(\alpha_2-\alpha_1)} |E_{\beta}(-t^{\beta}B)w|_{H^{\alpha_2,p}} = 0.
$$

Proof. Let $w \in H^{\alpha_1,p}$. According to Lemma 4, we consider

$$
\begin{aligned}
|E_{\beta}(-t^{\beta}B)w|_{H^{\alpha_2,p}} &\le \int_0^{\infty} \mathcal{M}_{\beta}(s) |e^{-st^{\beta}B} w|_{H^{\alpha_2,p}} ds \\
&\le \left(\mathfrak{C}\int_0^{\infty} \mathcal{M}_{\beta}(s) s^{-(\alpha_2-\alpha_1)} ds\right) t^{-\beta(\alpha_2-\alpha_1)} |w|_{H^{\alpha_1,p}} \\
&\le \mathfrak{C}_1 t^{-\beta(\alpha_2-\alpha_1)} |w|_{H^{\alpha_1,p}}.
\end{aligned}
$$

A well-known theorem, $\mathcal{L}ebesgue\mathcal{D}ominated\mathcal{C}onvergence$ theorem shows that

$$
\lim_{t\to 0} t^{\beta(\alpha_2-\alpha_1)} |E_{\beta}(-t^{\beta}B)w|_{H^{\alpha_2,p}} \le \int_0^{\infty} \mathcal{M}(s) \lim_{t\to 0} t^{\beta(\alpha_2-\alpha_1)} |E_{\beta}(-t^{\beta}B)w|_{H^{\alpha_2,p}} = 0.
$$

Similarly

$$
|E_{\beta,\beta}(-t^{\beta}B)w|_{H^{\alpha_2,p}} \le \int_0^{\infty} \beta s \mathcal{M}_{\beta}(s) |e^{-st^{\beta}B} w|_{H^{\alpha_2,p}} ds
$$

$$
\begin{aligned}
|E_{\beta,\beta}(-t^{\beta}B)w|_{H^{\alpha_2,p}} &\leq \left(\beta\mathfrak{C}\int_0^\infty \mathcal{M}_\beta(s)s^{1-(\alpha_2-\alpha_1)}ds\right)t^{-\beta(\alpha_2-\alpha_1)}|w|_{H^{\alpha_1,p}} \\
&\leq \mathfrak{C}_1 t^{-\beta(\alpha_2-\alpha_1)}|w|_{H^{\alpha_1,p}},
\end{aligned}
$$

where the constant term is $\mathfrak{C}_1 = \mathfrak{C}_1(\beta,\alpha_1,\alpha_2)$, such that

$$
\mathfrak{C}_1 \geq \mathfrak{C}\max\left\{\frac{\Gamma(1-\alpha_2+\alpha_1)}{\Gamma(1+\beta(\alpha_1-\alpha_2))}, \frac{\beta\Gamma(2-\alpha_2+\alpha_1)}{\Gamma(1+\beta(\alpha_1-\alpha_2))}\right\}.
$$

□

3.1. *Global Existence in $H^{\alpha,p}$*

The global mild solution of (2) in $H^{\alpha,p}$ is investigated in this subsection. For comfort, we signify

$$
\begin{aligned}
\mathcal{N}(t) &= \sup_{s\in(0,t]}\{s^{\beta(1-\alpha)}|Pg(s)|_p\}, \\
V_1 &= \mathfrak{C}_1\max\{V(\beta(1-\alpha),1-\beta(1-\alpha)),V(\beta(1-\xi),1-\beta(1-\alpha))\}, \\
K &\geq \mathcal{M}\mathfrak{C}_1\max\left\{V(\beta(1-\alpha),1-2\beta(\xi-\alpha)),V(\beta(1-\xi),1-2\beta(\xi-\alpha))\right\}.
\end{aligned}
$$

Theorem 2. *Suppose that $1<p<\infty, 0<\alpha<1$ and condition (H_1) holds. For each $\beta \in H^{\alpha,p}$. Let*

$$
\mathfrak{C}_1|b|_{H^{\alpha,p}} + V_1\mathcal{N}_\infty < \frac{1}{4K}, \tag{8}
$$

where $\mathcal{N}_\infty = \sup_{s\in(0,\infty)}\{s^{\beta(1-\alpha)}|Pg(s)|_p\}$. If $\frac{m}{2p}-\frac{1}{2}<\alpha$, then at that point there is $b\xi > \max\{\alpha,\frac{1}{2}\}$ and a unique function $v:[0,\infty)\to H^{\alpha,p}$ fulfils the conditions given below:

(i) $v:[0,\infty)\to H^{\alpha,p}$ is continuous as well as $v(0)=b$;

(ii) $v:[0,\infty)\to H^{\xi,p}$ is continuous as well as $\lim_{t\to 0} t^{\beta(\xi-\alpha)}|v(t)|_{H^{\xi,p}}=0$;

(iii) v fulfils (3) for $t\in[0,\infty)$.

Proof. The proof of this theorem is similar to that in [24] with a slight change according to our problem. □

3.2. *Local Existence in $H^{\alpha,p}$*

The local mild solution of (2) in $H^{\alpha,p}$ is discussed in this section.

Theorem 3. *Let $1<p<\infty, 0<\alpha<1$ and* **(H1)** *(the supposition is given in the beginning of Section 3) holds. Assume that*

$$
\frac{m}{2p}-\frac{1}{2}<\alpha.
$$

Then, there is $\xi > \max\{\alpha,\frac{1}{2}\}$ in such a way that for each $b\in H^{\alpha,p}$ there exist $\mathfrak{T}_*>0$ as well as $v:[0,\mathfrak{T}_*]\to H^{\alpha,p}$ is a unique function that fulfils the following properties:

(i) $v:[0,\mathfrak{T}_*]\to H^{\alpha,p}$ is continuous and $v(0)=b$;

(ii) $v:[0,\mathfrak{T}_*]\to H^{\xi,p}$ is continuous and $\lim_{t\to 0} t^{\beta(\xi-\alpha)}|v(t)|_{H^{\xi,p}}=0$;

(iii) For $t\in[0,\mathfrak{T}_*]$, v satisfy (3).

Proof. Suppose that $\xi = \frac{1+\alpha}{2}$ and the space of all curves is $Y = Y[\mathfrak{T}]\ v : (0, \mathfrak{T}] \to H^{\alpha,p}$ in such a way that:

(i) $v : [0, \mathfrak{T}_*] \to H^{\alpha,p}$ is continuous and $v(0) = b$;

(ii) $v : [0, \mathfrak{T}_*] \to H^{\xi,p}$ is continuous and $\lim_{t\to 0} t^{\beta(\xi-\alpha)}|v(t)|_{H^{\xi,p}} = 0$; with its neutral form

$$\|v\|_Y = \sup_{t\in[0,\mathfrak{T}]} \{t^{\beta(\xi-\alpha)}|v(t)|_{H^{\xi,p}}\}.$$

Alike the proof of Theorem 2, it is not difficult to claim that $\mathcal{U} : Y \times Y \to Y$ be continuous linear mapping as well as $\varphi(t) \in Y$.

$$\begin{aligned} E_\beta(-t^\beta B)b &\in \mathfrak{C}([0, \mathfrak{T}], H^{\alpha,p}), \\ E_\beta(-t^\beta B)b &\in \mathfrak{C}([0, \mathfrak{T}], H^{\xi,p}). \end{aligned}$$

By Lemma 5, it can easily be seen that

$$\begin{aligned} E_\beta(-t^\beta B)b &\in Y, \\ t^{\beta(\xi-\alpha)} E_\beta(-t^\beta B)b &\in \mathfrak{C}([0, \mathfrak{T}], H^{\xi,p}). \end{aligned}$$

Therefore, let $\mathfrak{T}_* > 0$ be small in such a way that

$$\|E_\beta(-t^\beta B)b + \varphi(t)\|_{Y[\mathfrak{T}_*]} \le \|E_\beta(-t^\beta B)b\|_{Y[\mathfrak{T}_*]} + \|\phi(t)\|_{Y[\mathfrak{T}_*]} < \frac{1}{4K}.$$

As a result of Lemma 3, $\mathcal{F}$ has a fixed point that is unique. □

4. Local Existence in Q_p

In this section, we discuss the local mild solution of (2) by using iteration method. Suppose that $\xi = \frac{1+\alpha}{2}$:

Theorem 4. *Suppose that $1 < p < \infty, 0 < \alpha < 1$ and (**H1**)(the supposition is given in the beginning of Section 3) holds. Assume that*

$$b \in H^{\alpha,p} with \frac{m}{2p} - \frac{1}{2} < \alpha.$$

Then, the problem (2) has mild solution v by Q_p for $b \in H^{\alpha,p}$. Furthermore, v must be continuous on $(0, \mathfrak{T}]$, $B^\xi v$, be continuous on $(0, \mathfrak{T}]$ and $t^{\beta(\xi-\alpha)} B^\xi v(t)$ is bounded as $t \to 0$.

Proof. Step 1: Describe

$$\mathfrak{R}(t) := \sup_{s\in(0,t]} s^{\beta(\xi-\alpha)}|B^\xi v(s)|_p,$$

and

$$\psi(t) := \mathcal{U}(v, w)(t) = \int_0^t (t-s)^{\beta-1} E_{\beta,\beta}(-(t-s)^\beta B) F(v(s) - w(s)) ds.$$

$$|B^\xi \psi(t)|_p \le \mathcal{N}\mathfrak{C}_1 V(\beta(1-\xi), 1 - 2\beta(\xi-\alpha))\mathfrak{R}^2(t) t^{-\beta(\xi-\alpha)},$$

considering the integral $\varphi(t)$. Thus

$$|Pg(s)|_p \le \mathcal{N}(t) s^{\beta(1-\alpha)},$$

where $\mathcal{N}$ is a continuous function. Using Theorem 2, we show that $B^{\xi}(\mathrm{t})$ is continuous in the interval $(0, \mathfrak{T}]$ by using

$$|B^{\xi}\varphi(\mathrm{t})|_p \leq \mathfrak{C}_1\mathcal{N}(\mathrm{t})V(\beta(1-\xi), 1-\beta(1-\alpha))\mathrm{t}^{-\beta(\xi-\alpha)}. \tag{9}$$

For $|Pg(\mathrm{t})|_p = s(\mathrm{t}^{-\beta(1-\alpha)})$ as $\mathrm{t} \to 0$, $\mathcal{N}(\mathrm{t}) = 0$ is the solution. Here, (9) denotes, $|B^{\xi}\varphi(\mathrm{t})|_p s(\mathrm{t}^{-\beta(1-\alpha)})$ as $\mathrm{t} \to 0$. In Q_p, we show that φ is continuous. In fact, if we take $0 \leq \mathrm{t}_0 < \mathrm{t} < \mathfrak{T}$, we get

$$\begin{aligned}
& |\varphi(\mathrm{t}) - \varphi(\mathrm{t}_0)|_p \\
\leq\; & \mathfrak{C}_3 \int_{\mathrm{t}_0}^{\mathrm{t}} (\mathrm{t}-s)^{\beta-1}|Pg(s)|_p ds + \mathfrak{C}_3 \int_0^{\mathrm{t}_0} ((\mathrm{t}_0-s)^{\beta-1} - (\mathrm{t}-s)^{\beta-1})|Pg(s)|_p ds \\
+\; & \mathfrak{C}_3 \int_0^{\mathrm{t}_0-\epsilon} (\mathrm{t}_0-s)^{\beta-1}\|E_{\beta,\beta}(-(\mathrm{t}-s)^{\beta}B) - E_{\beta,\beta}(-(\mathrm{t}_0-s)^{\beta}B\||Pg(s)|_p ds \\
+\; & 2\mathfrak{C}_3 \int_{\mathrm{t}_0-\epsilon}^{\mathrm{t}_0} (\mathrm{t}_0-s)^{\beta-1}|Pg(s)|_p ds \\
\leq\; & \mathfrak{C}_3\mathcal{N}(\mathrm{t}) \int_{\mathrm{t}_0}^{\mathrm{t}} (\mathrm{t}-s)^{\beta-1}s^{-\beta(1-\alpha)} ds + \mathfrak{C}_3\mathcal{N}(\mathrm{t}) \int_0^{\mathrm{t}} ((\mathrm{t}-s)^{\beta-1} - (\mathrm{t}_0-s)^{\beta-1})s^{-\beta(1-\alpha)} ds \\
+\; & \mathfrak{C}_3\mathcal{N}(\mathrm{t}) \int_0^{\mathrm{t}_0-\epsilon} (\mathrm{t}_0-s)^{\beta-1}s^{-\beta(1-\alpha)} ds \sup_{s\in[0,\mathrm{t}-\epsilon]} \|E_{\beta,\beta}(-(\mathrm{t}-s)^{\beta}B) - E_{\beta,\beta}(-(\mathrm{t}_0-s)^{\beta}B\| \\
+\; & 2\mathfrak{C}_3\mathcal{N}(\mathrm{t}) \int_{\mathrm{t}_0-\epsilon}^{\mathrm{t}_0} (\mathrm{t}_0-s)^{\beta-1}s^{-\beta(1-\alpha)} ds \to 0, as \mathrm{t} \to \mathrm{t}_0,
\end{aligned}$$

as a result of previous conversations.

We also consider the function $E_{\beta}(-\mathrm{t}^{\beta}B)b$. It is clear by Lemma 5 that

$$\begin{aligned}
|B^{\xi}E_{\beta}(-\mathrm{t}^{\beta}B)b|_p &\leq \mathfrak{C}_1\mathrm{t}^{-\beta(1-\alpha)}|B^{\alpha}b|_p = \mathfrak{C}_1\mathrm{t}^{-\beta(1-\alpha)}|b|_{H^{\alpha,p}}, \\
\lim_{\mathrm{t}\to 0} \mathrm{t}^{\beta(\xi-\alpha)}|B^{\xi}E_{\beta}(-\mathrm{t}^{\beta}B)b|_p &= \lim_{\mathrm{t}\to 0} \mathrm{t}^{\beta(\xi-\alpha)}|E_{\beta}(-\mathrm{t}^{\beta}B)b|_{H^{\alpha,p}} = 0.
\end{aligned}$$

Step 2: Now, we derive the result using successive approximations:

$$\begin{aligned}
v_0(\mathrm{t}) &= E_{\beta}(-\mathrm{t}^{\beta}B)b + \varphi(\mathrm{t}), \\
v_{m+1} &= v_0(\mathrm{t}) + \mathcal{U}(v_m, w_m)(\mathrm{t}), m = 0, 1, 2 \cdots.
\end{aligned} \tag{10}$$

Using the information presented above, we can deduce that

$$\mathfrak{R}_m(\mathrm{t}) := \sup_{s\in(0,\mathrm{t}]} s^{\beta(\xi-\alpha)}|B^{\xi}v_m(s)|_p$$

are increasing and continuous functions on $[0, \mathfrak{T}]$ with $\mathfrak{R}_m(0) = 0$. Furthermore, $\mathfrak{R}_m(t)$ fulfils the following inequality as a result of (9) and (10):

$$\mathfrak{R}_{m+1}(\mathrm{t}) \leq \mathfrak{R}_0(\mathrm{t}) + \mathcal{N}\mathfrak{C}_1 V(\beta(1-\xi), 1-2\beta(\xi-\alpha))\mathfrak{R}_m^2(\mathrm{t}). \tag{11}$$

We choose $\mathfrak{T} > 0$ such that $\mathfrak{R}_0(0) = 0$,

$$4\mathcal{N}\mathfrak{C}_1 V(\beta(1-\xi), 1-2\beta(\xi-\alpha))\mathfrak{R}_0(\mathfrak{T}) < 1. \tag{12}$$

The sequence $\mathfrak{R}_m(\mathfrak{T})$ is thus bounded, according to a fundamental consideration of (11).

$$\mathfrak{R}_m(\mathfrak{T}) \leq \varrho(\mathfrak{T}), m = 0, 1, 2...,$$

where

$$\varrho(t)=\frac{1-\sqrt{1-4\mathcal{N}\mathfrak{C}_1V(\beta(1-\xi),1-2\beta(\xi-\alpha))\mathfrak{K}_0(t)}}{2\mathcal{N}V\mathfrak{C}_1(\beta(1-\xi),1-2\beta(\xi-\alpha))}.$$

In the same way, $\mathfrak{R}_m(t)\leq\varrho(t)$ holds for any $t\in(0,\mathfrak{T})$. Similarly, we may see that

$$\varrho(t)\leq 2\mathfrak{R}_0(t).$$

Suppose that the equality

$$k_{m+1}(t)=\int_0^t(t-s)^{\beta-1}E_{\beta,\beta}(-(t-s)^{\beta}B)[F(v_{m+1}(s),w_{m+1}(s))-F(v_m(s),w_m(s))]ds,$$

where $k_m=v_{m+1}-v_n, m=0,1,...,$ as well as $t\in(0,\mathfrak{T}]$. Writing

$$\mathcal{W}_m(t):=\sup_{s\in(0,t]}s^{\beta(\xi-\alpha)}|B^{\xi}k_m(s)|_p.$$

By Equation (8), we get

$$|J(v_{m+1}(s),w_{m+1}(s))-J(v_m(s),w_m(s))|_p\leq\mathcal{N}(\mathfrak{R}_{m+1}(s)+\mathfrak{R}_m(t))J_m(s)s^{-2\beta(\xi-\alpha)},$$

by Theorem 2, we have

$$t^{\beta(\xi-\alpha)}|B^{\xi}k_{m+1}(t)|\leq 2\mathcal{N}\mathfrak{C}_1V(\beta(1-\xi),1-\beta(1-\alpha))\varrho(t)\mathcal{W}_m(t).$$

The above inequality gives

$$\begin{aligned}\mathcal{W}_{m+1}(\mathfrak{T}) &\leq 2\mathcal{N}\mathfrak{C}_1V(\beta(1-\xi),1-\beta(1-\alpha))\varrho(t)\mathcal{W}_m(t)\\ &\leq 4\mathcal{N}\mathfrak{C}_1V(\beta(1-\xi),1-\beta(1-\alpha))\varrho(t)\mathfrak{R}_0(t)\mathcal{W}_m(t).\end{aligned}\tag{13}$$

By Equations (12) and (13), it is not difficult to show that

$$\lim_{m\to 0}\frac{J_{m+1}(\mathfrak{T})}{J_m(\mathfrak{T})}<4\mathcal{N}\mathfrak{C}_1V(\beta(1-\xi),1-\beta(1-\alpha))<1,$$

as a result, the series $\Sigma_{m=0}^{\infty}J_m(\mathfrak{T})$ converge. It prove that for $t\in(0,\mathfrak{T}]$ the series

$$\Sigma_{m=0}^{\infty}t^{\beta(\xi-\alpha)}B^{\xi}k_m(t)$$

converge uniformly. As a result, the sequence $t^{\beta(\xi-\alpha)}B^{\xi}v_m(t)$ converge uniformly in $(0,\mathfrak{T}]$. This suggest that

$$\lim_{m\to 0}v_m(t)=v(t)\in D(B^{\xi})$$

as well as

$$\lim_{m\to 0}t^{\beta(\xi-\alpha)}B^{\xi}v_m(t)=t^{\beta(\xi-\alpha)}B^{\xi}v(t) uniformly,$$

since B^{ξ} is both bounded and $B^{-\xi}$ is closed. As a result, the function

$$\mathfrak{R}(t)=\sup_{s\in(0,\mathfrak{T}]}t^{\beta(\xi-\alpha)}|B^{\xi}v(s)|_p$$

also meets the condition

$$\mathfrak{R}(t)\leq\varrho(t)\leq 2\mathfrak{R}_0(t), t\in(0,t].\tag{14}$$

as well as

$$\begin{aligned} S_m &:= \sup_{s\in(0,\mathfrak{T}]} s^{2\beta(\xi-\alpha)}|F(v_m(s),w_m(s)) - F(v(s),w(s))|_p \\ &\le \mathcal{N}(\mathfrak{R}_m(\mathfrak{T}) + \mathfrak{R}(\mathfrak{T})) \sup_{s\in(0,\mathfrak{T}]} s^{\beta(\xi-\alpha)}|B^{\xi}(v_m(s) - v(s))|_p \to 0, as m \to \infty. \end{aligned}$$

Finally, make sure that v in $[0,\mathfrak{T}]$ is a mild solution to problem (2). Since

$$|\mathcal{U}(v_n,w_n)(t) - \mathcal{U}(v,w)|_p \le \int_0^t (t-s)^{\beta-1} S_m s^{-2\beta(\xi-\alpha)} ds = t^{\beta\alpha} S_m \to 0, (m\to\infty),$$

we have $\mathcal{U}(v_m,w_m)(t) \to \mathcal{U}(v,w)(t)$. We get (9) by taking the limits on both sides

$$v(t) = v_0(t) + \mathcal{U}(v,w)(t). \tag{15}$$

If we set $v(0) = b$, we get (15) for $t \in [0,\mathfrak{T}]$ and $v \in \mathfrak{C}([0,\mathfrak{T}], Q_p)$. Furthermore, the consistent convergence of $t^{\beta(\xi-\alpha)} B^{\xi} v_m(t) to t^{\beta(\xi-\alpha)} B^{\xi} v(t)$ drive the continuity of $B^{\xi}v(t) on (0,\mathfrak{T}]$. According to (14) and $\mathfrak{R}_0(0) = 0$, we have $|B^{\xi}v(t)|_p = s(t^{-\beta(\xi-\alpha)})$ is obvious.

Step 3: We show that the mild solution is unique. Assume that v and w are the mild solutions of problem (2). We consider the equality $k = v - w$

$$k(t) = \int_0^t (t-s)^{\beta-1} E_{\beta,\beta}(-(t-s)^{\beta}B)[F(v(s),v(s)) - F(w(s),w(s))]ds.$$

Introducing the function

$$\tilde{\mathfrak{R}}(t) := \max\{\sup_{s\in(0,t]} s^{\beta(\xi-\alpha)}|B^{\xi}v(s)|_p, \sup_{s\in(0,t]} s^{\beta(\xi-\alpha)}|B^{\xi}w(s)|_p\}.$$

By (8) and Lemma 5, we get

$$|B^{\xi}k(t)|_p \le \mathcal{N}\mathfrak{C}_1\tilde{\mathfrak{R}}(t) \int_0^t (t-s)^{\beta(1-\xi)-1} s^{-\beta(\xi-\alpha)} |B^{\xi}k(s)|_p ds.$$

For $t \in (0,\mathfrak{T})$, the Gronwall inequality demonstrates that $B^{\xi}k(t) = 0$. Since $t \in [0,\mathfrak{T}]$, this means that $k(t) = v(t) - w(t) = 0$. As a result, the mild solution is unique. □

5. Regularity

Considering the regularity of v which satisfy (2), overall in this section, we suppose that:

Hypothesis 2 (H2). *$Pg(t)$ be the Holďer continuous along the exponent $\theta \in (0, \beta(1-\xi))$, i.e,*

$$|Pg(t) - Pg(s)|_p \le K|t-s|^{\theta}, \forall t > 0, s \le \mathfrak{T}.$$

Definition 3. *The function $v : [0,\mathfrak{T}] \to Q_p$ is said to be the classical solution of (2), if $v \in \mathfrak{C}([0,\mathfrak{T}], Q_p)$ with ${}^C D_t^t v(t) \in \mathfrak{C}([0,\mathfrak{T}], Q_p)$, which takes the value of $D(B)$ and satisfy (2) for every $t \in (0,\mathfrak{T}]$.*

Lemma 6. *Let (**H2**) (the supposition is given in the beginning of Sec. 5) be fulfilled. If*

$$\varphi_1(t) := \int_0^t (t-s)^{\beta-1} E_{\beta,\beta}(-(t-s)^{\beta}B)(Pg(s) - Pg(t))ds, t \in (0,\mathfrak{T}],$$

then $\varphi_1(t) \in D(B)$ and $B\varphi_1(t)\mathfrak{C}^{\theta}([0,\mathfrak{T}], Q_p)$.

Proof. As

$$
\begin{aligned}
(\mathrm{t}-s)^{\beta-1}|BE_{\beta,\beta}(-(\mathrm{t}-s)^{\beta}B)(Pg(s)-Pg(\mathrm{t}))|_p &\leq (\mathrm{t}-s)^{-1}|(Pg(s)-Pg(\mathrm{t}))|_p \\
&\leq \mathfrak{C}_1K(\mathrm{t}-s)^{\theta-1}\in\mathcal{L}^1([0,\mathfrak{T}],Q_p), \quad (16)
\end{aligned}
$$

then

$$
\begin{aligned}
|B\varphi_1(\mathrm{t})|_p &\leq \int_0^{\mathrm{t}}(\mathrm{t}-s)^{\beta-1}|BE_{\beta,\beta}(-(\mathrm{t}-s)^{\beta}B)(Pg(s)-Pg(\mathrm{t}))|_p ds \\
&\leq \mathfrak{C}_1K\int_0^{\mathrm{t}}(\mathrm{t}-s)^{\theta-1}\leq\frac{\mathfrak{C}_1R}{\theta}\mathrm{t}^{\theta}<\infty.
\end{aligned}
$$

We must show that $B\varphi_1(\mathrm{t})$ is *Hölder* continuous.

$$
\frac{d}{d\mathrm{t}}(\mathrm{t}^{\beta-1}E_{\beta,\beta}(-\nu\mathrm{t}^{\beta}))=\mathrm{t}^{\beta-2}E_{\beta,\beta-1}(-\nu\mathrm{t}^{\beta}),
$$

then

$$
\begin{aligned}
\frac{d}{d\mathrm{t}}(\mathrm{t}^{\beta-1}E_{\beta,\beta}(-\nu\mathrm{t}^{\beta})) &= \frac{1}{2\pi i}\int_{\Gamma\theta}\mathrm{t}^{\beta-2}E_{\beta,\beta-1}(-\nu\mathrm{t}^{\beta})B(\nu I+B)^{-1}d\nu \\
&= \frac{1}{2\pi i}\int_{\Gamma\theta}\mathrm{t}^{\beta-2}E_{\beta,\beta-1}(-\nu\mathrm{t}^{\beta})d\nu-\frac{1}{2\pi i}\int_{\Gamma\theta}\mathrm{t}^{\beta-2}\nu E_{\beta,\beta-1}(-\nu\mathrm{t}^{\beta})(\nu I+B)^{-1}d\nu \\
&= \frac{1}{2\pi i}\int_{\Gamma\theta}-\mathrm{t}^{\beta-2}E_{\beta,\beta-1}(\zeta)\frac{1}{\mathrm{t}^{\beta}}d\zeta \\
&- \frac{1}{2\pi i}\int_{\Gamma\theta}-\mathrm{t}^{\beta-2}E_{\beta,\beta-1}(\zeta)\frac{\zeta}{\mathrm{t}^{\beta}}\left(-\frac{\zeta}{\mathrm{t}^{\beta}}I+B\right)^{-1}\frac{1}{\mathrm{t}^{\beta}}d\zeta.
\end{aligned}
$$

In view of $\|\nu I+B\|\leq\frac{\mathfrak{C}}{|\nu|}$, we derive that

$$
\left\|\frac{d}{d\mathrm{t}}(\mathrm{t}^{\beta-1}E_{\beta,\beta}(-\mathrm{t}^{\beta}B))\right\|\leq\mathfrak{C}_{\beta}\mathrm{t}^{-2}, 0<\mathrm{t}<\mathfrak{T}.
$$

By the Mean Value Theorem, for each $\mathfrak{T}\geq\mathrm{t}>s>0$, we get

$$
\begin{aligned}
\|\mathrm{t}^{\beta-1}E_{\beta,\beta}(-\mathrm{t}^{\beta}B)-s^{\beta-1}BE_{\beta,\beta}(-s^{\beta}B)\| &= \left\|\int_s^{\mathrm{t}}(\tau^{\beta-1}BE_{\beta,\beta}(\tau^{\beta}B))d\tau\right\| \\
&\leq \left\|\int_s^{\mathrm{t}}(\tau^{\beta-1}BE_{\beta,\beta}(\tau^{\beta}B))\right\|d\tau \\
&\leq \mathfrak{C}_{\beta}\int_s^{\mathrm{t}}\tau^{-2}d\tau=\mathfrak{C}+\beta(s^{-1}-\mathrm{t}^{-1}). \quad (17)
\end{aligned}
$$

Let $k>0$ in such a way that $0<\mathrm{t}<\mathrm{t}+k\leq\mathfrak{T}$, then

$$
\begin{aligned}
B\varphi_1(\mathrm{t}+k)-B\varphi_1(\mathrm{t}) &= \int_0^{\mathrm{t}}((\mathrm{t}+k-s)^{\beta-1}BE_{\beta,\beta}(-(\mathrm{t}+k-s)^{\beta}B)) \\
&- (\mathrm{t}-s)^{\beta-1}BE_{\beta,\beta}(-(\mathrm{t}+k-s)^{\beta}B)(Pg(s)-Pg(\mathrm{t}))ds \\
&+ \int_0^{\mathrm{t}}(\mathrm{t}+k-s)^{\beta-1}BE_{\beta,\beta}(-(\mathrm{t}+k-s)^{\beta}B)(Pg(\mathrm{t})-Pg(\mathrm{t}+k))ds \\
&+ \int_{\mathrm{t}}^{\mathrm{t}+k}(\mathrm{t}+k-s)^{\beta-1}BE_{\beta,\beta}(-(\mathrm{t}+k-s)^{\beta}B)(Pg(\mathrm{t})-Pg(\mathrm{t}+k))ds \\
&:= k_1(\mathrm{t})+k_2(\mathrm{t})+k_3(\mathrm{t}).
\end{aligned}
$$

We discuss these terms step by step. For $k_1(t)$, by (16) and (**H1**), we get

$$\begin{aligned} |k_1(t)|_p &\leq \int_0^t \|(t+k-s)^{\beta-1}BE_{\beta,\beta}(-(t+k-s)^{\beta}B) \\ &- (t-s)^{\beta-1}BE_{\beta,\beta}(-(t-s)^{\beta}B)\|\|(Pg(s)-Pg(t))|_p ds \\ &\leq K\mathfrak{C}_{\beta}k\int_0^t (t+k-s)^{-1}(t-s)^{\theta-1}ds \\ &\leq K\mathfrak{C}_{\beta}k\int_0^t (s+k)^{-1}(t-s)^{\theta-1}ds \\ &\leq \mathfrak{C}_{\beta}K\int_0^k \frac{k}{s+k}s^{\theta-1}ds + KC_{\beta}k\int_h^{\infty}\frac{s}{s+k}s^{\theta-1}ds, \end{aligned}$$

so

$$|k_1(t)|_p \leq K\mathfrak{C}_{\beta}k^{\theta}. \tag{18}$$

For $k_2(t)$, by using Lemma 5 and (**H2**),

$$\begin{aligned} |k_2(t)|_p &\leq \int_0^t (t+k-s)^{\beta-1}|BE_{\beta,\beta}(-(t+k-s)^{\beta}B)(Pg(t)-Pg(t+k))|_p ds \\ &\leq \mathfrak{C}_1\int_0^t (t+k-s)^{-1}|(Pg(t)-Pg(t+k))|_p ds \\ &\leq K\mathfrak{C}_1k^{\theta}\int_0^t (t+k-s)^{-1}ds \\ &= K\mathfrak{C}_1[\ln k - \ln(t+k)]k^{\theta}. \end{aligned} \tag{19}$$

Moreover, for $k_3(t)$, again we use (**H2**) and Lemma 5, we get

$$\begin{aligned} |k_3(t)|_p &\leq \int_t^{t+k} (t+k-s)^{\beta-1}|BE_{\beta,\beta}(-(t+k-s)^{\beta}B)(Pg(t)-Pg(t+k))|_p ds \\ &\leq \mathfrak{C}_1\int_t^{t+k} (t+k-s)^{-1}|(Pg(s)-Pg(t+k))|_p ds \\ &\leq \mathfrak{C}_1K\int_t^{t+k} (t+k-s)^{\theta-1}ds = \mathfrak{C}_1K\frac{k^{\theta}}{\theta}. \end{aligned} \tag{20}$$

Combining Equations (18), (19) and (20), we conclude that $B\varphi_1(t)$ is Hölder continuous. □

Theorem 5. *Assume that the suppositions of Theorem 4 are fulfilled. The mild solution of Theorem 4 is classic if for each $b \in D(B)$, (**H2**) holds.*

Proof. In the case of $b \in D(B)$, Part (ii) of Lemma 2 show that $v(t) = E_{\beta}(-t^{\beta}B)b(0 < t)$ the following problem has a classic solution:

$$\begin{cases} {}^{C}D_t^{\beta}v = -Bv, 0 < t, \\ v(0) = b. \end{cases}$$

Step 1: We show that

$$\varphi(t) = \int_0^t (t-s)^{\beta-1}E_{\beta,\beta}(-((t-s)^{\beta}B)Pg(s)ds,$$

is classic solution of the problem

$$\begin{cases} {}^{C}D_t^{\beta}v = -Bv + Pg(t), 0 < t, \\ v(0) = b. \end{cases}$$

From Theorem 4 $\varphi \in \mathfrak{C}([0,\mathfrak{T}],Q_p)$, we write $\varphi(t) = \varphi_1(t) + \varphi_2(t)$, where

$$\varphi_1(t) = \int_0^t (t-s)^{\beta-1} E_{\beta,\beta}(-(t-s)^\beta B)(Pg(t) - Pg(t+k))ds$$

$$\varphi_2(t) = \int_0^t (t-s)^{\beta-1} E_{\beta,\beta}(-(t-s)^\beta B)Pg(t)ds.$$

$$B\varphi_2(t) = Pg(t) - E_\beta(-t^\beta B)Pg(t).$$

Since (**H2**) hold, it observes that

$$|B\varphi_2(t)|_p \leq (1 + (\mathfrak{C}_1)|Pg(t)|_p,$$

as a result

$$\varphi_2(t) \in D(B) aswellas B\varphi_2(t) \in \mathfrak{C}^\mu((0,\mathfrak{T}],Q_p) fort \in (0,\mathfrak{T}].$$

We also explain that ${}^C D_t^\beta \varphi \in \mathfrak{C}((0,\mathfrak{T}],Q_p)$. By Lemma 2(iv), as well as $\varphi(0) = 0$, we get

$${}^C D_t^\beta \varphi(t) = \frac{d}{dt}(I_t^{1-\beta}\varphi(t)) = \frac{d}{dt}(E_\beta(-t^\beta B) * Pg).$$

It remains to show that $E_\beta(-t^\beta B) * Pg$ is continuously differentiable in Q_p. Suppose that $\mathfrak{T} - t \geq k > 0$, we have

$$\begin{aligned}\frac{1}{k}(E_\beta(-(t+k)^\beta B) * Pg - E_\beta(-t^\beta B) * Pg) &= \int_0^t \frac{1}{k}(E_\beta(-(t+k-s)^\beta B)Pg(s) \\ &- E_\beta(-(t-s)^\beta B)Pg(s))ds \\ &+ \frac{1}{k}\int_0^{t+k} E_\beta(-(t+k-s)^\beta B)Pg(s).\end{aligned}$$

Note that

$$\begin{aligned}&\int_0^t \frac{1}{k}|E_\beta(-(t+k-s)^\beta B)Pg(s) - E_\beta(-(t-s)^\beta B)Pg(s)|_p ds \\ \leq\ & \mathfrak{C}_1\frac{1}{k}\int_0^t |E_\beta(-(t-s)^\beta B)Pg(s)|_p \\ +\ & \mathfrak{C}_1\frac{1}{k}\int_0^t |E_\beta(-(t+k-s)^\beta B)Pg(s)|_p ds \\ \leq\ & \mathfrak{C}_1\mathcal{N}(t)\frac{1}{k}\int_0^t (t+k-s)^{-\beta}s^{-\beta(1-\alpha)}ds \\ +\ & \mathfrak{C}_1\mathcal{N}(t)\frac{1}{k}\int_0^t (t-s)^{-\beta}s^{-\beta(1-\alpha)}ds \\ \leq\ & \mathfrak{C}_1\mathcal{N}(t)\frac{1}{k}((t+k)^{1-\beta} + t^{1-\beta}) \\ & V(1-\beta, 1-\beta(1-\alpha)),\end{aligned}$$

according to Dominated Convergence Theorem, we note that

$$\begin{aligned}&\lim_{k\to 0}\int_0^t \frac{1}{k}(E_\beta(-(t+k-s)^\beta B)Pg(s) - E_\beta(-(t-s)^\beta B)Pg(s))ds \\ =\ & \int_0^t (t-s)^{\beta-1} B E_{\beta,\beta}(-(t-s)^\beta B)Pg(s)ds \\ =\ & B\varphi(t).\end{aligned}$$

Furthermore,

$$\begin{aligned}\frac{1}{k}\int_{t}^{t+k}E_{\beta}(-(t+k-s)^{\beta}B)Pg(s) &= \frac{1}{k}\int_{0}^{k}E_{\beta}(-s^{\beta}B)Pg(t+k-s)ds\\ &= \frac{1}{k}\int_{0}^{k}E_{\beta}(-s^{\beta}B)\big(Pg(t+k-s)ds-Pg(t-s)\big)ds\\ &+ \frac{1}{k}\int_{0}^{k}E_{\beta}(-s^{\beta}B)\big(Pg(t-s)-Pg(t)\big)ds\\ &+ \frac{1}{k}\int_{0}^{k}E_{\beta}(-s^{\beta}B)Pf(s)ds.\end{aligned}$$

By Lemma 1 and 5 and (**H2**), we get

$$\begin{aligned}\left|\frac{1}{k}\int_{0}^{k}E_{\beta}(-s^{\beta}B)\big(Pg(t+k-s)ds-Pg(t-s)\big)ds\right|_{p} &\leq \mathfrak{C}_1k^{\theta},\\ \left|\frac{1}{k}\int_{0}^{k}E_{\beta}(-s^{\beta}B)\big(Pg(t-s)-Pg(t)\big)ds\right|_{p} &\leq \mathfrak{C}_1K\frac{k^{\theta}}{\theta+1}.\end{aligned}$$

We conclude that $E_{\beta}(t^{\beta}B)*Pg$ is differentiable at t_{+} as well as $\frac{d}{dt}\big(E_{\beta}(t^{\beta}B)*Pg\big)_{+}=B\varphi(t)+Pg(t)$. Same as $E_{\beta}(t^{\beta}B)*Pg$ is differentiable at t_{-} as well as $\frac{d}{dt}\big(E_{\beta}(t^{\beta}B)*Pg\big)_{-}=B\varphi(t)+Pg(t)$.

We indicate $\varphi(t):=E_{\beta}(-t^{\beta}B)b$. According to Lemma 2(iv) and (5)

$$\begin{aligned}|B^{\xi}\varphi(t+k)-B^{\xi}\varphi(t)|_p &= \left|\int_{t}^{t+k}-s^{\beta-1}B^{\xi}E_{\beta,\beta}(-s^{\beta-1}B)bds\right|_{p}\\ &\leq \int_{t}^{t+k}s^{\beta-1}|B^{\xi-\alpha}E_{\beta,\beta}(-s^{\beta-1}B)B^{\beta}b|_{p}ds\\ &\leq L_1\int_{t}^{t+k}s^{\beta(1+\alpha-\xi)-1}ds|B^{\beta}b|_p\\ &= \frac{L_1|b|_{H^{\alpha,p}}}{\beta(1+\alpha-\xi)}k^{\beta(1+\alpha-\xi)}.\end{aligned}$$

Thus, $B^{\xi}\varphi\in\mathfrak{C}^{\theta}((0,\mathfrak{T}],Q_p)$.

For each small $\epsilon>0$, take k in such a way that $\epsilon\leq t<t+k\leq k$, since

$$\begin{aligned}|B^{\xi}\varphi(t+k)-B^{\xi}\varphi(t)|_p &\leq \left|\int_{t}^{t+k}(t+k-s)^{\beta-1}B^{\xi}E_{\beta,\beta}(-(t+k-s)^{\beta}B)Pg(s)ds\right|_{p}\\ &+ \Big|B^{\xi}\big((t+k-s)^{\beta-1}E_{\beta,\beta}(-(t+k-s)^{\beta}B)\\ &- (t-s)^{\beta-1}E_{\beta,\beta}(-(t-s)^{\beta}B)\big)Pg(s)ds\Big|_{p}\\ &= \varphi_1(t)+\varphi_2(t).\end{aligned}$$

By applying (**H1**) and Lemma 5, we have

$$\begin{aligned}\varphi_1(t) &\le \mathfrak{C}_1\int_t^{t+k}(t+k-s)^{\beta(1-\xi)-1}|Pg(s)|_p ds\\ &\le \mathfrak{C}_1\mathcal{N}(t)\int_t^{t+k}(t+k-s)^{\beta(1-\xi)-1}s^{-\beta(1-\alpha)}ds\\ &\le \mathcal{N}(t)\frac{\mathfrak{C}_1}{\beta(1-\xi)}k^{\beta(1-\xi)}t^{-\beta(1-\alpha)}\\ &\le \mathcal{N}(t)\frac{\mathfrak{C}_1}{\beta(1-\xi)}k^{\beta(1-\xi)}\epsilon^{-\beta(1-\alpha)}.\end{aligned}$$

To prove $\varphi_2(t)$, we consider the inequality

$$\begin{aligned}\frac{d}{dt}(t^{\beta-1}B^{\xi}E_{\beta,\beta}(-t^{\beta}B)) &= \frac{1}{2\pi\iota}\int_{\Gamma}\nu^{\xi}t^{\beta-2}E_{\beta,\beta-1}(-\nu t^{\beta})(\nu I+B)^{-1}d\nu\\ &= \frac{1}{2\pi\iota}\int_{\Gamma'}-\left(-\frac{\zeta}{t^{\beta}}\right)^{\xi}t^{\beta-2}E_{\beta,\beta-1}(\zeta)\left(-\frac{\zeta}{t^{\beta}}I+B\right)^{-1}\frac{1}{t^{\beta}}d\zeta.\end{aligned}$$

This gives that $\|\frac{d}{dt}(t^{\beta-1}B^{\xi}E_{\beta,\beta}(-t^{\beta}B))\| \le \mathfrak{C}_{\beta}t^{\beta(1-\xi)-2}$. By Mean Value Theorem

$$\begin{aligned}\|t^{\beta-1}B^{\xi}E_{\beta,\beta}(-t^{\beta}B)-s^{\beta-1}B^{\xi}E_{\beta,\beta}(-s^{\beta}B)\| &\le \int_s^t\left\|\frac{d}{d\tau}(\tau^{\beta-1}B^{\xi}E_{\beta,\beta}(-\tau^{\beta}B))\right\|d\tau\\ &\le \mathfrak{C}_{\beta}\int_s^t\tau^{\beta(1-\xi)-2}d\tau = \mathfrak{C}_{\beta}(s^{\beta(1-\xi)-1}-t^{\beta(1-\xi)-1}),\end{aligned}$$

thus

$$\begin{aligned}&\varphi_2(t)\\ \le\ &\int_0^t|B^{\xi}((t+k-s)^{\beta-1}E_{\beta,\beta}(-(t+k-s)^{\beta}B)-(t-s)^{\beta-1}E_{\beta,\beta}(-(t-s)^{\beta}B))Pg(s)ds|_p\\ \le\ &\int_0^t((t-s)^{\beta(1-\xi)-1}-(t+k-s)^{\beta(1-\xi)-1})|Pg(s)|_p ds\\ \le\ &\mathfrak{C}_{\beta}\mathcal{N}(t)\left(\int_0^t(t-s)^{\beta(1-\xi)-1}s^{-\beta(1-\alpha)}ds-\int_0^{t+k}(t-s+k)^{\beta(1-\xi)-1}s^{-\beta(1-\alpha)}ds\right)\\ +\ &\mathfrak{C}_{\beta}\mathcal{N}(t)\int_t^{t+k}(t-s+k)^{\beta(1-\xi)-1}s^{-\beta(1-\alpha)}ds\\ \le\ &\mathfrak{C}_{\beta}\mathcal{N}(t)(t^{\beta(\alpha-\xi)}-(t+k)^{\beta(\alpha-\xi)})B(\beta(1-\xi),1-\beta(1-\alpha))+\mathfrak{C}_{\beta}\mathcal{N}(t)k^{\beta(1-\xi)}t^{-\beta(1-\alpha)}\\ \le\ &\mathfrak{C}_{\beta}\mathcal{N}(t)k^{\beta(1-\xi)}[\epsilon(\epsilon+k)]^{\beta(\alpha-\xi)}+\mathfrak{C}_{\beta}\mathcal{N}(t)k^{\beta(1-\xi)}\epsilon^{-\beta(1-\alpha)},\end{aligned}$$

which shows that $B^{\xi}\varphi \in \mathfrak{C}^{\theta}([\epsilon,\mathfrak{T}],Q_p)$. Therefore $B^{\xi}\varphi \in \mathfrak{C}^{\theta}([0,\mathfrak{T}],Q_p)$, because of arbitrary ϵ.

Recall

$$\psi(t) = \int 0^t(t-s)^{\beta-1}E_{\beta,\beta}(-(t-s)^{\beta}B)F(v(s),v(s))ds.$$

Since $|F(v(s),w(s))|_p \le \mathcal{N}\mathfrak{R}^2(t)s^{-2\beta(\xi-\alpha)}$, where $\mathfrak{R}(t) := \sup_{s\in[0,t]} s^{\beta(\xi-\alpha)}|v(s)|_{H^{\xi,p}}$ in $(0,\mathfrak{T}]$, is bounded and continuous. A similar conversation made it possible to provide the Holder continuity of $B^{\xi}\psi$ in $\mathfrak{C}^{\theta}((0,\mathfrak{T}],Q_p)$. Hence, we have $B^{\xi}v(t) = B^{\xi}\varphi(t)+B^{\xi}\varphi(t)+B^{\xi}\psi(t) \in \mathfrak{C}^{\theta}((0,\mathfrak{T}],Q_p)$.

Since $F(v,w) \in \mathfrak{C}^{\theta}((0,\mathfrak{T}],Q_p)$, by Step 2, this proves that ${}^{C}D_t^{\beta}\psi \in \mathfrak{C}^{\theta}((0,\mathfrak{T}],Q_p)$, $B\psi \in \mathfrak{C}^{\theta}((0,\mathfrak{T}],Q_p)$. and ${}^{C}D_t^{\beta}\psi = -B\psi+F(v,w)$. We obtain ${}^{C}D_t^{\beta}v \in \mathfrak{C}^{\theta}((0,\mathfrak{T}],Q_p)$, $Bv \in \mathfrak{C}^{\theta}((0,\mathfrak{T}],Q_p)$ and ${}^{C}D_t^{\beta}v = -Bv+F(v,w)+Pg$.

Hence, we prove that v is a classical solution. □

6. Example

In this section, we present an example to indicate the applicability of our results:

Example 1. *Suppose that* $Y \in L^2(0, 2\pi)$ *as well as* $\mathfrak{e}_m(y) = 3\sqrt{\frac{3}{2}\pi}\cos x, m = 1, 2, ...$ *At that point, we define infinitesimal dimensional space* $\mathcal{U} = Y$ *and consider a system*

$$\begin{cases} {}^C D_t^{\frac{4}{5}} \mathfrak{Z}(t, y) = {}^C D_t^{\frac{2}{3}} \mathfrak{Z}(t, y) + f(t, \mathfrak{Z}(t, y)) + Qw(t, y), 0 < t < d, 0 < y < 2\pi, \\ \mathfrak{Z}(0, y) = \mathfrak{Z}_0(y), 0 \le y \le 2\pi, \\ \mathfrak{Z}(t, 0) = \mathfrak{Z}(t, 2\pi), 0 \le y \le d, \end{cases}$$

where **(H1)** *is satisfied by the nonlinear function* f *as an operator for every* $w \in L^2(0, d; \mathcal{U})$ *and* $\sum_{m=1}^{\infty} \hat{w}_m s(t)\mathfrak{e}_m$*. Consider*

$$Qw(t) = \sum_{m=1}^{\infty} \hat{w}_m s(t)\mathfrak{e}_m,$$

$$\hat{w}_m(t) - \begin{cases} 0, 0 \le t < d(1 - \frac{1}{m}), \\ w_m(t), d(1 - \frac{1}{m}) \le t \le d. \end{cases}$$

Because

$$\|Qw\|_{L^2(0,d;\mathcal{U})} \le \|w\|_{L^2(0,d;\mathcal{U})'},$$

from $\mathcal{U}$ *into* $L^2(\mathfrak{J}, Y)$*, the operator* Q *is bounded. However, it is not easy to see that* $\overline{Q\mathcal{U}} \ne L^2(\mathfrak{J}, Y)$*. Suppose that* φ *is an arbitrary element in* $L^2(0, d, Y)$ *and* $k \in Y$ *is defined as*

$$k = E_\beta(-d - s)^\beta \mathfrak{Z}(0)y + \int_0^d (d - s)^{\beta - 1} T_{\frac{4}{5}}(d - s)\varphi(s)ds.$$

Suppose that

$$\varphi(t) = \sum_{m=1}^{\infty} f_m(t)\mathfrak{e}_m,$$

as well as

$$k = \sum_{m=1}^{\infty} k_m(t)\mathfrak{e}_m.$$

Hence, we declare that for each given $\varphi \in L^2(0, d, Y)$*, there exist* $w \in \mathcal{U}$ *in such a way that*

$$\begin{aligned} & E_\beta(-d - s)^\beta \mathfrak{Z}(0)y + \int_0^t (d - s)^{\beta - 1} T_{\frac{4}{5}}(d - s)Qw(s)ds \\ = \; & E_\beta(-d - s)^\beta \mathfrak{Z}(0)y + \int_0^d (d - s)^{\beta - 1} T_{\frac{4}{5}}(d - s)\varphi(s)ds, \end{aligned}$$

this indicates that **(H2)** *is fulfilled.*

7. Conclusions

The purpose of this paper is to study the time fractional NS-equations using initial value problem with the Caputo derivative. We proved the global and local existence of mild solution in $H^{\alpha,p}$. We established sufficient conditions for the existence and uniqueness of the mild solution for problem (2) in $H^{\alpha,p}$. Moreover, we showed that classical solutions that satisfy problem (2) are regular. Furthermore, we presented the regularity of mild solutions for time fractional NS-equations. In the end, we presented an example.

Author Contributions: Methodology, M.A. and A.H.; validation, M.A.; formal analysis, M.A., A.H., F.H. and K.A.; investigation, K.A.; writing—original draft preparation, F.H.; writing—review and editing, M.A., A.H., F.H. and K.A.; supervision, A.H.; funding acquisition, M.A. All authors have read and agreed to the published version of the manuscript.

Funding: This work was supported by the Deanship of Scientific Research, Vice Presidency for Graduate Studies and Scientific Research, King Faisal University, Saudi Arabia [Grant No. 2248].

Data Availability Statement: Data sharing not applicable to this article as no data sets were generated or analyzed during the current study.

Conflicts of Interest: The authors declare no conflict of interest.

References

1. Varnhorn, W. *The Stokes Equations*; Akademie Verlag: Berlin, Germany, 1994.
2. Cannone, M. Nombres de Reynolds, stabilité et Navier–Stokes. *Banach Cent. Publ.* **2000**, *52*, 29–59.
3. Lemari-Rieusset, P.G. *Recent Developments in the Navier–Stokes Problem*; CRC Press: Boca Raton, FL, USA, 2002.
4. Wojciech, S.O.; Benjamin, C.P. Leray's Fundamental Work on the Navier–Stokes Equations: A Modern Review of (Sur le Mouvement d'un Liquide Visqueux Emplissant l'espace). In *Partial Differential Equations in Fluid Mechanics*; Cambridge University Press: Cambridge, UK, 2018.
5. Bermudez, B.; Huerta, A.R.; Guerrero-Sanchez, W.F.; Alans, J.D. Two Different Formulations for Solving the Navier–Stokes Equations with Moderate and High Reynolds Numbers. In *Computational Fluid Dynamics: Basic Instruments & Applications in Science*; BoD: Norderstedt, Germany, 2018; p. 319.
6. Chemin, J.Y.; Gallagher, I. Large, global solutions to the Navier–Stokes equations, slowly varying in one direction. *Trans. Am. Math. Soc.* **2010**, *362*, 2859–2873. [CrossRef]
7. Giga, Y.; Sohr, H. Abstract Lp estimates for the Cauchy problem with applications to the Navier–Stokes equations in exterior domains. *J. Funct. Anal.* **1991**, *102*, 72–94. [CrossRef]
8. Amrouche, C.; Rejaiba, A. Lp-theory for Stokes and Navier–Stokes equations with Navier boundary condition. *J. Differ. Equ.* **2014**, *256*, 1515–1547. [CrossRef]
9. Kozono, H.; Yamazaki, M. On a larger class of stable solutions to the Navier–Stokes equations in exterior domains. *Math. Z.* **1998**, *228*, 751–785. [CrossRef]
10. Raugel, G.; Sell, G.R. Navier–Stokes equations on thin 3D domains. I. Global attractors and global regularity of solutions. *J. Am. Math. Soc.* **1993**, *6*, 503–568.
11. Choe, H.J. Boundary regularity of suitable weak solution for the Navier–Stokes equations. *J. Funct. Anal.* **2015**, *268*, 2171–2187. [CrossRef]
12. Bazhlekova, E.; Jin, B.; Lazarov, R.; Zhou, Z. An analysis of the Rayleigh–Stokes problem for a generalized second-grade fluid. *Numer. Math.* **2015**, *131*, 1–31. [CrossRef] [PubMed]
13. Hilfer, R. *Applications of Fractional Calculus in Physics*; World Scientific: Singapore, 2000.
14. Herrmann, R. *Fractional Calculus: An Introduction for Physicists*; World Scientific: Singapore, 2011.
15. Zhou, Y.; Zhang, L.; Shen, X.H. Existence of mild solutions for fractional evolution equations. *J. Integral Equ. Appl.* **2013**, *25*, 557–586. [CrossRef]
16. Zhou, Y.; Vijayakumar, V.; Murugesu, R. Controllability for fractional evolution inclusions without compactness. *Evol. Equ. Control Theory* **2015**, *4*, 507. [CrossRef]
17. Zhou, Y.; Jiao, F.; Pecaric, J. Abstract Cauchy problem for fractional functional differential equations. *Topol. Methods Nonlinear Anal.* **2013**, *42*, 119–136.
18. Momani, S.; Odibat, Z. Analytical solution of a time-fractional Navier–Stokes equation by Adomian decomposition method. *Appl. Math. Comput.* **2006**, *177*, 488–494. [CrossRef]
19. Ganji, Z.Z.; Ganji, D.D.; Ganji, A.; Rostamian, M. Analytical solution of time-fractional Navier–Stokes equation in polar coordinate by homotopy perturbation method. *Numer. Methods Partial Differ. Equ. Int. J.* **2010**, *26*, 117–124. [CrossRef]
20. El-Shahed, M.; Salem, A. On the generalized Navier–Stokes equations. *Appl. Math. Comput.* **2004**, *156*, 287–293. [CrossRef]
21. Wahl, W.V. *The Equations of Navier–Stokes and Abstract Parabolic Equations*; Springer: Berlin/Heidelberg, Germany, 2013.
22. Kilbas, A.A.; Srivastava, H.M.; Trujillo, J.J. *Theory and Applications of Fractional Differential Equations*; Elsevier: Amsterdam, The Netherlands, 2006; Volume 204.
23. Galdi, G.P. *An Introduction to the Mathematical Theory of the Navier–Stokes Equations: Nonlinear Steady Problems*; Springer Tracts in Natural Philosophy; Springer: New York, NY, USA, 1998.
24. Shafqat, R.; Niazi, A.U.K.; Yavuz, M.; Jeelani, M.B.; Saleem, K. Mild Solution for the Time-Fractional Navier–Stokes Equation Incorporating MHD Effects. *Fractal Fract.* **2022**, *6*, 580. [CrossRef]

symmetry

MDPI

Article

Existence and Uniqueness Results for Different Orders Coupled System of Fractional Integro-Differential Equations with Anti-Periodic Nonlocal Integral Boundary Conditions

Ymnah Alruwaily [1], Shorog Aljoudi [2], Lamya Almaghamsi [3], Abdellatif Ben Makhlouf [1,*] and Najla Alghamdi [3]

1 Department of Mathematics, College of Science, Jouf University, P.O. Box 2014, Sakaka 72388, Saudi Arabia
2 Department of Mathematics and Statistics, Taif University, P.O. Box 11099, Taif 21944, Saudi Arabia
3 Department of Mathematics, College of Science, University of Jeddah, P.O. Box 80327, Jeddah 21589, Saudi Arabia
* Correspondence: abmakhlouf@ju.edu.sa

Abstract: This paper presents a new class of boundary value problems of integrodifferential fractional equations of different order equipped with coupled anti-periodic and nonlocal integral boundary conditions. We prove the existence and uniqueness criteria of the solutions by using the Leray-Schauder alternative and Banach contraction mapping principle. Examples are constructed for the illustration of our results.

Keywords: coupled system; fractional integro-differential equations; boundary conditions; existence and uniqueness; fixed point theorems

Citation: Alruwaily, Y.; Aljoudi, S.; Almaghamsi, L.; Ben Makhlouf, A.; Alghamdi, N. Existence and Uniqueness Results for Different Orders Coupled System of Fractional Integro-Differential Equations with Anti-Periodic Nonlocal Integral Boundary Conditions. *Symmetry* **2023**, *15*, 182. https://doi.org/10.3390/sym15010182

Academic Editors: Francisco Martínez González, Mohammed K. A. Kaabar, Jan Awrejcewicz and Sergei D. Odintsov

Received: 6 December 2022
Revised: 28 December 2022
Accepted: 4 January 2023
Published: 7 January 2023

1. Introduction

Fractional calculus has gained a rapid rise in popularity in the past few decades due to the nonlocal nature of the derivatives and integrals of fractional order [1]. As a matter of fact, this field incorporates the methods and concepts used to solve symmetrical differential equations with fractional derivatives. Thereby, it evolved in many theoretical and applications area. For application details in ecology, chaos and fractional dynamics, medical sciences, financial economics bio-engineering, immune system, etc., we refer the reader to the works [2–9]. For more theoretical aspects of fractional calculus, we refer the reader to the monographs [10–18].

During this development, nonlinear Fractional Differential Equations (FDEs) equipped with different kinds of Boundary Conditions (BCs) such as multi-point, periodic, anti-periodic, nonlocal, and integral conditions have also been widely studied and investigated. Many new results of variety boundary value problems were given in [19–25]. At the same time, fractional differential system subjects with different kinds of BCs also received the attention of such systems in the mathematical models with engineering and physical phenomena [26–31].

Recently, fractional Integro-Differential Equations (IDEs) with nonlocal conditions are considered a useful mathematical tool for the description of various real materials, for instance, see [32,33], and references therein. By side, several researchers have applied classical fixed point theorems to prove the existence and uniqueness results for such boundary value problems [19,31,34–42].

In addition, the authors in [43–45] investigated some coupled systems (CSs) of mixed-order FDEs with different kinds of BCs. To enrich the topic, we introduce and investigate a CS of fractional IDEs of Caputo type with different derivatives orders given by

$$\begin{cases} {}^{c}D^{q_1}[\kappa_1 \mathsf{v}(t) + \lambda_1 I_{\mathsf{x}_1}^{\theta_1}\phi(t,\mathsf{v}(t),\mathsf{u}(t))] = \mathsf{k}(t,\mathsf{v}(t),\mathsf{u}(t)), \ 2 < q_1 \leq 3, \ t \in [\mathsf{x}_1,\mathsf{x}_2], \\ {}^{c}D^{q_2}[\kappa_2 \mathsf{u}(t) + \lambda_2 I_{\mathsf{x}_1}^{\theta_2}\psi(t,\mathsf{v}(t),\mathsf{u}(t))] = \mathsf{p}(t,\mathsf{v}(t),\mathsf{u}(t)), \ 1 < q_2 \leq 2, \ t \in [\mathsf{x}_1,\mathsf{x}_2], \end{cases} \tag{1}$$

supplemented with coupled anti-periodic and nonlocal integral BCs:

$$\begin{cases} \mathsf{v}(\mathsf{x}_1) + \mathsf{v}(\mathsf{x}_2) = 0, \ \mathsf{v}'(\mathsf{x}_2) = 0, \ \mathsf{v}'(\mathsf{x}_1) = h\int_{\mathsf{x}_1}^{\xi} \mathsf{u}(s)ds, \\ \mathsf{u}(\mathsf{x}_1) + \mathsf{u}(\mathsf{x}_2) = 0, \ \mathsf{u}'(\mathsf{x}_2) = 0, \end{cases} \tag{2}$$

where ${}^{c}D^{\Upsilon}$ denotes the Caputo fractional differential operator of order $\Upsilon \in \{q_1, q_2\}$, $I_{\mathsf{x}_1}^{\bar{\Upsilon}}$ denotes the Riemann-Liouville fractional integral of order $\bar{\Upsilon} \in \{\theta_1, \theta_2\}$ such that $\theta_1, \theta_2 > 1$, $\kappa_i, \lambda_i, h, \ i = 1,2$ are real constants with $\kappa_i, h \neq 0, \phi, \psi, \mathsf{k}, \mathsf{p} : [\mathsf{x}_1, \mathsf{x}_2] \times \mathbb{R} \times \mathbb{R} \longrightarrow \mathbb{R}$ are given continuous functions and $\mathsf{x}_1 < \xi < \mathsf{x}_2$.

For usefulness, we emphasize that the current study is novel, and contributes extensively to the existing results on the topic. Furthermore, new results follow as special cases of the present work.

The structure of this paper is as follows. In Section 2, we give some important definitions of fractional calculus and establish an auxiliary lemma that helps to transform the system (1) into equivalent integral equations. In Section 3, the existence and uniqueness results for the given system (1) are derived. Two examples are also presented to illustrate the obtained outcomes.

2. Preliminary Material

First, we outline some main definitions of fractional calculus.

Definition 1 ([11]). *Let U be an integrable function on $\mathsf{x}_1 \leq z \leq \mathsf{x}_2$. The Riemann-Liouville fractional integral $I_{\mathsf{x}_1}^{\vartheta}$ of order $\vartheta \in \mathbb{R}$ ($\vartheta > 0$) for U is given by*

$$I_{\mathsf{x}_1}^{\vartheta} U(z) = \frac{1}{\Gamma(\vartheta)} \int_{\mathsf{x}_1}^{z} (z-s)^{\vartheta-1} U(s) ds,$$

where Γ is the Euler Gamma function.

Definition 2 ([11]). *The Caputo derivative for a function $U \in AC^{\mathfrak{r}}[\mathsf{x}_1, \mathsf{x}_2]$ of order $\vartheta \in (\mathfrak{r}-1, \mathfrak{r}]$, $\mathfrak{r} \in \mathbb{N}$ existing on $[\mathsf{x}_1, \mathsf{x}_2]$, is given by*

$${}^{c}D^{\vartheta} U(z) = \frac{1}{\Gamma(\mathfrak{r}-\vartheta)} \int_{\mathsf{x}_1}^{z} (z-l)^{\mathfrak{r}-\vartheta-1} U^{(\mathfrak{r})}(l) dl, z \in [\mathsf{x}_1, \mathsf{x}_2].$$

Lemma 1 ([11]). *The solution of the equation ${}^{c}D^{\vartheta}\mathsf{x}(z) = 0, \ \mathfrak{r}-1 < \vartheta < \mathfrak{r}, \ z \in [\mathsf{x}_1, \mathsf{x}_2]$, is*

$$\mathsf{x}(z) = m_0 + m_1(z-\mathsf{x}_1) + m_2(z-\mathsf{x}_1)^2 + \ldots + m_{r-1}(z-\mathsf{x}_1)^{\mathfrak{r}-1},$$

with $m_i \in \mathbb{R}, \ i = 0, 1, \ldots, \mathfrak{r}-1$. Moreover,

$$I_{\mathsf{x}_1}^{\vartheta}\, {}^{c}D^{\vartheta}\mathsf{x}(z) = \mathsf{x}(z) + \sum_{i=0}^{\mathfrak{r}-1} m_i (z-\mathsf{x}_1)^i.$$

Next, we introduce an important lemma related to our new results.

Lemma 2. *For $\Phi, \Psi, K, P \in C([\mathsf{x}_1, \mathsf{x}_2], \mathbb{R})$, the unique solution of the following linear system*

$$\begin{cases} {}^cD^{q_1}[\kappa_1 \mathrm{v}(t) + \lambda_1 I_{\mathrm{x}_1}^{\theta_1}\Phi(t)] = K(t), \ 2 < q_1 \leq 3, \ t \in [\mathrm{x}_1, \mathrm{x}_2], \\ {}^cD^{q_2}[\kappa_2 \mathrm{u}(t) + \lambda_2 I_{\mathrm{x}_1}^{\theta_2}\Psi(t)] = P(t), \ 1 < q_2 \leq 2, \ t \in [\mathrm{x}_1, \mathrm{x}_2], \end{cases} \tag{3}$$

equipped with the BCs (2) is given by:

$$\begin{aligned} \mathrm{v}(t) &= \frac{1}{\kappa_1}\int_{\mathrm{x}_1}^{t} \frac{(t-s)^{q_1-1}}{\Gamma(q_1)} K(s)ds - \frac{\lambda_1}{\kappa_1}\int_{\mathrm{x}_1}^{t} \frac{(t-s)^{\theta_1-1}}{\Gamma(\theta_1)}\Phi(s)ds \\ &+ \frac{\lambda_1}{2\kappa_1}\int_{\mathrm{x}_1}^{\mathrm{x}_2} \frac{(\mathrm{x}_2-s)^{\theta_1-1}}{\Gamma(\theta_1)}\Phi(s)ds - \frac{1}{2\kappa_1}\int_{\mathrm{x}_1}^{\mathrm{x}_2} \frac{(\mathrm{x}_2-s)^{q_1-1}}{\Gamma(q_1)} K(s)ds \\ &+ \rho_1(t)\Big[\lambda_2 \int_{\mathrm{x}_1}^{\mathrm{x}_2} \frac{(\mathrm{x}_2-s)^{\theta_2-1}}{\Gamma(\theta_2)}\Psi(s)ds - \int_{\mathrm{x}_1}^{\mathrm{x}_2} \frac{(\mathrm{x}_2-s)^{q_2-1}}{\Gamma(q_2)}P(s)ds\Big] \\ &+ \rho_2(t)\Big[\lambda_1 \int_{\mathrm{x}_1}^{\mathrm{x}_2} \frac{(\mathrm{x}_2-s)^{\theta_1-2}}{\Gamma(\theta_1-1)}\Phi(s)ds - \int_{\mathrm{x}_1}^{\mathrm{x}_2} \frac{(\mathrm{x}_2-s)^{q_1-2}}{\Gamma(q_1-1)}K(s)ds\Big] \\ &+ \rho_3(t)\Big[\lambda_2 \int_{\mathrm{x}_1}^{\mathrm{x}_2} \frac{(\mathrm{x}_2-s)^{\theta_2-2}}{\Gamma(\theta_2-1)}\Psi(s)ds - \int_{\mathrm{x}_1}^{\mathrm{x}_2} \frac{(\mathrm{x}_2-s)^{q_2-2}}{\Gamma(q_2-1)}P(s)ds\Big] \\ &+ \rho_4(t)\Big[\int_{\mathrm{x}_1}^{\xi}\Big(\frac{h\lambda_2}{\kappa_2}\int_{\mathrm{x}_1}^{s} \frac{(s-\tau)^{\theta_2-1}}{\Gamma(\theta_2)}\Psi(\tau)d\tau - \frac{h}{\kappa_2}\int_{\mathrm{x}_1}^{s} \frac{(s-\tau)^{q_2-1}}{\Gamma(q_2)}P(\tau)d\tau\Big)ds\Big], \end{aligned} \tag{4}$$

$$\begin{aligned} \mathrm{u}(t) &= \frac{1}{\kappa_2}\int_{\mathrm{x}_1}^{t} \frac{(t-s)^{q_2-1}}{\Gamma(q_2)} P(s)ds - \frac{\lambda_2}{\kappa_2}\int_{\mathrm{x}_1}^{t} \frac{(t-s)^{\theta_2-1}}{\Gamma(\theta_2)}\Psi(s)ds \\ &+ \frac{\lambda_2}{2\kappa_2}\int_{\mathrm{x}_1}^{\mathrm{x}_2} \frac{(\mathrm{x}_2-s)^{\theta_2-1}}{\Gamma(\theta_2)}\Psi(s)ds - \frac{1}{2\kappa_2}\int_{\mathrm{x}_1}^{\mathrm{x}_2} \frac{(\mathrm{x}_2-s)^{q_2-1}}{\Gamma(q_2)} P(s)ds \\ &+ \rho_5(t)\Big[\lambda_2 \int_{\mathrm{x}_1}^{\mathrm{x}_2} \frac{(\mathrm{x}_2-s)^{\theta_2-2}}{\Gamma(\theta_2-1)}\Psi(s)ds - \int_{\mathrm{x}_1}^{\mathrm{x}_2} \frac{(\mathrm{x}_2-s)^{q_2-2}}{\Gamma(q_2-1)}P(s)ds\Big], \end{aligned} \tag{5}$$

where

$$\begin{cases} \rho_1(t) = \dfrac{-\epsilon(\mathrm{x}_2-\mathrm{x}_1)}{4\kappa_1} + \dfrac{\epsilon(t-\mathrm{x}_1)}{\kappa_1} - \dfrac{\epsilon(t-\mathrm{x}_1)^2}{2\kappa_1(\mathrm{x}_2-\mathrm{x}_1)}, \\ \rho_2(t) = \dfrac{-(\mathrm{x}_2-\mathrm{x}_1)}{4\kappa_1} + \dfrac{(t-\mathrm{x}_1)^2}{2\kappa_1(\mathrm{x}_2-\mathrm{x}_1)}, \\ \rho_3(t) = \dfrac{-\epsilon(\xi-\mathrm{x}_2)(\mathrm{x}_2-\mathrm{x}_1)}{4\kappa_1} + \dfrac{\epsilon(\xi-\mathrm{x}_2)(t-\mathrm{x}_1)}{\kappa_1} - \dfrac{\epsilon(\xi-\mathrm{x}_2)(t-\mathrm{x}_1)^2}{2\kappa_1(\mathrm{x}_2-\mathrm{x}_1)}, \\ \rho_4(t) = \dfrac{(\mathrm{x}_2-\mathrm{x}_1)}{4} - (t-\mathrm{x}_1) + \dfrac{(t-\mathrm{x}_1)^2}{2(\mathrm{x}_2-\mathrm{x}_1)}, \\ \rho_5(t) = \dfrac{-(\mathrm{x}_2-\mathrm{x}_1)}{2\kappa_2} + \dfrac{(t-\mathrm{x}_1)}{\kappa_2}, \end{cases} \tag{6}$$

$$\epsilon = \frac{h\kappa_1(\xi-\mathrm{x}_1)}{2\kappa_2}. \tag{7}$$

Proof. Using Lemma 1 and applying the integral operators $I_{\mathrm{x}_1}^{q_1}, I_{\mathrm{x}_1}^{q_2}$ on both sides of the equations in (3), we get the general solution that can be written as

$$\begin{aligned} \mathrm{v}(t) &= \frac{1}{\kappa_1}\int_{\mathrm{x}_1}^{t} \frac{(t-s)^{q_1-1}}{\Gamma(q_1)} K(s)ds - \frac{\lambda_1}{\kappa_1}\int_{\mathrm{x}_1}^{t} \frac{(t-s)^{\theta_1-1}}{\Gamma(\theta_1)}\Phi(s)ds + \frac{c_1}{\kappa_1} + \frac{c_2}{\kappa_1}(t-\mathrm{x}_1) \\ &+ \tfrac{c_3}{\kappa_1}(t-\mathrm{x}_1)^2, \end{aligned} \tag{8}$$

$$
\begin{aligned}
v'(t) &= \frac{1}{\kappa_1}\int_{x_1}^{t}\frac{(t-s)^{q_1-2}}{\Gamma(q_1-1)}K(s)ds - \frac{\lambda_1}{\kappa_1}\int_{x_1}^{t}\frac{(t-s)^{\theta_1-2}}{\Gamma(\theta_1-1)}\Phi(s)ds + \frac{c_2}{\kappa_1} + 2\frac{c_3}{\kappa_1}(t-x_1), \quad (9)\\
u(t) &= \frac{1}{\kappa_2}\int_{x_1}^{t}\frac{(t-s)^{q_2-1}}{\Gamma(q_2)}P(s)ds - \frac{\lambda_2}{\kappa_2}\int_{x_1}^{t}\frac{(t-s)^{\theta_2-1}}{\Gamma(\theta_2)}\Psi(s)ds + \frac{c_4}{\kappa_2} + \frac{c_5}{\kappa_2}(t-x_1), \quad (10)\\
u'(t) &= \frac{1}{\kappa_2}\int_{x_1}^{t}\frac{(t-s)^{q_2-2}}{\Gamma(q_2-1)}P(s)ds - \frac{\lambda_2}{\kappa_2}\int_{x_1}^{t}\frac{(t-s)^{\theta_2-2}}{\Gamma(\theta_2-1)}\Psi(s)ds + \frac{c_5}{\kappa_2}, \quad (11)
\end{aligned}
$$

with $c_i \in \mathbb{R}$, $i = 1, ..., 5$ are unknown arbitrary constants.

Using the conditions (2) in Equations (8)–(11), we obtain a system of equations in c_i $(i = 1, ..., 5)$ given by

$$
\begin{cases}
2c_1 + (x_2 - x_1)c_2 + (x_2 - x_1)^2 c_3 = I_1,\\
2c_4 + (x_2 - x_1)c_5 = I_2,\\
c_2 + 2(x_2 - x_1)c_3 = I_3,\\
-\frac{c_2}{\kappa_1} + \frac{h(\xi - x_1)}{\kappa_2}c_4 + \frac{h(\xi - x_1)^2}{2\kappa_2}c_5 = I_4,\\
c_5 = I_5
\end{cases} \quad (12)
$$

where I_i; $(i = 1, ..., 5)$ are defined by

$$
\begin{aligned}
I_1 &= \lambda_1\int_{x_1}^{x_2}\frac{(x_2-s)^{\theta_1-1}}{\Gamma(\theta_1)}\Phi(s)ds - \int_{x_1}^{x_2}\frac{(x_2-s)^{q_1-1}}{\Gamma(q_1)}K(s)ds,\\
I_2 &= \lambda_2\int_{x_1}^{x_2}\frac{(x_2-s)^{\theta_2-1}}{\Gamma(\theta_2)}\Psi(s)ds - \int_{x_1}^{x_2}\frac{(x_2-s)^{q_2-1}}{\Gamma(q_2)}P(s)ds,\\
I_3 &= \lambda_1\int_{x_1}^{x_2}\frac{(x_2-s)^{\theta_1-2}}{\Gamma(\theta_1-1)}\Phi(s)ds - \int_{x_1}^{x_2}\frac{(x_2-s)^{q_1-2}}{\Gamma(q_1-1)}K(s)ds,\\
I_4 &= \int_{x_1}^{\xi}\Big(\frac{h\lambda_2}{\kappa_2}\int_{x_1}^{s}\frac{(s-\tau)^{\theta_2-1}}{\Gamma(\theta_2)}\Psi(\tau)d\tau - \frac{h}{\kappa_2}\int_{x_1}^{s}\frac{(s-\tau)^{q_2-1}}{\Gamma(q_2)}G(\tau)d\tau\Big)ds,\\
I_5 &= \lambda_2\int_{x_1}^{x_2}\frac{(x_2-s)^{\theta_2-2}}{\Gamma(\theta_2-1)}\Psi(s)ds - \int_{x_1}^{x_2}\frac{(x_2-s)^{q_2-2}}{\Gamma(q_2-1)}P(s)ds. \quad (13)
\end{aligned}
$$

Solving the system (12) for c_i $(i = 1, ..., 5)$, we get that

$$
\begin{aligned}
c_1 &= \frac{-\epsilon(\xi - x_2)(x_2 - x_1)}{4}c_5 + \frac{1}{2}I_1 - \frac{\epsilon(x_2 - x_1)}{4}I_2 - \frac{(x_2 - x_1)}{4}I_3 + \frac{\kappa_1(x_2 - x_1)}{4}I_4,\\
c_2 &= \epsilon(\xi - x_2)c_5 + \epsilon I_2 - \kappa_1 I_4,\\
c_3 &= \frac{-\epsilon(\xi - x_2)}{2(x_2 - x_1)}c_5 - \frac{\epsilon}{2(x_2 - x_1)}I_2 + \frac{1}{2(x_2 - x_1)}I_3 + \frac{\kappa_1}{2(x_2 - x_1)}I_4,\\
c_4 &= \frac{-(x_2 - x_1)}{2}c_5 + \frac{1}{2}I_2,\\
c_5 &= \lambda_2\int_{x_1}^{x_2}\frac{(x_2-s)^{\theta_2-2}}{\Gamma(\theta_2-1)}\Psi(s)ds - \int_{x_1}^{x_2}\frac{(x_2-s)^{q_2-2}}{\Gamma(q_2-1)}P(s)ds,
\end{aligned}
$$

where ϵ is given by (7). Inserting the values of c_i $(i = 1, ...5)$ in (8) and (9) together with notations (6), we get (4) and (5). The converse follows by direct computation. This completes the proof. □

3. Existence and Uniqueness Results

Let $\mathcal{V} = \{v | v \in C([x_1, x_2], \mathbb{R})\}$ be a Banach space endowed with the norm

$$\|v\| = \sup_{l \in [x_1, x_2]} |v(l)|.$$

Obviously the product space $(\mathcal{V} \times \mathcal{V}, \|.\|)$ is also a Banach space with norm $\|(\mathsf{v},\mathsf{u})\| = \|\mathsf{v}\| + \|\mathsf{u}\|$ for $(\mathsf{v},\mathsf{u}) \in \mathcal{V} \times \mathcal{V}$.

In view of Lemma 2, we define an operator $\mathcal{J} : \mathcal{V} \times \mathcal{V} \to \mathcal{V} \times \mathcal{V}$ as

$$\mathcal{J}(\mathsf{v},\mathsf{u})(t) := (\mathcal{J}_1(\mathsf{v},\mathsf{u})(t), \mathcal{J}_2(\mathsf{v},\mathsf{u})(t)), \tag{14}$$

where

$$\begin{aligned}
\mathcal{J}_1(\mathsf{v},\mathsf{u})(t) &= \frac{1}{\kappa_1}\int_{\mathsf{x}_1}^{t}\frac{(t-s)^{q_1-1}}{\Gamma(q_1)}\mathsf{k}(s,\mathsf{v}(s),\mathsf{u}(s))ds - \frac{\lambda_1}{\kappa_1}\int_{\mathsf{x}_1}^{t}\frac{(t-s)^{\theta_1-1}}{\Gamma(\theta_1)}\phi(s,\mathsf{v}(s),\mathsf{u}(s))ds \\
&+ \frac{\lambda_1}{2\kappa_1}\int_{\mathsf{x}_1}^{\mathsf{x}_2}\frac{(\mathsf{x}_2-s)^{\theta_1-1}}{\Gamma(\theta_1)}\phi(s,\mathsf{v}(s),\mathsf{u}(s))ds - \frac{1}{2\kappa_1}\int_{\mathsf{x}_1}^{\mathsf{x}_2}\frac{(\mathsf{x}_2-s)^{q_1-1}}{\Gamma(q_1)}\mathsf{k}(s,\mathsf{v}(s),\mathsf{u}(s))ds \\
&+ \rho_1(t)\Big[\lambda_2\int_{\mathsf{x}_1}^{\mathsf{x}_2}\frac{(\mathsf{x}_2-s)^{\theta_2-1}}{\Gamma(\theta_2)}\psi(s,\mathsf{v}(s),\mathsf{u}(s))ds - \int_{\mathsf{x}_1}^{\mathsf{x}_2}\frac{(\mathsf{x}_2-s)^{q_2-1}}{\Gamma(q_2)}\mathsf{p}(s,\mathsf{v}(s),\mathsf{u}(s))ds\Big] \\
&+ \rho_2(t)\Big[\lambda_1\int_{\mathsf{x}_1}^{\mathsf{x}_2}\frac{(\mathsf{x}_2-s)^{\theta_1-2}}{\Gamma(\theta_1-1)}\phi(s,\mathsf{v}(s),\mathsf{u}(s))ds - \int_{\mathsf{x}_1}^{\mathsf{x}_2}\frac{(\mathsf{x}_2-s)^{q_1-2}}{\Gamma(q_1-1)}\mathsf{k}(s,\mathsf{v}(s),\mathsf{u}(s))ds\Big] \\
&+ \rho_3(t)\Big[\lambda_2\int_{\mathsf{x}_1}^{\mathsf{x}_2}\frac{(\mathsf{x}_2-s)^{\theta_2-2}}{\Gamma(\theta_2-1)}\psi(s,\mathsf{v}(s),\mathsf{u}(s))ds - \int_{\mathsf{x}_1}^{\mathsf{x}_2}\frac{(\mathsf{x}_2-s)^{q_2-2}}{\Gamma(q_2-1)}\mathsf{p}(s,\mathsf{v}(s),\mathsf{u}(s))ds\Big] \\
&+ \rho_4(t)\Big[\int_{\mathsf{x}_1}^{\xi}\Big(\frac{h\lambda_2}{\kappa_2}\int_{\mathsf{x}_1}^{s}\frac{(s-\tau)^{\theta_2-1}}{\Gamma(\theta_2)}\psi(\tau,\mathsf{v}(\tau),\mathsf{u}(\tau))d\tau - \frac{h}{\kappa_2}\int_{\mathsf{x}_1}^{s}\frac{(s-\tau)^{q_2-1}}{\Gamma(q_2)}\mathsf{p}(\tau,\mathsf{v}(\tau),\mathsf{u}(\tau))d\tau\Big)ds\Big],
\end{aligned} \tag{15}$$

$$\begin{aligned}
\mathcal{J}_2(\mathsf{v},\mathsf{u})(t) &= \frac{1}{\kappa_2}\int_{\mathsf{x}_1}^{t}\frac{(t-s)^{q_2-1}}{\Gamma(q_2)}\mathsf{p}(s,\mathsf{v}(s),\mathsf{u}(s))ds - \frac{\lambda_2}{\kappa_2}\int_{\mathsf{x}_1}^{t}\frac{(t-s)^{\theta_2-1}}{\Gamma(\theta_2)}\psi(s,\mathsf{v}(s),\mathsf{u}(s))ds \\
&+ \frac{\lambda_2}{2\kappa_2}\int_{\mathsf{x}_1}^{\mathsf{x}_2}\frac{(\mathsf{x}_2-s)^{\theta_2-1}}{\Gamma(\theta_2)}\psi(s,\mathsf{v}(s),\mathsf{u}(s))ds - \frac{1}{2\kappa_2}\int_{\mathsf{x}_1}^{\mathsf{x}_2}\frac{(\mathsf{x}_2-s)^{q_2-1}}{\Gamma(q_2)}\mathsf{p}(s,\mathsf{v}(s),\mathsf{u}(s))ds \\
&+ \rho_5(t)\Big[\lambda_2\int_{\mathsf{x}_1}^{\mathsf{x}_2}\frac{(\mathsf{x}_2-s)^{\theta_2-2}}{\Gamma(\theta_2-1)}\psi(s,\mathsf{v}(s),\mathsf{u}(s))ds - \int_{\mathsf{x}_1}^{\mathsf{x}_2}\frac{(\mathsf{x}_2-s)^{q_2-2}}{\Gamma(q_2-1)}\mathsf{p}(s,\mathsf{v}(s),\mathsf{u}(s))ds\Big],
\end{aligned} \tag{16}$$

where $\rho_i(t),\ i = 1,....,5$ are given by (6). For brevity, we use the subsequent notations.

$$M_1 = \frac{3(\mathsf{x}_2-\mathsf{x}_1)^{q_1}}{2|\kappa_1||\Gamma(q_1+1)} + \widetilde{\rho}_2\frac{(\mathsf{x}_2-\mathsf{x}_1)^{q_1-1}}{\Gamma(q_1)}, \tag{17}$$

$$M_2 = \widetilde{\rho}_1\frac{(\mathsf{x}_2-\mathsf{x}_1)^{q_2}}{\Gamma(q_2+1)} + \widetilde{\rho}_3\frac{(\mathsf{x}_2-\mathsf{x}_1)^{q_2-1}}{\Gamma(q_2)} + \widetilde{\rho}_4\frac{|h|(\xi-\mathsf{x}_1)^{q_2+1}}{|\kappa_2|\Gamma(q_2+2)}, \tag{18}$$

$$M_3 = \frac{3|\lambda_1|(\mathsf{x}_2-\mathsf{x}_1)^{\theta_1}}{2|\kappa_1||\Gamma(\theta_1+1)} + \widetilde{\rho}_2\frac{|\lambda_1|(\mathsf{x}_2-\mathsf{x}_1)^{\theta_1-1}}{\Gamma(\theta_1)}, \tag{19}$$

$$M_4 = \widetilde{\rho}_1\frac{|\lambda_2|(\mathsf{x}_2-\mathsf{x}_1)^{\theta_2}}{\Gamma(\theta_2+1)} + \widetilde{\rho}_3\frac{|\lambda_2|(\mathsf{x}_2-\mathsf{x}_1)^{\theta_2-1}}{\Gamma(\theta_2)} + \widetilde{\rho}_4\frac{|h||\lambda_2|(\xi-\mathsf{x}_1)^{\theta_2+1}}{|\kappa_2|\Gamma(\theta_2+2)}, \tag{20}$$

$$M_5 = \frac{3(\mathsf{x}_2-\mathsf{x}_1)^{q_2}}{2|\kappa_2||\Gamma(q_2+1)} + \widetilde{\rho}_5\frac{(\mathsf{x}_2-\mathsf{x}_1)^{q_2-1}}{\Gamma(q_2)}, \tag{21}$$

$$M_6 = \frac{3|\lambda_2|(\mathsf{x}_2-\mathsf{x}_1)^{\theta_2}}{2|\kappa_2||\Gamma(\theta_2+1)} + \widetilde{\rho}_5\frac{|\lambda_2|(\mathsf{x}_2-\mathsf{x}_1)^{\theta_2-1}}{\Gamma(\theta_2)}, \tag{22}$$

where $\widetilde{\rho}_i = \sup_{t\in[\mathsf{x}_1,\mathsf{x}_2]} |\rho_i(t)|,\ i = 1,\cdots,5.$

3.1. Existence Result via Leray-Schauder Alternative

Lemma 3 ([46]). *(Leray-Schauder alternative) Let $\mathfrak{L} : \mathfrak{E} \to \mathfrak{E}$ be a completely continuous operator. Let $\mathfrak{X}(\mathfrak{L}) = \{\mathsf{x} \in \mathfrak{E} : \mathsf{x} = \delta\mathfrak{L}(\mathsf{x})$ for some $0 < \delta < 1\}$. Then either the set $\mathfrak{X}(\mathfrak{L})$ is unbounded or $\mathfrak{L}$ has at least one fixed point.*

Theorem 1. *Assume the following assumption holds*

(H_1) $\mathsf{k}, \mathsf{p}, \phi, \psi : [\mathsf{x}_1, \mathsf{x}_2] \times \mathbb{R}^2 \to \mathbb{R}$ *are continuous functions and there exist real constants* $\gamma_i, \nu_i, \mu_i, \varpi_i \geq 0$ $(i = 1, 2)$ *and* $\gamma_0, \nu_0, \mu_0, \varpi_0 > 0$ *such that, for all* $t \in [\mathsf{x}_1, \mathsf{x}_2]$ *and* $\mathsf{v}, \mathsf{u} \in \mathbb{R}$,

$$
\begin{aligned}
|\mathsf{k}(t, \mathsf{v}, \mathsf{u})| &\leq \gamma_0 + \gamma_1|\mathsf{v}| + \gamma_2|\mathsf{u}|, \ \ |\mathsf{p}(t, \mathsf{v}, \mathsf{u})| \leq \nu_0 + \nu_1|\mathsf{v}| + \nu_2|\mathsf{u}|, \\
|\phi(t, \mathsf{v}, \mathsf{u})| &\leq \mu_0 + \mu_1|\mathsf{v}| + \mu_2|\mathsf{u}|, \ \ |\psi(t, \mathsf{v}, \mathsf{u})| \leq \varpi_0 + \varpi_1|\mathsf{v}| + \varpi_2|\mathsf{u}|,
\end{aligned}
$$

then the CS (1) and (2) has at least one solution on $[\mathsf{x}_1, \mathsf{x}_2]$ *if*

$$\mathfrak{N}_1 = \gamma_1 M_1 + \nu_1(M_2 + M_5) + \mu_1 M_3 + \varpi_1(M_4 + M_6) < 1, \tag{23}$$

$$\mathfrak{N}_2 = \gamma_2 M_1 + \nu_2(M_2 + M_5) + \mu_2 M_3 + \varpi_2(M_4 + M_6) < 1. \tag{24}$$

where M_j, $j = 1, \cdots, 6$ *are given by (17)–(22) respectively.*

Proof. First, we demonstrate that the operator $\mathcal{J} : \mathcal{V} \times \mathcal{V} \to \mathcal{V} \times \mathcal{V}$ is completely continuous. By continuity of the functions $\mathsf{k}, \mathsf{p}, \phi$ and ψ, it follows that the operators $\mathcal{J}_1$ and $\mathcal{J}_2$ are continuous. In consequence, the operator $\mathcal{J}$ is continuous. Let $\mathfrak{G} \subset \mathcal{V} \times \mathcal{V}$ be a bounded set. Then $\forall(\mathsf{v}, \mathsf{u}) \in \mathfrak{G}$, there exist positive constants L_n, $n = 1, 2, 3, 4$ such that:

$$
\begin{cases}
|\mathsf{k}(t, \mathsf{v}(t), \mathsf{u}(t))| \leq L_1, \ \ |\mathsf{p}(t, \mathsf{v}(t), \mathsf{u}(t))| \leq L_2, \\
|\phi(t, \mathsf{v}(t), \mathsf{u}(t))| \leq L_3, \ \ |\psi(t, \mathsf{v}(t), \mathsf{u}(t))| \leq L_4.
\end{cases} \tag{25}
$$

Then, for any $(\mathsf{v}, \mathsf{u}) \in \mathfrak{G}$, we have

$$
\begin{aligned}
|\mathcal{J}_1(\mathsf{v},\mathsf{u})(t)| \leq{}& \frac{1}{|\kappa_1|}\int_{\mathsf{x}_1}^{t}\frac{(t-s)^{q_1-1}}{\Gamma(q_1)}\Big|\mathsf{k}(s,\mathsf{v}(s),\mathsf{u}(s))\Big|ds + \frac{|\lambda_1|}{|\kappa_1|}\int_{\mathsf{x}_1}^{t}\frac{(t-s)^{\theta_1-1}}{\Gamma(\theta_1)}\Big|\phi(s,\mathsf{v}(s),\mathsf{u}(s))\Big|ds \\
+{}& \frac{|\lambda_1|}{2|\kappa_1|}\int_{\mathsf{x}_1}^{\mathsf{x}_2}\frac{(\mathsf{x}_2-s)^{\theta_1-1}}{\Gamma(\theta_1)}\Big|\phi(s,\mathsf{v}(s),\mathsf{u}(s))\Big|ds + \frac{1}{2|\kappa_1|}\int_{\mathsf{x}_1}^{\mathsf{x}_2}\frac{(\mathsf{x}_2-s)^{q_1-1}}{\Gamma(q_1)}\Big|\mathsf{k}(s,\mathsf{v}(s),\mathsf{u}(s))\Big|ds \\
+{}& |\rho_1(t)|\Big[|\lambda_2|\int_{\mathsf{x}_1}^{\mathsf{x}_2}\frac{(\mathsf{x}_2-s)^{\theta_2-1}}{\Gamma(\theta_2)}\Big|\psi(s,\mathsf{v}(s),\mathsf{u}(s))\Big|ds + \int_{\mathsf{x}_1}^{\mathsf{x}_2}\frac{(\mathsf{x}_2-s)^{q_2-1}}{\Gamma(q_2)}\Big|\mathsf{p}(s,\mathsf{v}(s),\mathsf{u}(s))\Big|ds\Big] \\
+{}& |\rho_2(t)|\Big[|\lambda_1|\int_{\mathsf{x}_1}^{\mathsf{x}_2}\frac{(\mathsf{x}_2-s)^{\theta_1-2}}{\Gamma(\theta_1-1)}\Big|\phi(s,\mathsf{v}(s),\mathsf{u}(s))\Big|ds + \int_{\mathsf{x}_1}^{\mathsf{x}_2}\frac{(\mathsf{x}_2-s)^{q_1-2}}{\Gamma(q_1-1)}\Big|\mathsf{k}(s,\mathsf{v}(s),\mathsf{u}(s))\Big|ds\Big] \\
+{}& |\rho_3(t)|\Big[|\lambda_2|\int_{\mathsf{x}_1}^{\mathsf{x}_2}\frac{(\mathsf{x}_2-s)^{\theta_2-2}}{\Gamma(\theta_2-1)}\Big|\psi(s,\mathsf{v}(s),\mathsf{u}(s))\Big|ds + \int_{\mathsf{x}_1}^{\mathsf{x}_2}\frac{(\mathsf{x}_2-s)^{q_2-2}}{\Gamma(q_2-1)}\Big|\mathsf{p}(s,\mathsf{v}(s),\mathsf{u}(s))\Big|ds\Big] \\
+{}& |\rho_4(t)|\Big[\int_{\mathsf{x}_1}^{\xi}\Big(\frac{|h||\lambda_2|}{|\kappa_2|}\int_{\mathsf{x}_1}^{s}\frac{(s-\tau)^{\theta_2-1}}{\Gamma(\theta_2)}\Big|\psi(\tau,\mathsf{v}(\tau),\mathsf{u}(\tau))\Big|d\tau \\
+{}& \frac{|h|}{|\kappa_2|}\int_{\mathsf{x}_1}^{s}\frac{(s-\tau)^{q_2-1}}{\Gamma(q_2)}\Big|\mathsf{p}(\tau,\mathsf{v}(\tau),\mathsf{u}(\tau))\Big|d\tau\Big)ds\Big] \\
\leq{}& L_1\Bigg\{\frac{1}{|\kappa_1|}\int_{\mathsf{x}_1}^{t}\frac{(t-s)^{q_1-1}}{\Gamma(q_1)}ds + \frac{1}{2|\kappa_1|}\int_{\mathsf{x}_1}^{\mathsf{x}_2}\frac{(\mathsf{x}_2-s)^{q_1-1}}{\Gamma(q_1)}ds + |\rho_2(t)|\int_{\mathsf{x}_1}^{\mathsf{x}_2}\frac{(\mathsf{x}_2-s)^{q_1-2}}{\Gamma(q_1-1)}ds\Bigg\} \\
+{}& L_2\Bigg\{|\rho_1(t)|\int_{\mathsf{x}_1}^{\mathsf{x}_2}\frac{(\mathsf{x}_2-s)^{q_2-1}}{\Gamma(q_2)}ds + |\rho_3(t)|\int_{\mathsf{x}_1}^{\mathsf{x}_2}\frac{(\mathsf{x}_2-s)^{q_2-2}}{\Gamma(q_2-1)}ds + \frac{|\rho_4(t)||h|}{|\kappa_2|}\int_{\mathsf{x}_1}^{\xi}\Big(\int_{\mathsf{x}_1}^{s}\frac{(s-\tau)^{q_2-1}}{\Gamma(q_2)}d\tau\Big)ds\Bigg\} \\
+{}& L_3\Bigg\{\frac{|\lambda_1|}{|\kappa_1|}\int_{\mathsf{x}_1}^{t}\frac{(t-s)^{\theta_1-1}}{\Gamma(\theta_1)}ds + \frac{|\lambda_1|}{2|\kappa_1|}\int_{\mathsf{x}_1}^{\mathsf{x}_2}\frac{(\mathsf{x}_2-s)^{\theta_1-1}}{\Gamma(\theta_1)}ds + |\rho_2(t)||\lambda_1|\int_{\mathsf{x}_1}^{\mathsf{x}_2}\frac{(\mathsf{x}_2-s)^{\theta_1-2}}{\Gamma(\theta_1-1)}ds\Bigg\} \\
+{}& L_4\Bigg\{|\rho_1(t)||\lambda_2|\int_{\mathsf{x}_1}^{\mathsf{x}_2}\frac{(\mathsf{x}_2-s)^{\theta_2-1}}{\Gamma(\theta_2)}ds + |\rho_3(t)||\lambda_2|\int_{\mathsf{x}_1}^{\mathsf{x}_2}\frac{(\mathsf{x}_2-s)^{\theta_2-2}}{\Gamma(\theta_2-1)}ds \\
+{}& \frac{|\rho_4(t)||h||\lambda_2|}{|\kappa_2|}\int_{\mathsf{x}_1}^{\xi}\Big(\int_{\mathsf{x}_1}^{s}\frac{(s-\tau)^{\theta_2-1}}{\Gamma(\theta_2)}d\tau\Big)ds\Bigg\},
\end{aligned}
$$

taking the norm for $t \in [\mathrm{x}_1, \mathrm{x}_2]$ and using the notations (17)–(20) yields

$$\|\mathcal{J}_1(\mathrm{v},\mathrm{u})\| \leq L_1 M_1 + L_2 M_2 + L_3 M_3 + L_4 M_4. \tag{26}$$

Similarly, we have

$$\|\mathcal{J}_2(\mathrm{v},\mathrm{u})\| \leq L_2 M_5 + L_4 M_6. \tag{27}$$

From above inequalities (26) and (27), we deduce that $\mathcal{J}_1$ and $\mathcal{J}_2$ are uniformly bounded, which implies that

$$\|\mathcal{J}(\mathrm{v},\mathrm{u})\| \leq L_1 M_1 + L_2(M_2 + M_5) + L_3 M_3 + L_4(M_4 + M_6). \tag{28}$$

Hence the operator $\mathcal{J}$ is uniformly bounded.

Next, we prove that $\mathcal{J}$ is equicontinuous. Let $t_1,\ t_2 \in [\mathrm{x}_1, \mathrm{x}_2]$ with $t_1 < t_2$. Then we get

$$\begin{aligned}
&|\mathcal{J}_1(\mathrm{v},\mathrm{u})(t_2) - \mathcal{J}_1(\mathrm{v},\mathrm{u})(t_1)| \leq \frac{1}{|\kappa_1|}\Big[\int_{\mathrm{x}_1}^{t_1} \frac{|(t_2-s)^{q_1-1} - (t_1-s)^{q_1-1}|}{\Gamma(q_1)}\Big|\mathrm{k}(s,\mathrm{v}(s),\mathrm{u}(s))\Big|ds \\
&+ \int_{t_1}^{t_2} \frac{|(t_2-s)^{q_1-1}|}{\Gamma(q_1)}\Big|\mathrm{k}(s,\mathrm{v}(s),\mathrm{u}(s))\Big|\Big]ds + \frac{|\lambda_1|}{|\kappa_1|}\Big[\int_{\mathrm{x}_1}^{t_1} \frac{|(t_2-s)^{\theta_1-1} - (t_1-s)^{\theta_1-1}|}{\Gamma(\theta_1)}\Big|\phi(s,\mathrm{v}(s),\mathrm{u}(s))\Big|ds \\
&+ \int_{t_1}^{t_2} \frac{|(t_2-s)^{\theta_1-1}|}{\Gamma(\theta_1)}\Big|\phi(s,\mathrm{v}(s),\mathrm{u}(s))\Big|ds\Big] \\
&+ |\rho_1(t_2) - \rho_1(t_1)|\Big[|\lambda_2|\int_{\mathrm{x}_1}^{\mathrm{x}_2} \frac{(\mathrm{x}_2-s)^{\theta_2-1}}{\Gamma(\theta_2)}\Big|\psi(s,\mathrm{v}(s),\mathrm{u}(s))\Big|ds + \int_{\mathrm{x}_1}^{\mathrm{x}_2} \frac{(\mathrm{x}_2-s)^{q_2-1}}{\Gamma(q_2)}\Big|\mathrm{p}(s,\mathrm{v}(s),\mathrm{u}(s))\Big|ds\Big] \\
&+ |\rho_2(t_2) - \rho_2(t_1)|\Big[|\lambda_1|\int_{\mathrm{x}_1}^{\mathrm{x}_2} \frac{(\mathrm{x}_2-s)^{\theta_1-2}}{\Gamma(\theta_1-1)}\Big|\phi(s,\mathrm{v}(s),\mathrm{u}(s))\Big|ds + \int_{\mathrm{x}_1}^{\mathrm{x}_2} \frac{(\mathrm{x}_2-s)^{q_1-2}}{\Gamma(q_1-1)}\Big|\mathrm{k}(s,\mathrm{v}(s),\mathrm{u}(s))\Big|ds\Big] \\
&+ |\rho_3(t_2) - \rho_3(t_1)|\Big[|\lambda_2|\int_{\mathrm{x}_1}^{\mathrm{x}_2} \frac{(\mathrm{x}_2-s)^{\theta_2-2}}{\Gamma(\theta_2-1)}\Big|\psi(s,\mathrm{v}(s),\mathrm{u}(s))\Big|ds + \int_{\mathrm{x}_1}^{\mathrm{x}_2} \frac{(\mathrm{x}_2-s)^{q_2-2}}{\Gamma(q_2-1)}\Big|\mathrm{p}(s,\mathrm{v}(s),\mathrm{u}(s))\Big|ds\Big] \\
&+ |\rho_4(t_2) - \rho_4(t_1)|\Big[\int_{\mathrm{x}_1}^{\xi}\Big(\frac{|h||\lambda_2|}{|\kappa_2|}\int_{\mathrm{x}_1}^{s} \frac{(s-\tau)^{\theta_2-1}}{\Gamma(\theta_2)}\Big|\psi(\tau,\mathrm{v}(\tau),\mathrm{u}(\tau))\Big|d\tau \\
&+ \frac{|h|}{|\kappa_2|}\int_{\mathrm{x}_1}^{s} \frac{(s-\tau)^{q_2-1}}{\Gamma(q_2)}\Big|\mathrm{p}(\tau,\mathrm{v}(\tau),\mathrm{u}(\tau))\Big|d\tau\Big)ds\Big] \\
&\leq L_1\Bigg\{\frac{1}{|\kappa_1|\Gamma(q_1+1)}\Big[2(t_2-t_1)^{q_1} + \Big|(t_2-\mathrm{x}_1)^{q_1} - (t_1-\mathrm{x}_1)^{q_1}\Big|\Big] \\
&+ |\rho_2(t_2) - \rho_2(t_1)|\int_{\mathrm{x}_1}^{\mathrm{x}_2} \frac{(\mathrm{x}_2-s)^{q_1-2}}{\Gamma(q_1-1)}ds\Bigg\} \\
&+ L_2\Bigg\{|\rho_1(t_2) - \rho_1(t_1)|\int_{\mathrm{x}_1}^{\mathrm{x}_2} \frac{(\mathrm{x}_2-s)^{q_2-1}}{\Gamma(q_2)}ds + |\rho_3(t_2) - \rho_3(t_1)|\int_{\mathrm{x}_1}^{\mathrm{x}_2} \frac{(\mathrm{x}_2-s)^{q_2-2}}{\Gamma(q_2-1)}ds \\
&+ \frac{|\rho_4(t_2) - \rho_4(t_1)||h|}{|\kappa_2|}\int_{\mathrm{x}_1}^{\xi}\Big(\int_{\mathrm{x}_1}^{s} \frac{(s-\tau)^{q_2-1}}{\Gamma(q_2)}d\tau\Big)ds\Bigg\} \\
&+ L_3\Bigg\{\frac{|\lambda_1|}{|\kappa_1|\Gamma(\theta_1+1)}\Big[2(t_2-t_1)^{\theta_1} + \Big|(t_2-\mathrm{x}_1)^{\theta_1} - (t_1-\mathrm{x}_1)^{\theta_1-1}\Big|\Big] \\
&+ \Big|\bar{\rho}_2(t_2) - \bar{\rho}_2(t_1)\Big||\lambda_1|\int_{\mathrm{x}_1}^{\mathrm{x}_2} \frac{(\mathrm{x}_2-s)^{\theta_1-2}}{\Gamma(\theta_1-1)}ds\Bigg\} \\
&+ L_4\Bigg\{|\rho_1(t_2) - \rho_1(t_1)||\lambda_2|\int_{\mathrm{x}_1}^{\mathrm{x}_2} \frac{(\mathrm{x}_2-s)^{\theta_2-1}}{\Gamma(\theta_2)}ds + |\rho_3(t_2) - \rho_3(t_1)||\lambda_2|\int_{\mathrm{x}_1}^{\mathrm{x}_2} \frac{(\mathrm{x}_2-s)^{\theta_2-2}}{\Gamma(\theta_2-1)}ds \\
&+ \frac{|\rho_4(t_2) - \rho_4(t_1)||h||\lambda_2|}{|\kappa_2|}\int_{\mathrm{x}_1}^{\xi}\Big(\int_{\mathrm{x}_1}^{s} \frac{(s-\tau)^{\theta_2-1}}{\Gamma(\theta_2)}d\tau\Big)ds\Bigg\},
\end{aligned}$$

which imply that $|\mathcal{J}_1(\mathsf{v},\mathsf{u})(t_2) - \mathcal{J}_1(\mathsf{v},\mathsf{u})(t_1)| \to 0$ independent of $(\mathsf{v},\mathsf{u}) \in \mathfrak{G}$ as $t_2 \to t_1$. In a similar way, we get

$$|\mathcal{J}_2(\mathsf{v},\mathsf{u})(t_2) - \mathcal{J}_2(\mathsf{v},\mathsf{u})(t_1)| \to 0$$

as $t_2 \to t_1$. Thus $\mathcal{J}$ is equicontinuous. Therefore, by Arzela-Ascoli's theorem, it follows that $\mathcal{J}$ is compact (completely continuous).

Finally, we ought to prove that $\mathfrak{Z}(\mathcal{J}) = \{(\mathsf{v},\mathsf{u}) \in \mathcal{V} \times \mathcal{V} : (\mathsf{v},\mathsf{u}) = \delta\mathcal{J}(\mathsf{v},\mathsf{u})\,;\, 0 \le \delta \le 1\}$ is bounded. Let $(\mathsf{v},\mathsf{u}) \in \mathfrak{Z}(\mathcal{J})$. Then $(\mathsf{v},\mathsf{u}) = \delta\mathcal{J}(\mathsf{v},\mathsf{u})$. For every $t \in [\mathsf{x}_1,\mathsf{x}_2]$, we have

$$\mathsf{v}(t) = \delta\mathcal{J}_1(\mathsf{v},\mathsf{u})(t), \; \mathsf{u}(t) = \delta\mathcal{J}_2(\mathsf{v},\mathsf{u})(t).$$

Using (H_1) in (1), we get

$$\begin{aligned}
|\mathcal{J}_1(\mathsf{v},\mathsf{u})(t)| \le\; & \frac{1}{|\kappa_1|}\int_{\mathsf{x}_1}^{t} \frac{(t-s)^{q_1-1}}{\Gamma(q_1)}\Big[\gamma_0 + \gamma_1|\mathsf{v}(s)| + \gamma_2|\mathsf{u}(s)|\Big]ds \\
+ & \frac{|\lambda_1|}{|\kappa_1|}\int_{\mathsf{x}_1}^{t} \frac{(t-s)^{\theta_1-1}}{\Gamma(\theta_1)}\Big[\mu_0 + \mu_1|\mathsf{v}(s)| + \mu_2|\mathsf{u}(s)|\Big]ds \\
+ & \frac{|\lambda_1|}{2|\kappa_1|}\int_{\mathsf{x}_1}^{\mathsf{x}_2} \frac{(\mathsf{x}_2-s)^{\theta_1-1}}{\Gamma(\theta_1)}\Big[\mu_0 + \mu_1|\mathsf{v}(s)| + \mu_2|\mathsf{u}(s)|\Big]ds \\
+ & \frac{1}{2|\kappa_1|}\int_{\mathsf{x}_1}^{\mathsf{x}_2} \frac{(\mathsf{x}_2-s)^{q_1-1}}{\Gamma(q_1)}\Big[\gamma_0 + \gamma_1|\mathsf{v}(s)| + \gamma_2|\mathsf{u}(s)|\Big]ds \\
+ & |\rho_1(t)|\bigg\{|\lambda_2|\int_{\mathsf{x}_1}^{\mathsf{x}_2} \frac{(\mathsf{x}_2-s)^{\theta_2-1}}{\Gamma(\theta_2)}\Big[\varpi_0 + \varpi_1|\mathsf{v}(s)| + \varpi_2|\mathsf{u}(s)|\Big]ds \\
+ & \int_{\mathsf{x}_1}^{\mathsf{x}_2} \frac{(\mathsf{x}_2-s)^{q_2-1}}{\Gamma(q_2)}\Big[\nu_0 + \nu_1|\mathsf{v}(s)| + \nu_2|\mathsf{u}(s)|\Big]ds\bigg\} \\
+ & |\rho_2(t)|\bigg\{|\lambda_1|\int_{\mathsf{x}_1}^{\mathsf{x}_2} \frac{(\mathsf{x}_2-s)^{\theta_1-2}}{\Gamma(\theta_1-1)}\Big[\mu_0 + \mu_1|\mathsf{v}(s)| + \mu_2|\mathsf{u}(s)|\Big]ds \\
+ & \int_{\mathsf{x}_1}^{\mathsf{x}_2} \frac{(\mathsf{x}_2-s)^{q_1-2}}{\Gamma(q_1-1)}\Big[\gamma_0 + \gamma_1|\mathsf{v}(s)| + \gamma_2|\mathsf{u}(s)|\Big]ds\bigg\} \\
+ & |\rho_3(t)|\bigg\{|\lambda_2|\int_{\mathsf{x}_1}^{\mathsf{x}_2} \frac{(\mathsf{x}_2-s)^{\theta_2-2}}{\Gamma(\theta_2-1)}\Big[\varpi_0 + \varpi_1|\mathsf{v}(s)| + \varpi_2|\mathsf{u}(s)|\Big]ds \\
+ & \int_{\mathsf{x}_1}^{\mathsf{x}_2} \frac{(\mathsf{x}_2-s)^{q_2-2}}{\Gamma(q_2-1)}\Big[\nu_0 + \nu_1|\mathsf{v}(s)| + \nu_2|\mathsf{u}(s)|\Big]ds\bigg\} \\
+ & |\rho_4(t)|\bigg\{\int_{\mathsf{x}_1}^{\xi}\bigg(\frac{|h||\lambda_2|}{|\kappa_2|}\int_{\mathsf{x}_1}^{s} \frac{(s-\tau)^{\theta_2-1}}{\Gamma(\theta_2)}\Big[\varpi_0 + \varpi_1|\mathsf{v}(\tau)| + \varpi_2|\mathsf{u}(\tau)|\Big]d\tau \\
+ & \frac{|h|}{|\kappa_2|}\int_{\mathsf{x}_1}^{s} \frac{(s-\tau)^{q_2-1}}{\Gamma(q_2)}\Big[\nu_0 + \nu_1|\mathsf{v}(\tau)| + \nu_2|\mathsf{u}(\tau)|\Big]d\tau\bigg)ds\bigg\},
\end{aligned}$$

which implies that

$$\begin{aligned}
\|\mathsf{v}\| \le\; & \gamma_0 M_1 + \nu_0 M_2 + \mu_0 M_3 + \varpi_0 M_4 + \Big[\gamma_1 M_1 + \nu_1 M_2 + \mu_1 M_3 + \varpi_1 M_4\Big]\|\mathsf{v}\| \\
+ & \Big[\gamma_2 M_1 + \nu_2 M_2 + \mu_2 M_3 + \varpi_2 M_4\Big]\|\mathsf{u}\|. \qquad (29)
\end{aligned}$$

Similarly, we get

$$\begin{aligned} \|\mathsf{u}\| &\leq \nu_0 M_5 + \varpi_0 M_6 + \Big[\nu_1 M_5 + \varpi_1 M_6\Big]\|\mathsf{v}\| \\ &+ \Big[\nu_2 M_5 + \varpi_2 M_6\Big]\|\mathsf{u}\|. \end{aligned} \tag{30}$$

From inequalities (29) and (30), we have

$$\|(\mathsf{v},\mathsf{u})\| \leq \frac{1}{\mathfrak{N}}\Big[\gamma_0 M_1 + \nu_0(M_2+M_5) + \mu_0 M_3 + \varpi_0(M_4+M_6)\Big], \tag{31}$$

with $\mathfrak{N} = \min\Big\{1-\mathfrak{N}_1, 1-\mathfrak{N}_2\Big\}$. The inequality (31) shows that $\mathfrak{Z}(\mathcal{J})$ is bounded. Hence, $\mathcal{J}$ has at least one fixed point according to Lemma 3. Thus, there is at least one solution on $[\mathsf{x}_1,\mathsf{x}_2]$ for the CS (1) and (2). □

Example 1. *Consider the CS of fractional differential equations given by*

$$\begin{cases} {}^cD^{\frac{7}{3}}\Big[\frac{1}{3}\mathsf{v}(t) + \frac{2}{110}I^{\frac{23}{5}}\phi(t,\mathsf{v}(t),\mathsf{u}(t))\Big] = \mathsf{k}(t,\mathsf{v}(t),\mathsf{u}(t)), \\ {}^cD^{\frac{5}{4}}\Big[\frac{7}{9}\mathsf{u}(t) + \frac{3}{70}I^{\frac{11}{3}}\psi(t,\mathsf{v}(t),\mathsf{u}(t))\Big] = \mathsf{p}(t,\mathsf{v}(t),\mathsf{u}(t)),\ t\in[0,1], \end{cases} \tag{32}$$

with the BCs

$$\begin{cases} \mathsf{v}(0)+\mathsf{v}(1)=0,\ \mathsf{v}'(1)=0,\ \mathsf{v}'(0)=\dfrac{3}{125}\displaystyle\int_0^{2/5}\mathsf{u}(s)ds, \\ \mathsf{u}(0)+\mathsf{u}(1)=0,\ \mathsf{u}'(1)=0. \end{cases} \tag{33}$$

Here $q_1 = 7/3,\ q_2 = 5/4,\ \theta_1 = 23/5,\ \theta_2 = 11/3, h = 3/125,\ \xi = 2/5$ *with*

$$\begin{aligned} \mathsf{k}(t,\mathsf{v}(t),\mathsf{u}(t)) &= \frac{3t^2}{6+t^4} + \frac{\sin\mathsf{v}(t)}{\sqrt{49+t^2}} + \frac{\mathsf{u}(t)|\mathsf{v}(t)|}{70(1+|\mathsf{v}(t)|)}, \\ \mathsf{p}(t,\mathsf{v}(t),\mathsf{u}(t)) &= \frac{2t}{3} + \frac{8\sin\mathsf{v}(t)|\tan^{-1}\mathsf{u}(t)|}{16\pi(t^3+1)} + \frac{\mathsf{u}(t)}{\sqrt{19}}, \\ \phi(t,\mathsf{v}(t),\mathsf{u}(t)) &= \frac{3}{11} + \frac{20\mathsf{v}(t)}{(t^2+8)^2} + \frac{4}{\sqrt[3]{125+t^2}}\mathsf{u}(t), \\ \psi(t,\mathsf{v}(t),\mathsf{u}(t)) &= (\frac{t+6}{70\pi})\tan^{-1}\mathsf{v}(t) + \frac{\mathsf{v}(t)|\cos\mathsf{u}(t)|}{t^5+22} + \frac{\sqrt{t^2+8}}{9}\sin\mathsf{u}(t). \end{aligned}$$

Clearly,

$$\begin{aligned} |\mathsf{k}(t,\mathsf{v}(t),\mathsf{u}(t))| &\leq \frac{1}{2} + \frac{1}{7}\|\mathsf{v}\| + \frac{1}{70}\|\mathsf{u}\|, \\ |\mathsf{p}(t,\mathsf{v}(t),\mathsf{u}(t))| &\leq \frac{2}{3} + \frac{1}{4}\|\mathsf{v}\| + \frac{1}{\sqrt{19}}\|\mathsf{u}\|, \\ |\phi(t,\mathsf{v}(t),\mathsf{u}(t))| &\leq \frac{3}{11} + \frac{5}{16}\|\mathsf{v}\| + \frac{4}{5}\|\mathsf{u}\|, \\ |\psi(t,\mathsf{v}(t),\mathsf{u}(t))| &\leq \frac{1}{20} + \frac{1}{22}\|\mathsf{v}\| + \frac{1}{3}\|\mathsf{u}\|, \end{aligned}$$

and hence $\gamma_0 = \frac{1}{2},\ \gamma_1 = \frac{1}{7},\ \gamma_2 = \frac{1}{70},\ \nu_0 = \frac{2}{3},\ \nu_1 = \frac{1}{4},\ \nu_2 = \frac{1}{\sqrt{19}},\ \mu_0 = \frac{3}{11},\ \mu_1 = \frac{5}{16},\ \mu_2 = \frac{4}{5},\ \varpi_0 = \frac{1}{20},\ \varpi_1 = \frac{1}{22}$ *and* $\varpi_2 = \frac{1}{3}$. *Using (23) and (24) with the given data we find that* $\mathfrak{N}_1 \simeq 0.836430 < 1,\ \mathfrak{N}_2 \simeq 0.583054 < 1$. *Therefore, by Theorem 1, the problem (32) and (33) has at least one solution on* $[0,1]$.

3.2. Uniqueness Result via Banach's Fixed Point Theorem

Theorem 2. *Assume the following assumption holds*

(H_2) $\mathsf{k}, \mathsf{p}, \phi, \psi : [\mathsf{x}_1, \mathsf{x}_2] \times \mathbb{R}^2 \to \mathbb{R}$ *are continuous functions and there exist positive constants* $l_m, m = 1, \cdots, 4$ *such that* $\forall t \in [\mathsf{x}_1, \mathsf{x}_2]$, v_i, u_i, $i = 1, 2 \in \mathbb{R}$ *we have*

$$\begin{aligned} |\mathsf{k}(t, \mathsf{v}_1, \mathsf{u}_1) - \mathsf{k}(t, \mathsf{v}_2, \mathsf{u}_2)| &\leq l_1(|\mathsf{v}_1 - \mathsf{v}_2| + |\mathsf{u}_1 - \mathsf{u}_2|), \quad (34)\\ |\mathsf{p}(t, \mathsf{v}_1, \mathsf{u}_1) - \mathsf{p}(t, \mathsf{v}_2, \mathsf{u}_2)| &\leq l_2(|\mathsf{v}_1 - \mathsf{v}_2| + |\mathsf{u}_1 - \mathsf{u}_2|), \quad (35)\\ |\phi(t, \mathsf{v}_1, \mathsf{u}_1) - \phi(t, \mathsf{v}_2, \mathsf{u}_2)| &\leq l_3(|\mathsf{v}_1 - \mathsf{v}_2| + |\mathsf{u}_1 - \mathsf{u}_2|), \quad (36)\\ |\psi(t, \mathsf{v}_1, \mathsf{u}_1) - \psi(t, \mathsf{v}_2, \mathsf{u}_2)| &\leq l_4(|\mathsf{v}_1 - \mathsf{v}_2| + |\mathsf{u}_1 - \mathsf{u}_2|), \quad (37) \end{aligned}$$

then the CS (1) and (2) has a unique solution on $[\mathsf{x}_1, \mathsf{x}_2]$, *provided that*

$$\mathcal{M}^* = M_1 l_1 + (M_2 + M_5) l_2 + M_3 l_3 + (M_4 + M_6) l_4 < 1, \quad (38)$$

where M_j, $(j = 1, ..., 6)$ *are given by (17)–(22).*

Proof. Define $l_1^* = \sup_{t \in [\mathsf{x}_1, \mathsf{x}_2]} |\mathsf{k}(t, 0, 0)| < \infty$, $l_2^* = \sup_{t \in [\mathsf{x}_1, \mathsf{x}_2]} |\mathsf{p}(t, 0, 0)| < \infty$, $l_3^* = \sup_{t \in [\mathsf{x}_1, \mathsf{x}_2]} |\phi(t, 0, 0)| < \infty$, $l_4^* = \sup_{t \in [\mathsf{x}_1, \mathsf{x}_2]} |\psi(t, 0, 0)| < \infty$ and $\mathfrak{K} > 0$ such that

$$\mathfrak{K} > \frac{M_1 l_1^* + (M_2 + M_5) l_2^* + M_3 l_3^* + (M_4 + M_6) l_4^*}{1 - (M_1 l_1 + (M_2 + M_5) l_2 + M_3 l_3 + (M_4 + M_6) l_4)}$$

Firstly, we show that $\mathcal{J} \mathcal{B}_{\mathfrak{K}} \subset \mathcal{B}_{\mathfrak{K}}$, where

$$\mathcal{B}_{\mathfrak{K}} = \{(\mathsf{v}, \mathsf{u}) \in \mathcal{V} \times \mathcal{V} : \|(\mathsf{v}, \mathsf{u})\| \leq \mathfrak{K}\}$$

For $(\mathsf{v}, \mathsf{u}) \in \mathcal{B}_{\mathfrak{K}}$, $t \in [\mathsf{x}_1, \mathsf{x}_2]$ and by the assumption (H_2), we have

$$\begin{aligned} |\mathsf{k}(t, \mathsf{v}(t), \mathsf{u}(t)| &\leq |\mathsf{k}(t, \mathsf{v}(t), \mathsf{u}(t)) - \mathsf{k}(t, 0, 0)| + |\mathsf{k}(t, 0, 0)| \\ &\leq l_1\Big(|\mathsf{v}(t)| + |\mathsf{u}(t)|\Big) + l_1^* \\ &\leq l_1\Big(\|\mathsf{v}\|_{\mathcal{V}} + \|\mathsf{u}\|_{\mathcal{U}}\Big) + l_1^* \leq l_1 \mathfrak{K} + l_1^*. \end{aligned}$$

In the same manner, we can get,

$$|\mathsf{p}(t, \mathsf{v}(t), \mathsf{u}(t)| \leq l_2 \mathfrak{K} + l_2^*,\ |\phi(t, \mathsf{v}(t), \mathsf{u}(t)| \leq l_3 \mathfrak{K} + l_3^*,\ |\psi(t, \mathsf{v}(t), \mathsf{u}(t)| \leq l_4 \mathfrak{K} + l_4^*.$$

Therefore, we have

$$\begin{aligned} |\mathcal{J}_1(\mathsf{v}, \mathsf{u})(t)| \leq{}& \Big(l_1 \mathfrak{K} + l_1^*\Big) \Bigg\{ \frac{1}{|\kappa_1|} \int_{\mathsf{x}_1}^{t} \frac{(t - s)^{q_1 - 1}}{\Gamma(q_1)} ds + \frac{1}{2|\kappa_1|} \int_{\mathsf{x}_1}^{\mathsf{x}_2} \frac{(\mathsf{x}_2 - s)^{q_1 - 1}}{\Gamma(q_1)} ds \\ &+ |\rho_2(t)| \int_{\mathsf{x}_1}^{\mathsf{x}_2} \frac{(\mathsf{x}_2 - s)^{q_1 - 2}}{\Gamma(q_1 - 1)} ds \Bigg\} + \Big(l_2 \mathfrak{K} + l_2^*\Big) \Bigg\{ |\rho_1(t)| \int_{\mathsf{x}_1}^{\mathsf{x}_2} \frac{(\mathsf{x}_2 - s)^{q_2 - 1}}{\Gamma(q_2)} ds \\ &+ |\rho_3(t)| \int_{\mathsf{x}_1}^{\mathsf{x}_2} \frac{(\mathsf{x}_2 - s)^{q_2 - 2}}{\Gamma(q_2 - 1)} ds + \frac{|\rho_4(t)||h|}{|\kappa_2|} \int_{\mathsf{x}_1}^{\xi} \Big(\int_{\mathsf{x}_1}^{s} \frac{(s - \tau)^{q_2 - 1}}{\Gamma(q_2)} d\tau \Big) ds \Bigg\} \\ &+ \Big(l_3 \mathfrak{K} + l_3^*\Big) \Bigg\{ \frac{|\lambda_1|}{|\kappa_1|} \int_{\mathsf{x}_1}^{t} \frac{(t - s)^{\theta_1 - 1}}{\Gamma(\theta_1)} ds + \frac{|\lambda_1|}{2|\kappa_1|} \int_{\mathsf{x}_1}^{\mathsf{x}_2} \frac{(\mathsf{x}_2 - s)^{\theta_1 - 1}}{\Gamma(\theta_1)} ds \\ &+ |\rho_2(t)||\lambda_1| \int_{\mathsf{x}_1}^{\mathsf{x}_2} \frac{(\mathsf{x}_2 - s)^{\theta_1 - 2}}{\Gamma(\theta_1 - 1)} ds \Bigg\} + \Big(l_4 \mathfrak{K} + l_4^*\Big) \Bigg\{ |\rho_1(t)||\lambda_2| \int_{\mathsf{x}_1}^{\mathsf{x}_2} \frac{(\mathsf{x}_2 - s)^{\theta_2 - 1}}{\Gamma(\theta_2)} ds \end{aligned}$$

$$+|\rho_3(t)||\lambda_2|\int_{x_1}^{x_2}\frac{(x_2-s)^{\theta_2-2}}{\Gamma(\theta_2-1)}ds+\frac{|\rho_4(t)||h||\lambda_2|}{|\kappa_2|}\int_{x_1}^{\xi}\Big(\int_{x_1}^{s}\frac{(s-\tau)^{\theta_2-1}}{\Gamma(\theta_2)}d\tau\Big)ds\Big\},$$
$$\leq\Big(M_1l_1+M_2l_2+M_3l_3+M_4l_4\Big)\mathfrak{K}+M_1l_1^*+M_2l_2^*+M_3l_3^*+M_4l_4^*.$$

In consequence, we get

$$\|\mathcal{J}_1(v,u)\|\leq\Big(M_1l_1+M_2l_2+M_3l_3+M_4l_4\Big)\mathfrak{K}+M_1l_1^*+M_2l_2^*+M_3l_3^*+M_4l_4^*.$$

Likewise, we can find that

$$\|\mathcal{J}_2(v,u)\|\leq(M_5l_2+M_6l_4)\mathfrak{K}+M_5l_2^*+M_6l_4^*,$$

and consequently, we get

$$\begin{aligned}\|\mathcal{J}(v,u)\| &\leq \Big(M_1l_1+(M_2+M_5)l_2+M_3l_3+(M_4+M_6)l_4\Big)\mathfrak{K}\\ &+ \Big(M_1l_1^*+(M_2+M_5)l_2^*+M_3l_3^*+(M_4+M_6)l_4^*\Big)\leq\mathfrak{K}.\end{aligned}$$

which implies that $\mathcal{J}\mathcal{B}_{\mathfrak{K}}\subset\mathcal{B}_{\mathfrak{K}}$.

Next, we show that the operator $\mathcal{J}$ is a contraction. For that, let v_i, $u_i\in\mathcal{B}_{\mathfrak{K}}$; $i=1,2$ and for each $t\in[x_1,x_2]$. Then we have

$$\begin{aligned}&|\mathcal{J}_1(v_1,u_1)(t)-\mathcal{J}_1(v_2,u_2)(t)|\leq\frac{1}{|\kappa_1|}\int_{x_1}^{t}\frac{(t-s)^{q_1-1}}{\Gamma(q_1)}\Big|k(s,v_1(s),u_1(s))-k(s,v_2(s),u_2(s))\Big|ds\\
&+\frac{|\lambda_1|}{|\kappa_1|}\int_{x_1}^{t}\frac{(t-s)^{\theta_1-1}}{\Gamma(\theta_1)}\Big|\phi(s,v_1(s),u_1(s))-\phi(s,v_2(s),u_2(s))\Big|ds\\
&+\frac{|\lambda_1|}{2|\kappa_1|}\int_{x_1}^{x_2}\frac{(x_2-s)^{\theta_1-1}}{\Gamma(\theta_1)}\Big|\phi(s,v_1(s),u_1(s))-\phi(s,v_2(s),u_2(s))\Big|ds\\
&+\frac{1}{2|\kappa_1|}\int_{x_1}^{x_2}\frac{(x_2-s)^{q_1-1}}{\Gamma(q_1)}\Big|k(s,v_1(s),u_1(s))-k(s,v_2(s),u_2(s))\Big|ds\\
&+|\rho_1(t)|\Big\{|\lambda_2|\int_{x_1}^{x_2}\frac{(x_2-s)^{\theta_2-1}}{\Gamma(\theta_2)}\Big|\psi(s,v_1(s),u_1(s))-\psi(s,v_2(s),u_2(s))\Big|ds\\
&+\int_{x_1}^{x_2}\frac{(x_2-s)^{q_2-1}}{\Gamma(q_2)}\Big|p(s,v_1(s),u_1(s))-p(s,v_2(s),u_2(s))\Big|ds\Big\}\\
&+|\rho_2(t)|\Big\{|\lambda_1|\int_{x_1}^{x_2}\frac{(x_2-s)^{\theta_1-2}}{\Gamma(\theta_1-1)}\Big|\phi(s,v_1(s),u_1(s))-\phi(s,v_2(s),u_2(s))\Big|ds\\
&+\int_{x_1}^{x_2}\frac{(x_2-s)^{q_1-2}}{\Gamma(q_1-1)}\Big|k(s,v_1(s),u_1(s))-k(s,v_2(s),u_2(s))\Big|ds\Big\}\\
&+|\rho_3(t)|\Big\{|\lambda_2|\int_{x_1}^{x_2}\frac{(x_2-s)^{\theta_2-2}}{\Gamma(\theta_2-1)}\Big|\psi(s,v_1(s),u_1(s))-\psi(s,v_2(s),u_2(s))\Big|ds\\
&+\int_{x_1}^{x_2}\frac{(x_2-s)^{q_2-2}}{\Gamma(q_2-1)}\Big|p(s,v_1(s),u_1(s))-p(s,v_2(s),u_2(s))\Big|ds\Big\}\\
&+|\rho_4(t)|\Big\{\int_{x_1}^{\xi}\Big(\frac{|h||\lambda_2|}{|\kappa_2|}\int_{x_1}^{s}\frac{(s-\tau)^{\theta_2-1}}{\Gamma(\theta_2)}\Big|\psi(\tau,v_1(\tau),u_1(\tau))-\psi(\tau,v_2(\tau),u_2(\tau))\Big|d\tau\\
&+\frac{|h|}{|\kappa_2|}\int_{x_1}^{s}\frac{(s-\tau)^{q_2-1}}{\Gamma(q_2)}\Big|p(\tau,x_1(\tau),y_1(\tau))-p(\tau,v_2(\tau),u_2(\tau))\Big|d\tau\Big)ds\Big\}.\end{aligned}$$

By applying (H_2), we have

$$\|\mathcal{J}_1(\mathsf{v}_1,\mathsf{u}_1)-\mathcal{J}_1(\mathsf{v}_2,\mathsf{u}_2)\| \leq [M_1l_1+M_2l_2+M_3l_3+M_4l_4](\|\mathsf{v}_1-\mathsf{v}_2\|+\|\mathsf{u}_1-\mathsf{u}_2\|). \quad (39)$$

Similarly, we find

$$\|\mathcal{J}_2(\mathsf{v}_1,\mathsf{u}_1)-\mathcal{J}_1(\mathsf{v}_2,\mathsf{u}_2)\| \leq [M_5l_2+M_6l_4](\|\mathsf{v}_1-\mathsf{v}_2\|+\|\mathsf{u}_1-\mathsf{u}_2\|). \quad (40)$$

It follows from (39) and (40) that

$$\|\mathcal{J}(\mathsf{v}_1,\mathsf{u}_1)-\mathcal{J}(\mathsf{v}_2,\mathsf{u}_2)\| \leq [M_1l_1+(M_2+M_5)l_2+M_3l_3+(M_4+M_6)l_4](\|\mathsf{v}_1-\mathsf{v}_2\|+\|\mathsf{u}_1-\mathsf{u}_2\|). \quad (41)$$

The inequalities (38) and (41) shows that $\mathcal{J}$ is a contraction. Due to the Banach fixed point theorem, the operator $\mathcal{J}$ has a unique fixed point that corresponds to the unique solution of the system (1) and (2) on $[\mathsf{x}_1,\mathsf{x}_2]$. □

Example 2. *Consider the same system in Example (3.2) with*

$$\begin{aligned}
\mathsf{k}(t,\mathsf{v}(t),\mathsf{u}(t)) &= \frac{2t^3}{\sqrt{225+t^8}}\Big(\frac{|\mathsf{v}(t)|}{1+|\mathsf{v}(t)|}+\cos\mathsf{u}(t)\Big),\\
\mathsf{p}(t,\mathsf{v}(t),\mathsf{u}(t)) &= \frac{e^{-4t}}{13}\Big(\sin\mathsf{v}(t)+\mathsf{u}(t)+\ln 7\Big),\\
\phi(t,\mathsf{v}(t),\mathsf{u}(t)) &= \tan^{-1}(t)+\frac{1}{12\pi}\sin 2\pi\mathsf{v}(t)+\frac{|\mathsf{u}(t)|}{6(1+|\mathsf{u}(t)|)},\\
\psi(t,\mathsf{v}(t),\mathsf{u}(t)) &= \frac{1}{240}\sin\mathsf{u}(t)+\frac{3e^{-t}}{720}\mathsf{v}(t).
\end{aligned}$$

Clearly,

$$\begin{aligned}
|\mathsf{k}(t,\mathsf{v}_1,\mathsf{u}_1)-\mathsf{k}(t,\mathsf{v}_2,\mathsf{u}_2)| &\leq l_1(\|\mathsf{v}_1-\mathsf{v}_2\|+\|\mathsf{u}_1-\mathsf{u}_2\|) \text{ with } l_1=2/15\\
|\mathsf{p}(t,\mathsf{v}_1,\mathsf{u}_1)-\mathsf{p}(t,\mathsf{v}_2,\mathsf{u}_2)| &\leq l_2(\|\mathsf{v}_1-\mathsf{v}_2\|+\|\mathsf{u}_1-\mathsf{u}_2\|) \text{ with } l_2=1/13\\
|\phi(t,\mathsf{v}_1,\mathsf{u}_1)-\phi(t,\mathsf{v}_2,\mathsf{u}_2)| &\leq l_3(\|\mathsf{v}_1-\mathsf{v}_2\|+\|\mathsf{u}_1-\mathsf{u}_2\|) \text{ with } l_3=1/6\\
|\psi(t,\mathsf{v}_1,\mathsf{u}_1)-\psi(t,\mathsf{v}_2,\mathsf{u}_2)| &\leq l_4(\|\mathsf{v}_1-\mathsf{v}_2\|+\|\mathsf{u}_1-\mathsf{u}_2\|) \text{ with } l_4=1/240
\end{aligned}$$

Moreover, it is found that $\mathcal{M}^* \simeq 0.402293 < 1$. So, the hypothesis of Theorem 2 is satisfied. Based on Theorem 2, there is a unique solution for the system (32) equipped with the conditions (33) on $[0,1]$.

4. Conclusions

In this work, we have successfully proved the existence and uniqueness results for a CS of nonlinear fractional IDEs of different orders type Caputo complemented with coupled anti-periodic and nonlocal integral BCs by using the Leray Schauder alternative and Banach fixed point theorem. As a special case, if we take $\lambda_1=\lambda_2=0$, consequently, our outcomes correspond to the solutions of the form:

$$\begin{aligned}
\mathcal{J}_1^*(\mathsf{v},\mathsf{u})(t) &= \frac{1}{\kappa_1}\int_{\mathsf{x}_1}^{t}\frac{(t-s)^{q_1-1}}{\Gamma(q_1)}\mathsf{k}(s,\mathsf{v}(s),\mathsf{u}(s))ds-\frac{1}{2\kappa_1}\int_{\mathsf{x}_1}^{\mathsf{x}_2}\frac{(\mathsf{x}_2-s)^{q_1-1}}{\Gamma(q_1)}\mathsf{k}(s,\mathsf{v}(s),\mathsf{u}(s))ds\\
&- \rho_1(t)\int_{\mathsf{x}_1}^{\mathsf{x}_2}\frac{(\mathsf{x}_2-s)^{q_2-1}}{\Gamma(q_2)}\mathsf{p}(s,\mathsf{v}(s),\mathsf{u}(s))ds-\rho_2(t)\int_{\mathsf{x}_1}^{\mathsf{x}_2}\frac{(\mathsf{x}_2-s)^{q_1-2}}{\Gamma(q_1-1)}\mathsf{k}(s,\mathsf{v}(s),\mathsf{u}(s))ds \quad (42)\\
&- \rho_3(t)\int_{\mathsf{x}_1}^{\mathsf{x}_2}\frac{(\mathsf{x}_2-s)^{q_2-2}}{\Gamma(q_2-1)}\mathsf{p}(s,\mathsf{v}(s),\mathsf{u}(s))ds-\rho_4(t)\int_{\mathsf{x}_1}^{\xi}\Big(\frac{h}{\kappa_2}\int_{\mathsf{x}_1}^{s}\frac{(s-\tau)^{q_2-1}}{\Gamma(q_2)}\mathsf{p}(\tau,\mathsf{v}(\tau),\mathsf{u}(\tau))d\tau\Big)ds,
\end{aligned}$$

$$\mathcal{J}_2^*(\mathsf{v},\mathsf{u})(t) = \frac{1}{\kappa_2}\int_{\mathsf{x}_1}^{t}\frac{(t-s)^{q_2-1}}{\Gamma(q_2)}\mathsf{p}(s,\mathsf{v}(s),\mathsf{u}(s))ds - \frac{1}{2\kappa_2}\int_{\mathsf{x}_1}^{\mathsf{x}_2}\frac{(\mathsf{x}_2-s)^{q_2-1}}{\Gamma(q_2)}\mathsf{p}(s,\mathsf{v}(s),\mathsf{u}(s))ds$$
$$+ \quad \rho_5(t) - \int_{\mathsf{x}_1}^{\mathsf{x}_2}\frac{(\mathsf{x}_2-s)^{q_2-2}}{\Gamma(q_2-1)}\mathsf{p}(s,\mathsf{v}(s),\mathsf{u}(s))ds, \tag{43}$$

and the values of $M_i,\ i = 1,...,6$ given by (17)–(22) takes the following form in this situations:

$$\begin{aligned}
M_1^* &= \frac{3(\mathsf{x}_2-\mathsf{x}_1)^{q_1}}{2|\kappa_1||\Gamma(q_1+1)} + \widetilde{\rho}_2\frac{(\mathsf{x}_2-\mathsf{x}_1)^{q_1-1}}{\Gamma(q_1)},\\
M_2^* &= \widetilde{\rho}_1\frac{(\mathsf{x}_2-\mathsf{x}_1)^{q_2}}{\Gamma(q_2+1)} + \widetilde{\rho}_3\frac{(\mathsf{x}_2-\mathsf{x}_1)^{q_2-1}}{\Gamma(q_2)} + \widetilde{\rho}_4\frac{|h|(\xi-\mathsf{x}_1)^{q_2+1}}{|\kappa_2|\Gamma(q_2+2)},\\
M_5^* &= \frac{3(\mathsf{x}_2-\mathsf{x}_1)^{q_2}}{2|\kappa_2||\Gamma(q_2+1)} + \widetilde{\rho}_5\frac{(\mathsf{x}_2-\mathsf{x}_1)^{q_2-1}}{\Gamma(q_2)},\\
M_6^* &= \frac{3|\lambda_2|(\mathsf{x}_2-\mathsf{x}_1)^{\theta_2}}{2|\kappa_2||\Gamma(\theta_2+1)} + \widetilde{\rho}_5\frac{|\lambda_2|(\mathsf{x}_2-\mathsf{x}_1)^{\theta_2-1}}{\Gamma(\theta_2)},
\end{aligned}$$

In addition, the methods presented in this study can be utilized to solve the system of FDEs type Riemann-Liouville with the BCs (2).

The simulation results of such an equation are the goal of a numerical study which could be interesting for future work.

Author Contributions: Conceptualization, Y.A. and S.A.; writing—original draft preparation, Y.A. and S.A.; validation, L.A. and N.A.; investigation, L.A. and N.A.; writing—review and editing, A.B.M. All authors have read and agreed to the published version of the manuscript.

Funding: Taif University Researchers Supporting Project number (TURSP-2020/218), Taif University, Taif, Saudi Arabia.

Data Availability Statement: Not applicable.

Acknowledgments: The second author thanks Taif University Researchers Supporting Program (Project number: TURSP-2020/218), Taif University, Saudi Arabia for technical and financial support.

Conflicts of Interest: The authors declare no conflict of interest.

References

1. Vazquez, L. A Fruitful Interplay: From Nonlocality to Fractional Calculus. In *Nonlinear Waves: Classical and Quantum Aspects*; Springer: Dordrecht, The Netherlands, 2004; pp. 129–133.
2. Zhang, F.; Chen, G.; Li, C.; Kurths, J. Chaos synchronization in fractional differential systems. *Phil. Trans. R. Soc. A* **2013**, *371*, 20120155. [CrossRef]
3. Qureshi, S. Effects of vaccination on measles dynamics under fractional conformable derivative with Liouville–Caputo operator. *Eur. Phys. J. Plus* **2020**, *135*, 63. [CrossRef]
4. Fallahgoul, H.A.; Focardi, S.M.; Fabozzi, F.J. *Fractional Calculus and Fractional Processes with Applications to Financial Economics: Theory and Application*; Elsevier/Academic Press: London, UK, 2017.
5. Javidi, M.; Ahmad, B. Dynamic analysis of time fractional order phytoplankton-toxic phytoplankton–zooplankton system. *Ecol. Model.* **2015**, *318*, 8–18. [CrossRef]
6. Magin, R.L. *Fractional Calculus in Bioengineering*; Begell House Publishers: Danbury, CT, USA, 2006.
7. Dingl, Y.; Wang, Z.; Ye, H. Optimal control of a fractional-order HIV-immune system with memory. *IEEE Trans. Control Syst. Technol.* **2012**, *20*, 763–769. [CrossRef]
8. Xu, Y.; Li, W. Finite-time synchronization of fractional-order complex-valued coupled system. *Phys. A* **2020**, *549*, 123903. [CrossRef]
9. Zaslavsky, G.M. *Hamiltonian Chaos and Fractional Dynamics*; Oxford University Press: Oxford, UK, 2005.
10. Sabatier, J.; Agrawal, O.P.; Machado, J.A.T. (Eds.) *Advances in Fractional Calculus: Theoretical Developments and Applications in Physics and Engineering*; Springer: Dordrecht, The Netherlands, 2007.
11. Kilbas, A.A.; Srivastava, H.M.; Trujillo, J.J. *Theory and Applications of Fractional Differential Equations*; North-Holland Mathematics Studies; Elsevier Science B.V.: Amsterdam, The Netherlands, 2006; p. 204.
12. Podlubny, I. *Fractional Differential Equations*; Academic Press: San Diego, CA, USA, 1999.

13. Miller, K.S.; Ross, B. *An Introduction to the Fractional Calculus and Fractional Differential Equations*; Wiley and Sons: New York, NY, USA, 1993.
14. Samko, S.; Kilbas, A.A.; Marichev, O. *Fractional Integrals and Derivatives: Theory and Applications*; Gordon and Breach: London, UK, 1993.
15. Lakshimikantham, V.; Leela, S.; Devi, J.V. *Theory of Fractional Dynamic Systems*; Cambridge Academic Publishers: Cambridge, UK, 2009.
16. Ghafoor, A.; Khan, N.; Hussain, M.; Ullah, R. A hybrid collocation method for the computational study of multi-term time fractional partial differential equations. *Comput. Math. Appl.* **2022**, *128*, 130–144. [CrossRef]
17. Ghafoor, A.; Haq, S.; Hussain, M.; Abdeljawad, T.; Alqudah, M.A. Numerical Solutions of Variable Coefficient Higher-Order Partial Differential Equations Arising in Beam Models. *Entropy* **2022**, *24*, 567. [CrossRef]
18. Shaheen, S.; Haq, S.; Ghafoor, A. A meshfree technique for the numerical solutions of nonlinear Fornberg-Whitham and Degasperis-Procesi equations with their modified forms. *Comput. Appl. Math.* **2022**, *41*, 183. [CrossRef]
19. Ahmad, B.; Alruwaily, Y.; Alsaedi, A.; Nieto, J.J. Fractional integro-differential equations with dual anti-periodic boundary conditions. *Differ. Integral Equ.* **2020**, *33*, 181–206. [CrossRef]
20. Ahmad, B.; Alsaedi, A.; Salem, S.; Ntouyas, S.K. Fractional differential equation involving mixed nonlinearities with nonlocal multi-point and Riemann-Stieltjes integral-multi-strip conditions. *Fractal Fract.* **2019**, *3*, 34. [CrossRef]
21. Ahmad, B.; Alsaedi, A.; Alruwaily, Y.; Ntouyas, S.K. Nonlinear multi-term fractional differential equations with Riemann-Stieltjes integro-multipoint boundary conditions. *AIMS Math.* **2020**, *5*, 1446–1461. [CrossRef]
22. Abbas, M.I. Existence and uniqueness of solution for a boundary value problem of fractional order involving two Caputo's fractional derivatives. *Adv. Differ. Equ.* 2015, 1–19. [CrossRef]
23. Agarwal, R.P.; Ahmad, B.; Garout, D.; Alsaedi, A. Existence results for coupled nonlinear fractional differential equations equipped with nonlocal coupled flux and multi-point boundary conditions. *Chaos Solitons Fracts* **2017**, *102*, 149–161. [CrossRef]
24. Su, X.; Zhang, S.; Zhang, L. Periodic boundary value problem involving sequential fractional derivatives in Banach space. *AIMS Math.* **2020**, *5*, 7510–7530. [CrossRef]
25. Wang, Y.; Liang, S.; Wang, Q. Existence results for fractional differential equations with integral and multi-point boundary conditions. *Bound. Value Probl.* **2018**, *4*, 1–11. [CrossRef]
26. Ahmad, B.; Alsaedi, A.; Ntouyas, S.K. Nonlinear coupled Fractional order systems with integro-multistrip-multipoint boundary conditions. *Int. J. Anal. Appl.* **2019**, *17*, 940–957.
27. Ahmad, B.; Alghanmi, M.; Alsaedi, A. Existence results for a nonlinear coupled systems involving both Caputo and Riemann-Liouville generalized fractional derivatives and coupled integral boundary conditions. *Rocty Mt. J. Math.* **2020** *50*, 1901–1922. [CrossRef]
28. Ahmad, B.; Alghanmi, M.; Alsaedi, A.; Nieto, J.J. Existence and uniqueness results for a nonlinear coupled system involving Caputo fractional derivatives with a new kind of coupled boundary conditions. *Appl. Math. Lett.* **2021**, *116*, 107018. [CrossRef]
29. Ahmad, B.; Alghamdi, N.; Alsaedi, A.; Notouyas, S.K. A system of coupled multi-term fractional differential equations with three-point coupled boundary conditions. *Fract. Calc. Appl. Anal.* **2019**, *22*, 601–618. [CrossRef]
30. Ntouyas, S.K.; Sulami, H.H. A study of coupled systems of mixed order fractional differential equations and inclusions with coupled integral fractional boundary conditions. *Adv. Differ. Equ.* **2020**, *2020*. [CrossRef]
31. Alsaedi, A.; Ahmad, B.; Aljoudi, S.; Ntouyas, S.K. A study of a fully coupled two-parameter system of sequential fractional integro-differential equations with nonlocal integro-multipoint boundary conditions. *Acta Math. Sci.* **2019**, *39*, 927–944. [CrossRef]
32. Butkovskii, A.G.; Postnov, S.S.; Postnova, E.A. Fractional integro-differential calculus and its control-theoretical applications. II. Fractional dynamic systems: Modeling and hardware implementation. *Autom. Remote Control.* **2013**, *74*, 725–749. [CrossRef]
33. Abro, K.A.; Atangana, A.; Gómez-Aguilar, J.F. A comparative analysis of plasma dilution based on fractional integro-differential equation: An application to biological science. *Int. J. Model. Simul.* **2022**, *2022*, 1–10. [CrossRef]
34. Ahmad, B. On nonlocal boundary value problems for nonlinear integro-differential equations of arbitrary fractional order. *Results Math.* **2013**, *63*, 183–194. [CrossRef]
35. Ahmad, B.; Agarwal, R.P.; Alsaedi, A.; Ntouyas, S.K.; Alruwaily, Y. Fractional order Coupled systems for mixed fractional derivatives with nonlocal multi-point and Riemann–Stieltjes integral-multi-strip conditions. *Dynam. System Appl.* **2020**, *29*, 71–86. [CrossRef]
36. Alsaedi, A.; Aljoudi, S.; Ahmad, B. Existence of solutions for Riemann-Liouville type coupled systems of fractional integro-differential equations and boundary conditions. *Electron. J. Differ. Equ.* **2016**, *211*, 1–14.
37. Ahmad, B.; Alsaedi, A.; Aljoudi, S.; Ntouyas, S.K. A six-point nonlocal boundary value problem of nonlinear coupled sequential fractional integro-differential equations and coupled integral boundary conditions. *J. Appl. Math. Comput.* **2018**, *56*, 367–389. [CrossRef]
38. Ahmad, B.; Broom, A.; Alsaedi, A.; Ntouyas, S.K. Nonlinear integro-differential equations involving mixed right and left fractional derivatives and integrals with nonlocal boundary data. *Mathematics* **2020**, *8*, 336. [CrossRef]
39. Ahmad, B.; Alghamdi, R.P.; Agarwal, P.; Alsaedi, A. Riemann-Liouville fractional integro-differential equations with fractional nonlocal multi-point boundary conditions. *Fractals* **2022**, *30*, 2240002. [CrossRef]
40. Smart, D.R. *Fixed Point Theorems*; Cambridge University Press: Cambridge, UK, 1980.

41. Henderson, J.; Luca, R.; Tudorache, A. On a system of fractional differential equations with coupled integral boundary conditions. *Fract. Calc. Appl. Anal* **2015**, *18*, 361–386. [CrossRef]
42. Wang, J.R.; Zhou, Y.; Feckan, M. On recent developments in the theory of boundary value problems for impulsive fractional differential equations. *Comput. Math. Appl.* **2012**, *64*, 3008–3020. [CrossRef]
43. Alsaedi, A.; Hamdan, S.; Ahmad, B.; Ntouyas, S.K. Existence results for coupled nonlinear fractional differential equations of different orders with nonlocal coupled boundary conditions. *J. Inequalities Appl.* **2021**, *95*, 759. [CrossRef]
44. Ahmad, B.; Ntouyas, S.K.; Alsaedi, A. On fully coupled nonlocal multi-point boundary value problem of nonlinear mixed-order fractional differential equations on an arbitrary domain. *Filomat* **2018**, *32*, 4503–4511. [CrossRef]
45. Ahmad, B.; Hamdan, S.; Alsaedi, A.; Ntouyas, S.K. On a nonlinear mixed-order coupled fractional differential system with new integral boundary conditions. *AIMS Math.* **2021**, *6*, 5801–5816. [CrossRef]
46. Granas, A.; Dugundji, J. *Fixed Point Theory*; Springer: New York, NY, USA, 2005.

symmetry

MDPI

Article

Analysis of Controllability of Fractional Functional Random Integroevolution Equations with Delay

Kinda Abuasbeh [1,*], Ramsha Shafqat [2,*], Ammar Alsinai [3,*] and Muath Awadalla [1]

1 Department of Mathematics and Statistics, College of Science, King Faisal University, Hafuf 31982, Al Ahsa, Saudi Arabia
2 Department of Mathematics and Statistics, The University of Lahore, Sargodha 40100, Pakistan
3 Department of Studies in Mathematics, University of Mysore, Manasagangotri, Mysore 570006, India
* Correspondence: kabuasbeh@kfu.edu.sa (K.A.); ramshawarriach@gmail.com (R.S.); aliiammar1985@gmail.com (A.A.)

Abstract: Various scholars have lately employed a wide range of strategies to resolve two specific types of symmetrical fractional differential equations. The evolution of a number of real-world systems in the physical and biological sciences exhibits impulsive dynamical features that can be represented via impulsive differential equations. In this paper, we explore some existence and controllability theories for the Caputo order $q \in (1, 2)$ of delay- and random-effect-affected fractional functional integroevolution equations (FFIEEs). In order to prove that random solutions exist, we must prove a random fixed point theorem using a stochastic domain and the mild solution. Then we demonstrate that our solutions are controllable. At the end, applications and example is illustrated which indicates the applicability of this manuscript.

Keywords: random fixed point; state dependent delay; controllability; functional differential equation; mild solution; finite delay; cosine and sine family

MSC: 26A33; 34K37

Citation: Abuasbeh, K.; Shafqat, R.; Alsinai, A.; Awadalla, M. Analysis of Controllability of Fractional Functional Random Integroevolution Equations with Delay. *Symmetry* **2023**, *15*, 290. https://doi.org/10.3390/sym15020290

Academic Editor: Dumitru Baleanu

Received: 26 December 2022
Revised: 11 January 2023
Accepted: 17 January 2023
Published: 20 January 2023

1. Introduction

Many different applications have been investigated through the theory of impulsive fractional differential equations (IFDEs) in the accurate mathematical representation of a wide variety of practical problems. It is acknowledged as a crucial area for research, as much as the modelling of impulsive issues in population dynamics, ecology, biotechnology, and other fields. In real-world situations, many processes and phenomena are characterised by rapid shifts in their states. The mentioned quick modifications are called impulsive effects within the system. Instantaneous and noninstantaneous impulses are the two main forms of impulses discussed in the literature to date. In contrast to the length of a whole evolution, such as that of shocks and natural disasters, the period of these fluctuations in instantaneous impulses is insignificant; in the case of noninstantaneous impulses, on the other hand, the duration of the changes exists throughout a finite time period.

Over the past three decades, the field of mathematical analysis has incorporated fractional calculus, FDEs, and integrodifferential equations, and the qualitative theory of these equations on both a theoretical and a practical level. Fundamentally, fractional calculus theory, the qualitative theory of FDEs and fractional integrodifferential equations, numerical simulations, and symmetry analysis are mathematical analytical tools used to study arbitrary-order integrals and derivatives that unify and generalise the conventional ideas of differentiation and integration. Compared to classical formulations, nonlinear operators with a fractional order are more useful. Throughout the development of emerging control theory, the controllability of DEs problems has played a major role. Typically, it means that a dynamical system may be moved from any initial state to the desired terminal state using a set of legal controls. Control theory places much emphasis on the qualitative characteristics of control systems. There has been particular focus on the controllability of linear and nonlinear systems in a finite-dimensional space that are described by ordinary DEs; see [1–4] for a list of researchers who have extended the idea to infinite-dimensional systems with bounded operators in Banach spaces (BS). The controllability problem was converted into a fixed-point problem by the authors of [5]. We advise reading [6,7] for additional information. The authors of [8,9] investigated a variety of functional DEs and inclusions, and proposed various controllability findings. A family of integrodifferential evolution equations' controllability was examined by Dilao et al. [10].

It is often advantageous to handle second-order abstract DEs explicitly rather than always reducing them to first-order systems. For the investigation of second-order issues, the theory of strongly continuous cosine families is an invaluable resource. We use some of the core ideas in cosine family theory [11]. Typically, this means that a dynamical system may be moved from any initial state to the desired terminal state using a set of legal controls. Control theory places much emphasis on the qualitative characteristics of control systems. There has been particular focus on the controllability of linear and nonlinear systems in finite-dimensional space that are described by ordinary DEs [12,13].

The reader is recommended to read [14–16] for more information on random differential equations, which are natural generalisations of deterministic DEs and appear in a variety of applications. The accuracy of our knowledge about the system's characteristics determines the nature of a dynamic system. When knowledge about a dynamic system is exact, a deterministic dynamical system emerges. Moreover, many of the available details for identifying and assessing dynamic system characteristics are incorrect, uncertain, or imprecise. To put it another way, determining the parameters of a dynamic system is highly risky. However, when we have probable knowledge and an understanding of statistical characteristics, we can use stochastic DEs in mathematically modelling such systems.

Ji-Huan He [17] studied fractal calculus. Wang et al. [18–20] worked on nondifferentiable exact solutions, the modification of the unsteady model, and diverse exact and explicit solutions. Mehmood et al. [21] worked on a partial DE. Niazi et al. [22], Shafqat et al. [23], Alnahdi [24], and Abuasbeh et al. [25] investigated the existence and uniqueness of FEEs. Inspired by the above studies [26], this paper deals with the controllability of the fractional functional integroevolution equation with delay and random effects:

$$\begin{aligned} &{}_0^cD_v^q U(\chi,\xi) = B_1 U(\chi,\xi) + \varphi(\chi, U_\chi(.,\xi),\xi) + \textstyle\int_0^v B_2 f(\chi,\xi) dC_v + Bx(v)Cx(v)dv,\ \xi \in \Theta := [0,\kappa],\ v \in [0,T] \\ &U(\chi,\xi) + m(U) = \varrho_1(\chi,\xi);\ \ \xi \in (-\infty, 0], \\ &U'(\chi,\xi) = \varrho_2(\xi) \end{aligned} \tag{1}$$

Knowing that complete probability space $(\Phi, F, \wp)$ given functions $\varphi : \Theta \times D \times \Psi \to \Xi, \sigma_1 \in D \in D \times \Phi$, and infinitesimal generator $B_1 : D(B_1) \subset \Xi \to \Xi$ of a strongly continuous cosine family, the phase space is $(H_q(\chi))_{\chi \in \mathbf{R^m}}$ on Ξ, D, and a real BS is $(\Xi, |.|)$. Control function $\mathcal{P}(.,\xi)$ is specified in $L^2(\Theta, \Omega)$, a BS of possible control functions with Ω as a BS, and B_2 is a bounded linear operator (LO) from Ω into Ξ.

The component of $D \times \Phi$ determined with $D \times \Phi$, given by $U_\xi(\iota,\xi) = U(\xi + \iota, \xi), \iota \in (-\infty, 0]$ is denoted by $U_\chi(.,\xi)$. Here, the state's existence from the year $-\infty$ to the current day ξ is represented by the string $U_\chi(.,\xi)$. Eras $U_\chi(.,\xi)$ were presumptively part of some abstract phases D.

First, we suppose random issue

$$\begin{aligned} &{}_0^cD_v^q U(\chi,\xi) = B_1 U(\chi,\xi) + \varphi(\chi, U_{\vartheta(\chi,U_\chi)}(.,\xi),\xi) + \textstyle\int_0^v B_2 f(\chi,\xi) dC_v + Bx(v)Cx(v)dv,\ \xi \in \Theta := [0,\kappa],\ v \in [0,T] \\ &U(\chi,\xi) + m(U) = \varrho_1(\chi,\xi);\ \ \xi \in (-\infty, 0], \\ &U'(\chi,\xi) = \varrho_2(\xi) \end{aligned} \tag{2}$$

where $\varphi : \Theta \times D \times \Psi \to \Xi, \sigma_1 \in D \in D \times \Phi$ are given random functions, $B_1 : D(B_1) \subset \Xi \to \Xi$ is as in problem (1), D is the phase space, $\psi; \Theta \times D \to (-\infty, \kappa]$, and $(\Xi, |.|)$ is a real

BS. For the key conclusions on Schauder's fixed theorem [27], and random fixed-point theorem paired with the family of cosine operators, we employ our' arguments.

The layout of this article is as follows. Section 2 contains some needed preliminaries and fundamental results. Sections 3 and 4 present our main results in two cases: infinite fixed delay and state-dependent delay, respectively. In Sections 5 and 6, we give applications and an example, respectively. In Section 7, we present the conclusion.

Motivation and Novelties

The incorporation of fractional-order derivatives in delay DEs provides a range of advantages, including hereditary properties, additional degrees of freedom, and other advantages of fractional modelling. As these equations are primarily used in control theory and robotics, the stability and asymptotics of these equations are of vital importance. However, stability and asymptotic analyses of fractional delay DEs are still in their early stages. Most of the current stability results on autonomous equations of this type are based on the root locus of their corresponding characteristic equations, and do not offer a universal and reliable way of assessing the stability of a given fractional delay DE.

FDEs with a time delay are widely used in natural phenomena, and the fields of science and engineering. To capture the dynamic behavior of travelling wave solutions on the basis of these equations, researchers have created algorithms with high performance for various spatial and time fractional delay DEs. However, there are still challenges to be addressed in the field of fractional delay DEs, such as the stability analysis of numerical time integration schemes and the numerical theory of the numerical scheme. Additionally, there is a need for stability and numerical simulations of travelling wave solutions, critical travelling wave solutions, and the design of compact fourth- and sixth-order schemes for fractional delay DEs with strong nonlinearity.

This paper aims to investigate the existence and controllability of solutions to FDEs with delay and random effects. While the majority of results in the literature have focused on first-order equations, some researchers produced FDE results. In our study, we obtained findings for Caputo derivatives of order (1,2) using a mild solution. Stability is a major area of research in DE theory, and over the past 20 years, stability for FDE has been a major focus of research. In order to illustrate this, we consider the prerequisites for solution stability and FDE asymptotic stability. We also examine delay fractional functional random integroevolution equations.

2. Preliminaries

We discuss a few of the abbreviations, definitions, and theorems that are used throughout the work in this part. Considering the BS $D(\Xi)$ of bounded LOs from Ξ into Ξ, where $\Theta := [0,\kappa], \kappa > 0$,

$$||\aleph||_{D(\Xi)} = \sup_{||\chi||=1} ||\aleph(U)||.$$

Let $C := C(I,\Xi)$ be the Banach space of continuous functions $U : \Theta \to \Xi$ with the norm

$$||U||_C = \sup_{\chi\in\Theta} |U(\chi)|.$$

We follow to the methodology described in [28] and apply the axiomatic description of the phase space D given in [29]. Once $(D, ||.||_D)$ is defined as a seminormed linear space of functions translating $(-\infty, 0]$ into Ξ, we have

(J_1) Let $U : (-\infty,\kappa) \to \Xi, \kappa > 0$, is a continuous function on Θ and $U_0 \in D$, then, for every $\chi \in \Theta$, the following hold.

- (a) $U_\chi \in D$;
- (b) There $\exists$ a positive constant ρ, $|U(\chi)| \leq \varpi||U_\chi||_D$.

(c) There $\exists$ two functions $\beta(.), \omega(.) : \mathbf{R}^{\mathbf{m}}_{+} \rightarrow \mathbf{R}^{\mathbf{m}}_{+}$ independent of U with β continuous and bounded and ω locally bounded where:

$$||U_{\chi}||_{D} \leq \beta(\chi) \sup\{|U(\rho)| : 0 \leq \rho \leq \rho\} + \omega(\chi)||U_{0}||_{D}.$$

(J_2) For function U in (A_1), U_{χ} is a D-valued continuous function on Θ.
(J_3) The space D is complete.
Set

$$\varsigma = \sup\{\beta(\chi) : \chi \in \Theta\}, \text{ and } \omega = \sup\{\omega(\chi) : \chi \in \Theta\}.$$

Remark 1. *1.* (2) *is equivalent to* $|\varrho_1||_D \leq \varpi||\varrho_1||_D \forall \varrho_1 \in D$.
2. $||.||_D$ *is a seminorm, this implies that the two elements* $\varrho_1, \eta \in D$ *satisfy* $||\varrho_1 - \eta||_D = 0$ *not necessarily that* $\varrho_1(\iota) = \eta(\iota) \forall \iota \leq 0$.
3. *For all* $\varrho_1, \eta \in D$ *where* $||\varrho_1 - \chi||_D = 0. \Rightarrow \varrho_1(0) = \eta(0)$.
Let us present the space

$$\Xi := \{U : (-\infty, \kappa] : U|_{(\infty,0]} \in D \text{ and } U|_{\Theta} \in C\},$$

and let $||U||_{\Xi}$ *be the seminorm in* Ξ *given by*

$$||U||_{\Xi} = ||\varrho_1||_D + ||U||_C.$$

Definition 1. *Let* $\{H_q(\chi) : \chi \in \mathbf{R}^{\mathbf{m}}\}$ *be a family of bounded LOs in the Banach space* Ψ*, which is a strongly continuous cosine family if*

- $H_q(0) = I$.
- $H_q(\chi)_{\eta}$ *is strongly continuous in* χ *on* $\mathbf{R}^{\mathbf{m}}$ *for each fixed* $\eta \in \Psi$.
- $H_q(\chi - \rho) = 2H_q(\chi)H_q(\rho) \forall \chi, \rho \in \mathbf{R}^{\mathbf{m}}$.

Let $\{H_q(\chi) : \chi \in \mathbf{R}^{\mathbf{m}}\}$ *be a strongly continuous cosine family in* Ψ. *Define the sine family* $\{K_q(\chi) : \chi \in \mathbf{R}^{\mathbf{m}}\}$ *with*

$$K_q(\chi)\eta = \int_0^{\chi} H_q(\rho)\eta d\rho, \ \eta \in \Xi, \ \chi \in \mathbf{R}^{\mathbf{m}}.$$

The infinitesimal generator $B_1 : \Xi \rightarrow \Xi$ *of the cosine family* $\{S_{(\chi)} : \chi \in \mathbf{R}^{\mathbf{m}}\}$ *is defined by*

$$B_1\eta = \frac{d^2}{d\chi^2} H_q(\chi)\eta|_{\chi=0}, \ \eta \in D(B_1),$$

where

$$D(B_1) = \{\eta \in \Xi : H_q(.)\eta \in C^2(\mathbf{R}^{\mathbf{m}}, \Xi)\}.$$

Definition 2. *Consider the map* $\phi : \Theta \times D \times \psi \rightarrow \Xi$ *is random Caratheodory if*
(i) $\chi \rightarrow \phi(\chi, U, \Delta)$*, this map measurable* $\forall\, U \in D$ *and for all* $\Delta \in \psi$.
(ii) $U \rightarrow \phi(\chi, U, \Delta)$ *is measurable* $\forall\, U \in D$ *and for all* $\Delta \in \psi$.
(iii) $\Delta \rightarrow \phi(\chi, U, \Delta)$ *is measurable* $\forall\, U \in D$*, and almost* $\chi \in \Theta$.

Let D_{Ξ} *be the Borel* σ*-algebra in separable BS* Ξ. *If, for each* $\Pi \in D_{\Xi}, p^{-1}(\Pi) \in F$*, then the map* $p : \psi \rightarrow \Xi$ *is a random variable. If* $G(., p)$*, written as* $G(\Delta, p) = G(\Delta)p$*, is measurable for each* $p \in \Xi$*, then* $G : \psi \times \Xi \rightarrow \Xi$ *is a random operator.*

Definition 3 ([30])**.** *Let* $\acute{G}$ *be a mapping from* ψ *into* 2^{Ξ}. *A mapping* $G : \{(\Delta, p) : \Delta \in \psi \wedge p \in \acute{G}(\Delta)\} \rightarrow \Xi$ *is a random operator with stochastic domain* $\acute{G}$ *if and only if, for all closed* $\Pi_1 \subseteq \Xi, \{\Delta \in \psi : \acute{G}(\Delta) \cap \acute{G}_1 \neq \emptyset\} \in F$*, and for all open* $\Pi_2 \subseteq \Xi$ *and all* $p \in \Xi, \{\Delta \in \psi : p \in \acute{G}(\Delta) \wedge G(\Delta, p) \in \Pi_2\} \in F$. *G is continuous if every* $G(\Delta)$ *is continuous. A mapping* $p : \psi \rightarrow \Xi$ *is a random fixed point of G if and only if for all* $\Delta \in \psi, p(\Delta) \in \acute{G}(\Delta)$ *and* $G(\Delta)p(\Delta) = p(\Delta)$ *and p is measurable if for all open* $\Pi_2 \subseteq \Xi, \{\Delta \in \psi : p(\Delta) \in \Pi_2\} \in F$.

Lemma 1 ([30]). *Let $\acute{G} : \psi \to 2^{\Xi}$ be measurable for every $\Delta \in \psi$ with $\acute{G}(\Delta)$ closed, convex, and solid (i.e., $\int G(\Delta) \neq \emptyset$). We assumed the existence of a measurable $p_0 : \psi \to \Xi$ with $p_0 \in \int \acute{G}(\Delta)$ for all $\Delta \in \psi$. Assume that G is a continuous random operator with the stochastic domain $\acute{G}$; as such, $G(\Delta)p = p \neq \emptyset$ for any $\Delta \in \psi, \{p \in \acute{G}(\Delta)$. Once this happens, G has a stochastic fixed point. If the function $p(\chi, .)$ is measurable for each $\chi \in \Theta$, then the mapping of p of $\Theta \times \psi$ into Ξ is stochastic.*

Definition 4 ([31]). *Assume that U is a BS, and ϕ_U is the bounded subsets of Ξ. The Kuratowski measure of noncompactness is map $\mu : \psi_U \to [0, \infty)$ given by $\mu(\Pi) = \inf\{\epsilon > 0 : \Pi \subseteq \cup_{i=1}^{n}$ and $diam(\Pi_i) \leq \epsilon\}$; here $\Pi \in \psi_U$ and verifies the following properties:*

(a) $\mu(\Pi) = 0 \Leftrightarrow \bar{\Pi}$ *is compact.*
(b) $\mu(\Pi) = \mu(\bar{\Pi})$.
(c) $\tilde{\Pi} \subset \Pi \Rightarrow \mu(\tilde{\Pi}) \leq (\Pi)$.
(d) $\mu(\tilde{\Pi} + \Pi) \leq \mu(\tilde{\Pi} + \mu(\Pi))$.
(e) $\mu(\epsilon\Pi) = |\epsilon|\mu(\Pi); \epsilon \in \mathbf{R^m}$.
(f) $\mu(conv\Pi) = \mu(B)$.

Lemma 2 ([32]). *$\mu(g(\chi))$ is continuous on theta if and only if $g \subset C(\Theta, \Xi)$ is bounded and equicontinuous:*

$$\mu\left(\left\{\int_{\Theta} \eta(\rho)d\rho : \eta \in g\right\}\right) \leq \int_{\Theta} \mu(g(\rho))d\rho,$$

where $g(\chi) = \{\eta(\chi) : \eta \in g\}, \chi \in \Theta$.

Lemma 3 (Gronwall lemma [28]). *Assume $\mu, y \in \mathcal{H}([0,1], \mathbb{R}_+)$ and let μ be increasing. If $u \in \mathcal{H}([0,1], \mathbb{R}_+)$ satisfies*

$$u(\omega) \leqslant \mu(\omega) + \int_0^{\omega} y(s)u(s)ds, \ \omega \in [0,1],$$

then

$$u(\omega) \leqslant \mu(\omega) \exp \int_0^{\omega} y(s)u(s)ds, \ \omega \in [0,1].$$

Definition 5 ([30]). *The fractional Riemann–Liouville (RL) derivative is defined as follows.*

$$_aD^p_\omega\chi(\omega) = \frac{1}{\Gamma(n-p+1)}\left(\frac{d}{d\omega}\right)^{n+1}$$
$$\int_a^{\omega} (\omega - \tau)^{n-p}\chi(\tau)d\tau, \ n \leqslant p \leqslant n+1.$$

Definition 6 ([30]). *Caputo fractional derivatives ${}^C_aD^\alpha_\omega\chi(\omega)$ of order $\alpha \in \mathbb{R}^+$ are defined by*

$${}^C_aD^\alpha_\omega\chi(\omega) = {}_aD^\alpha_\omega(\chi(\omega) - \sum_{j=0}^{k-1} \frac{\chi^{(j)}(a)}{j!}(\omega - a)^j),$$

in which $k = [\alpha] + 1$.

Definition 7 ([31]). *Wright function ψ_α is defined by*

$$\begin{aligned}\psi_\alpha(\kappa) &= \sum_{j=0}^{\infty} \frac{(-\kappa)^j}{j!\Gamma(-\alpha j + 1 - \alpha)} \\ &= \frac{1}{\pi}\sum_{j=1}^{\infty} \frac{(-\kappa)^j}{(j-1)!}\Gamma(j\alpha)\sin(j\pi\alpha),\end{aligned}$$

$\alpha \in (0,1), \kappa \in \mathbb{C}$.

3. Results of Controllability for the Steady Delay Case

Definition 8. *Equation (1) is controllable on the interval $(-\infty, \kappa]$ if, for all final state $U^1(\xi)$, there $\exists$ a control $\mathcal{P}(.,\xi)$ in $L^2(\Theta,\Omega)$, such that the solution $U(\chi,\xi)$ of (1) satisfies $U(\kappa,\xi) = U^1(\xi)$.*

Definition 9. *A stochastic process $U : (-\infty,\kappa] \times \Phi \to \Xi$ is a random mild solution of Problem (1) if $U(\chi,\xi) = \varrho_1(\chi,\xi); \chi \in (-\infty,\chi], U^{\infty}(0,\xi) = \varrho_2(\xi)$, and the restriction of $U(.,\xi)$ to the interval Θ is continuous and verifies:*

$$U(\chi,\xi) = H_q(\chi)(\varrho_1(\chi,\xi) - m(U)) + K_q(\chi)\varrho_2(\chi) + \int_0^V (\chi-\rho)P_q(\chi-\rho)B_1U(\chi,\xi)d\rho + \int_0^V (\chi-\rho)P_q(\chi-\rho)$$
$$[\varphi(\chi, U_\chi(.,\xi),\xi)]d\rho + \int_0^\chi \left((\chi-\rho)P_q(\chi-\rho)\int_0^V B_2 f(\chi,\xi)dC_v + Bx(\rho)Cx(\rho)\right)d\rho$$

Let

$$\omega = \sup\{||H_q(\chi)||_{D(\Xi)} : \chi \geq 0\}$$

and

$$\omega = \sup\{||K_q(\chi)||_{D(\Xi)} : \chi \geq 0\}.$$

The following hypotheses must be introduced:

(H_1) $H_q(\chi)$ is compact for $\chi > 0$,
(H_2) The function $\phi : \Theta \times D \times \psi \to \Psi$ is random Caratheodory.
(H_3) There $\exists$ functions $\eta : \Theta \times \phi \to \mathbf{R}^{\mathbf{m}}_{+}$ and $p : \Theta \times \psi \to \mathbf{R}^{\mathbf{m}}_{+}$ for each $\Delta \in \psi, \eta(.,\Delta)$ is continuous nondecreasing and $p(.,\Delta)$ integrable with:

$$|\phi(\chi,\mathcal{P},\Delta)| \leq p(\chi,\Delta)\eta(||\mathcal{P}||_D,\Delta) \; \textit{for a.e.}\; \chi \in \Theta \; \textit{and each}\; \mathcal{P} \in D,$$

(H_4) There $\exists$ a random function $Q : \psi \to \mathbf{R}^{\mathbf{m}}_{+}\ \{0\}$ where:

$$\omega(1+\kappa\omega\zeta(||\varrho_1||_D + \eta(D,\Delta||p||_{L^1}) + \kappa\omega\zeta||\eta^1|| + \omega'(1+\kappa\omega\zeta)|\varrho_2| \leq Q(\Delta)$$

where

$$D := \zeta Q(\Delta) + \sigma||\varrho_1||_D,$$

(H_5) The linear $\beth : L^2(\Theta,\Omega) \to \Psi$ given by

$$\beth\mathcal{P} = \int_0^\kappa H_q(\kappa-\rho)B_2\mathcal{P}(\rho,\Delta)d\rho$$

has an inverse operator $\beth^{-1}$ in $L^2(\Theta,\Omega)/\ker\beth$, and there $\exists$ a positive constant ζ, such that $||B_2\beth^{-1}|| \leq \zeta$,
(H_6) for each $\Delta \in \psi, \varrho(.,\Delta)$ is continuous and $\chi, \varrho_1(\chi,.)$ and $\Delta \in \psi, \varrho_2(\Delta)$ are measurable.

Theorem 1. *Assume that (H_1)–(H_2) are met; then Problem (1) is controllable on Θ.*

Proof. Define the control:

$$\mathcal{P}(\chi,\Delta) = \beth^{-1}\Big(p^1(\Theta) - H_q(\chi)(\varrho_1(\chi,\xi) - m(U)) - K_q(\chi)\varrho_2(\chi) - \int_0^V (\chi-\rho)P_q(\chi-\rho)B_1U(\chi,\xi)d\rho$$
$$- \int_0^V (\chi-\rho)P_q(\chi-\rho)[\varphi(\chi,U_\chi(.,\xi),\xi)]d\rho\Big).$$

The operator $I : \psi \times \Xi \to \Xi$ defined by $(I(\xi)p)(\chi) = \varrho_1(\chi,\xi)$, if $\chi \in (-\infty,0]$, and for $\chi \in \Theta$:

$$\begin{aligned}U(\chi,\xi) &= H_q(\chi)(\varrho_1(\chi,\xi)-m(U))+K_q(\chi)\varrho_2(\chi)+\int_0^V(\chi-\rho)P_q(\chi-\rho)B_1U(\chi,\xi)d\rho+\int_0^V(\chi-\rho)P_q(\chi-\rho)\\ &\quad [\varphi(\chi,U_\chi(.,\xi),\xi)]d\rho+\int_0^\chi\Big((\chi-\rho)P_q(\chi-\rho)B_\Xi^{-1}\Big(U^1(\Theta)-H_q(\chi)(\varrho_1(\chi,\xi)-m(U))\\ &\quad -K_q(\chi)\varrho_2(\chi)-\int_0^V(\chi-\rho)P_q(\chi-s)B_1U(\chi,\xi)d\rho-\int_0^V(\chi-\rho)P_q(\chi-\rho)\\ &\quad [\varphi(\chi,U_\chi(.,\xi),\xi)]dC_\rho\Big)+Bx(\rho)Cx(\rho)\Big)d\rho. \end{aligned} \tag{3}$$

We use (H_5) to show that I has a fixed point $U(\chi,\xi)$ that is a mild solution of (1). This suggests that Issue (1) is manageable on Θ. Additionally, we establish that I is a random operator. To prove this, we show that $I(.)(U):\psi\to\Xi$ is a random variable for any $U\in\Xi$. The measurement of $I(.)(U):\psi\to\Xi$ is then shown. Because of the assumptions (H_2) and (H_6), the mapping $\varphi(\chi,U,.),\chi\in\Theta,U\in\Xi$ is measurable. Assume that $D:\psi\to 2^\Xi$ is provided by:

$$D(\xi)=\{U\in\Xi:\|U\|_\Xi\le Q(\xi)\}.$$

$D(\chi)$ is bounded, convex, closed, and solid for all $\xi\in\psi$. So, D is measurable. Suppose $\xi\in\psi$ is fixed; then, $U\in D(\xi)$ and by (A_1), we obtain:

$$\begin{aligned}\|U_\rho\|_D &\le \beta(\rho)|U(\rho)|W+\omega(\rho)\|U_0\|_D\\ &\le \zeta_\kappa|U(\rho)|+\omega_\kappa\|\varrho_1\|_D,\end{aligned}$$

and via (H_3) and (H_4), we have

$$\begin{aligned}|(I(\xi)U)(\chi)| &\le \omega\|\varrho_1\|_D+\omega'|\varrho_2|+\omega\int_0^\chi|\varphi(\rho,U_\rho,\xi)|d\rho+\omega\zeta\int_0^\chi|U^1(\xi)|+\omega\|\varrho_1\|_D\\ &\quad +\omega'|\varrho_2|d\rho\omega\zeta\int_0^\chi\int_0^\kappa\|H_q(\epsilon-\rho)\||\varphi(\epsilon,U_\epsilon(.,\xi),\xi)|d\epsilon d\rho\\ &\le \omega\|\varrho_1\|_D+\omega'|\varrho_2|+\omega\int_0^\kappa p(\varrho,\xi)\chi(\|U_\chi\|_D,\xi)d\rho+\kappa\omega\zeta|U^1(\xi)|+\kappa\omega^2\zeta\|\varrho_1\|_D+\kappa\omega\omega'\zeta|\varrho_2|\\ &\quad +\kappa\omega^2\zeta\int_0^\kappa p(\epsilon,\xi)U(\|U_\epsilon\|_D,\omega)d\epsilon\\ &\le \omega(1+\kappa\omega\zeta)\|\varrho_1|_D+\kappa\omega\zeta|U^1(\xi)|+\omega'(1+\kappa\omega\zeta)|\varrho_2|+\omega(1+\kappa\omega\zeta)\int_0^\kappa p(\rho,\xi)U(\|p_\rho\|_D,\xi)d\rho\\ &\le \omega(1+\kappa\omega\zeta)\Big(\|\varrho_1|_D+U(D_\kappa,\xi)\int_0^\kappa p(\rho,\xi)d\rho\Big)\kappa\omega\zeta\|U^1(\xi)\|+\omega'(1+\kappa\omega\zeta)|\varrho_2|.\end{aligned}$$

Set

$$D_\kappa:=\zeta_\kappa Q(\xi)+\rho_\kappa\|\varrho_1\|_D.$$

Then, we have

$$|(I(\xi)U(\chi)| \le \omega(1+\kappa\omega\zeta)\Big(\|\varrho_1|_D+U(D_\kappa,\xi)\int_0^\kappa p(\rho,\xi)d\rho\Big)\kappa\omega\zeta\|p^1(\xi)\|+\omega'|\varrho_2|(1+\kappa\omega\zeta).$$

Thus

$$\begin{aligned}\|I(\xi)U\|_\Xi &\le \omega(1+\kappa\omega\zeta)(\|\varrho_1\|_D+U(D_\kappa,\omega)\|\varrho\|_L^{\frac{1}{L}})\kappa\omega\zeta|U^1(\xi)|+\omega'(1+\kappa\omega\zeta)|\varrho_2|\\ &\le Q(\omega).\end{aligned}$$

Thus, we deduce that, with stochastic domain D, I is a random operator and $I(\xi):D(\xi)\to D(\xi)$ for each $\xi\in\psi$.

Claim 1: I is continuous.

Assume that U^n is a sequence where $U^n\to U$ in Y. Then,

$$
\begin{aligned}
|(I(\xi)U^n)(\chi)-(I(\xi)U(\chi)| &\le \omega\int_0^{\chi}|\varphi(\rho,U^n_\rho,\xi)-\varphi(\rho,U_\rho,\xi)|d\epsilon d\rho+\zeta\omega\int_0^{\chi}\int_0^{\kappa}\|H_q(\kappa-\epsilon)\| \\
&\quad |\varphi(\epsilon,U^n_\epsilon(.,\xi)-\varphi(\epsilon,U_\epsilon,\xi)|d\epsilon d\rho \\
&\le \omega\int_0^{\chi}|\varphi(\rho,U^n_\rho,\xi)-\varphi(\rho,U_\rho,\xi)|d\epsilon d\rho+\kappa\omega^2\zeta\int_0^{\kappa}|\varphi(\epsilon,U^n_\epsilon(.,\xi)-\varphi(\epsilon,U_\epsilon,\xi)|d\epsilon \\
&\le \omega(1+\kappa\omega\zeta)\int_0^{\kappa}|\varphi(\epsilon,U^n_\epsilon(.,\xi)-\varphi(\epsilon,U_\epsilon,\xi)|d\epsilon
\end{aligned}
$$

As $\varphi(\chi,.,\xi)$ is continuous, we obtain

$$\|\varphi(.,U^n,\xi)-\varphi(.,U,\xi)\|_{L^1}\to 0\;as\;n\to+\infty.$$

I is continuous.

Claim 2: we show that $\xi\in\psi,\{U\in D(\xi):I(\xi)U=U\}\ne\varnothing$ by applying Schauder's theorem.

(a) I maps bounded sets into equicontinuous sets in $D(\xi)$.
Assume that $\epsilon_1,\epsilon_2\in[0,\kappa]$ with $\epsilon_2>\epsilon_1,D(\xi)$ are a bounded set, as in Claim 2, and $U\in D(\xi)$. Now,

$$
\begin{aligned}
|(I(\xi)U)(\epsilon_2)-(I(\xi)U)(\epsilon_1)| &\le \|H_q(\epsilon_2)-H_q(\epsilon_1)\|_{D(\Psi)}\|\varrho_1\|_D+\|K_q(\epsilon_2)-K_q(\epsilon_1)\|_{D(\Psi)}|\varrho+\int_0^{\epsilon_1}\|H_q(\epsilon_2 \\
&\quad -\rho)-H_q(\epsilon_1-\rho)\|_{D(\Psi)}|\varphi(\rho,U_\rho,\xi)|d\rho+\int_{\epsilon_1}^{\epsilon_2}\|C(\epsilon_2-\rho)\|_{D(\Psi)}|\varphi(\rho,U_\rho,\xi)|d\rho \\
&\quad +\zeta\int_0^{\epsilon_1}\|H_q(\epsilon_2-\rho)-H_q(\epsilon_1-\rho)\|_{D(\Psi)}\times[|U^1(\xi)|+\|H_q(\kappa)\|_{D(\Psi)}\|\varrho_1\|_D+ \\
&\quad \|K_q(\kappa)\|_{D(\Psi)}|\varrho_2|]d\rho+\zeta\int_0^{\epsilon_1}\|H_q(\epsilon_2-\rho)-H_q(\epsilon_1-\rho)\|_{D(\Psi)}\int_0^{\kappa}\|H_q(\kappa-\epsilon)\|_{D(\Psi)}| \\
&\varphi(\epsilon,U_\epsilon(.,\xi),\xi)|d\epsilon d\rho+\zeta\int_{\epsilon_1}^{\epsilon_2}\|C(\epsilon_2-\rho)\|_{D(\Psi)}[|U^1(\xi)|+\|H_q(\kappa)\|_{D(\Psi)}\|\varrho_1\|_D+\|H_q(\kappa)\|_{D(\Psi)}|\varrho_2|]d\rho \\
&+\zeta\int_{\epsilon_1}^{\epsilon_2}\|C(\epsilon_2-\rho)\|_{D(\Psi)}\int_0^{\kappa}\|H_q(\kappa-\epsilon)\|_{D(\Psi)}|\varphi(\epsilon,U_\epsilon(.,\xi)\xi)|d\epsilon d\rho \\
\le\;& \|H_q(\epsilon-\rho)-H_q(\epsilon_1-\rho)\|_{D(\Psi)}\|\varrho_1\|_D+\|K_q(\epsilon_2)-K_q(\epsilon_1)\|_{D(\Psi)}|\varrho_2|U(D_\kappa,\xi)\int_0^{\epsilon_1}\|H_q(\epsilon_2-\rho) \\
&-H_q(\epsilon_1-\rho)\|_{D(\Psi)}U(\rho,\xi)d\rho+\omega x(D_\kappa,\xi)\int_{\epsilon_1}^{\epsilon_2}p(\rho,\xi)d\rho+\zeta\int_0^{\epsilon_1}\|H_q(\epsilon_2-\rho)-H_q(\epsilon_1-\rho)\|_{D(\Psi)} \\
&\times[|U^1(\xi)|+\|H_q(\kappa)\|_{D(\Psi)}\|\varrho_1\|_D+\|K_q(\kappa)\|_{D(\Psi)}|\varrho_2|]d\rho+\zeta\omega U(D_\kappa,\xi)\int_0^{\epsilon_1}\|H_q(\epsilon_2-\rho) \\
&-H_q(\epsilon_1-\rho\|_{D(\Psi)}\int_0^{\kappa}U(\epsilon,\xi)d\epsilon d\rho\zeta\omega\int_{\epsilon_1}^{\epsilon_2}(|U^1(\xi)|+\|H_q(\kappa)\|_{D(\Psi)}\|\varrho_1\|_D+\|K_q(\kappa)\|_{D(\Psi)}|\varrho_2| \\
&+\omega U(D_\kappa,\xi)\int_0^{\kappa}U(\epsilon,\xi)d\epsilon d\rho.
\end{aligned}
$$

In the above inequality, right-hand side tends to zero as $\epsilon_2-\epsilon_1\to 0$, since $H_q(\chi),K_q(\chi)$ are compact for $\chi>0$ and strongly continuous; then, we obtain the continuity in the uniform operator topology [12,33].

(b) Assume that $\chi\in[0,\kappa]$ is, fixed and $U\in D(\xi)$: by assumption $(H_3),(H_5)$; since $H_q(\chi)$ is compact, the set

$$\left\{\int_0^{\chi}H_q(\chi-\rho)\varphi(\rho,U_\rho(.,\xi),\xi)d\rho\int_0^{\chi}H_q(\chi-\rho)B_2\mathfrak{p}(\chi,\xi)d\rho\right\}$$

is precompact in Ψ; then, the set

$$\Big\{H_q(\chi)(\varrho_1(\chi,\xi)-m(U))+K_q(\chi)\varrho_2(\chi)+\int_0^\chi(\chi-\rho)P_q(\chi-s)B_1U(\chi,\xi)d\rho+\int_0^\chi(\chi-\rho)P_q(\chi-s)$$
$$[\varphi(\chi,U_2(.,\xi),\xi)]d\rho+\int_0^\chi\Big((\chi-\rho)P_q(\chi-\rho)\int_0^v B_2f(\chi,\xi)dC_v+Bx(\rho)Cx(\rho)\Big)d\rho\Big\}$$

is precompact in Ψ. Thus, $I(\xi):D(\xi)\to D(\xi)$ is continuous. Through compact Schauder's theorem, we obtain that $I(\xi)$ has a fixed point $U(\xi)$ in $D(\xi)$. Since $\cap_{\xi\in\psi}D(\xi)\neq\varnothing$, and a measurable selector of $\int D$ exists, then via Lemma 4, I has a stochastic fixed point $U^*(\xi)$, which is a random mild solution of (1).

□

4. Results for State-Dependent Delay Case Controllability

Definition 10. *A stochastic process $U:(-\infty,\kappa]\times\psi\to\Psi$ is a random mild solution of Problem (2) if $U(\chi,\xi)=\varrho(\chi,\xi);\chi\in(-\infty,0], U'(0,\xi)=\varpi_2(\xi)$, and the restriction of $U(.,\xi)$ to the interval Θ is continuous and verifies the following equation:*

$$U(\chi,\xi) = H_q(\chi)(\varrho_1(\chi,\xi)-m(U))+K_q(\chi)\varrho_2(\chi)+\int_0^\chi(\chi-\rho)P_q(\chi-s)B_1U(\chi,\xi)d\rho+\int_0^\chi(\chi-\rho)$$
$$P_q(\chi-\rho)[\varphi(\chi,U_2(.,\xi),\xi)]d\rho+\int_0^\chi\Big((\chi-\rho)P_q(\chi-\rho)\int_0^v B_2f(\chi,\xi)dC_v+BU(\rho)CU(\rho)\Big)d\rho$$

Set

$$Q(\theta^{-1})=\{\theta(\rho,\varrho_2):(\rho,\varrho_2)\in\Theta\times D,\theta(\rho,\varrho_2)\leq 0\}.$$

Suppose that $\theta:\Theta\to(-\infty,\kappa]$ is continuous. (H_{ϱ_1}) the function $\chi\to\varrho_{1\chi}$ is continuous from $Q(\theta^{-1})$ into D, and there exists a continuous and bounded function $\beta^{\varrho_1}:Q(\theta^-)\to(0,\infty)$ where $\beta^{\varrho_1}(\chi)||\varrho_1||_D$ for every $\chi\in Q(\theta^-)$.

Remark 2 ([28]). *Hypothesis H_{ϱ_1} is satisfied through continuous and bounded functions.*

Lemma 4 ([34]). *If $U:(-\infty,\kappa]\to\Psi$ is a function, such that $U_0=\varrho_1$, then*

$$||U_\varrho||_D\leq(\omega_\kappa+\beta^{\varrho_1})||\varrho_1||_D+\zeta_\kappa\sup\{|U(i)|;I\in[0,\max\{0,\rho\}]\},\varrho\in Q(\theta^-)\bigcup\Theta.$$

where $\beta^{\varrho_1}=\sup_{\chi\in Q(\theta^{-1})}\beta^{\varrho_1}(\chi)$.

The hypotheses

(H_1') $H_q(\chi)$ is compact for $\chi>0$ in Ψ.

(H_2') The function $\varphi:\Theta\times D\times\psi\to\Psi$ is random Caratheodory.

(H_3') There $\exists$ a function $\eta:\Theta\times\psi\to\mathbf{R}_+^{\mathbf{m}}$ and $p:\Theta\times\to\mathbf{R}_+^{\mathbf{m}}$, such that $\xi\in\psi, U(.,\xi)$ is a continuous nondecreasing function and $p(.,\xi)$ integrable with:

$$|\phi(\chi,\mathcal{P},\Delta)|\leq p(\chi,\Delta)\eta(||\mathcal{P}||_D,\Delta)\ \textit{for a.e.}\ \chi\in\Theta\ \textit{and each}\ \mathcal{P}\in D,$$

(H_4') There $\exists$ a random function $\alpha:\Theta\times\psi\to\mathbf{R}_+^{\mathbf{m}}$ with $\alpha(.,\chi)\in L^1(\Theta,\mathbf{R}_+^{\mathbf{m}})$ for each $\xi\in\psi$ such that for any bounded $B\subseteq\Psi$.

$$\mu(\varphi(\chi,B,\chi))\leq\alpha(\chi,\xi)\mu(B).$$

(H_5') There $\exists$ a random function $Q:\psi\to\mathbf{R}_+^{\mathbf{m}}\ \{0\}$ where:

$$\omega(1+\kappa\omega\lambda)\Big(||\varrho_1||_D+\eta(\omega_\kappa+\beta^{\varrho_1})||\varrho_1||_D+\zeta_\kappa Q(\chi),\chi)\int_0^\kappa p(\rho,\chi)d\rho\Big)+\kappa\omega\lambda||U^1(\chi)||+\omega'(1+\kappa\omega\lambda)|\varrho_2|\leq Q(\xi).$$

(H_6') The linear LO $\daleth : L^2(\Theta,\Omega) \to \Psi$ defined by:

$$\daleth U = \int_0^{\kappa} H_q(\kappa-\rho)B_2U(\rho,\xi)d\rho$$

has an inverse operator $\daleth^{-1}$ that takes values in $L^2(\Theta,\Omega)/ker\daleth$, and there $\exists$ a positive constant λ, such that $\|B_2\daleth^{-1}\| \leq \lambda$.

(H_7') For each $\Delta \in \psi, \varrho(.,\Delta)$ is continuous and, for each $\chi, \varrho_1(\chi,.)$, is measurable, and, for each $\Delta \in \psi, \varrho_2(\Delta)$, is measurable.

Theorem 2. *Suppose that (H_1')–(H_7') and (H_{ϱ_1}) hold. If*

$$\omega(1+\omega\lambda\kappa)\int_0^{\kappa}\alpha(\rho)\xi(\rho)d\rho < 1. \tag{4}$$

Therefore, Theta can be used to control Random Problem (2).

Proof. Using (H_6), the control is

$$U(\chi,\xi) = \daleth^{-1}\Big(U^1(\xi) - H_q(\kappa)\varrho_1(0,\xi) - K_q(\kappa)\varrho_2(\xi) - \int_0^{\kappa} H_q(\kappa-\rho)B_2U(\chi,\xi)d\rho - \int_0^{\kappa} H_q(\kappa-\rho)\varphi(\rho,U_{\theta(\rho,U_\rho)}(.,\xi),\xi)d\rho\Big).$$

The operator $I : \psi \times \Xi \to \Xi$ given by: $(I(\xi)U)(\chi) = \varrho_1(\chi,\xi)$, if $\chi \in (-\infty,0]$, and for $\chi \in \Theta$:

$$\begin{aligned} U(\chi,\xi) &= H_q(\chi)(\varrho_1(\chi,\xi)-m(U)) + K_q(\chi)\varrho_2(\chi) + \int_0^{\chi}(\chi-\rho)P_q(\chi-s)B_1U(\chi,\xi)d\rho + \int_0^{\chi}(\chi-\rho)P_q(\chi-\rho) \\ &\quad [\varphi(\chi,U_2(.,\xi),\xi)]d\rho + \int_0^{\chi}\Big((\chi-\rho)P_q(\chi-\rho)B_{\daleth}^{-1}\Big(p^1(\Theta) - H_q(\chi)(\varrho_1(\chi,\xi)-m(U)) \\ &\quad -K_q(\chi)\varrho_2(\chi) - \int_0^{\chi}(\chi-\rho)P_q(\chi-s)[\varphi(\chi,U_2(.,\xi),\xi)]dC_\rho\Big) + BU(\rho)CU(\rho)\Big)d\rho \end{aligned} \tag{5}$$

This proves that I has a fixed point $U(\chi,\xi)$, and that (2) is controllable. Moreover, we demonstrate that I is a random operator by showing that, for any $U \in \Xi, I(.)(U) : \psi \to \Xi$ is a random variable. We also show that $I(.)(U) : \psi \to \Xi$ is measurable, as a mapping $\varphi(\chi,U,.), \chi \in \Theta, U \in \Xi$ is measurable through assumptions (H_2') and (H_6'). Assume that $D : \psi \to 2^{\Xi}$ is given by:

$$D(\xi) = \{U \in \Xi : \|U\|_{\Xi} \leq Q(\xi)\}.$$

$D(\chi)$ is bounded, convex, closed and solid for all $\xi \in \psi$. Then, D is measurable. Let $\xi \in \psi$ be fixed; if $p \in D(\xi)$, then

$$\|U_{\varrho(\chi,U_\chi)}\|_D = (\omega_\kappa + L^{\varrho_1})\|\varrho_1\|_D + \zeta_\kappa Q(\xi),$$

For each $U \in D(\xi)$, (H_3'), and (H_4'), for each $\chi \in \Theta$, we have

$$
\begin{aligned}
|(I(\xi)U)(\chi)| &\leq \omega\|\varrho_1\|_D + \omega'|\varrho_2| + \omega\int_0^\chi |\varphi(\rho, U_{\varrho(\chi,U_\chi)}, \xi)|d\rho + \omega\zeta\int_0^\chi |U^1(\xi)| + \omega\|\varrho_1\|_D \\
&\quad +\omega'|\varrho_2|d\rho\omega\zeta\int_0^\chi\int_0^\kappa \|H_q(\epsilon-\rho)\||\varphi(\epsilon, U_{\varrho(\chi,U_\chi)}(.,\xi),\xi)|d\epsilon d\rho \\
&\leq \omega\|\varrho_1\|_D + \omega'|\varrho_2| + \omega\int_0^\kappa p(\varrho,\xi)\eta(\|U_\chi\|_D,\xi)d\rho + \kappa\omega\zeta|U^1(\xi)| + \kappa\omega^2\zeta\|\varrho_1\|_D + \kappa\omega\omega'\zeta|\varrho_2| \\
&\quad +\kappa\omega^2\zeta\int_0^\kappa p(\epsilon,\xi)\eta(\|p_\epsilon\|_D,\omega)d\epsilon \\
&\leq \omega(1+\kappa\omega\lambda)\|\varrho_1|_D + \kappa\omega\lambda|U^1(\xi)| + \omega'(1+\kappa\omega\lambda)|\varrho_2| + \omega(1+\kappa\omega\lambda)\int_0^\kappa p(\rho,\xi)\eta(\|U_{\varrho(\chi,U_\chi)}\|_D,\xi)d\rho \\
&\leq \omega(1+\kappa\omega\lambda)\times\left(\|\varrho_1\|_D + \eta(\omega_\kappa+\beta^{\varrho_1})\|\varrho_1\|_D + \zeta_\kappa Q(\xi),\xi)\int_0^\kappa p(\rho,\xi)d\rho\right)\kappa\omega\lambda\|U^1(\xi)\| \\
&\quad +\omega'(1+\kappa\omega\lambda)|\varrho_2|.
\end{aligned}
$$

Thus, with stochastic domain D, I is a random operator and $I(\xi) : D(\xi) \to D(\xi)$ for each $\xi \in \psi$.

Claim 1: I is continuous.

Suppose that U^n is a sequence where $U^n \to U$ in Ξ. Then,

$$
\begin{aligned}
|(I(\xi)U^n)(\chi) - (I(\xi)U(\chi)| &\leq \omega\int_0^\chi |\varphi(\rho, U_\vartheta(\chi, U_\chi^n)^n, \xi) - \varphi(\rho, U_{\vartheta(\chi,U_\chi)}, \xi)|d\epsilon d\rho \\
&\quad +\zeta\omega\int_0^\chi\int_0^\kappa \|H_q(\kappa-\epsilon)\||\varphi(\epsilon, p_\epsilon^n(.,\xi) - \varphi(\epsilon, p_\epsilon,\xi))|d\epsilon d\rho \\
&\leq \omega\int_0^\chi |\varphi(\rho, U_\vartheta(\chi, U_\chi^n),\xi)^n) - \varphi(\rho, U_\vartheta(\chi, U_\chi),\xi))|d\epsilon d\rho \\
&\quad \kappa\omega^2\zeta\int_0^\kappa |\varphi(\epsilon, U_\vartheta(\chi, U_\chi^n)^n(.,\xi)) - \varphi(\epsilon U_\vartheta(\chi, U_\chi),\xi)|d\epsilon \\
&\leq \omega(1+\kappa\omega\zeta)\int_0^\kappa |\varphi(\epsilon, U^n_{\vartheta(\chi,U_\chi^n)}(.,\xi) - \varphi(\epsilon U_\vartheta(\chi, U_\chi),\xi)|d\epsilon
\end{aligned}
$$

As $\varphi(\chi, ., \xi)$ is continuous, we have

$$
\|\varphi(., U^n, \xi) - \varphi(., U, \xi)\|_\Xi \to 0 \; as \; n \to +\infty.
$$

I is continuous.

Claim 2: We show that $\xi \in \psi, \{U \in D(\xi) : I(\xi)U = U\} \neq \varnothing$. We apply Mönch fixed point theorem [35,36].

(a) In $D(\xi)$, I transforms bounded sets into equicontinuous sets.
Let $\epsilon_1, \epsilon_2 \in [0, \kappa]$ with $\epsilon_2 > \epsilon_1$, $D(\xi)$ be a bounded set as in Claim 2, and $U \in D(\xi)$. Then,

$$
\begin{aligned}
|(I(\xi)U)(\epsilon_2) - (I(\xi)U)(\epsilon_1)| &\leq \|H_q(\epsilon_2) - H_q(\epsilon_1)\|_{D(\Psi)}\|\varrho_1\|_D + \|K_q(\epsilon_2) - K_q(\epsilon_1)\|_{D(\Psi)}|\varrho \\
&\quad + \int_0^{\epsilon_1} \|H_q(\epsilon_2-\rho) - H_q(\epsilon_1-\rho)\|_{D(\Psi)}|\varphi(\rho, U_{\vartheta(\chi,U_\chi)},\xi)|d\rho \\
&\quad + \int_{\epsilon_1}^{\epsilon_2} \|C(\epsilon_2-\rho)\|_{D(\Psi)}|\varphi(\chi, U_{\vartheta(\chi,U_\chi)},\xi)|d\rho + \zeta\int_0^{\epsilon_1} \|H_q(\epsilon_2-\rho) - H_q(\epsilon_1-\rho)\|_{D(\Psi)} \\
&\quad \times[|p^1(\xi)| + \|H_q(\kappa)\|_{D(\Psi)}\|\varrho_1\|_D + \|K_q(\kappa)\|_{D(\Psi)}|\varrho_2|]d\rho \\
&\quad +\zeta\int_0^{\epsilon_1} \|H_q(\epsilon_2-\rho) - S_1(\epsilon_1-\rho)\|_{D(\Psi)}\int_0^\kappa \|H_q(\kappa-\epsilon)\|_{D(\Psi)}|\varphi(\epsilon, U_{\vartheta(\chi,U_\chi)},\xi)|d\epsilon d\rho \\
&\quad +\zeta\int_{\epsilon_1}^{\epsilon_2} \|C(\epsilon_2-\rho)\|_{D(\Psi)}[|U^1(\xi)| + \|H_q(\kappa)\|_{D(\Psi)}\|\varrho_1\|_D + \|H_q(\kappa)\|_{D(\Psi)}|\varrho_2|]d\rho \\
&\quad +\zeta\int_{\epsilon_1}^{\epsilon_2} \|C(\epsilon_2-\rho)\|_{D(\Psi)}\int_0^\kappa \|H_q(\kappa-\epsilon)\|_{D(\Psi)}|\varphi(\epsilon, U_{\vartheta(\chi,U_\chi)},\xi)|d\epsilon d\rho
\end{aligned}
$$

Thus,

$$
\begin{aligned}
|(I(\xi)U)(\epsilon_2)-(I(\xi)U)(\epsilon_1)| \leq\ & |H_q(\epsilon_2)-H_q(\epsilon_1)|\|\varrho_1\|_D+\|K_q(\epsilon_2)-K_q(\epsilon_1)\|_{D(\psi)}|\varrho_2| \\
& +\int_0^{\epsilon_1}\|H_q(\epsilon_2-\rho)-H_q(\epsilon_1-\rho)\|_{D(\psi)}\varphi(\rho,U_{\vartheta(\chi,U_\chi)},\xi)d\rho+\int_{\epsilon_1}^{\epsilon_2}\|H_q(\epsilon_2-\rho)\|_{D(\psi)}\varphi(\rho,U_{\vartheta(\chi,U_\chi)},\xi)d\rho \\
& +\lambda\int_0^{\epsilon_1}\|H_q(\epsilon_2-\rho)-H_q(\epsilon_1-\rho)\|_{D(\psi)}[\|p^1(\xi)\|+\|H_q(\kappa)\|_{D(\psi)}|\varrho_1(0,\xi)|]d\rho \\
& +\lambda\int_0^{\epsilon_1}\|H_q(\epsilon_2-\rho)-H_q(\epsilon_1-\rho)\|_{D(\psi)}\eta((\omega_\kappa+\beta^{\varrho_1})\|\varrho_1\|_D+\zeta_\kappa Q(\xi))\times\int_0^\kappa p(\epsilon,\xi)d\epsilon d\rho+ \\
& \lambda\omega\int_{\epsilon_1}^{\epsilon_2}\|U^1\|+\|H_q(\kappa)\|_{D(\psi)}|\varrho_1(0,\xi)|+\omega\eta((\omega_\kappa+\beta^{\varrho_1})\|\varrho_1\|_D+\zeta_\kappa Q(\xi))\times\int_0^\kappa p(\epsilon,\xi)d\epsilon d\rho
\end{aligned}
$$

Hence,

$$
\begin{aligned}
|(I(\xi)U)(\epsilon_2)-(I(\xi)U)(\epsilon_1)| \leq\ & |H_q(\epsilon_2)-H_q(\epsilon_1)|_{D(\psi)}\|\varrho_1\|_D+\|K_q(\epsilon_2)-K_q(\epsilon_1)\|_{D(\psi)}|\varrho_2| \\
& +\eta(\omega_\kappa+\beta^{\varrho_1}\|\varrho_1\|_D+\zeta_\kappa Q(\varpi))\int_0^{\epsilon_1}\|H_q(\epsilon_2-\rho)-H_q(\epsilon_1-\rho)\|_{D(\psi)}p(\chi,\xi)d\rho \\
& +\eta((\omega_\kappa+\beta^{\varrho_1}\|\varrho_1\|_D+\zeta_\kappa Q(\varpi),\varpi)\int_{\epsilon_1}^{\epsilon_2}p(\chi,\xi)d\rho \\
& +\lambda\int_0^{\epsilon_1}\|H_q(\epsilon_2-\rho)-H_q(\epsilon_1-\rho)\|_{D(\psi)}[\|U^1(\xi)\|+\|H_q(\kappa)\|_{D(\psi)}|\varrho_1(0,\xi)|]d\rho \\
& +\lambda\int_0^{\epsilon_1}\|H_q(\epsilon_2-\rho)-H_q(\epsilon_1-\rho)\|_{D(\psi)}\eta((\omega_\kappa+\beta^{\varrho_1})\|\varrho_1\|_D+\zeta_\kappa Q(\xi)) \\
& \times\int_0^\kappa p(\epsilon,\xi)d\epsilon d\rho+\lambda\omega\int_{\epsilon_1}^{\epsilon_2}\|U^1(\varpi)\|+\|H_q(\kappa)\|_{D(\psi)}|\varrho_1(0,\xi)|+ \\
& \omega\eta((\omega_\kappa+\beta^{\varrho_1})\|\varrho_1\|_D+\zeta_\kappa Q(\xi))\times\int_0^\kappa p(\epsilon,\xi)d\epsilon d\rho
\end{aligned}
$$

In the previous inequality, the right-hand side went to zero as $\epsilon_2-\epsilon_1\to 0$, $H_q(\chi), K_q(\chi)$ are a strongly continuous operator, and $H_q(\chi)$ and $K_q(\chi)$ for $\chi>0$ are compact, which implies that uniform operator topology is continuous. Suppose that $\xi\in\psi$ is fixed.

(b) Suppose that Λ is a subset of $D(\xi)$ where $\Lambda\subset\overline{conv}(I(\Lambda)\bigcup\{0\})$. Λ is bounded and equicontinuous, and function $\chi\to v(\chi)=\varsigma(\Lambda(\chi))$ is continuous on $(-\infty,\kappa]$. Via (H_2), and by considering the characteristics of the measure Λ, we have $\chi\in(-\infty,\kappa]$:

$$
\begin{aligned}
v &\leq \varsigma(I(\Lambda))(\chi)\bigcup\{0\}) \\
&\leq \varsigma(I(\Lambda)(\chi)) \\
&\leq \varsigma(H_q(\chi)\varrho_1(0,\xi)) + \varsigma(K_q(\chi)\varrho_2(\xi)) + \varsigma\Big(\int_0^{\chi} H_q(\chi-\rho)\varphi(\epsilon, U_{\vartheta(\chi,U_\chi)}(.,\xi)d\rho\Big) + \omega\lambda\int_0^{\chi}\varsigma(U^1(\xi) \\
&\quad -H_q(\kappa)\varrho_1(0,\xi) - K_q(\kappa)\varrho_2(\xi)) + \varsigma\Big(\int_0^{\kappa} H_q(\kappa-\epsilon)\varphi(\epsilon, U_{\vartheta(\chi,U_\chi)}(.,\xi),\xi\Big)d\rho \\
&\leq \omega\int_0^{\chi}\varsigma(\varphi(\rho, U_{\vartheta(\chi,U_\chi)}(.,\xi),\xi))d\rho\omega\lambda\int_0^{\chi}\int_0^{\kappa}\varsigma(H_q(\kappa-\epsilon)\varphi(\epsilon, U_{\vartheta(\chi,U_\chi)}(.,\xi),\xi)d\epsilon d\rho \\
&\leq \omega\int_0^{\chi}\alpha(\rho)\varsigma(\{U_{\vartheta(\chi,p_\chi)} : p\in\Lambda\})d\rho\omega\lambda\int_0^{\chi}\int_0^{\kappa}\varsigma(H_q(\kappa-\epsilon)\varphi(\epsilon, U_{\vartheta(\chi,U_\chi)}(.,\xi),\xi)d\epsilon d\rho \\
&\leq \omega\int_0^{\chi}\gamma(\rho)\zeta(\rho)\sup_{0\leq\epsilon\leq\rho}\varsigma(\Lambda(\epsilon))\rho + \omega^2\lambda\int_0^{\chi}\int_0^{\kappa}\varsigma(\varphi(\epsilon, U_{\vartheta(\chi,U_\chi)},\xi)d\epsilon d\rho \\
&\leq \omega\int_0^{\chi}\gamma(\rho)\zeta(\rho)\varsigma(\Lambda(\rho))d\rho + \omega^2\lambda\kappa\int_0^{\kappa}\alpha(\epsilon)\varsigma(\varphi(\{U_{\vartheta(\chi,U_\chi)} : U\in\Lambda)d\epsilon \\
&\leq \omega\int_0^{\chi}v(\rho)\alpha(\rho)\zeta(\rho)d\rho + \omega^2\lambda\kappa\int_0^{\kappa}\alpha(\epsilon)\zeta(\epsilon)\varsigma(\Lambda(\epsilon))d\epsilon \\
&= \omega\int_0^{\chi}\alpha(\rho)\zeta(\rho)v(\rho)d\rho + \omega^2\lambda\kappa\int_0^{\kappa}\alpha(\epsilon)\zeta(\epsilon)v(\epsilon))d\epsilon \\
&\leq \omega\int_0^{\chi}\alpha(\rho)\zeta(\rho)v(\rho)d\rho + \omega^2\lambda\kappa\int_0^{\kappa}\alpha(\epsilon)\zeta(\epsilon)v(\epsilon))d\epsilon \\
&\leq \omega(1+\omega\lambda\kappa)\int_0^{\kappa}\alpha(\rho)\zeta(\rho)v(\rho))d\rho \\
&\leq \omega(1+\omega\lambda\kappa)\int_0^{\kappa}\alpha(\rho)\zeta(\rho)\sup_{0\leq\epsilon\leq\rho}v(\epsilon))d\rho \\
&\leq \omega(1+\omega\lambda\kappa)\|v\|_\infty\int_0^{\kappa}\alpha(\rho)\zeta(\rho)d\rho.
\end{aligned}
$$

Thus,

$$\|v\|_\infty \leq \omega(1+\omega\lambda\kappa)\|v\|_\infty\int_0^{\kappa}\alpha(\rho)\zeta(\rho)d\rho$$

Then,

$$\|v\|_\infty\Big(1-\omega(1+\omega\lambda\kappa)\int_0^{\kappa}\alpha(\rho)\zeta(\rho)d\rho\Big) \leq 0.$$

Hereby, $\|v\|_\infty = 0$; thus, $v(\chi) = 0$ for each $\chi \in \Theta$, this implies $\Lambda(\chi)$ is relatively compact in Ψ. Through the result of Ascoli-Arzel *à* theorem, Λ is relatively compact in $D(\xi)$. Via *Mönch* fixed-point theorem, we show that I has a fixed point $U(\xi) \in D(\xi)$. As $\bigcap_{\xi\in\varphi} D(\xi) \neq \emptyset$; moreover, a measurable selector of $\int D$ exists. Lemma implies that I has a stochastic fixed point $U^*(\xi)$, which is a mild solution of (2).

□

5. Applications

The qualitative theory of FDEs, fractional integrodifferential equations, and fractional-order operators can be applied to a wide range of scientific fields, including fluid mechanics, viscoelasticity, physics, biology, chemistry, dynamical systems, signal processing, and entropy theory. Due to this, academics from all over the world have become interested in the applications of the theory of fractional calculus and the qualitative theory of the aforementioned equations, and many researchers have included them into their most recent research.

For a very long time, DEs driven by a Brownian motion (or Wiener process) have been the focus of study on the qualitative characteristics of stochastic DEs and their applications. Furthermore, applications from a variety of domains, including storage, queueing, economic, and neurophysiological systems, can be found frequently in stochastic DEs driven by a Poisson process. Additionally, stochastic DEs with Poisson jumps have gained much traction in modelling phenomena from a variety of disciplines, especially economics, where jump processes are frequently used to describe asset and commodity price dynamics. These factors are sufficient for the existence and uniqueness of non-Lipschitz stochastic neutral delay DEEs driven by Poisson jumps.

Levy procedures are becoming increasingly significant in the world of banking. While Levy processes are often employed in newer models to accommodate jumps (which can be regarded as external shocks) and achieve a better fit to empirical data, Brownian motion is still frequently used in older models as a source of randomness. As a result, Levy process applications in finance are simple to locate. There have been numerous applications of the theory of impulsive DEs of an integer order in accurate mathematical modelling. It has recently become a crucial subject of research due to the large range of practical problems. This is because many evolutionary systems' states are frequently exposed to rapid disturbances and undergo abrupt shifts from time to time. These changes have a very brief and insignificant length when compared to the lifespan of the process under consideration, and can be viewed as impulses. Due to the lack of effective methods, the control analysis of problems, including the impulse effect, fractional calculus, and white noise, is challenging.

6. Example

Consider

$$\begin{aligned} &{}_0^cD_\nu^q U(\chi,\xi,\varsigma) = \varphi(\chi, U(\chi,\xi,\varsigma),\varsigma) + \textstyle\int_0^\nu B_2 f(\chi,\varsigma)dC_v,\ \xi \in \Theta := [0,\kappa],\ \nu \in [0,T] \\ &U(\chi,\pi,\varsigma) + m(U) = U_1(\chi,2\pi,\varsigma);\ \ \xi \in [0,\kappa], \\ &U'(\chi,\xi,\varsigma) = U_2(\xi), \end{aligned} \tag{6}$$

where $\Phi : \Theta \times R \times \zeta \to \mathbf{R^m}$ is a given function. If $\Xi = L^2[\pi, 2\pi]$, and $B_1 : \Xi \to \Xi$ given by $B_1U = U'$ with domain $D(B_1) = \{U \in \Phi; U, U'\ \textit{are absolutely continuous},\ U' \in \Xi, U(\pi) = U(2\pi) = 0\}$. Let the strongly continuous cosine function $(H_q(\chi))_{\chi\in\mathbf{R^m}}$ on Φ be infinitesimally generated by the operator B_1. Furthermore, B_1 has a discrete spectrum, and the eigenvalues are $-n^2, n \in IN$ with corresponding normalized eigenvectors

$$U_n(\varepsilon) := \left(\frac{2}{2\pi}\right)^{\frac{1}{2}} cos(n\varepsilon),$$

and

(i) $\{U_n : n \in IN\}$ is an orthonormal basis of Φ,
(ii) If $x \in \Phi$, then $B_1x = -\sum_{n=1}^{\infty} n^2 \langle x, U_n\rangle U_n$,
(iii) For $x \in \Phi, H_q(\vartheta)x = \sum_{n=1}^{\infty} \sin(nt)\langle x, U_n\rangle U_n$, and the associated cosine family is

$$K_q(\vartheta)x = \sum_{n=1}^{\infty} \frac{\cos(nt)}{n}\langle x, U_n\rangle U_n.$$

Consequently, $K_q(\chi)$ is compact for all $\chi > 0$ and

$$\|H_q(\vartheta)\| = \|K_q(\chi)\| \le 1, \forall \chi \ge 0.$$

(iv) Let the group of translation be denoted by Φ:

$$\overline{\psi}(\chi)x(U,\varsigma) = \widetilde{x}(U+\chi,\varsigma),$$

where $\widetilde{x}$ is the extension of x with period 4π. Then,

$$H_q(\chi) = \frac{1}{2}(\overline{\psi} + \psi(-\chi)); U_1 = D,$$

where D is the infinitesimal generator of the group on

$$X = \{x(.,\varsigma) \in H^1(\pi, 2\pi) : x(\pi, \varsigma) = x(2\pi, \varsigma) = 0\}.$$

Suppose that B_2 is a bounded LO from Ω into Ξ and the linear operator $K : L^2(\Theta, \Omega) \to \Xi$ given by:

$$Kf = \int_0^k H_q(k - \varrho) B_2 f(\varrho, \varsigma) d\rho,$$

has an inverse operator K^{-1} in $L^2(\Theta, \Omega) / \ker K$. We deduce that Equation (1) is an abstract formulation of Equation (6) if H_1 to H_6 are met. Via Theorem 1, we conclude that Equation (6) is controllable.

7. Conclusions

Existence and controllability results were presented for a couple of classes of second-order fractional functional differential equations. A stochastic random fixed-point theorem established the basis for our claims. Then, we demonstrated that our problems were controllable. Some of the findings in this area form the basis of our future research plans. New results can be obtained by either changing or generalising the conditions and the functional spaces, or even by involving some fractional differential problems.

Author Contributions: Conceptualization, R.S.; Methodology, R.S.; Software, A.A.; Formal analysis, M.A.; Investigation, R.S. and A.A.; Resources, A.A. and M.A.; Writing—original draft, R.S.; Writing—review and editing, R.S.; Supervision, A.A.; Project administration, K.A.; Funding acquisition, K.A. All authors have read and agreed to the published version of the manuscript.

Funding: This work was supported by the Deanship of Scientific Research, Vice Presidency for Graduate Studies and Scientific Research, King Faisal University, Saudi Arabia (grant no. 2384).

Data Availability Statement: The data are original, and references are given where required.

Conflicts of Interest: the authors declare that they have no known competing financial interest or personal relationships that could have appeared to influence the work reported in this paper.

References

1. Abbas, S.; Benchohra, M. *Advanced Functional Evolution Equations and Inclusions*; Springer: Cham, Switzerland, 2015; Volume 39.
2. Naito, K. On controllability for a nonlinear Volterra equation. In Proceedings of the 29th IEEE Conference on Decision and Control, Honolulu, HI, USA, 5–7 December 1990; pp. 2817–2818.
3. Nakagiri, S.I.; Yamamoto, M. Controllability and observability of linear retarded systems in Banach spaces. *Int. J. Control* **1989**, *49*, 1489–1504. [CrossRef]
4. Triggiani, R. On the stabilizability problem in Banach space. *J. Math. Anal. Appl.* **1975**, *52*, 383–403. [CrossRef]
5. Quinn, M.D.; Carmichael, N. An approach to non-linear control problems using fixed-point methods, degree theory and pseudo-inverses. *Numer. Funct. Anal. Optim.* **1985**, *7*, 197–219. [CrossRef]
6. Fu, X.; Ezzinbi, K. Existence of solutions for neutral functional differential evolution equations with nonlocal conditions. *Nonlinear Anal. Theory Methods Appl.* **2003**, *54*, 215–227. [CrossRef]
7. Kwun, Y.C.; Park, J.Y.; Ryu, J.W. Approximate controllability and controllability for delay Volterra system. *Bull. Korean Math. Soc.* **1991**, *28*, 131–145.
8. Balachandran, K.; Dauer, J.P. Controllability of nonlinear systems in Banach spaces: A survey. *J. Optim. Theory Appl.* **2002**, *115*, 7–28. [CrossRef]
9. Abada, N.; Benchohra, M.; Hammouche, H.; Ouahab, A. Controllability of impulsive semilinear functional differential inclusions with finite delay in Fréchet spaces. *Discuss. Math. Differ. Incl. Control Optim.* **2007**, *27*, 329–347. [CrossRef]
10. Dieye, M.; Diop, M.A.; Ezzinbi, K. Controllability for some integrodifferential equations driven by vector measures. *Math. Methods Appl. Sci.* **2017**, *40*, 2090–2106. [CrossRef]
11. Travis, C.C.; Webb, G.F. Cosine families and abstract nonlinear second order differential equations. *Acta Math. Hung.* **1978**, *32*, 75–96. [CrossRef]
12. Balachandran, K.; Anthoni, S.M. Controllability of second-order semilinear neutral functional differential systems in Banach spaces. *Comput. Math. Appl.* **2001**, *41*, 1223–1235. [CrossRef]
13. Henríquez, H.R.; Hernández, E. Approximate controllability of second-order distributed implicit functional systems. *Nonlinear Anal. Theory Methods Appl.* **2009**, *70*, 1023–1039. [CrossRef]
14. Bharucha-Reid, A.T. *Random Integral Equations*; Academic Press: New York, NY, USA, 1972; Volume 96.

15. Ladde, G.S.; Lakshmikantham, V. Stochastic differential inequalities of Itô type. In *Applied Stochastic Processes*; Academic Press: New York, NY, USA, 1980; pp. 109–120.
16. Tsokos, C.P.; Padgett, W.J. *Random Integral Equations with Applications to Life Sciences and Engineering*; Academic Press: New York, NY, USA, 1974.
17. He, J.H. Fractal calculus and its geometrical explanation. *Results Phys.* **2018**, *10*, 272–276. [CrossRef]
18. Wang, K.J.; Si, J. On the non-differentiable exact solutions of the (2 + 1)-dimensional local fractional breaking soliton equation on Cantor sets. *Math. Methods Appl. Sci.* **2022**, *46*, 1456–1465. [CrossRef]
19. Wang, K.J. A fractal modification of the unsteady Korteweg–de Vries model and its generalized fractal variational principle and diverse exact solutions. *Fractals* **2022**, *30*, 2250192. [CrossRef]
20. Wang, K.J. BÄCKLUND Transformation and Diverse Exact Explicit Solutions of the Fractal Combined KdV-mKdV Equation. *Fractals* **2022**, *30*, 2250189. [CrossRef]
21. Mehmood, Y.; Shafqat, R.; Sarris, I.E.; Bilal, M.; Sajid, T.; Akhtar, T. Numerical Investigation of MWCNT and SWCNT Fluid Flow along with the Activation Energy Effects over Quartic Auto Catalytic Endothermic and Exothermic Chemical Reactions. *Mathematics* **2022**, *10*, 4636. [CrossRef]
22. Niazi, A.U.K.; He, J.; Shafqat, R.; Ahmed, B. Existence, Uniqueness, and Eq–Ulam-Type Stability of Fuzzy Fractional Differential Equation. *Fractal Fract.* **2021**, *5*, 66. [CrossRef]
23. Shafqat, R.; Niazi, A.U.K.; Jeelani, M.B.; Alharthi, N.H. Existence and Uniqueness of Mild Solution Where $\alpha \in (1,2)$ for Fuzzy Fractional Evolution Equations with Uncertainty. *Fractal Fract.* **2022**, *6*, 65. [CrossRef]
24. Alnahdi, A.S.; Shafqat, R.; Niazi, A.U.K.; Jeelani, M.B. Pattern Formation Induced by Fuzzy Fractional-Order Model of COVID-19. *Axioms* **2022**, *11*, 313. [CrossRef]
25. Abuasbeh, K.; Shafqat, R.; Niazi, A.U.K.; Awadalla, M. Nonlocal fuzzy fractional stochastic evolution equations with fractional Brownian motion of order (1,2). *AIMS Math.* **2022**, *7*, 19344–19358. [CrossRef]
26. Benchohra, M.; Bouazzaoui, F.; Karapinar, E.; Salim, A. Controllability of Second Order Functional Random Differential Equations with Delay. *Mathematics* **2022**, *10*, 1120. [CrossRef]
27. Granas, A.; Dugundji, J. *Fixed Point Theory*; Springer: New York, NY, USA, 2003; Volume 14, pp. 15–16.
28. Hino, Y.; Murakami, S.; Naito, T. *Functional Differential Equations with Infinite Delay*; Springer: Berlin/Heidelberg, Germany, 2006.
29. Hale, J.K. Retarded equations with infinite delays. In *Functional Differential Equations and Approximation of Fixed Points*; Springer: Berlin/Heidelberg, Germany, 1979; pp. 157–193.
30. Engl, H.W. A general stochastic fixed-point theorem for continuous random operators on stochastic domains. *J. Math. Anal. Appl.* **1978**, *66*, 220–231. [CrossRef]
31. Banaś, J. On measures of noncompactness in Banach spaces. *Comment. Math. Univ. Carol.* **1980**, *21*, 131–143.
32. Guo, D.J.; Lakshmikantham, V.; Liu, X.Z. *Nonlinear Integral Equations in Abstract Spaces*; Kluwer Academic Publishers: Dordrecht, The Netherlands, 1996.
33. Travis, C.C.; Webb, G.F. Compactness, regularity, and uniform continuity properties of strongly continuous cosine families. *Houst. J. Math.* **1977**, *3*, 555–567.
34. Hernandez, E.; Sakthivel, R.; Aki, S.T. Existence results for impulsive evolution differential equations with state-dependent delay. *Electron. J. Differ. Equ. (EJDE)* **2008**, *2008*, 1–11.
35. Agarwal, R.P.; Meehan, M.; O'regan, D. *Fixed Point Theory and Applications*; Cambridge University Press: Cambridge, UK, 2001; Volume 141.
36. Mönch, H. Boundary value problems for nonlinear ordinary differential equations of second order in Banach spaces. *Nonlinear Anal. Theory Methods Appl.* **1980**, *4*, 985–999. [CrossRef]

symmetry

MDPI

Article

The Exact Solutions of Fractional Differential Systems with *n* Sinusoidal Terms under Physical Conditions

Laila F. Seddek [1,2,*], Essam R. El-Zahar [1,3] and Abdelhalim Ebaid [4]

1 Department of Mathematics, College of Science and Humanities in Al-Kharj, Prince Sattam Bin Abdul-Aziz University, Al-Kharj 11942, Saudi Arabia
2 Department of Engineering Mathematics and Physics, Faculty of Engineering, Zagazig University, Zagazig 44519, Egypt
3 Department of Basic Engineering Science, Faculty of Engineering, Menofia University, Shebin El-Kom 32511, Egypt
4 Department of Mathematics, Faculty of Science, University of Tabuk, P.O. Box 741, Tabuk 71491, Saudi Arabia
* Correspondence: l.morad@psau.edu.sa

Abstract: This paper considers the classes of the first-order fractional differential systems containing a finite number n of sinusoidal terms. The fractional derivative employs the Riemann–Liouville fractional definition. As a method of solution, the Laplace transform is an efficient tool to solve linear fractional differential equations. However, this method requires to express the initial conditions in certain fractional forms which have no physical meaning currently. This issue formulated a challenge to solve fractional systems under real/physical conditions when applying the Riemann–Liouville fractional definition. The principal incentive of this work is to overcome such difficulties via presenting a simple but effective approach. The proposed approach is successfully applied in this paper to solve linear fractional systems of an oscillatory nature. The exact solutions of the present fractional systems under physical initial conditions are derived in a straightforward manner. In addition, the obtained solutions are given in terms of the entire exponential and periodic functions with arguments of a fractional order. The symmetric/asymmetric behaviors/properties of the obtained solutions are illustrated. Moreover, the exact solutions of the classical/ordinary versions of the undertaken fractional systems are determined smoothly. In addition, the properties and the behaviors of the present solutions are discussed and interpreted.

Keywords: Riemann–Liouville fractional derivative; fractional differential equation; sinusoidal; exact solution

Citation: Seddek, L.F.; El-Zahar, E.R.; Ebaid, A. The Exact Solutions of Fractional Differential Systems with *n* Sinusoidal Terms under Physical Conditions. *Symmetry* **2022**, *14*, 2539. https://doi.org/10.3390/sym14122539

Academic Editors: Francisco Martínez González and Mohammed K. A. Kaabar

Received: 4 November 2022
Accepted: 15 November 2022
Published: 1 December 2022

1. Introduction

Unlike the classical calculus (CC) with integer derivatives, the fractional calculus (FC) implements the derivatives of an arbitrary order (non-integer) [1–3]. So, the FC is considered as a generalization of the CC. During the past decades, numerous physical, engineering, and biological problems have been investigated by means of the FC ([4–9]). There are several definitions for the derivatives of an arbitrary order, such as the Caputo fractional derivative (CFD) [10–22], the Riemann–Liouville fractional derivative (RLFD) [23–25], and the conformable derivative [26–29]. However, some difficulties arise when applying the RLFD to solve fractional models under real physical conditions. The present paper is an attempt to face such an issue by considering the following class of first-order fractional ordinary equations (FODEs):

$$\begin{aligned} {}^{RL}_{-\infty}D^{\alpha}_{t}y(t)+\omega^2 y(t) &= b_1\sin(\Omega_1 t)+b_2\sin(\Omega_2 t)+\cdots+b_n\sin(\Omega_n t),\\ &= \sum_{j=1}^{n} b_j\sin(\Omega_j t), \qquad y(0)=A, \quad \alpha\in(0,1], \end{aligned} \tag{1}$$

where α is the non-integer order of the RLFD. The constant A is real while ω, b_j, and Ω_j may be real or complex $\forall\, j = 1, 2, 3, \ldots, n$.

The applications of the class (1) may arise in oscillatory models in engineering when the FC is incorporated. This class splits to other physical classes. As examples, for complex ω, i.e., $\omega = i\mu$ (μ is real), where i is the imaginary number, the model (1) becomes

$$ {}_{-\infty}^{RL}D_t^{\alpha} y(t) - \mu^2 y(t) = \sum_{j=1}^{n} b_j \sin(\Omega_j t), \;\; y(0) = A, \;\; \alpha \in (0, 1]. \tag{2} $$

In addition, if $\Omega_j = i\sigma_j$ and $b_j = -id_j$, the classes (1) and (2) take the form:

$$ {}_{-\infty}^{RL}D_t^{\alpha} y(t) + \omega^2 y(t) = \sum_{j=1}^{n} d_j \sinh(\sigma_j t), \;\; y(0) = A, \;\; \alpha \in (0, 1], \tag{3} $$

and

$$ {}_{-\infty}^{RL}D_t^{\alpha} y(t) - \mu^2 y(t) = \sum_{j=1}^{n} d_j \sinh(\sigma_j t), \;\; y(0) = A, \;\; \alpha \in (0, 1], \tag{4} $$

in terms of hyperbolic functions, respectively.

In Refs. [1–3], the RLFD of order $\alpha \in \mathbb{R}_0^+$ of function $f : [c, d] \to \mathbb{R}$ $(-\infty < c < d < \infty)$ is defined as

$$ {}_{c}^{RL}D_t^{\alpha} f(t) = \frac{1}{\Gamma(n-\alpha)} \frac{d^n}{dt^n} \left(\int_c^t \frac{f(\tau)}{(t-\tau)^{\alpha-n+1}} d\tau \right), \quad n = [\alpha] + 1, \;\; t > c, \tag{5} $$

where $[\alpha]$ is the integral part of α. If $0 < \alpha \le 1$ and $c \to -\infty$, then

$$ {}_{-\infty}^{RL}D_t^{\alpha} f(t) = \frac{1}{\Gamma(1-\alpha)} \frac{d}{dt} \left(\int_{-\infty}^t \frac{f(\tau)}{(t-\tau)^{\alpha}} d\tau \right). \tag{6} $$

It is important to refer to the initial condition (IC) $y(0) = A$ being physical, unlike the nonphysical condition $D_t^{\alpha-1} y(0) = A$ that has been considered by the authors [30]. In fact, the IC in the last fractional form is required when solving an FODE via the Laplace transform (LT). This is, simply, because the LT of the RLFD as $c \to 0$, i.e., ${}_{0}^{RL}D_t^{\alpha}$, is [1–3,23,30]

$$ \mathcal{L}\left[{}_{0}^{RL}D_t^{\alpha} y(t) \right] = s^{\alpha} Y(s) - D_t^{\alpha-1} y(0), \tag{7} $$

which is given in terms of $D_t^{\alpha-1} y(0)$. Really, the main difference between ${}_{-\infty}^{RL}D_t^{\alpha}$ and ${}_{0}^{RL}D_t^{\alpha}$ lies in the nature of the considered IC of the problem. In the literature, one can see that the obtained solutions of the physical models depend on both the nature of the given classical/fractional ICs along with the implemented method of solution.

In this regard, Ebaid and Al-Jeaid [30] applied the RLFDs ${}_{-\infty}^{RL}D_t^{\alpha}$ and ${}_{0}^{RL}D_t^{\alpha}$ to obtain a dual solution for a similar model under the nonphysical IC $D_t^{\alpha-1} y(0)$ using the LT. Although the LT was shown as an effective tool to exactly investigate several models [31–37], it may not be appropriate to deal with the class (1) under the physical IC $y(0) = A$ by means of the RLFD operator ${}_{0}^{RL}D_t^{\alpha}$. However, the solution is still available under this physical condition via the RLFD operator ${}_{-\infty}^{RL}D_t^{\alpha}$ along with avoiding the LT, as will be shown through this paper.

Therefore, the main incentive of the present work is to introduce a new approach to obtain the real solution of the current model under the physical IC $y(0) = A$ through the following properties (see Refs. [30,38]):

$$ {}^{RL}_{-\infty}D^{\alpha}_{t} e^{i\omega t} = (i\omega)^{\alpha} e^{i\omega t}, \tag{8} $$

$$ {}^{RL}_{-\infty}D^{\alpha}_{t} \cos(\omega t) = \omega^{\alpha} \cos\Big(\omega t + \frac{\alpha\pi}{2}\Big), \tag{9} $$

$$ {}^{RL}_{-\infty}D^{\alpha}_{t} \sin(\omega t) = \omega^{\alpha} \sin\Big(\omega t + \frac{\alpha\pi}{2}\Big). \tag{10} $$

By using the above properties, it will be shown that the real solution of class (1) exists at specific values of the fractional-order α. The symmetric/asymmetric behaviors/properties of the obtained solutions will be demonstrated. Furthermore, it will be declared that the solution of the class (2) is real at any arbitrary value α. In addition, the solutions of the corresponding classes with the classical/ordinary derivative, i.e., as $\alpha \to 1$, will be evaluated.

A brief description of the structure of this paper is as follows. In Section 2, an analysis of the complementary and particular solutions is presented. Section 3 is devoted to obtaining the exact solutions for the fractional classes. In Section 4, the exact solutions for the ordinary classes are obtained. The behaviors/properties of the solution are introduced in Section 5. The paper is concluded in Section 6.

2. Analysis

The complementary solution $y_c(t)$ of Equation (1) can be obtained in the form, see [30]:

$$ y_c(t) = c\, e^{i\delta t}, \quad \delta = -i\Big(-\omega^2\Big)^{1/\alpha}, \tag{11} $$

which satisfies the homogeneous equation:

$$ {}^{RL}_{-\infty}D^{\alpha}_{t} y(t) + \omega^2 y(t) = 0. \tag{12} $$

In order to evaluate the constant c, the given IC will be applied on the general solution $y(t) = y_c(t) + y_p(t)$ in a subsequent section where $y_p(t)$ is a particular solution of the non-homogeneous Equation (1). A simple method to calculate $y_p(t)$ is explained through the following theorem.

Theorem 1. *The $y_p(t)$ of the class (1) is in the form:*

$$ y_p(t) = \sum_{j=1}^{n} b_j \left(\frac{\omega^2 \sin(\Omega_j t) + \Omega_j^{\alpha} \sin(\Omega_j t - \frac{\pi\alpha}{2})}{\omega^4 + \Omega_j^{2\alpha} + 2\omega^2 \Omega_j^{\alpha} \cos(\frac{\pi\alpha}{2})} \right), \tag{13} $$

Proof. Let us assume that

$$ y_p(t) = \sum_{j=1}^{n} (\rho_{1j} \cos(\Omega_j t) + \rho_{2j} \sin(\Omega_j t)). \tag{14} $$

Using the preceding properties of the RLFD operator ${}^{RL}_{-\infty}D^{\alpha}_{t}$, we have

$$ \begin{aligned} {}^{RL}_{-\infty}D^{\alpha}_{t} y_p &= \sum_{j=1}^{n} \Big(\rho_{1j}\, {}^{RL}_{-\infty}D^{\alpha}_{t} \cos(\Omega_j t) + \rho_{2j}(\alpha)\, {}^{RL}_{-\infty}D^{\alpha}_{t} \sin(\Omega_j t) \Big), \\ &= \sum_{j=1}^{n} \Omega_j^{\alpha} \cos(\Omega_j t) \Big(\rho_{1j} \cos\Big(\frac{\pi\alpha}{2}\Big) + \rho_{2j} \sin\Big(\frac{\pi\alpha}{2}\Big) \Big) + \\ & \quad \sum_{j=1}^{n} \Omega_j^{\alpha} \sin(\Omega_j t) \Big(\rho_{2j} \cos\Big(\frac{\pi\alpha}{2}\Big) - \rho_{1j} \sin\Big(\frac{\pi\alpha}{2}\Big) \Big). \end{aligned} \tag{15} $$

Thus,

$$\begin{aligned}{}^{RL}_{-\infty}D^{\alpha}_{t}y_p+\omega^2y_p &= \sum_{j=1}^{n}\Big[\Big(\Omega_j^{\alpha}\cos\Big(\frac{\pi\alpha}{2}\Big)+\omega^2\Big)\rho_{1j}+\Omega_j^{\alpha}\sin\Big(\frac{\pi\alpha}{2}\Big)\rho_{2j}\Big]\cos(\Omega_j t)+\\ &\sum_{j=1}^{n}\Big[\Big(\Omega_j^{\alpha}\cos\Big(\frac{\pi\alpha}{2}\Big)+\omega^2\Big)\rho_{2j}-\Omega_j^{\alpha}\sin\Big(\frac{\pi\alpha}{2}\Big)\rho_{1j}\Big]\sin(\Omega_j t). \quad (16)\end{aligned}$$

Inserting the last result into Equation (1) yields

$$\begin{aligned}&\Big(\Omega_j^{\alpha}\cos\Big(\frac{\pi\alpha}{2}\Big)+\omega^2\Big)\rho_{1j}+\Omega_j^{\alpha}\sin\Big(\frac{\pi\alpha}{2}\Big)\rho_{2i}=0,\\ &\Big(\Omega_j^{\alpha}\cos\Big(\frac{\pi\alpha}{2}\Big)+\omega^2\Big)\rho_{2j}-\Omega_j^{\alpha}\sin\Big(\frac{\pi\alpha}{2}\Big)\rho_{1j}=b_j,\end{aligned} \quad (17)$$

which can be easily solved to obtain ρ_{1j} and ρ_{2j} in the forms:

$$\rho_{1j}=-\frac{\Omega^{\alpha}b_j\sin\big(\frac{\pi\alpha}{2}\big)}{\omega^4+\Omega_j^{2\alpha}+2\omega^2\Omega_j^{\alpha}\cos\big(\frac{\pi\alpha}{2}\big)},\quad \rho_{2j}=\frac{b_j\omega^2+\Omega_j^{\alpha}b_j\cos\big(\frac{\pi\alpha}{2}\big)}{\omega^4+\Omega_j^{2\alpha}+2\omega^2\Omega_j^{\alpha}\cos\big(\frac{\pi\alpha}{2}\big)}. \quad (18)$$

Employing (18) into (14), we find

$$y_p(t)=\sum_{j=1}^{n}b_j\left(\frac{\omega^2\sin(\Omega_j t)+\Omega_j^{\alpha}\sin\big(\Omega_j t-\frac{\pi\alpha}{2}\big)}{\omega^4+\Omega_j^{2\alpha}+2\omega^2\Omega_j^{\alpha}\cos\big(\frac{\pi\alpha}{2}\big)}\right), \quad (19)$$

which completes the proof. □

3. Solution of the Fractional Models: $\alpha \in (0,1)$

Lemma 1. *The solution of the fractional class (1) is*

$$y(t)=\left(A+\sum_{j=1}^{n}\frac{\Omega_j^{\alpha}b_j\sin\big(\frac{\pi\alpha}{2}\big)}{\omega^4+\Omega_j^{2\alpha}+2\omega^2\Omega_j^{\alpha}\cos\big(\frac{\pi\alpha}{2}\big)}\right)e^{(-\omega^2)^{\frac{1}{\alpha}}t}+\sum_{j=1}^{n}b_j\left(\frac{\omega^2\sin(\Omega_j t)+\Omega_j^{\alpha}\sin\big(\Omega_j t-\frac{\pi\alpha}{2}\big)}{\omega^4+\Omega_j^{2\alpha}+2\omega^2\Omega_j^{\alpha}\cos\big(\frac{\pi\alpha}{2}\big)}\right). \quad (20)$$

Proof. The preceding analysis reveals that the general solution of the class (1) is in the form:

$$y(t)=c\,e^{i\delta t}+\sum_{j=1}^{n}b_j\left(\frac{\omega^2\sin(\Omega_j t)+\Omega_j^{\alpha}\sin\big(\Omega_j t-\frac{\pi\alpha}{2}\big)}{\omega^4+\Omega_j^{2\alpha}+2\omega^2\Omega_j^{\alpha}\cos\big(\frac{\pi\alpha}{2}\big)}\right). \quad (21)$$

From this equation, at $t=0$, we obtain

$$y(0)=c-\sum_{j=1}^{n}\frac{\Omega_j^{\alpha}b_j\sin\big(\frac{\pi\alpha}{2}\big)}{\omega^4+\Omega_j^{2\alpha}+2\omega^2\Omega_j^{\alpha}\cos\big(\frac{\pi\alpha}{2}\big)}, \quad (22)$$

and hence the IC can be applied to give

$$c=A+\sum_{j=1}^{n}\frac{\Omega_j^{\alpha}b_j\sin\big(\frac{\pi\alpha}{2}\big)}{\omega^4+\Omega_j^{2\alpha}+2\omega^2\Omega_j^{\alpha}\cos\big(\frac{\pi\alpha}{2}\big)}. \quad (23)$$

Substituting (23) into (21), the solution reads

$$y(t)=\left(A+\sum_{j=1}^{n}\frac{\Omega_j^{\alpha}b_j\sin\big(\frac{\pi\alpha}{2}\big)}{\omega^4+\Omega_j^{2\alpha}+2\omega^2\Omega_j^{\alpha}\cos\big(\frac{\pi\alpha}{2}\big)}\right)e^{(-\omega^2)^{\frac{1}{\alpha}}t}+\sum_{j=1}^{n}b_j\left(\frac{\omega^2\sin(\Omega_j t)+\Omega_j^{\alpha}\sin\big(\Omega_j t-\frac{\pi\alpha}{2}\big)}{\omega^4+\Omega_j^{2\alpha}+2\omega^2\Omega_j^{\alpha}\cos\big(\frac{\pi\alpha}{2}\big)}\right). \quad (24)$$

It can be seen that the above solution satisfies the IC. In addition, the solution (24) is real at specific values of α; this point will be discussed later. □

Lemma 2. *The solution of the fractional class (2) is*

$$y(t)=\left(A+\sum_{j=1}^{n}\frac{\Omega_j^{\alpha}b_j\sin\left(\frac{\pi\alpha}{2}\right)}{\mu^4+\Omega_j^{2\alpha}-2\mu^2\Omega_j^{\alpha}\cos\left(\frac{\pi\alpha}{2}\right)}\right)e^{\mu^{\frac{2}{\alpha}}t}-\sum_{j=1}^{n}b_j\left(\frac{\mu^2\sin(\Omega_j t)-\Omega_j^{\alpha}\sin\left(\Omega_j t-\frac{\pi\alpha}{2}\right)}{\mu^4+\Omega_j^{2\alpha}-2\mu^2\Omega_j^{\alpha}\cos\left(\frac{\pi\alpha}{2}\right)}\right). \tag{25}$$

Proof. As mentioned in Section 1, the class (2) is a transformed version of the class (1) when $\omega=i\mu$. Hence, the solution of the class (2) can be directly obtained from the solution of the class (1), given in lemma 1, with the aide of the substitution $\omega=i\mu$, which yields

$$y(t)=\left(A+\sum_{j=1}^{n}\frac{\Omega_j^{\alpha}b_j\sin\left(\frac{\pi\alpha}{2}\right)}{\mu^4+\Omega_j^{2\alpha}-2\mu^2\Omega_j^{\alpha}\cos\left(\frac{\pi\alpha}{2}\right)}\right)e^{\mu^{\frac{2}{\alpha}}t}+\sum_{j=1}^{n}b_j\left(\frac{-\mu^2\sin(\Omega_j t)+\Omega_j^{\alpha}\sin\left(\Omega_j t-\frac{\pi\alpha}{2}\right)}{\mu^4+\Omega_j^{2\alpha}-2\mu^2\Omega_j^{\alpha}\cos\left(\frac{\pi\alpha}{2}\right)}\right), \tag{26}$$

or

$$y(t)=\left(A+\sum_{j=1}^{n}\frac{\Omega_j^{\alpha}b_j\sin\left(\frac{\pi\alpha}{2}\right)}{\mu^4+\Omega_j^{2\alpha}-2\mu^2\Omega_j^{\alpha}\cos\left(\frac{\pi\alpha}{2}\right)}\right)e^{\mu^{\frac{2}{\alpha}}t}-\sum_{j=1}^{n}b_j\left(\frac{\mu^2\sin(\Omega_j t)-\Omega_j^{\alpha}\sin\left(\Omega_j t-\frac{\pi\alpha}{2}\right)}{\mu^4+\Omega_j^{2\alpha}-2\mu^2\Omega_j^{\alpha}\cos\left(\frac{\pi\alpha}{2}\right)}\right), \tag{27}$$

which completes the proof. □

Remark 1. *The analytic method used to obtain the exact solutions of the fractional classes (1) and (2) is shown in this section. The other fractional classes (3) and (4) can also be obtained similarly. It can be seen from the solution (20) of the fractional class (1) that it is not always a real solution for $\alpha\in(0,1)$. This is simply because $(-\omega^2)^{1/\alpha}\notin\mathbb{R}\ \forall\ \alpha\in(0,1)$, but there are certain values of the fractional-order α at which the solution (20) is real, $y(t)\in\mathbb{R}$. Such values of α will be addressed in a subsequent section.*

However, the solution (25) of the fractional class (2) is always a real solution $\forall\ \alpha\in(0,1)$ where $\mu^{2/\alpha}\in\mathbb{R}$ for $\mu\in\mathbb{R}$. In the case of the ordinary/classical derivative, i.e., as $\alpha\to1$, then the solutions (20) and (25) are real. The solution of the fractional classes (3) and (4) can be obtained via substituting $\Omega_j=i\sigma_j$ and $b_j=-id_j$ into the solutions (20) and (25), respectively. Although, the resulting solutions of fractional classes (3) and (4) are not real at any value of α. In fact, the solutions of classes (3) and (4) are only real when $\alpha\to1$. The solutions of the four classes (1)–(4), as $\alpha\to1$, are determined in the next section.

4. Solution of the Classical/Ordinary Models: $\alpha\to1$

This section focuses on obtaining the exact solutions of the classical/ordinary versions of the classes (1)–(4) when $\alpha\to1$,

4.1. Class (1)

As $\alpha\to1$, the class (1) is transformed to the following class of ODEs:

$$y'(t)+\omega^2y(t)=\sum_{j=1}^{n}b_j\sin(\Omega_j t),\ \ y(0)=A. \tag{28}$$

The solution of this class can be derived from Equation (20) by letting $\alpha\to1$, and accordingly, we have

$$y(t)=\left(A+\sum_{j=1}^{n}\frac{\Omega_j b_j}{\omega^4+\Omega_j^2}\right)e^{-\omega^2t}+\sum_{j=1}^{n}b_j\left(\frac{\omega^2\sin(\Omega_j t)+\Omega_j\sin\left(\Omega_j t-\frac{\pi}{2}\right)}{\omega^4+\Omega_j^2}\right), \tag{29}$$

which is equivalent to

$$y(t)=\left(A+\sum_{j=1}^{n}\frac{\Omega_j b_j}{\omega^4+\Omega_j^2}\right)e^{-\omega^2t}+\sum_{j=1}^{n}b_j\left(\frac{\omega^2\sin(\Omega_j t)-\Omega_j\cos(\Omega_j t)}{\omega^4+\Omega_j^2}\right). \tag{30}$$

The validity of the solution (30) can be easily verified by direct substitution into (28). Moreover, this solution satisfies the given IC.

4.2. Class (2)

The class (2), as $\alpha \to 1$, reduces to ODEs:

$$y'(t) - \mu^2 y(t) = \sum_{j=1}^{n} b_j \sin(\Omega_j t), \; y(0) = A. \tag{31}$$

From Equation (24), we obtain as $\alpha \to 1$ that

$$y(t) = \left(A + \sum_{j=1}^{n} \frac{\Omega_j b_j}{\mu^4 + \Omega_j^2}\right) e^{\mu^2 t} - \sum_{j=1}^{n} b_j \left(\frac{\mu^2 \sin(\Omega_j t) - \Omega_j \sin\left(\Omega_j t - \frac{\pi}{2}\right)}{\mu^4 + \Omega_j^2}\right), \tag{32}$$

or

$$y(t) = \left(A + \sum_{j=1}^{n} \frac{\Omega_j b_j}{\mu^4 + \Omega_j^2}\right) e^{\mu^2 t} - \sum_{j=1}^{n} b_j \left(\frac{\mu^2 \sin(\Omega_j t) + \Omega_j \cos(\Omega_j t)}{\mu^4 + \Omega_j^2}\right). \tag{33}$$

4.3. Class (3)

The class (3) as $\alpha \to 1$ becomes

$$y'(t) + \omega^2 y(t) = \sum_{j=1}^{n} d_j \sinh(\sigma_j t), \; y(0) = A. \tag{34}$$

Because this class is transformed from the class (1) when $\Omega_j = i\sigma_j$, and $b_j = -id_j$, then the solution of the current class is determined from Equation (30) as

$$y(t) = \left(A + \sum_{j=1}^{n} \frac{\sigma_j d_j}{\omega^4 - \sigma_j^2}\right) e^{-\omega^2 t} - \sum_{j=1}^{n} id_j \left(\frac{\omega^2 \sin(i\sigma_j t) - i\sigma_j \cos(i\sigma_j t)}{\omega^4 - \sigma_j^2}\right), \tag{35}$$

i.e.,

$$y(t) = \left(A + \sum_{j=1}^{n} \frac{\sigma_j d_j}{\omega^4 - \sigma_j^2}\right) e^{-\omega^2 t} + \sum_{j=1}^{n} d_j \left(\frac{\omega^2 \sinh(\sigma_j t) - \sigma_j \cosh(\sigma_j t)}{\omega^4 - \sigma_j^2}\right). \tag{36}$$

4.4. Class (4)

If $\omega = i\mu$, $\Omega_j = i\sigma_j$, and $b_j = -id_j$, then the class (1) as $\alpha \to 1$ is equivalent to the following class of ODEs:

$$y'(t) - \mu^2 y(t) = \sum_{j=1}^{n} d_j \sinh(\sigma_j t), \; y(0) = A. \tag{37}$$

In this case, we have three possible ways to obtain the solution of the current class. The first way is to substitute $\omega = i\mu$, $\Omega_j = i\sigma_j$, and $b_j = -id_j$ into Equation (30). The second is to substitute $\Omega_j = i\sigma_j$ and $b_j = -id_j$ into Equation (33). The third way is the simplest one, by substituting only $\omega = i\mu$ into Equation (36). Following the third option, one can obtain the exact solution:

$$y(t) = \left(A + \sum_{j=1}^{n} \frac{\sigma_j d_j}{\mu^4 - \sigma_j^2}\right) e^{\mu^2 t} + \sum_{j=1}^{n} d_j \left(\frac{-\mu^2 \sinh(\sigma_j t) - \sigma_j \cosh(\sigma_j t)}{\mu^4 - \sigma_j^2}\right), \tag{38}$$

or

$$y(t) = \left(A + \sum_{j=1}^{n} \frac{\sigma_j d_j}{\mu^4 - \sigma_j^2}\right) e^{\mu^2 t} - \sum_{j=1}^{n} d_j \left(\frac{\mu^2 \sinh(\sigma_j t) + \sigma_j \cosh(\sigma_j t)}{\mu^4 - \sigma_j^2}\right), \tag{39}$$

for the present class of ODEs.

Remark 2. *The obtained exact solutions for the four classes of ODEs satisfy the condition $y(0) = A$. On the other hand, the validity of the obtained solutions can be easily checked through direct substitutions into the governing ODEs of these classes. We can say that the FC is of great importance and benefits. This is because the FC not only gives the solutions of fractional models but also helps in deriving the solutions of corresponding classical/ordinary models.*

5. Behavior of Solution

It is seen from the previous sections that the fractional systems (1) and (2) have the exact solutions given by Equation (20) and Equation (24), respectively. The main observation is that the solution (20) of the class (1) is real if the quantity $(-\omega^2)^{1/\alpha}$ is real. For real ω, we note that $(-\omega^2)^{1/\alpha} = \nu\omega^{2/\alpha}$ where $\nu = (-1)^{1/\alpha}$. So, the solution (20) is real when ν is real. The authors [31] were able to specify the α-values such that $\nu = (-1)^{1/\alpha}$ is real and this occurs that the α-values follow the next theorem [30].

Theorem 2. *For $n, k \in \mathbb{N}^+$, the solution (20) is real when $\alpha = \frac{2n-1}{2(k+n-1)}$ $(\nu = 1)$ and $\alpha = \frac{2n-1}{2(k+n)-1}$ $(\nu = -1)$.*

Based on the above theorem, the solution (20) for the fractional class (1) is plotted in Figure 1 for $\alpha = \frac{1}{2}$ at different numbers of the sinusoidal terms. Figure 2 shows the variation in the solution (20) for the fractional class (1) with two sinusoidal terms at different values of the initial condition A. In addition, Figure 3 indicates the behavior of the solution at various values of the fractional-order α when ten sinusoidal terms are incorporated in the fractional class (1). Furthermore, the solution is depicted in Figure 4 at some selected values α close to unity. This figure declares that the fractional solution becomes identical to the ordinary/classical solution as $\alpha \to 1$ which validates the present results.

For the fractional class (2), the solution (25) is displayed in Figure 5 when $\alpha = \frac{1}{2}$ at different numbers of the sinusoidal terms. The behavior of the solution of this class is similar to Figure 1 but with a slightly higher magnitude of the oscillations for the same numbers of the sinusoidal terms. Figure 6 gives us a picture of the solution profile as the fractional-order α varies regarding the fractional class (2). Moreover, Figure 7 displays the profile of the solution (25) at various values of the parameter μ. The current results reveal the oscillatory nature of the obtained solutions for the fractional systems (1) and (2). Finally, the present analysis may be extended to effectively analyze higher-order fractional systems containing a finite number of sinusoidal terms.

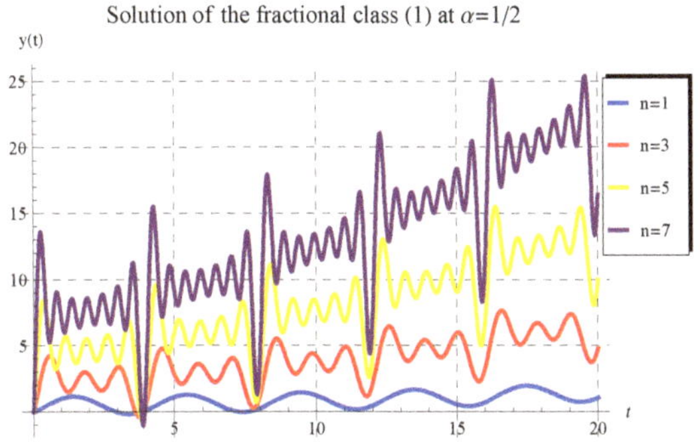

Figure 1. Plots of the solution for the fractional class (1) when $\alpha = \frac{1}{2}$, $A = 0$, $\omega = \frac{1}{2}$, $b_j = j$, and $\Omega_j = j\pi/2$ at different values of n (number of sinusoidal terms).

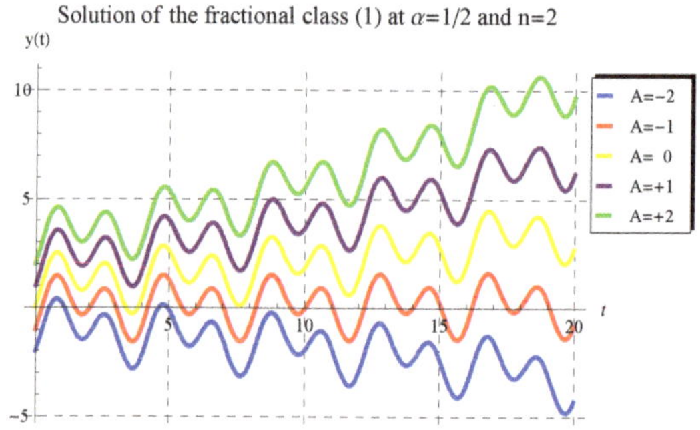

Figure 2. Plots of the solution for the fractional class (1) when $\alpha = \frac{1}{2}, \omega = \frac{1}{2}, b_j = j$, and $\Omega_j = j\pi/2$ at different values of $A = -2, -1, 0, 1, 2$ for two sinusoidal terms ($n = 2$).

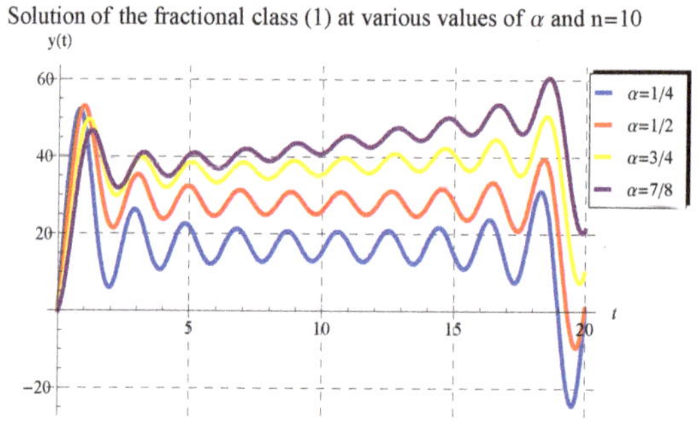

Figure 3. Plots of the solution for the fractional class (1) when $\alpha = \frac{1}{2}$, $A = 0$, $\omega = \frac{1}{5}$, $b_j = j$, and $\Omega_j = j\pi/10$ at different values of $\alpha = \frac{1}{4}, \frac{1}{2}, \frac{3}{4}, \frac{7}{8}$ for ten sinusoidal terms ($n = 10$).

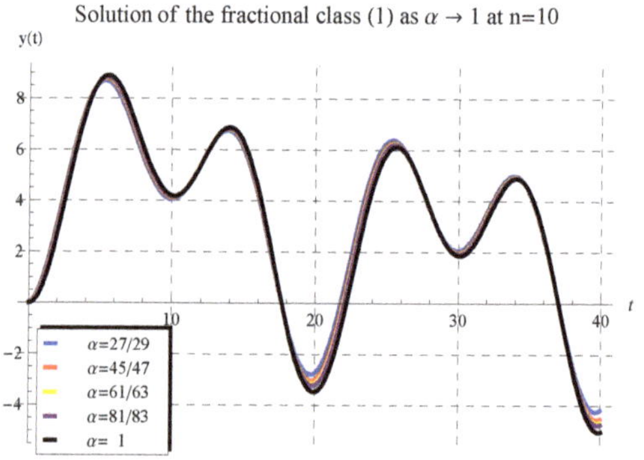

Figure 4. Plots of the solution for the fractional class (1) when $A = 0, \omega = \frac{1}{5}, b_j = j$, and $\Omega_j = j\pi/10$ at different values of $\alpha = \frac{27}{29}, \frac{45}{47}, \frac{61}{63}, \frac{81}{83}, 1$ for ten sinusoidal terms ($n = 10$).

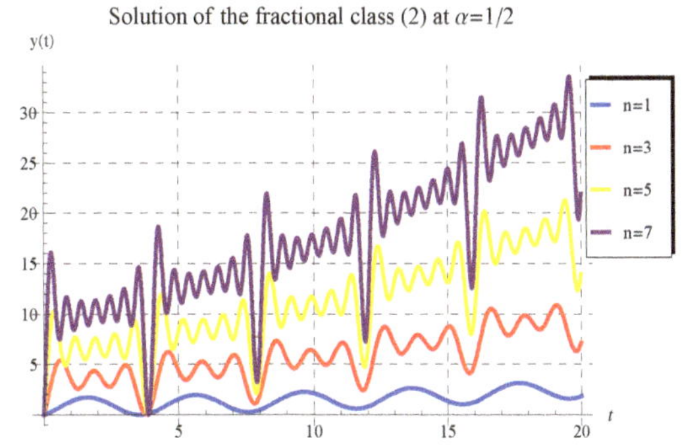

Figure 5. Plots of the solution for the fractional class (2) when $\alpha = \frac{1}{2}$, $A = 0$, $\mu = \frac{1}{2}$, $b_j = j$, and $\Omega_j = j\pi/2$ at different values of n (number of sinusoidal terms).

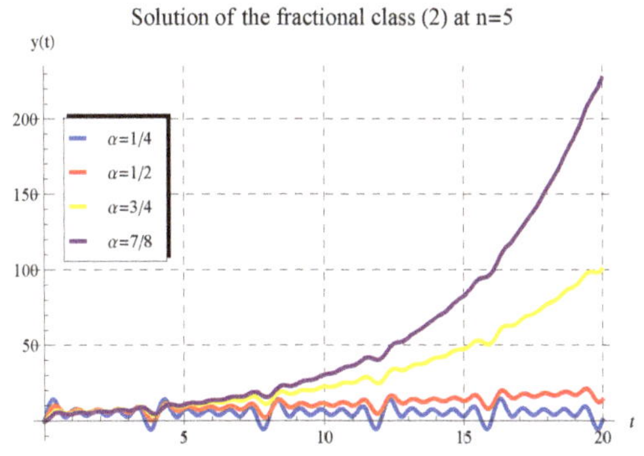

Figure 6. Plots of the solution for the fractional class (2) when $\mu = \frac{1}{2}$, $A = 0$, $b_j = j$, and $\Omega_j = j\pi/2$ at different values of α for five sinusoidal terms ($n = 5$).

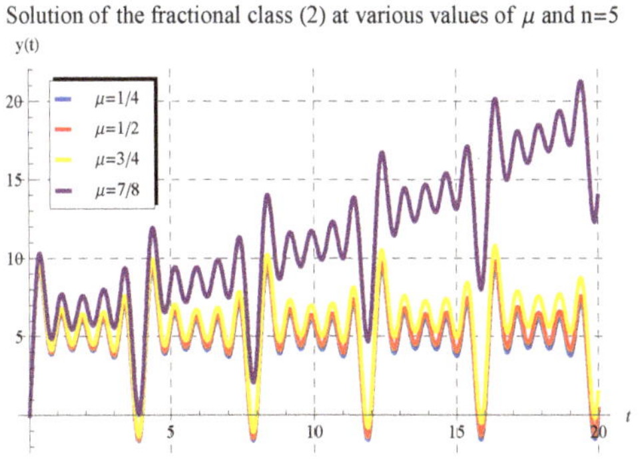

Figure 7. Plots of the solution for the fractional class (2) when $\alpha = \frac{1}{2}$, $A = 0$, $b_j = j$, and $\Omega_j = j\pi/2$ at different values of μ.

6. Conclusions

In this paper, a class of first-order fractional differential systems containing a finite number n of sinusoidal terms was analyzed by means of the Riemann–Liouville fractional definition. The difficulties in solving fractional systems under real/physical initial conditions using the Riemann–Liouville fractional definition are overcome in this paper. This task was achieved via a straightforward method. The suggested method was successfully applied to extract the exact solutions of the considered fractional systems. In addition, the corresponding exact solutions of the classical/ordinary versions were determined. The obtained results reveal the oscillatory nature of the present fractional systems. Moreover, the properties/behaviors of the obtained solutions were investigated graphically and hence interpreted. Accordingly, the current approach may deserve a further extension to include fractional systems of a higher order when the sinusoidal terms of a finite number are incorporated. Finally, the current approach may be applied to include other ideas [39–47].

Author Contributions: Conceptualization, L.F.S., E.R.E.-Z. and A.E.; methodology, L.F.S., E.R.E.-Z. and A.E.; software, L.F.S., E.R.E.-Z. and A.E.; validation, L.F.S., E.R.E.-Z. and A.E.; formal analysis, L.F.S., E.R.E.-Z. and A.E.; investigation, L.F.S., E.R.E.-Z. and A.E.; resources, L.F.S., E.R.E.-Z. and A.E.; data curation, L.F.S., E.R.E.-Z. and A.E.; writing—original draft preparation, L.F.S., E.R.E.-Z. and A.E.; writing—review and editing, L.F.S., E.R.E.-Z. and A.E.; visualization, L.F.S., E.R.E.-Z. and A.E.; supervision, L.F.S. and E.R.E.-Z.; project administration, L.F.S. and E.R.E.-Z.; funding acquisition, L.F.S. and E.R.E.-Z. All authors have read and agreed to the published version of the manuscript.

Funding: The authors extend their appreciation to the Deputyship for Research and Innovation, the Ministry of Education in Saudi Arabia, for funding this research work through the project number (IF2/PSAU/2022/01/22726)

Data Availability Statement: Not applicable.

Acknowledgments: The authors extend their appreciation to the Deputyship for Research and Innovation, the Ministry of Education in Saudi Arabia, for funding this research work through the project number (IF2/PSAU/2022/01/22726). Moreover, they would like to thank the referees for their valuable comments and suggestions which helped to improve the manuscript.

Conflicts of Interest: The authors have no competing interest regarding the publication of this paper.

References

1. Miller, K.S.; Ross, B. *An Introduction to the Fractional Calculus and Fractional Differential Equations*; John Wiley & Sons: New York, NY, USA, 1993.
2. Podlubny, I. *Fractional Differential Equations*; Academic Press: San Diego, CA, USA, 1999.
3. Hilfer, R. *Applications of Fractional Calculus in Physics*; World Scientific Publishing Company: Singapore, 2000.
4. Achar, B.N.N.; Hanneken, J.W.; Enck, T.; Clarke, T. Dynamics of the fractional oscillator. *Phys. A* **2001**, *297*, 361–367. [CrossRef]
5. Sebaa, N.; Fellah, Z.E.A.; Lauriks, W.; Depollier, C. Application of fractional calculus to ultrasonic wave propagation in human cancellous bone. *Signal Process.* **2006**, *86*, 2668–2677. [CrossRef]
6. Tarasov, V.E. Fractional Heisenberg equation. *Phys. Lett. A* **2008**, *372*, 2984–2988. [CrossRef]
7. Ding, Y.; Yea, H. A fractional-order differential equation model of HIV infection of CD4+T-cells. *Math. Comput. Model.* **2009**, *50*, 386–392. [CrossRef]
8. Wang, S.; Xu, M.; Li, X. Green's function of time fractional diffusion equation and its applications in fractional quantum mechanics. *Nonlinear Anal. Real World Appl.* **2009**, *10*, 1081–1086. [CrossRef]
9. Song, L.; Xu, S.; Yang, J. Dynamical models of happiness with fractional order. *Commun. Nonlinear Sci. Numer. Simul.* **2010**, *15*, 616–628. [CrossRef]
10. Ebaid, A. Analysis of projectile motion in view of the fractional calculus. *Appl. Math. Model.* **2011**, *35*, 1231–1239. [CrossRef]
11. Machado, J.T.; Kiryakova, V.; Mainardi, F. Recent history of fractional calculus. *Commun. Nonlinear Sci. Numer. Simul.* **2011**, *16*, 1140–1153. [CrossRef]
12. Ebaid, A.; El-Sayed, D.M.M.; Aljoufi, M.D. Fractional calculus model for damped Mathieu equation: Approximate analytical solution. *Appl. Math. Sci.* **2012**, *6*, 4075–4080.
13. Gómez-Aguilara, J.F.; Rosales-García, J.J.; Bernal-Alvarado, J.J. Fractional mechanical oscillators. *Rev. Mex. De Física* **2012**, *58*, 348–352.
14. Garcia, J.J.R.; Calderon, M.G.; Ortiz, J.M.; Baleanu, D. Motion of a particle in a resisting medium using fractional calculus approach. *Proc. Rom. Acad. Ser. A* **2013**, *14*, 42–47.
15. Kumar, D.; Singh, J.; Baleanu, D.; Rathore, S. Analysis of a fractional model of the Ambartsumian equation. *Eur. Phys. J. Plus* **2018**, *133*, 133–259. [CrossRef]
16. Ebaid, A.; El-Zahar, E.R.; Aljohani, A.F.; Salah, B.; Krid, M.; Machado, J.T. Analysis of the two-dimensional fractional projectile motion in view of the experimental data. *Nonlinear Dyn.* **2019**, *97*, 1711–1720. [CrossRef]
17. Ebaid, A.; Cattani, C.; Juhani1, A.S.A.; El-Zahar, E.R. A novel exact solution for the fractional Ambartsumian equation. *Adv. Differ. Equations* **2021**, *2021*, 88. [CrossRef]
18. Kaur, D.; Agarwal, P.; Rakshit, M.; Chand, M. Fractional Calculus involving (p,q)-Mathieu Type Series. *Appl. Math. Nonlinear Sci.* **2020**, *5*, 15–34. [CrossRef]

19. Agarwal, P.; Mondal, S.R.; Nisar, K.S. On fractional integration of generalized struve functions of first kind. *Thai J. Math.* **2020**, *to appear*.
20. Agarwal, P.; Singh, R. Modelling of transmission dynamics of Nipah virus (Niv): A fractional order approach. *Phys. A Stat. Mech. Its Appl.* **2020**, *547*, 124243. [CrossRef]
21. Alderremy, A.A.; Saad, K.M.; Agarwal, P.; Aly, S.; Jain, S. Certain new models of the multi space-fractional Gardner equation. *Phys. A Stat. Mech. Its Appl.* **2020**, *545*, 123806. [CrossRef]
22. Aljohani, A.F.; Ebaid, A.; Algehyne, E.A.; Mahrous, Y.M.; Cattani, C.; Al-Jeaid, H.K. The Mittag-Leffler function for re-evaluating the chlorine transport model: Comparative analysis. *Fractal Fract.* **2022**, *6*, 125. [CrossRef]
23. Ahmad, B.; Batarfi, H.; Nieto, J.J.; Oscar, O.-Z.; Shammakh, W. Projectile motion via Riemann-Liouville calculus. *Adv. Differ. Equ.* **2015**, *2015*, 63. [CrossRef]
24. Elzahar, E.R.; Gaber, A.A.; Aljohani, A.F.; Machado, J.T.; Ebaid, A. Generalized Newtonian fractional model for the vertical motion of a particle. *Appl. Math. Model* **2020**, *88*, 652–660. [CrossRef]
25. El-Zahar, E.R.; Alotaibi, A.M.; Ebaid, A.; Aljohani, A.F.; Gomez Aguilar, J.F. The Riemann-Liouville fractional derivative for Ambartsumian equation. *Results Phys.* **2020**, *19*, 103551. [CrossRef]
26. Ebaid, A.; Masaedeh, B.; El-Zahar, E. A new fractional model for the falling body problem. *Chin. Phys. Lett.* **2017**, *34*, 020201. [CrossRef]
27. Khaled, S.M.; El-Zahar, E.R.; Ebaid, A. Solution of Ambartsumian delay differential equation with conformable derivative. *Mathematics* **2019**, *7*, 425. [CrossRef]
28. Alharbi, F.M.; Baleanu, D.; Ebaid, A. Physical properties of the projectile motion using the conformable derivative. *Chin. J. Phys.* **2019**, *58*, 18–28. [CrossRef]
29. Algehyne, E.A.; El-Zahar, E.R.; Alharbi, F.M.; Ebaid, A. Development of analytical solution for a generalized Ambartsumian equation. *AIMS Math.* **2020**, *5*, 249–258. [CrossRef]
30. Ebaid, A.; Al-Jeaid, H.K. The Mittag–Leffler Functions for a Class of First-Order Fractional Initial Value Problems: Dual Solution via Riemann–Liouville Fractional Derivative. *Fractal Fract.* **2022**, *6*, 85. [CrossRef]
31. Khaled, S.M.; Ebaid, A.; Mutairi, F.A. The exact endoscopic effect on the peristaltic flow of a nanofluid. *J. Appl. Math.* **2014**, *2014*, 367526. [CrossRef]
32. Ebaid, A.; Sharif, M.A. Application of Laplace transform for the exact effect of a magnetic field on heat transfer of carbon-nanotubes suspended nanofluids. *Z. Nature. A* **2015**, *70*, 471–475. [CrossRef]
33. Saleh, H.; Alali, E.; Ebaid, A. Medical applications for the flow of carbon-nanotubes suspended nanofluids in the presence of convective condition using Laplace transform. *J. Assoc. Arab Univ. Basic Appl. Sci.* **2017**, *24*, 206–212. [CrossRef]
34. Ebaid, A.; Wazwaz, A.M.; Alali, E.; Masaedeh, B. Hypergeometric Series Solution to a Class of Second-Order Boundary Value Problems via Laplace Transform with Applications to Nanofuids. *Commun. Theor. Phys.* **2017**, *67*, 231. [CrossRef]
35. Ebaid, A.; Alali, E.; Saleh, H. The exact solution of a class of boundary value problems with polynomial coefficients and its applications on nanofluids. *J. Assoc. Arab Univ. Basi Appl. Sci.* **2017**, *24*, 156–159. [CrossRef]
36. Ali, H.S.; Alali, E.; Ebaid, A.; Alharbi, F.M. Analytic solution of a class of singular second-order boundary value problems with applications. *Mathematics* **2019**, *7*, 172. [CrossRef]
37. Ebaid, A.; Alharbi, W.; Aljoufi, M.D.; El-Zahar, E.R. The exact solution of the falling body problem in three-dimensions: Comparative study. *Mathematics* **2020**, *8*, 1726. [CrossRef]
38. El-Dib, Y.O.; Elgazery, N.S. Effect of Fractional Derivative Properties on the Periodic Solution of the Nonlinear Oscillations. *Fractals* **2020**, *28*, 2050095. [CrossRef]
39. Azam, M. Effects of Cattaneo-Christov heat flux and nonlinear thermal radiation on MHD Maxwell nanofluid with Arrhenius activation energy. *Case Stud. Therm. Eng.* **2022**, *34*, 102048. [CrossRef]
40. Azam, M. Bioconvection and nonlinear thermal extrusion in development ofchemically reactive Sutterby nano-material due to gyrotactic microorganisms. *Int. Commun. Heat Mass Transf.* **2022**, *130*, 105820. [CrossRef]
41. Azam, M.; Abbas, N.; Ganesh, K.K.; Wali, S. Transient bioconvection and activation energy impacts on Casson nanofluid with gyrotactic microorganisms and nonlinear radiation. *Waves Random Complex Media* **2022**, 1–20. [CrossRef]
42. Azam, M.; Nayak, M.K.; Khan, W.A.; Khan, M. Significance of bioconvection and variable thermal properties on dissipative Maxwell nanofluid due to gyrotactic microorganisms and partial slip. *Waves Random Complex Media* **2022**, 1–21. [CrossRef]
43. Azam, M.; Xu, T.; Nayak, M.K.; Khan, W.A.; Khan, M. Gyrotactic microorganisms and viscous dissipation features on radiative Casson nanoliquid over a moving cylinder with activation energy. *Waves Random Complex Media* **2022**, 1–23. [CrossRef]
44. Oderinu, R.A.; Owolabi, J.A.; Taiwo, M. Approximate solutions of linear time-fractional differential equations. *J. Math. Comput. Sci.* **2022**, *29*, 60–72. [CrossRef]
45. Akram, T.; Abbas, M.; Ali, A. A numerical study on time fractional Fisher equation using an extended cubic B-spline approximation. *J. Math. Comput. Sci.* **2021**, *22*, 85–96. [CrossRef]
46. AlAhmad, R.; AlAhmad, Q.; Abdelhadi, A. Solution of fractional autonomous ordinary differential Equations. *J. Math. Comput. Sci.* **2022**, *27*, 59–64. [CrossRef]
47. Nikan, O.; Avazzadeh, Z.; Machado, J.T. An efficient local meshless approach for solving nonlinear time-fractional fourth-order diffusion model. *J. King Saud Univ.-Sci.* **2021**, *33*, 101243. [CrossRef]

symmetry

MDPI

Article

Some Existence and Uniqueness Results for a Class of Fractional Stochastic Differential Equations

Omar Kahouli [1,*], Abdellatif Ben Makhlouf [2], Lassaad Mchiri [3,4], Pushpendra Kumar [5], Naim Ben Ali [6] and Ali Aloui [1]

1 Department of Electronics Engineering, Community College, University of Ha'il, Ha'il P.O. Box 2440, Saudi Arabia
2 Mathematics Department, College of Science, Jouf University, Sakaka P.O. Box 2014, Saudi Arabia
3 Mathematics Department, Faculty of Sciences of Sfax, University of Sfax, Route Soukra, BP 1171, Sfax 3000, Tunisia
4 ENSIIE, University of Evry-Val-d'Essonne, 1 Square de la Résistance, CEDEX, 91025 Évry-Courcouronnes, France
5 Institute for the Future of Knowledge, University of Johannesburg, P.O. Box 524, Auckland Park 2006, South Africa
6 Department of Industrial Engineering, College of Engineering, University of Ha'il, Ha'il P.O. Box 2440, Saudi Arabia
* Correspondence: a.kahouli@uoh.edu.sa

Abstract: Many techniques have been recently used by various researchers to solve some types of symmetrical fractional differential equations. In this article, we show the existence and uniqueness to the solution of ς-Caputo stochastic fractional differential equations (CSFDE) using the Banach fixed point technique (BFPT). We analyze the Hyers–Ulam stability of CSFDE using the stochastic calculus techniques. We illustrate our results with three examples.

Keywords: fractional calculus; fixed-point theory

Citation: Kahouli, O.; Ben Makhlouf, A.; Mchiri, L.; Kumar, P.; Ben Ali, N.; Aloui, A. Some Existence and Uniqueness Results for a Class of Fractional Stochastic Differential Equations. *Symmetry* **2022**, *14*, 2336. https://doi.org/10.3390/sym14112336

Academic Editors: Francisco Martínez González, Mohammed K. A. Kaabar and Luis Vázquez

Received: 30 September 2022
Accepted: 4 November 2022
Published: 7 November 2022

1. Introduction

Fractional calculus is a mathematical axis studying the characterizations of non-integer order derivatives and integrals [1,2]. In fact, this field contains the methods and notions of solving symmetrical differential equations with fractional derivatives. The theory of fractional calculus began almost in the same decade as the definition of classical calculus was decided. It was first defined in Leibniz's letter to L'Hospital in 1695, where the notion of semi-derivative was presented. During this period, fractional derivative was founded by many famous scientists, e.g., Riemann, Lagrange, Liouville, Fourier, Grünwald, Euler, Heaviside, Abel, etc. The fractional calculus has been used to describe many real-world phenomena: control theory, electrical networks, fluid flow, optics and signal processing, dynamical processes, etc. (see [1,3–7]). Particularly, in [8], the authors analyzed a system of neural networks in the sense of fractional derivatives. In [4], some novel applications of the non-integer order operators in the theory of viscoelasticity were derived. The authors of ref. [9] have proposed a scheme of approximate non-integer order differentiation, including noise immunity. A fruitful discussion on the Adams method in the fractional-order sense was given in the ref. [10]. In the last few decades, some new fractional derivatives have been introduced by various researchers to improve the literature on fractional calculus. In [11], Almeida suggested a new fractional derivative with respect to a kernel function called ς-Caputo fractional derivative, and generalized the work of several researchers [1,12]. In this context, several research papers showed interest in the ς-Caputo fractional derivative; for instance, see [11,13,14]. In [15], a numerical study on the non-integer order relaxation–oscillation equations in terms of ς-Caputo fractional derivatives are proposed. In [16], a study on the Ulam stability for Langevin non-integer order differential equations in the sense of two different fractional orders of ς-Caputo derivative has been given. In [17], the authors explored an initial value problem for differential equations in the sense of ς-Caputo derivative via a monotone iterative approach.

Recently, the theory of Hyers–Ulam stability (HUS) has attracted the attention of several famous scientists due to its real-world applications in biology and fluid flow, where identifying the explicit solutions is a very hard task. Some novel research studies on this topic have been proposed in the following references [18–20]. In [21], the authors discussed the results regarding the existence and HUS of solutions for almost periodic stochastic differential equations in a fractional sense. In [22], some novel results on the existence and HUS of random stochastic impulsive functional differential equations with delay have been established. In [23], Ulam stability for partial integro-differential equations with uncertainty in a fractional-order sense has been explored. Most of the existing papers consider the Caputo fractional derivative for the existence, uniqueness and HUS of the solutions of fractional differential equations. There are a lot of papers which discuss the ψ-Caputo fractional derivative (see [24–26]) for the deterministic case. In this paper, we have studied this concept for the stochastic case.

In this work, the existence and uniqueness of CSFDE are provided. The HUS for the proposed problem with the help of the novel features of stochastic calculus is simulated.

This paper extends the work on [27–29] for the Caputo and Caputo–Hadamard fractional derivative.

We highlight the main advantages of our article as follows:

- To investigate the existence and uniqueness of the solution of CSFDE via BFPT.
- To investigate the HUS of CSFDE by using the stochastic calculus techniques.

We summarize the content of the article: Section 2 presents the basic definitions of ς-CFD and some fundamental notations. Section 3 investigates the global existence and uniqueness of the solution of CSFDE. In Section 4, we analyze the HUS of CSFDE. In Section 5, we give three illustrative examples.

2. Basic Notions

Denote by $\{\Sigma, \mathcal{F}, \mathbb{F}_\Pi, \mathbb{P}\}$, where $\mathbb{F}_\Pi = \{\mathbb{F}_\eta\}_{\eta\in[1,\Pi]}$ and $\Pi > 1$, the complete probability space; $W(\eta)$ is the standard Brownian motion.

Let $\mathcal{X}_\eta = L^2(\Sigma, \mathbb{F}_\eta, \mathbb{P})$ (for every $\eta \in [1, \Pi]$) be the family of all $\mathbb{F}_\eta$-measurable and mean square integrable functions $\lambda = (\lambda_1, \dots, \lambda_p)^T : \Sigma \to \mathbb{R}^p$ satisfies

$$||\lambda||_{ms} = \sqrt{\sum_{l=1}^{p} \mathbb{E}(|\lambda_l|^2)} = \sqrt{\mathbb{E}||\lambda||^2},$$

where $||\cdot||$ is the usual Euclidian norm.

Definition 1 ([14])**.** *Denote by $\varphi > 0$ and let $\varsigma \in C^1[c, b]$ the function satisfying $\varsigma'(\sigma) \neq 0$, $\forall\sigma \in [c, b]$. The ς-fractional integral of order φ for an integrable function g is defined as*

$$I_{c^+}^{\varphi,\varsigma} g(x) = \frac{1}{\Gamma(\varphi)} \int_c^x \varsigma'(\sigma)(\varsigma(x) - \varsigma(\sigma))^{\varphi-1} g(\sigma) d\sigma. \tag{1}$$

Definition 2 ([14])**.** *Denote by $\varphi > 0$ and let $\varsigma \in C^1[c, b]$ the function satisfying $\varsigma'(\sigma) \neq 0$, $\forall\sigma \in [c, b]$. The ς-Riemann–Liouville fractional derivative of order φ of a function g is defined by*

$$D_{c^+}^{\varphi,\varsigma} g(x) = \left(\frac{1}{\varsigma'(x)} \frac{d}{dx}\right) I_{c^+}^{1-\varphi,\varsigma} g(x). \tag{2}$$

Definition 3 ([14]). *Let $\varphi > 0$ and $\varsigma \in C^1[c,b]$ the functions satisfying $\varsigma'(\sigma) \neq 0, \forall \sigma \in [c,b]$. The ς-Caputo fractional derivative of order φ of a function g is defined by*

$$^{C}D^{\varphi,\varsigma}_{c^+}g(t) = D^{\varphi,\varsigma}_{c^+}\Big[g(t) - g(c)\Big]. \tag{3}$$

Definition 4 ([1]). *$E_{\rho,\kappa}(y)$ is called a Mittag–Leffler function with two parameters if:*

$$E_{\rho,\kappa}(y) = \sum_{m=0}^{+\infty} \frac{y^m}{\Gamma(m\rho + \kappa)},$$

where $\rho > 0, \kappa > 0, y \in \mathbb{C}$.

Theorem 1 ([30]). *Let $(\mathbb{E}, d)$ be a complete metric space and let $\mathcal{B} : \mathbb{E} \to \mathbb{E}$ (with $z \in [0,1)$) be a contraction. Assume that $j \in \mathbb{E}$, $d(j, \mathcal{B}(j)) \leq v$ and $v > 0$. Then, there is a unique $u \in \mathbb{E}$ such that $\mathcal{B}(u) = u$.*

Let the following CSFDE:

$$^{C}D^{\varphi,\varsigma}_{a^+}\xi(\eta) = f_1(\eta, \xi(\eta)) + f_2(\eta, \xi(\eta))\frac{dW(\eta)}{d\eta}, \tag{4}$$

where the initial condition is $\xi(a) = \delta$, $\varsigma : [a, \Pi] \to \mathbb{R}$ be a C^1-increasing function with $\varsigma'(\eta) \neq 0, \forall \eta \in [a, \Pi], 0 < \varphi < 1$, $f_1 : [a, \Pi] \times \mathbb{R}^p \longrightarrow \mathbb{R}^p$ and $f_2 : [a, \Pi] \times \mathbb{R}^p \longrightarrow \mathbb{R}^p$ are measurable functions.

Let the following hypothesis:

$\mathcal{H}_1$: There is $L > 0$ satisfying

$$||f_1(\eta, \xi_1) - f_1(\eta, \xi_2)|| \vee ||f_2(\eta, \xi_1) - f_2(\eta, \xi_2)|| \leq L||\xi_1 - \xi_2||, \tag{5}$$

for all $(\eta, \xi_1, \xi_2) \in [a, \Pi] \times \mathbb{R}^p \times \mathbb{R}^p$.

$\mathcal{H}_2$: $f_1(\cdot, 0)$ and $f_2(\cdot, 0)$ satisfying

$$||f_2(\cdot, 0)||_\infty = \text{ess} \sup_{\eta \in [a,\Pi]} ||f_2(\eta, 0)|| < \infty, \tag{6}$$

$$\int_a^\Pi ||f_1(\sigma, 0)||^2 d\sigma < \infty.$$

3. Existence and Uniqueness of Solutions

Denote by $\mathbb{H}^2([a, \Pi])$ the family of all the processes ξ which are $\mathbb{F}_\Pi$-adapted, measurable such that

$$||\xi||_{\mathbb{H}^2} = \sup_{a \leq r \leq \Pi} ||\xi(r)||_{ms} < \infty.$$

It is not hard to prove that $(\mathbb{H}^2([a, \Pi]), || \cdot ||_{\mathbb{H}^2})$ is a Banach space. Let the operator N_δ : $\mathbb{H}^2([a, \Pi]) \to \mathbb{H}^2([a, \Pi])$, for $\delta \in \mathcal{X}_a$, given by:

$$\begin{aligned} N_\delta y(\eta) &= \delta + \frac{1}{\Gamma(\varphi)}\left[\int_a^\eta \varsigma'(\sigma)(\varsigma(\eta) - \varsigma(\sigma))^{\varphi-1} f_1(\sigma, y(\sigma))d\sigma\right] \\ &+ \frac{1}{\Gamma(\varphi)}\left[\int_a^\eta \varsigma'(\sigma)(\varsigma(\eta) - \varsigma(\sigma))^{\varphi-1} f_2(\sigma, y(\sigma))dW(\sigma)\right]. \end{aligned} \tag{7}$$

Lemma 1. *N_δ, for every $\sigma \in \mathcal{X}_a$, is well defined.*

Proof. Let $q \in \mathbb{H}^2([a,\Pi])$. Then, one has

$$\begin{aligned} ||N_\delta q(\eta)||^2_{ms} &\leq 3||\delta||^2_{ms} + \frac{3}{\Gamma(\varphi)^2}\mathbb{E}\left(\left|\left|\int_a^\eta \varsigma'(\sigma)(\varsigma(\eta)-\varsigma(\sigma))^{\varphi-1} f_1(\sigma,q(\sigma))d\sigma\right|\right|^2\right) \\ &+ \frac{3}{\Gamma(\varphi)^2}\mathbb{E}\left(\left|\left|\int_a^\eta \varsigma'(\sigma)(\varsigma(\eta)-\varsigma(\sigma))^{\varphi-1} f_2(\sigma,q(\sigma))dW(\sigma)\right|\right|^2\right). \end{aligned} \quad (8)$$

Using the Cauchy–Schwartz inequality, one gets

$$\begin{aligned} &\mathbb{E}\left(\left|\left|\int_a^\eta \varsigma'(\sigma)(\varsigma(\eta)-\varsigma(\sigma))^{\varphi-1} f_1(\sigma,q(\sigma))d\sigma\right|\right|^2\right) \\ &\leq \left(\int_a^\eta (\varsigma'(\sigma))^2(\varsigma(\eta)-\varsigma(\sigma))^{2\varphi-2}d\sigma\right)\mathbb{E}\left(\int_a^\eta ||f_1(\sigma,q(\sigma))||^2 d\sigma\right) \\ &\leq M\left(\int_a^\eta \varsigma'(\sigma)(\varsigma(\eta)-\varsigma(\sigma))^{2\varphi-2}d\sigma\right)\mathbb{E}\left(\int_a^\eta ||f_1(\sigma,q(\sigma))||^2 d\sigma\right) \\ &\leq \frac{M}{2\varphi-1}(\varsigma(\eta)-\varsigma(a))^{2\varphi-1}E\left(\int_a^\eta ||f_1(\sigma,q(\sigma))||^2 d\sigma\right), \end{aligned} \quad (9)$$

where $M = \sup_{\sigma\in[a,\Pi]} \varsigma'(\sigma)$. By $\mathcal{H}_1$, one can derive that

$$||f_1(\sigma,q(\sigma))||^2 \leq 2L^2||q(\sigma)||^2 + 2||f_1(\sigma,0)||^2. \quad (10)$$

Thus,

$$E\left(\int_a^\eta ||f_1(\sigma,q(\sigma))||^2 d\sigma\right) \leq 2L^2(\Pi-a) \sup_{\sigma\in[a,\Pi]} E\left(||q(\sigma)||^2\right) + 2\int_a^\Pi ||f_1(\sigma,0)||^2 d\sigma. \quad (11)$$

Then,

$$\begin{aligned} &\mathbb{E}\left(\left|\left|\int_a^\eta \varsigma'(\sigma)(\varsigma(\eta)-\varsigma(\sigma))^{\varphi-1} f_1(\sigma,q(\sigma))d\sigma\right|\right|^2\right) \\ &\leq \frac{M(\varsigma(\Pi)-\varsigma(a))^{2\varphi-1}}{2\varphi-1}\left[2L^2(\Pi-a)\sup_{\sigma\in[a,\Pi]} E\left(||q(\sigma)||^2\right) + 2\int_a^\Pi ||f_1(\sigma,0)||^2 d\sigma\right]. \end{aligned} \quad (12)$$

Using Itô's isometry formula, one gets

$$\begin{aligned} &\mathbb{E}\left(\left|\left|\int_a^\eta \varsigma'(\sigma)(\varsigma(\eta)-\varsigma(\sigma))^{\varphi-1} f_2(\sigma,q(\sigma))dW(\sigma)\right|\right|^2\right) \\ &= \mathbb{E}\left(\int_a^\eta (\varsigma'(\sigma))^2(\varsigma(\eta)-\varsigma(\sigma))^{2\varphi-2}||f_2(\sigma,q(\sigma))||^2 d\sigma\right). \end{aligned} \quad (13)$$

Using $\mathcal{H}_1$, one has

$$||f_2(\sigma,q(\sigma))||^2 \leq 2L^2||q(\sigma)||^2 + 2||f_2(\cdot,0)||^2_\infty. \quad (14)$$

Hence,

$$\mathbb{E}\left(\left|\left|\int_a^\eta \varsigma'(\sigma)(\varsigma(\eta)-\varsigma(\sigma))^{\varphi-1} f_2(\sigma,q(\sigma))dW(\sigma)\right|\right|^2\right)$$

$$\begin{aligned}
&\leq 2ML^2\mathbb{E}\Big(\int_a^\eta \varsigma'(\sigma)(\varsigma(\eta)-\varsigma(\sigma))^{2\varphi-2}||q(\sigma)||^2 d\sigma\Big)\\
&+ 2M||f_2(\cdot,0)||_\infty^2\int_a^\eta \varsigma'(\sigma)(\varsigma(\eta)-\varsigma(\sigma))^{2\varphi-2}d\sigma\\
&\leq \frac{2ML^2}{2\varphi-1}(\varsigma(\Pi)-\varsigma(a))^{2\varphi-1}||q||_{\mathbb{H}_2}^2+\frac{2M}{2\varphi-1}(\varsigma(\Pi)-\varsigma(a))^{2\varphi-1}||f_2(\cdot,0)||_\infty^2. \quad (15)
\end{aligned}$$

Therefore, N_δ is well defined. □

Theorem 2. *Under $\mathcal{H}_1$ and $\mathcal{H}_2$, for every $\sigma \in \mathcal{X}_a$, Equation (4) has a unique global solution $\xi(\cdot,\sigma)$ on $[a,\Pi]$.*

Proof. Let $\Pi > a$ be arbitrary. Let $\theta > 0$, such that $\theta^{2\varphi-1} > 2L^2M(\Pi+1)\dfrac{\Gamma(2\varphi-1)}{\Gamma(\varphi)^2}$. We define a norm $||\cdot||$ on the space $\mathbb{H}^2([a,\Pi])$ by

$$||\xi||_\theta = \sup_{\eta\in[a,\Pi]}\sqrt{\frac{\mathbb{E}\Big(||\xi(\eta)||^2\Big)}{e^{\theta(\varsigma(\eta)-\varsigma(a))}}}, \quad \forall\xi\in\mathbb{H}^2([a,\Pi]). \quad (16)$$

It is not hard to show that $||\cdot||_{\mathbb{H}^2}$ and $||\cdot||_\theta$ are equivalent. Consequently, $(\mathbb{H}^2([a,\Pi]),||\cdot||_\theta)$ is a Banach space.

Let $\xi_1,\xi_2\in\mathbb{H}^2([a,\Pi])$. Using (7), we get $\forall\eta\in[a,\Pi]$

$$\begin{aligned}
&\mathbb{E}\Big(||N_\delta\xi_1(\eta)-N_\delta\xi_2(\eta)||^2\Big)\\
&\leq \frac{2}{\Gamma(\varphi)^2}\mathbb{E}\Bigg(\Bigg|\Bigg|\int_a^\eta \varsigma'(\sigma)(\varsigma(\eta)-\varsigma(\sigma))^{\varphi-1}(f_1(\sigma,\xi_1(\sigma))-f_1(\sigma,\xi_2(\sigma)))d\sigma\Bigg|\Bigg|^2\Bigg)\\
&+ \frac{2}{\Gamma(\varphi)^2}\mathbb{E}\Bigg(\Bigg|\Bigg|\int_a^\eta \varsigma'(\sigma)(\varsigma(\eta)-\varsigma(\sigma))^{\varphi-1}(f_2(\sigma,\xi_1(\sigma))-f_2(\sigma,\xi_2(\sigma)))dW(\sigma)\Bigg|\Bigg|^2\Bigg).
\end{aligned}$$

Using Hölder inequality, one has

$$\begin{aligned}
&\mathbb{E}\Bigg(\Bigg|\Bigg|\int_a^\eta \varsigma'(\sigma)(\varsigma(\eta)-\varsigma(\sigma))^{\varphi-1}(f_1(\sigma,\xi_1(\sigma))-f_1(\sigma,\xi_2(\sigma)))d\sigma\Bigg|\Bigg|^2\Bigg)\\
&\leq L^2M(\eta-a)\int_a^\eta \varsigma'(\sigma)(\varsigma(\eta)-\varsigma(\sigma))^{2\varphi-2}\mathbb{E}\Big(||\xi_1(\sigma)-\xi_2(\sigma)||^2\Big)d\sigma.
\end{aligned}$$

Moreover, using Itô isometry, we have

$$\begin{aligned}
&\mathbb{E}\Bigg(\Bigg|\Bigg|\int_a^\eta \varsigma'(\sigma)(\varsigma(\eta)-\varsigma(\sigma))^{\varphi-1}(f_2(\sigma,\xi_1(\sigma))-f_2(\sigma,\xi_2(\sigma)))dW(\sigma)\Bigg|\Bigg|^2\Bigg)\\
&= \mathbb{E}\Big(\int_a^\eta (\varsigma'(\sigma))^2(\varsigma(\eta)-\varsigma(\sigma))^{2\varphi-2}||f_2(\sigma,\xi_1(\sigma))-f_2(\sigma,\xi_2(\sigma))||^2 d\sigma\Big)\\
&\leq L^2M\int_a^\eta \varsigma'(\sigma)(\varsigma(\eta)-\varsigma(\sigma))^{2\varphi-2}\mathbb{E}\Big(||\xi_1(\sigma)-\xi_2(\sigma)||^2\Big)d\sigma. \quad (17)
\end{aligned}$$

Then,

$$\mathbb{E}\Big(||N_\delta\xi_1(\eta)-N_\delta\xi_2(\eta)||^2\Big)$$

$$\begin{aligned}
&\leq \frac{2L^2M}{\Gamma(\varphi)^2}(\Pi+1)\int_a^\eta \varsigma'(\sigma)(\varsigma(\eta)-\varsigma(\sigma))^{2\varphi-2}\mathbb{E}\Big(||\xi_1(\sigma)-\xi_2(\sigma)||^2\Big)d\sigma \\
&= \frac{2L^2M}{\Gamma(\varphi)^2}(\Pi+1)\int_a^\eta \varsigma'(\sigma)(\varsigma(\eta)-\varsigma(\sigma))^{2\varphi-2}\frac{\mathbb{E}\Big(||\xi_1(\sigma)-\xi_2(\sigma)||^2\Big)}{e^{\theta(\varsigma(\sigma)-\varsigma(a))}}e^{\theta(\varsigma(\sigma)-\varsigma(a))}d\sigma \\
&\leq \frac{2L^2M}{\Gamma(\varphi)^2}(\Pi+1)||\xi_1-\xi_2||_\theta^2\int_a^\eta \varsigma'(\sigma)(\varsigma(\eta)-\varsigma(\sigma))^{2\varphi-2}e^{\theta(\varsigma(\sigma)-\varsigma(a))}d\sigma. \qquad (18)
\end{aligned}$$

Set $J=\int_a^\eta \varsigma'(\sigma)(\varsigma(\eta)-\varsigma(\sigma))^{2\varphi-2}e^{\theta(\varsigma(\sigma)-\varsigma(a))}d\sigma$. Thus, by using Lemma 2.6 in [16], we get

$$J \leq \frac{\Gamma(2\varphi-1)}{\theta^{2\varphi-1}}e^{\theta(\varsigma(\eta)-\varsigma(a))}. \qquad (19)$$

Therefore, we have

$$\frac{\mathbb{E}\Big(||N_\delta\xi_1(\eta)-N_\delta\xi_2(\eta)||^2\Big)}{e^{\theta(\varsigma(\eta)-\varsigma(a))}} \leq \frac{2L^2M}{\Gamma(\varphi)^2}(\Pi+1)\frac{\Gamma(2\varphi-1)}{\theta^{2\varphi-1}}||\xi_1-\xi_2||_\theta^2. \qquad (20)$$

Hence,

$$||N_\delta\xi_1-N_\delta\xi_2||_\theta \leq C||\xi_1-\xi_2||_\theta, \qquad (21)$$

where $C=\sqrt{\frac{2L^2M}{\Gamma(\varphi)^2}(\Pi+1)\frac{\Gamma(2\varphi-1)}{\theta^{2\varphi-1}}}$. Therefore, there is a unique solution of (4) such that $\xi(a)=\delta$. □

4. Hyers–Ulam Stability

In this section, we study the Hyers–Ulam stability of Equation (4) using the generalized Gronwall inequality and the stochastic calculus techniques.

Definition 5. *Equation (4) is Hyers–Ulam stable with respect to ϵ if there is a number $M_1>0$ satisfying for each $\epsilon>0$, and for each solution $y\in\mathbb{H}^2([a,\Pi])$, with $y(a)=\delta$, of the following inequality:*

$$\mathbb{E}\left|\left|y(\eta)-y(a)-\left(\int_a^\eta \frac{\varsigma'(\sigma)(\varsigma(\eta)-\varsigma(\sigma))^{\varphi-1}}{\Gamma(\varphi)}(f_1(\sigma,y(\sigma))d\sigma+f_2(\sigma,y(\sigma))dW(\sigma))\right)\right|\right|^2 \leq \epsilon, \qquad (22)$$

for all $\eta\in[a,\Pi]$, there exists a solution $\xi\in\mathbb{H}^2([a,\Pi])$ of (4), with $\xi(a)=\delta$, such that

$$\mathbb{E}||y(\eta)-\xi(\eta)||^2 \leq M_1\epsilon,\ \forall\eta\in[a,\Pi].$$

Theorem 3. *Under Assumptions $\mathcal{H}_1$-$\mathcal{H}_2$, the ς-Caputo stochastic fractional differential Equation (4) are Hyers–Ulam stable with respect to ϵ on $[a,\Pi]$.*

Proof. Let $\epsilon>0$ and $y\in\mathbb{H}^2([a,\Pi])$ be a function satisfying (22) and denote by $\xi\in\mathbb{H}^2([a,\Pi])$ the solution of (4) with initial data $y(a)$; thus

$$\xi(\eta)=y(a)+\frac{1}{\Gamma(\varphi)}\left[\int_a^\eta \varsigma'(\sigma)(\varsigma(\eta)-\varsigma(\sigma))^{\varphi-1}(f_1(\sigma,\xi(\sigma))d\sigma+f_2(\sigma,\xi(\sigma))dW(\sigma))\right]. \qquad (23)$$

Thus,

$$\begin{aligned}
&\mathbb{E}||y(\eta)-\xi(\eta)||^2 \\
&\leq 2\mathbb{E}||y(\eta)-y(a)-\frac{1}{\Gamma(\varphi)}(\int_a^\eta \varsigma'(\sigma)(\varsigma(\eta)-\varsigma(\sigma))^{\varphi-1}[f_1(\sigma,y(\sigma))d\sigma \\
&+f_2(\sigma,y(\sigma))dW(\sigma)])||^2
\end{aligned}$$

$$+2\mathbb{E}||\frac{1}{\Gamma(\varphi)}(\int_a^\eta \varsigma'(\sigma)(\varsigma(\eta)-\varsigma(\sigma))^{\varphi-1}[(f_1(\sigma,y(\sigma))-f_1(\sigma,\xi(\sigma)))d\sigma$$
$$+(f_2(\sigma,y(\sigma))-f_2(\sigma,\xi(\sigma)))dW(\sigma)])||^2.$$

Then, applying assumptions $\mathcal{H}_1$-$\mathcal{H}_2$ and Cauchy–Schwartz inequality, we have

$$\mathbb{E}||y(\eta)-\xi(\eta)||^2$$

$$\begin{aligned}
&\leq 2\epsilon+4\mathbb{E}\left|\left|\frac{1}{\Gamma(\varphi)}\int_a^\eta \varsigma'(\sigma)(\varsigma(\eta)-\varsigma(\sigma))^{\varphi-1}(f_1(\sigma,y(\sigma))-f_1(\sigma,\xi(\sigma)))d\sigma\right|\right|^2\\
&+ 4\mathbb{E}\left|\left|\frac{1}{\Gamma(\varphi)}\int_a^\eta \varsigma'(\sigma)(\varsigma(\eta)-\varsigma(\sigma))^{\varphi-1}(f_2(\sigma,y(\sigma))-f_2(\sigma,\xi(\sigma)))dW(\sigma)\right|\right|^2\\
&\leq 2\epsilon+\frac{4L^2M(\varsigma(\eta)-\varsigma(a))^{2\varphi-1}}{(2\varphi-1)\Gamma(\varphi)^2}\mathbb{E}\left(\int_a^\eta ||y(\sigma)-\xi(\sigma)||^2d\sigma\right)\\
&+ \frac{4L^2M}{\Gamma(\varphi)^2}\mathbb{E}\left(\int_a^\eta \varsigma'(\sigma)(\varsigma(\eta)-\varsigma(\sigma))^{2\varphi-2}||y(\sigma)-\xi(\sigma)||^2d\sigma\right).
\end{aligned}$$

Then,

$$\begin{aligned}
\mathbb{E}||y(\eta)-\xi(\eta)||^2 &\leq 2\epsilon+\frac{4L^2M(\varsigma(\Pi)-\varsigma(a))^{2\varphi-1}}{(2\varphi-1)\Gamma(\varphi)^2}\int_a^\eta \mathbb{E}||y(\sigma)-\xi(\sigma)||^2d\sigma\\
&+ \frac{4L^2M}{\Gamma(\varphi)^2}\int_a^\eta \varsigma'(\sigma)(\varsigma(\eta)-\varsigma(\sigma))^{2\varphi-2}\mathbb{E}||y(\sigma)-\xi(\sigma)||^2d\sigma. \quad (24)
\end{aligned}$$

Set $z(\eta)=\mathbb{E}||y(\eta)-\xi(\eta)||^2$. Thus, one gets

$$z(\eta)\leq \alpha_1+\alpha_2\int_a^\eta z(\sigma)d\sigma+\alpha_3\int_a^\eta \varsigma'(\sigma)(\varsigma(\eta)-\varsigma(\sigma))^{2\varphi-2}z(\sigma)d\sigma, \quad (25)$$

where $\alpha_1=2\epsilon$, $\alpha_2=\frac{4L^2M(\varsigma(\Pi)-\varsigma(a))^{2\varphi-1}}{(2\varphi-1)\Gamma(\varphi)^2}$ and $\alpha_3=\frac{4L^2M}{\Gamma(\varphi)^2}$.

Applying the generalized Gronwall inequality (see [31]), we have

$$\begin{aligned}
z(\eta) &\leq \left[\alpha_1+\alpha_2\int_a^\eta z(\sigma)d\sigma\right]E_{2\varphi-1}\left(\alpha_3\Gamma(2\varphi-1)(\varsigma(\eta)-\varsigma(a))^{2\varphi-1}\right)\\
&\leq \alpha_4+\alpha_5\int_a^\eta z(\sigma)d\sigma, \quad (26)
\end{aligned}$$

where $\alpha_4=2\epsilon E_{2\varphi-1}\left(\alpha_3\Gamma(2\varphi-1)(\varsigma(\Pi)-\varsigma(a))^{2\varphi-1}\right)$ and $\alpha_5=\frac{4L^2M(\varsigma(\Pi)-\varsigma(a))^{2\varphi-1}}{(2\varphi-1)\Gamma(\varphi)^2}$ $E_{2\varphi-1}\left(\alpha_3\Gamma(2\varphi-1)(\varsigma(\Pi)-\varsigma(a))^{2\varphi-1}\right)$.

Applying the classical Gronwall inequality, we can derive that

$$z(\eta)\leq \alpha_4e^{\alpha_5(\eta-a)}\leq \alpha_4e^{\alpha_5(\Pi-a)}. \quad (27)$$

Hence,

$$z(\eta)\leq M_1\epsilon, \quad (28)$$

where $M_1=2E_{2\varphi-1}\left(\alpha_3\Gamma(2\varphi-1)(\varsigma(\Pi)-\varsigma(a))^{2\varphi-1}\right)e^{\alpha_5(\Pi-a)}$.

Therefore, Equation (4) is Hyers–Ulam stable with respect to ϵ. □

5. Examples

This section is devoted to show our results in three examples.

Example 1. *Let the CSFDE for each $\epsilon > 0$ and for $\eta \in [1, e^2]$, given by*

$$\begin{cases} {}^{C}D_{1^+}^{\frac{2}{3},\varsigma}\xi(\eta) = f_1(\eta, \xi(\eta)) + f_2(\eta, \xi(\eta))\frac{dW(\eta)}{d\eta}, \\ \mathbb{E}\left|y(\eta) - y(1) - \frac{1}{\Gamma(\varphi)}\left(\int_1^\eta \varsigma'(\sigma)(\varsigma(\eta) - \varsigma(\sigma))^{-\frac{1}{3}}(f_1(\sigma, y(\sigma))d\sigma + f_2(\sigma, y(\sigma))dW(\sigma))\right)\right|^2 \leq \epsilon, \\ y(1) = \delta, \end{cases} \tag{29}$$

where $\varphi = \frac{2}{3}$, $\varsigma(\eta) = \ln(\eta)$ and

$$\begin{aligned} \xi(\eta) &\in \mathbb{H}^2([1, e^2], \mathbb{R}) \\ f_1(\eta, \xi(\eta)) &= \sqrt{\ln(\eta)}(\arctan(\xi(\eta)) + \cos(\xi(\eta))) \\ f_2(\eta, \xi(\eta)) &= \sqrt{\eta}\cos(\xi(\eta)). \end{aligned}$$

We will prove that Equation (29) is Hyers–Ulam stable with respect to ϵ.

Let $(\eta, \xi_1, \xi_2) \in [1, e^2] \times \mathbb{R} \times \mathbb{R}$, thus

$$|f_1(\eta, \xi_1) - f_1(\eta, \xi_2)| \leq 4|\xi_1 - \xi_2|,$$

and

$$|f_2(\eta, \xi_1) - f_2(\eta, \xi_2)| \leq e|\xi_1 - \xi_2|.$$

Hence, assumption $\mathcal{H}_1$ fulfilled. Moreover,

$$||f_2(\cdot, 0)||_\infty = \text{ess} \sup_{\eta \in [1, e^2]} |f_2(\eta, 0)| \leq e,$$

and

$$\int_1^{e^2} |f_1(\eta, 0)|^2 d\eta \leq 2(e^2 + 1).$$

Thus, assumptions $\mathcal{H}_1$-$\mathcal{H}_2$ fulfilled. Hence, applying Theorem 3, Equation (29) has a unique solution, and it is Hyers–Ulam stable with respect to ϵ on $[1, e^2]$.

Example 2. *Let the CSFDE for each $\epsilon > 0$ and for $\eta \in [0.5, 6]$, given by*

$$\begin{cases} {}^{C}D_{1^+}^{\frac{3}{4},\varsigma}\xi(\eta) = f_1(\eta, \xi(\eta)) + f_2(\eta, \xi(\eta))\frac{dW(\eta)}{d\eta}, \\ \mathbb{E}\left|y(\eta) - y(0.5) - \frac{1}{\Gamma(\varphi)}\left(\int_{0.5}^\eta \varsigma'(\sigma)(\varsigma(\eta) - \varsigma(\sigma))^{-\frac{1}{4}}(f_1(\sigma, y(\sigma))d\sigma + f_2(\sigma, y(\sigma))dW(\sigma))\right)\right|^2 \leq \epsilon, \\ y(0.5) = \delta, \end{cases} \tag{30}$$

where $\varphi = \frac{3}{4}$, $\varsigma(\eta) = \sqrt{\eta}$ and

$$\begin{aligned} \xi(\eta) &\in \mathbb{H}^2([0.5, 6], \mathbb{R}) \\ f_1(\eta, \xi(\eta)) &= \frac{e^\eta}{1 + e^\eta}(1 + \xi(\eta)) \\ f_2(\eta, \xi(\eta)) &= \frac{1 + \sin(\xi(\eta))}{(1 + \eta)^2}. \end{aligned}$$

We will prove that Equation (31) is Hyers–Ulam stable with respect to ϵ.

$(\eta, \xi_1, \xi_2) \in [0.5, 6] \times \mathbb{R} \times \mathbb{R}$, then

$$|f_1(\eta, \xi_1) - f_1(\eta, \xi_2)| \leq |\xi_1 - \xi_2|,$$

and

$$|f_2(\eta, \xi_1) - f_2(\eta, \xi_2)| \leq |\xi_1 - \xi_2|.$$

Thus, assumption $\mathcal{H}_1$ holds. On the other hand,

$$||f_2(\cdot, 0)||_\infty = \text{ess} \sup_{\eta \in [0.5, 6]} |f_2(\eta, 0)| \leq 1,$$

and

$$\int_{0.5}^{6} |f_1(\eta, 0)|^2 d\eta \leq \ln(1 + e^6).$$

Then, assumptions $\mathcal{H}_1$-$\mathcal{H}_2$ are fulfilled. Hence, applying Theorem 3, Equation (31) has a unique solution, and it is Hyers–Ulam stable with respect to ϵ on $[0.5, 6]$.

Example 3. *Let the CSFDE, for each $\epsilon > 0$ and for $\eta \in [0, 5]$, given by*

$$\begin{cases} {}^{C}D_{0^+}^{\frac{1}{5}, \varsigma} \xi(\eta) = f_1(\eta, \xi(\eta)) + f_2(\eta, \xi(\eta)) \frac{dW(\eta)}{d\eta}, \\ \mathbb{E}\Big| y(\eta) - y(0) - \frac{1}{\Gamma(\varphi)} \Big(\int_0^\eta \varsigma'(\sigma)(\varsigma(\eta) - \varsigma(\sigma))^{-\frac{4}{5}} (f_1(\sigma, y(\sigma)) d\sigma + f_2(\sigma, y(\sigma)) dW(\sigma)) \Big) \Big|^2 \leq \epsilon, \\ y(0) = \delta, \end{cases} \tag{31}$$

where $\varphi = \frac{1}{5}$, $\varsigma(\eta) = \eta$ and

$$\begin{aligned} \xi(\eta) &\in \mathbb{H}^2([0, 5], \mathbb{R}) \\ f_1(\eta, \xi(\eta)) &= 2e^{-\eta} \xi(\eta) \\ f_2(\eta, \xi(\eta)) &= 3 \sin(\xi(\eta)). \end{aligned}$$

We will prove that Equation (31) is Hyers–Ulam stable with respect to ϵ.

$(\eta, \xi_1, \xi_2) \in [0, 5] \times \mathbb{R} \times \mathbb{R}$, then

$$|f_1(\eta, \xi_1) - f_1(\eta, \xi_2)| \leq 2|\xi_1 - \xi_2|,$$

and

$$|f_2(\eta, \xi_1) - f_2(\eta, \xi_2)| \leq 3|\xi_1 - \xi_2|.$$

Thus, assumption $\mathcal{H}_1$ hold. On the other hand,

$$||f_2(\cdot, 0)||_\infty = \text{ess} \sup_{\eta \in [0, 5]} |f_2(\eta, 0)| = 0,$$

and

$$\int_0^5 |f_1(\eta, 0)|^2 d\eta = 0.$$

Then, assumptions $\mathcal{H}_1$-$\mathcal{H}_2$ are fulfilled. Hence, applying Theorem 3, Equation (31) has a unique solution, and it is Hyers–Ulam stable with respect to ϵ on $[0, 5]$.

6. Conclusions

In this research paper, we have proved the existence and uniqueness of CSFDE. We have simulated the HUS for the proposed problem with the help of the novel features of stochastic calculus. We have illustrated three examples to justify the correctness and applicability of the proposed results. The applications of some well-known terms of functional analysis, such as the Cauchy–Schwarz inequality, properties of measurable functions, supremum norm, Itô's isometry formula, Hölder inequality, and generalized Gronwall inequality make the study more visible to the literature. The proposed results will be very useful to prove the existence of a unique solution and Hyers–Ulam stability of ς-Caputo type fractional stochastic differential equations.

Author Contributions: Formal analysis, L.M. and A.B.M.; writing—original draft preparation, O.K. and N.B.A.; visualization, P.K. and A.A. All authors have read and agreed to the published version of the manuscript.

Funding: This research has been funded by the Scientific Research Deanship at the University of Ha'il—Saudi Arabia through project number RG-21 159.

Institutional Review Board Statement: Not applicable.

Informed Consent Statement: Not applicable.

Data Availability Statement: Not applicable.

Conflicts of Interest: The authors declare no conflict of interest.

References

1. Kilbas, A.A.; Srivastava, H.M.; Trujillo, J.J. *Theory and Applications of Fractional Differential Equations*; Elsevier: Amsterdam, The Netherlands, 2006.
2. Podlubny, I. *Fractional Differential Equations*; Academic Press: San Diego, CA, USA, 1999.
3. Baleanu, D.; Machado, J.A.; Luo, A.C. *Fractional Dynamics and Control*; Springer Science and Business Media: New York, NY, USA, 2011.
4. Koeller, R. Applications of fractional calculus to the theory of viscoelasticity. *ASME J. Appl. Mech.* **1984**, *51*, 299–307. [CrossRef]
5. Li, C.P.; Zeng, F.H. *Numerical Methods for Fractional Calculus*; Chapman and Hall/CRC Press: Boca Raton, FL, USA, 2015.
6. Oussaeif, T.E.; Antara, B.; Ouannas, A.; Batiha, I.M.; Saad, K.M.; Jahanshahi, H.; Aljuaid, A.M.; Aly, A.A. Existence and Uniqueness of the Solution for an Inverse Problem of a Fractional Diffusion Equation with Integral Condition. *J. Funct. Spaces* **2020**, *2020*, 7667370. [CrossRef]
7. Batiha, I.M.; Ouannas, A.; Albadarneh, R.; Al-Nana, A.A.; Momani, S. Existence and uniqueness of solutions for generalized Sturm–Liouville and Langevin equations via Caputo–Hadamard fractional-order operator. *Eng. Comput.* **2022**, *39*, 2581–2603. [CrossRef]
8. Chen, L.; Chai, Y.; Wu, R.C.; Ma, T.D.; Zha, H.Z. Dynamic analysis of a class of fractional-order neural networks with delay. *Neurocomputing* **2013**, *111*, 190–194. [CrossRef]
9. Li, Y.L.; Pan, C.; Meng, X.; Ding, Y.Q.; Chen, X.H. A method of approximate fractional order differentiation with noise immunity. *Chemom. Intell. Lab. Syst.* **2015**, *144*, 31–38. [CrossRef]
10. Li, C.P.; Tao, C.X. On the fractional Adams method. *Comput. Math. Appl.* **2009**, *58*, 1573–1588. [CrossRef]
11. Almeida, R.; Malinowska, A.B.; Monteiro, M.T.T. Fractional differential equations with a Caputo derivative with respect to a kernel function and their applications. *Math. Methods Appl. Sci.* **2018**, *41*, 336–352. [CrossRef]
12. Jarad, F.; Abdeljawad, T.; Baleanu, D. Caputo-type modification of the Hadamard fractional derivatives. *Adv. Differ. Equ.* **2012**, *142*, 1–10. [CrossRef]
13. Abdo, M.S.; Panchal, S.K.; Saeed, A.M. Fractional boundary value problem with ψ-Caputo fractional derivative. *Proc. Indian Acad. Sci. Math. Sci.* **2019**, *129*, 157–169. [CrossRef]
14. Almeida, R. Further properties of Osler's generalized fractional integrals and derivatives with respect to another function. *Rocky Mt. J. Math.* **2019**, *49*, 2459–2493. [CrossRef]
15. Almeida, R.; Jleli, M.; Samet, B. A numerical study of fractional relaxation-oscillation equations involving ψ-Caputo fractional derivative. *Rev. R. Acad. Cienc. Exactas Fís Nat. Ser. Mat.* **2019**, *113*, 1873–1891. [CrossRef]
16. Baitiche, Z.; Derbazi, C.; Matar, M.M. Ulam stability for nonlinear-Langevin fractional differential equations involving two fractional orders in the ψ-Caputo sense. *Appl. Anal.* **2022**, *101*, 4866–4881. [CrossRef]
17. Derbazi, C.; Baitiche, Z.; Benchohra, M.; Cabada, A. Initial value problem for nonlinear fractional differential equations with ψ-Caputo derivative via monotone iterative technique. *Axioms* **2020**, *9*, 57. [CrossRef]
18. Abbas, S.; Benchohra, M.; Lazreg, J.E.; Zhou, Y. A survey on Hadamard and Hilfer fractional differential equations: Analysis and stability. *Chaos Solitons Fractals* **2017**, *102*, 47–71. [CrossRef]
19. Guo, Y.; Shu, X.B.; Li, Y.; Xu, F. The existence and Hyers–Ulam stability of solution for an impulsive Riemann–Liouville fractional neutral functional stochastic differential equation with infinite delay of order $1 < \beta < 2$. *Bound. Value Probl.* **2019**, *59*, 1–11.
20. Hyers, D.H. On the stability of the linear functional equation. *Proc. Natl. Acad. Sci. USA* **1941**, *27*, 222–224. [CrossRef] [PubMed]

21. Guo, Y.; Chen, M.; Shu, X.B.; Xu, F. The existence and Hyers-Ulam stability of solution for almost periodical fractional stochastic differential equation with fBm. *Stoch. Anal. Appl.* **2021**, *39*, 643–666. [CrossRef]
22. Li, S.; Shu, L.; Shu, X.B.; Xu, F. Existence and Hyers-Ulam stability of random impulsive stochastic functional differential equations with finite delays. *Stochastics* **2019**, *91*, 857–872. [CrossRef]
23. Long, H.V.; Son, N.T.K.; Tam, H.T.T.; Yao, J.C. Ulam stability for fractional partial integrodi differential equation with uncertainty. *Acta Math. Vietnam.* **2017**, *42*, 675–700. [CrossRef]
24. Derbazi, C.; Baitiche, Z.; Benchohra, M.; N'guérékata, G. Existence, uniqueness, approximation of solutions and Ealpha-Ulam stability results for a class of nonlinear fractional differential equations involving psi-Caputo derivative with initial conditions. *Math. Moravica* **2021**, *25*, 1–30. [CrossRef]
25. Boutiara, A.; Abdo, M.S.; Benbachir, M. Existence results for ψ-Caputo fractional neutral functional integro-differential equations with finite delay. *Turk. J. Math.* **2020**, *44*, 2380–2401. [CrossRef]
26. Derbazi, C.; Baitiche, Z. Uniqueness and Ulam–Hyers–Mittag–Leffler stability results for the delayed fractional multiterm differential equation involving the ϕ-Caputo fractional derivative. *Rocky Mt. J. Math.* **2020**, *52*, 887–897.
27. Ahmadova, A.; Mahmudov, N.I. Ulam-Hyers stability of Caputo type fractional stochastic neutral differential equations. *Stat. Probab. Lett.* **2021**, *168*, 108949. [CrossRef]
28. Ben Makhlouf, A.; Mchiri, L. Some results on the study of Caputo-Hadamard fractional stochastic differential equations. *Chaos Solitons Fractals* **2022**, *155*, 111757. [CrossRef]
29. Doan, T.S.; Huong, P.T.; Kloeden, P.; Tuan, H.T. Asymptotic separation between solutions of Caputo fractional stochastic differential equations. *Stoch. Anal. Appl.* **2018**, *36*, 1–11.
30. Diaz, J.B.; Margolis, B. A fixed point theorem of the alternative, for contractions on a generalized complete metric space. *Bull. Am. Math. Soc.* **1968**, *74*, 305–309. [CrossRef]
31. Vanterler da Sousa, J.; Capelas de Oliveira, E. A Gronwall inequality and the Cauchy-type problem by means of ψ-Hilfer operator. *arXiv* **2017**, arXiv:1709.03634.

symmetry MDPI

Article

On the Generalized Liouville–Caputo Type Fractional Differential Equations Supplemented with Katugampola Integral Boundary Conditions

Muath Awadalla [1,*], Muthaiah Subramanian [2], Kinda Abuasbeh [1] and Murugesan Manigandan [3]

1 Department of Mathematics and Statistics, College of Science, King Faisal University, Hafuf, Al Ahsa 31982, Saudi Arabia
2 Department of Mathematics, KPR Institute of Engineering and Technology, Coimbatore 641407, Tamil Nadu, India
3 Department of Mathematics, Sri Ramakrishna Mission Vidyalaya College of Arts and Science, Coimbatore 641020, Tamil Nadu, India
* Correspondence: mawadalla@kfu.edu.sa

Abstract: In this study, we examine the existence and Hyers–Ulam stability of a coupled system of generalized Liouville–Caputo fractional order differential equations with integral boundary conditions and a connection to Katugampola integrals. In the first and third theorems, the Leray–Schauder alternative and Krasnoselskii's fixed point theorem are used to demonstrate the existence of a solution. The Banach fixed point theorem's concept of contraction mapping is used in the second theorem to emphasise the analysis of uniqueness, and the results for Hyers–Ulam stability are established in the next theorem. We establish the stability of Ulam–Hyers using conventional functional analysis. Finally, examples are used to support the results. When a generalized Liouville–Caputo (ρ) parameter is modified, asymmetric results are obtained. This study presents novel results that significantly contribute to the literature on this topic.

Keywords: generalized fractional derivatives; generalized fractional integrals; coupled system; existence; fixed point

MSC: 34A08; 34B10; 34D10

Citation: Awadalla, M.; Subramanian, M.; Abuasbeh, K.; Manigandan, M. On the Generalized Liouville–Caputo Type Fractional Differential Equations Supplemented with Katugampola Integral Boundary Conditions. *Symmetry* **2022**, *14*, 2273. https://doi.org/10.3390/sym14112273

Academic Editor: Anton V. Krysko

Received: 10 October 2022
Accepted: 26 October 2022
Published: 29 October 2022

1. Introduction

We consider the nonlinear coupled fractional differential equations with generalized Liouville–Caputo derivatives

$$\begin{cases} {}^{\rho}_{C}\mathcal{D}^{\xi}_{0+}p(\tau) = f(\tau, p(\tau), q(\tau)), \tau \in \mathcal{G} := [0, \mathcal{T}], \\ {}^{\rho}_{C}\mathcal{D}^{\zeta}_{0+}q(\tau) = g(\tau, p(\tau), q(\tau)), \tau \in \mathcal{G} := [0, \mathcal{T}], \end{cases} \quad (1)$$

enhanced with boundary conditions which are defined by:

$$\begin{cases} p(0) = 0, \quad q(0) = 0, \\ p(\mathcal{T}) = \epsilon^{\rho}\mathcal{I}^{\varsigma}_{0+}q(\varpi) = \frac{\epsilon\rho^{1-\varsigma}}{\Gamma(\varsigma)}\int_0^{\varpi}\frac{\theta^{\rho-1}}{(\varpi^{\rho}-\theta^{\rho})^{1-\varsigma}}q(\theta)d\theta, \\ q(\mathcal{T}) = \pi^{\rho}\mathcal{I}^{\varrho}_{0+}p(\sigma) = \frac{\pi\rho^{1-\varrho}}{\Gamma(\varrho)}\int_0^{\sigma}\frac{\theta^{\rho-1}}{(\sigma^{\rho}-\theta^{\rho})^{1-\varrho}}p(\theta)d\theta, \\ 0 < \sigma < \varpi < \mathcal{T}, \end{cases} \quad (2)$$

where ${}^{\rho}_{C}\mathcal{D}^{\xi}_{0+}$, ${}^{\rho}_{C}\mathcal{D}^{\zeta}_{0+}$ are the Liouville–Caputo-type generalized fractional derivative of order $1 < \xi, \zeta \le 2$, ${}^{\rho}_{C}\mathcal{I}^{\varsigma}_{0+}$, ${}^{\rho}_{C}\mathcal{I}^{\varrho}_{0+}$ are the generalized fractional integral of order (Katugampola type) $\varrho, \varsigma > 0, \rho > 0, f, g : \mathcal{G} \times \mathbb{R} \times \mathbb{R} \to \mathbb{R}$ are continuous functions, $\epsilon, \pi \in \mathbb{R}$. The strip conditions states that the value of the unknown function at the right end point $\tau = T$ of the given interval is proportional to the values of the unknown function on the strips of varying lengths. When $\rho = 1$, the generalized Liouville–Caputo equation is changed to the Caputo sense, which leads to asymmetric results. In a similar way, when $\rho = 1$, the Katugampola integrals are changed to Riemann-Liouville integrals, which leads to cases that are not symmetric. To the best of our knowledge, the stability analysis of boundary value problems (BVPs) is still in its early stages. This paper's primary contribution is to study existence and Ulam-Hyers stability analysis. In addition, we demonstrate the problem (1)–(2) employed by Leray–Schauder, Banach and Krasnoselskii's fixed point theorems to prove the existence and uniqueness of solutions. The system (1) is the well-known fractional-order coupled logistic system [1]:

$$\begin{cases} \mathcal{D}^{\alpha}u(\tau) = r_1 u(\tau) - \frac{r_1}{k_1}u(\tau)(u(\tau)+v(\tau)), \tau \in I, \\ \mathcal{D}^{\beta}v(\tau) = r_2 v(\tau) - \frac{r_2}{k_2}v(\tau)(v(\tau)+u(\tau)), \end{cases}$$

and the Lotka–Volterra prey-predator system [1]:

$$\begin{cases} \mathcal{D}^{\alpha}u(\tau) = u(\tau)(a - u(\tau)E - \gamma v(\tau)), \tau \in I, \\ \mathcal{D}^{\beta}v(\tau) = v(\tau)(-b + \gamma E v(\tau) - \beta E). \end{cases}$$

We now provide some recent results related to our problem (1)–(2). In [2], the authors discussed the existence results for coupled system of fractional differential equations Riemann–Liouville derivatives

$$\begin{cases} \mathcal{D}^{\alpha_1}_{0+}(\mathcal{D}^{\beta_1}_{0+}x(t)) + f(t,x(t),y(t)), t \in [0,1], \\ \mathcal{D}^{\alpha_2}_{0+}(\mathcal{D}^{\beta_2}_{0+}y(t)) + f(t,x(t),y(t)), t \in [0,1], \end{cases} \tag{3}$$

with the Riemann–Stieltjes integral boundary conditions:

$$\begin{cases} \mathcal{D}^{\beta_1}_{0+}x(0) = 0,\ x(0) = 0,\ \mathcal{D}^{\beta_2}_{0+}y(0) = 0,\ y(0) = 0, \\ x(1) = \gamma_1 \mathcal{I}^{\delta_1}_{0+}y(\xi) + \sum_{i=1}^{p}\int_0^1 y(\tau)d\mathcal{H}_i(\tau), \\ y(1) = \gamma_2 \mathcal{I}^{\delta_2}_{0+}x(\eta) + \sum_{j=1}^{q}\int_0^1 x(\tau)d\mathcal{K}_i(\tau), \end{cases} \tag{4}$$

where α_1 is in the interval $(0,1)$, β_1 is in the interval $(1,2)$, α_2 is in the interval $(0,1]$, β_2 is in the interval $(1,2]$, $p,q \in N$, and $\gamma_1,\ \gamma_2,\ \delta_1,\ \delta_2 > 0$, $0 < \xi,\eta < 1$ $\mathcal{K}_j(t)$, $j = 1,\ldots,q$, $\mathcal{H}_i(t), i = 1,\ldots,p$ are bounded variation functions. Both function f and g are nonlinear. They used several theorems from fixed point index theory to prove the main results. In [3], the authors investigated existence of solutions for coupled system of fractional differential equations with Hilfer derivatives

$$\begin{cases} ({}^H\mathcal{D}^{\alpha_1,\beta_1}_{0+}x)(t) + \lambda_1({}^H\mathcal{D}^{\alpha_1-1,\beta_1}_{0+}x)(t) = f(t,x(t),R^{(\delta_q,\ldots,\delta_1)}x(t),y(t)), t \in [0,T], \\ ({}^H\mathcal{D}^{\alpha_2,\beta_2}_{0+}y)(t) + \lambda_2({}^H\mathcal{D}^{\alpha_2-1,\beta_2}_{0+}y)(t) = f(t,x(t),y(t),R^{(\zeta_q,\ldots,\zeta_1)}y(t)), t \in [0,T], \end{cases} \tag{5}$$

with Riemann–Liouville and Hadamard-type iterated integral boundary conditions:

$$\begin{cases} x(0) = 0,\ y(0) = 0, \\ x(T) = \sum_{i=1}^{m}\epsilon_i R^{(\mu_p,\ldots,\mu_1)}y(\eta_i)\ \eta_i \in (0,T), \\ y(T) = \sum_{j=1}^{n}\theta_j R^{(\nu_p,\ldots,\nu_1)}x(\xi_j)\ \xi_i \in (0,T), \end{cases} \tag{6}$$

where ${}^H\mathcal{D}^{\alpha_l,\beta_l}$ is the Hilfer fractional derivative operator of order α_l with parameters β_l, $l \in 1,2$, $1 < \alpha_l < 2$, $0 \le \beta_l \le 1$, $\lambda_1,\lambda_2,\epsilon_i,\theta_j \in \mathcal{R}\backslash\{0\}$, $i = 1,2,\ldots,m$, $j = 1,2,\ldots,n$, $f,g : [0,T] \times \mathcal{R} \times \mathcal{R} \times \mathcal{R} \to \times\mathcal{R}$ are nonlinear continuous functions and $R^{(\phi_\tau,\ldots,\phi_1)}$, $\phi_r \in \{\delta,\zeta,\mu,\nu\}$, $r \in \{q,p,\rho | q,p,\rho \in \mathcal{N}\}$, involves the iterated Riemann–Liouville and

Hadamard fractional integral operators. They used several theorems from fixed point index theory to prove the main results. Numerous scientific and engineering phenomena are mathematically modelled using fractional order differential and integral operators. The main benefit of adopting these operators is their nonlocality, which enables the description of the materials and processes involved in the history of the phenomenon. As a result, compared to their integer-order counterparts, fractional-order models are more precise and informative. As a result of the extensive use of fractional calculus techniques in a range of real-world occurrences, such as those described in the texts cited [4–8] numerous researchers developed this significant branch of mathematical study. In recent years, a lot of research has been done on fractional differential equations with different boundary conditions. Nonlocal nonlinear fractional-order boundary value problems, in particular, have attracted a lot of attention (BVPs). The idea of nonlocal circumstances, which help to describe physical processes occurring inside the confines of a specific domain, was originally introduced in the work of Bitsadze and Samarski [9]. It is challenging to defend the assumption of a circular cross-section in computational fluid dynamics investigations of blood flow problems because to the changing shape of a blood vessel throughout the vessel. To solve that problem, integral boundary conditions have been developed. In addition, the ill-posed parabolic backward problems are solved under integral boundary conditions. Integral boundary conditions are also essential in mathematical models of bacterial self-regularization, as shown in [10]. Fractional order differential equations, as well as inclusions including Riemann–Liouville, Liouville–Caputo (Caputo), and Hadamard-type derivatives, among others, have all been included in the literature on the topic recently. For some recent works on the topic, we point the reader to several papers [11–15] and the references listed therein. The use of fractional differential systems in mathematical representations of physical and engineering processes has drawn considerable interest. See [16–22] for additional details on the theoretical evolution of such systems. The following is the remainder of the article: Section 2 introduces some fundamental definitions, lemmas, and theorems that support our main results. For the existence and uniqueness of solutions to the given system (1) and (2), we use various conditions and some standard fixed-point theorems in Section 3. Section 4 discusses the Ulam–Hyers stability of the given system (1) and (2) under certain conditions. In Section 6, examples are provided to demonstrate the main results. Finally, the consequences of existence, uniqueness, and stability for the problem (1) and (75) are provided.

2. Preliminaries

For our research, we recall some preliminary definitions of generalized Liouville–Caputo fractional derivatives and Katugampola fractional integrals.

The space of all complex-valued Lebesgue measurable functions ϕ on (c,d) equipped with the norm is denoted by $\mathcal{Z}^q_b(c,d)$:

$$||\phi||_{\mathcal{Z}^q_b} = \left(\int_c^d |z^b\phi(z)|^q \frac{dz}{z}\right)^{\frac{1}{q}} < \infty, b \in \mathbb{R}, 1 \leq q \leq \infty.$$

Let $\mathcal{L}^1(c,d)$ represent the space of all Lebesgue measurable functions φ on (c,d) endowed with the norm:

$$||\varphi||_{\mathcal{L}^1} = \int_c^d |\varphi(z)|dz < \infty.$$

We further recall that $\mathcal{AC}^n(\mathcal{E},\mathbb{R}) = \{p : \mathcal{E} \to \mathbb{R} : p, p', \ldots, p^{(n-1)} \in C(\mathcal{E},\mathbb{R})$ and $p^{(n-1)}$ is absolutely continuous. For $0 \leq \epsilon < 1$, we define $C_{\epsilon,\rho}(\mathcal{E},\mathbb{R}) = \{f : \mathcal{E} \to \mathbb{R} : (\tau^\rho - a^\rho)^\epsilon f(\tau) \in C(\mathcal{E},\mathbb{R})$ endowed with the norm $\|f\|_{C_{\epsilon,\rho}} = \|(\tau^\rho - a^\rho)^\epsilon f(\tau)\|_C$. Moreover, we define the class of functions f that have absolute continuous δ^{n-1} derivative, denoted by

$\mathcal{AC}^n_\gamma(\mathcal{E},\mathbb{R})$, as follows: $\mathcal{AC}^n_\gamma(\mathcal{E},\mathbb{R}) = \{f : \mathcal{E} \to \mathbb{R} : \gamma^{n-1} f \in \mathcal{AC}(\mathcal{E},\mathbb{R}), \gamma = \tau^{1-\rho}\frac{d}{d\tau}\}$, which is equipped with the norm $\|f\|_{\mathcal{C}^n_{\gamma,\epsilon}} = \sum_{k=0}^{n-1} \|\gamma^k f\|_{\mathcal{C}} + \|\gamma^n f\|_{\mathcal{C}_{\epsilon,\rho}}$ is defined by

$$\mathcal{C}^n_{\gamma,\epsilon}(\mathcal{E},\mathbb{R}) = \left\{f : \mathcal{E} \to \mathbb{R} : \gamma^{n-1} f \in \mathcal{C}(\mathcal{E},\mathbb{R}), \gamma^n f \in \mathcal{C}_{\epsilon,\rho}(\mathcal{E},\mathbb{R}), \gamma = \tau^{1-\rho}\frac{d}{d\tau}\right\}.$$

Notice that $\mathcal{C}^n_{\gamma,0} = \mathcal{C}^n_\gamma$. We define space $\mathcal{P} = \{p(\tau) : p(\tau) \in \mathcal{C}(\mathcal{E},\mathbb{R})\}$ equipped with the norm $||p|| = \sup\{|p(\tau)|, \tau \in \mathcal{E}\}$- this is a Banach space. Furthermore $\mathcal{Q} = \{q(\tau) : q(\tau) \in \mathcal{C}(\mathcal{E},\mathbb{R})\}$ equipped with the norm is $||q|| = \sup\{|q(\tau)|, \tau \in \mathcal{E}\}$ is a Banach space. Then the product space $(\mathcal{P} \times \mathcal{Q}, ||(p,q)||)$ is also a Banach space with norm $||(p,q)|| = ||p|| + ||q||$.

Definition 1 ([23]). *The left and right-sided generalized fractional integrals (GFIs) of* $f \in \mathcal{Z}^q_b(c,d)$ *of order* $\xi > 0$ *and* $\rho > 0$ *for* $-\infty < c < \tau < d < \infty$*, are defined as follows:*

$$({}^\rho\mathcal{I}^\xi_{c+} f)(\tau) = \frac{\rho^{1-\xi}}{\Gamma(\xi)} \int_c^\tau \frac{\theta^{\rho-1}}{(\tau^\rho - \theta^\rho)^{1-\xi}} f(\theta)d\theta, \tag{7}$$

$$({}^\rho\mathcal{I}^\xi_{d-} f)(\tau) = \frac{\rho^{1-\xi}}{\Gamma(\xi)} \int_\tau^d \frac{\theta^{\rho-1}}{(\theta^\rho - \tau^\rho)^{1-\xi}} f(\theta)d\theta. \tag{8}$$

Definition 2 ([24]). *The generalized fractional derivatives (GFDs) which are associated with GFIs (7) and (8) for* $0 \le c < \tau < d < \infty$*, are defined as follows:*

$$\begin{aligned}({}^\rho\mathcal{D}^\xi_{c+} f)(\tau) &= \left(\tau^{1-\rho}\frac{d}{d\tau}\right)^n ({}^\rho\mathcal{I}^{n-\xi}_{c+} f)(\tau) \\ &= \frac{\rho^{\xi-n+1}}{\Gamma(n-\xi)} \left(\tau^{1-\rho}\frac{d}{d\tau}\right)^n \int_c^\tau \frac{\theta^{\rho-1}}{(\tau^\rho - \theta^\rho)^{\xi-n+1}} f(\theta)d\theta,\end{aligned} \tag{9}$$

$$\begin{aligned}({}^\rho\mathcal{D}^\xi_{d-} f)(\tau) &= \left(-\tau^{1-\rho}\frac{d}{d\tau}\right)^n ({}^\rho\mathcal{I}^{n-\xi}_{d-} f)(\tau) \\ &= \frac{\rho^{\xi-n+1}}{\Gamma(n-\xi)} \left(-\tau^{1-\rho}\frac{d}{d\tau}\right)^n \int_\tau^d \frac{\theta^{\rho-1}}{(\tau^\rho - \theta^\rho)^{\xi-n+1}} f(\theta)d\theta,\end{aligned} \tag{10}$$

if the integrals exist.

Definition 3 ([25]). *The above GFDs define the left and right-sided generalized Liouville–Caputo type fractional derivatives of* $f \in \mathcal{AC}^n_\gamma[c,d]$ *of order* $\xi \ge 0$

$${}^\rho_C\mathcal{D}^\xi_{c+} f(z) = {}^\rho\mathcal{D}^\xi_{c+}\left[f(\tau) - \sum_{k=0}^{n-1} \frac{\gamma^k f(c)}{k!}\left(\frac{\tau^\rho - c^\rho}{\rho}\right)^k\right](z), \gamma = z^{1-\rho}\frac{d}{dz}, \tag{11}$$

$${}^\rho_C\mathcal{D}^\xi_{d-} f(z) = {}^\rho\mathcal{D}^\xi_{d-}\left[f(\tau) - \sum_{k=0}^{n-1} \frac{(-1)^k\gamma^k f(d)}{k!}\left(\frac{d^\rho - \tau^\rho}{\rho}\right)^k\right](z), \gamma = z^{1-\rho}\frac{d}{dz}, \tag{12}$$

when $n = [\xi] + 1$.

Lemma 1 ([25]). *Let* $\xi \ge 0, n = [\xi] + 1$ *and* $f \in \mathcal{AC}^n_\gamma[c,d]$*, where* $0 < c < d < \infty$*. Then,*

1. *if* $\xi \notin \mathbb{N}$

$$ {}^{\rho}_{C}\mathcal{D}^{\xi}_{c+}f(\tau) = \frac{1}{\Gamma(n-\xi)}\int_{c}^{\tau}\left(\frac{\tau^{\rho}-\theta^{\rho}}{\rho}\right)^{n-\xi-1}\frac{(\gamma^{n}f)(\theta)d\theta}{\theta^{1-\rho}} = {}^{\rho}\mathcal{I}^{n-\xi}_{c^{+}}(\gamma^{n}f)(\tau), \tag{13} $$

$$ {}^{\rho}_{C}\mathcal{D}^{\xi}_{d-}f(\tau) = \frac{1}{\Gamma(n-\xi)}\int_{\tau}^{d}\left(\frac{\theta^{\rho}-\tau^{\rho}}{\rho}\right)^{n-\xi-1}\frac{(-1)^{n}(\gamma^{n}f)(\theta)d\theta}{\theta^{1-\rho}} = {}^{\rho}\mathcal{I}^{n-\xi}_{d^{-}}(\gamma^{n}f)(\tau). \tag{14} $$

2. *if* $\xi \in \mathbb{N}$

$$ {}^{\rho}_{C}\mathcal{D}^{\xi}_{c+}f = \gamma^{n}f, \qquad {}^{\rho}_{C}\mathcal{D}^{\xi}_{d-}f = (-1)^{n}\gamma^{n}f. \tag{15} $$

Lemma 2 ([25]). *Let $f \in \mathcal{AC}^{n}_{\gamma}[c,d]$ or $C^{n}_{\gamma}[c,d]$ and $\xi \in \mathbb{R}$. Then,*

$$ {}^{\rho}\mathcal{I}^{\xi}_{c+}\,{}^{\rho}_{C}\mathcal{D}^{\xi}_{c+}f(z) = f(z) - \sum_{k=0}^{n-1}\frac{\gamma^{k}f(c)}{k!}\left(\frac{z^{\rho}-c^{\rho}}{\rho}\right)^{k}, $$

$$ {}^{\rho}\mathcal{I}^{\xi}_{d-}\,{}^{\rho}_{C}\mathcal{D}^{\xi}_{d-}f(z) = f(z) - \sum_{k=0}^{n-1}\frac{(-1)^{k}\gamma^{k}f(d)}{k!}\left(\frac{d^{\rho}-z^{\rho}}{\rho}\right)^{k}. $$

In particular, for $0 < \xi \leq 1$, we have

$$ {}^{\rho}\mathcal{I}^{\xi}_{c+}\,{}^{\rho}_{C}\mathcal{D}^{\xi}_{c+}f(z) = f(z) - f(c), \qquad {}^{\rho}\mathcal{I}^{\xi}_{d-}\,{}^{\rho}_{C}\mathcal{D}^{\xi}_{d-}f(z) = f(z) - f(d). $$

We introduce the following notations for computational ease:

$$ \mathcal{E}_{1} = \epsilon\frac{\varpi^{\rho(\varsigma+1)}}{\rho^{\varsigma+1}\Gamma(\varsigma+2)},\ \mathcal{E}_{2} = \pi\frac{\sigma^{\rho(\varrho+1)}}{\rho^{\varrho+1}\Gamma(\varrho+2)},\ \widehat{\mathcal{E}} = \frac{\mathcal{T}^{\rho}}{\rho}, \tag{16} $$

$$ \mathcal{G} = \widehat{\mathcal{E}}^{2} - \mathcal{E}_{1}\mathcal{E}_{2} \neq 0, \tag{17} $$

$$ \delta(\tau) = \left(\frac{\tau^{\rho}}{\rho\mathcal{G}}\right). \tag{18} $$

Next, we are proving a lemma, which is vital in converting the given problem to a fixed-point problem.

Lemma 3. *Given the functions $\hat{f}, \hat{g} \in C(0,\mathcal{T}) \cup \mathcal{L}(0,\mathcal{T})$, $p, q \in \mathcal{AC}^{2}_{\gamma}(\mathcal{E})$ and $\Lambda \neq 0$. Then the solution of the coupled BVP:*

$$ \begin{cases} {}^{\rho}_{C}\mathcal{D}^{\xi}_{0+}p(\tau) = \hat{f}(\tau), \tau \in \mathcal{E} := [0,\mathcal{T}], \\ {}^{\rho}_{C}\mathcal{D}^{\zeta}_{0+}q(\tau) = \hat{g}(\tau), \tau \in \mathcal{E} := [0,\mathcal{T}], \\ p(0) = 0, \quad q(0) = 0, \quad p(\mathcal{T}) = \epsilon\,{}^{\rho}\mathcal{I}^{\varsigma}_{0+}q(\varpi), \quad q(\mathcal{T}) = \pi\,{}^{\rho}\mathcal{I}^{\varrho}_{0+}p(\sigma) \quad 0 < \sigma < \varpi < \mathcal{T}, \end{cases} \tag{19} $$

is given by

$$ p(\tau) = {}^{\rho}\mathcal{I}^{\xi}_{0+}\hat{f}(\tau) + \delta(\tau)\Big[\widehat{\mathcal{E}}\Big(\epsilon\,{}^{\rho}\mathcal{I}^{\zeta+\varsigma}_{0^{+}}\hat{g}(\varpi) - {}^{\rho}\mathcal{I}^{\xi}_{0+}\hat{f}(\mathcal{T})\Big) + \mathcal{E}_{1}\Big(\pi\,{}^{\rho}\mathcal{I}^{\xi+\varrho}_{0^{+}}\hat{f}(\sigma) - {}^{\rho}\mathcal{I}^{\zeta}_{0+}\hat{g}(\mathcal{T})\Big)\Big] \tag{20} $$

and

$$ q(\tau) = {}^{\rho}\mathcal{I}^{\zeta}_{0+}\hat{g}(\tau) + \delta(\tau)\Big[\widehat{\mathcal{E}}\Big(\pi\,{}^{\rho}\mathcal{I}^{\xi+\varrho}_{0^{+}}\hat{f}(\sigma) - {}^{\rho}\mathcal{I}^{\zeta}_{0+}\hat{g}(\mathcal{T})\Big) + \mathcal{E}_{2}\Big(\epsilon\,{}^{\rho}\mathcal{I}^{\zeta+\varsigma}_{0^{+}}\hat{g}(\varpi) - {}^{\rho}\mathcal{I}^{\xi}_{0+}\hat{f}(\mathcal{T})\Big)\Big]. \tag{21} $$

Proof. When ${}^{\rho}\mathcal{I}_{0+}^{\xi}, {}^{\rho}\mathcal{I}_{0+}^{\zeta}$ are applied to the FDEs in (19) and Lemma 2 is used, the solution of the FDEs in (19) for $\tau \in \mathcal{E}$ is

$$p(\tau) = {}^{\rho}\mathcal{I}_{0+}^{\xi}\hat{f}(\tau) + a_1 + a_2\frac{\tau^{\rho}}{\rho} = \frac{\rho^{1-\xi}}{\Gamma(\xi)}\int_0^{\tau}\theta^{\rho-1}(\tau^{\rho}-\theta^{\rho})^{\xi-1}\hat{f}(\theta)d\theta + a_1 + a_2\frac{\tau^{\rho}}{\rho}, \quad (22)$$

$$q(\tau) = {}^{\rho}\mathcal{I}_{0+}^{\zeta}\hat{g}(\tau) + b_1 + b_2\frac{\tau^{\rho}}{\rho} = \frac{\rho^{1-\zeta}}{\Gamma(\zeta)}\int_0^{\tau}\theta^{\rho-1}(\tau^{\rho}-\theta^{\rho})^{\zeta-1}\hat{g}(\theta)d\theta + b_1 + b_2\frac{\tau^{\rho}}{\rho}, \quad (23)$$

respectively, for some $a_1, a_2, b_1, b_2 \in \mathcal{R}$. Making use of the boundary conditions $p(0) = q(0) = 0$ in (22) and (23) respectively, we get $a_1 = b_1 = 0$. Next, we obtain by using the generalized integral operators ${}^{\rho}\mathcal{I}_{0+}^{\xi}, {}^{\rho}\mathcal{I}_{0+}^{\zeta}$ (22) and (23) respectively,

$${}^{\rho}\mathcal{I}_{0+}^{\varrho}p(\tau) = {}^{\rho}\mathcal{I}_{0+}^{\xi+\varrho}\hat{f}(\tau) + a_1\frac{\tau^{\rho\varrho}}{\rho^{\varrho}\Gamma(\varrho+1)} + a_2\frac{\tau^{\rho(\varrho+1)}}{\rho^{\varrho+1}\Gamma(\varrho+2)}, \quad (24)$$

$${}^{\rho}\mathcal{I}_{0+}^{\varsigma}q(\tau) = {}^{\rho}\mathcal{I}_{0+}^{\zeta+\varsigma}\hat{g}(\tau) + b_1\frac{\tau^{\rho\varsigma}}{\rho^{\varsigma}\Gamma(\varsigma+1)} + b_2\frac{\tau^{\rho(\varsigma+1)}}{\rho^{\varsigma+1}\Gamma(\varsigma+2)}, \quad (25)$$

which, when combined with the boundary conditions $p(\mathcal{T}) = \epsilon\,{}^{\rho}\mathcal{I}_{0+}^{\varsigma}q(\varpi)$, $q(\mathcal{T}) = \pi{}^{\rho}\mathcal{I}_{0+}^{\varrho}p(\sigma)$, gives the following results:

$${}^{\rho}\mathcal{I}_{0+}^{\xi}\hat{f}(\mathcal{T}) + a_1 + a_2\frac{\mathcal{T}^{\rho}}{\rho} = \epsilon{}^{\rho}\mathcal{I}_{0+}^{\zeta+\varsigma}\hat{g}(\varpi) + b_1\frac{\epsilon\varpi^{\rho\varsigma}}{\rho^{\varsigma}\Gamma(\varsigma+1)} + b_2\frac{\epsilon\varpi^{\rho(\varsigma+1)}}{\rho^{\varsigma+1}\Gamma(\varsigma+2)}, \quad (26)$$

$${}^{\rho}\mathcal{I}_{0+}^{\zeta}\hat{g}(\mathcal{T}) + b_1 + b_2\frac{\mathcal{T}^{\rho}}{\rho} = \pi{}^{\rho}\mathcal{I}_{0+}^{\xi+\varrho}\hat{f}(\sigma) + a_1\frac{\pi\sigma^{\rho\varrho}}{\rho^{\varrho}\Gamma(\varrho+1)} + a_2\frac{\pi\sigma^{\rho(\varrho+1)}}{\rho^{\varrho+1}\Gamma(\varrho+2)}. \quad (27)$$

Next, we obtain

$$a_2\widehat{\mathcal{E}} - b_2\mathcal{E}_1 = \epsilon\,{}^{\rho}\mathcal{I}_{0+}^{\zeta+\varsigma}\hat{g}(\varpi) - {}^{\rho}\mathcal{I}_{0+}^{\xi}\hat{f}(\mathcal{T}), \quad (28)$$

$$b_2\widehat{\mathcal{E}} - a_2\mathcal{E}_2 = \pi\,{}^{\rho}\mathcal{I}_{0+}^{\xi+\varrho}\hat{f}(\sigma) - {}^{\rho}\mathcal{I}_{0+}^{\zeta}\hat{g}(\mathcal{T}), \quad (29)$$

by employing the notations (16) in (26) and (27) respectively. We find that when we solve the system of Equations (28) and (29) for a_2 and b_2,

$$a_2 = \frac{1}{\mathcal{G}}\Big[\widehat{\mathcal{E}}\Big(\epsilon\,{}^{\rho}\mathcal{I}_{0+}^{\zeta+\varsigma}\hat{g}(\varpi) - {}^{\rho}\mathcal{I}_{0+}^{\xi}\hat{f}(\mathcal{T})\Big) + \mathcal{E}_1\Big(\pi\,{}^{\rho}\mathcal{I}_{0+}^{\xi+\varrho}\hat{f}(\sigma) - {}^{\rho}\mathcal{I}_{0+}^{\zeta}\hat{g}(\mathcal{T})\Big)\Big], \quad (30)$$

$$b_2 = \frac{1}{\mathcal{G}}\Big[\mathcal{E}_2\Big(\epsilon\,{}^{\rho}\mathcal{I}_{0+}^{\zeta+\varsigma}\hat{g}(\varpi) - {}^{\rho}\mathcal{I}_{0+}^{\xi}\hat{f}(\mathcal{T})\Big) + \widehat{\mathcal{E}}\Big(\pi\,{}^{\rho}\mathcal{I}_{0+}^{\xi+\varrho}\hat{f}(\sigma) - {}^{\rho}\mathcal{I}_{0+}^{\zeta}\hat{g}(\mathcal{T})\Big)\Big]. \quad (31)$$

Substituting the values of a_1, a_2, b_1, b_2 in (22) and (23) respectively, we get the solution for the BVP (19). □

3. Existence Results for the Problem (1) and (2)

As a result of Lemma 3, we define an operator $\Delta : \mathcal{P} \times \mathcal{Q} \to \mathcal{P} \times \mathcal{Q}$ by

$$\Delta(p,q)(\tau) = (\Delta_1(p,q)(\tau), \Delta_2(p,q)(\tau)), \quad (32)$$

where

$$\Delta_1(p,q)(\tau) = {}^{\rho}\mathcal{I}^{\xi}_{0+}f(\tau,p(\tau),q(\tau)) + \delta(\tau)\Bigg[\widehat{\mathcal{E}}\Big(\epsilon\,{}^{\rho}\mathcal{I}^{\zeta+\varsigma}_{0^+}g(\varpi,p(\varpi),q(\varpi)) - {}^{\rho}\mathcal{I}^{\xi}_{0+}f(\mathcal{T},p(\mathcal{T}),q(\mathcal{T}))\Big)$$
$$+\mathcal{E}_1\Big(\pi\,{}^{\rho}\mathcal{I}^{\xi+\varrho}_{0^+}f(\sigma,p(\sigma),q(\sigma)) - {}^{\rho}\mathcal{I}^{\zeta}_{0+}g(\mathcal{T},p(\mathcal{T}),q(\mathcal{T}))\Big)\Bigg], \tag{33}$$

$$\Delta_2(p,q)(\tau) = {}^{\rho}\mathcal{I}^{\zeta}_{0+}g(\tau,p(\tau),q(\tau)) + \delta(\tau)\Bigg[\widehat{\mathcal{E}}\Big(\pi\,{}^{\rho}\mathcal{I}^{\xi+\varrho}_{0^+}f(\sigma,p(\sigma),q(\sigma)) - {}^{\rho}\mathcal{I}^{\zeta}_{0+}g(\mathcal{T},p(\mathcal{T}),q(\mathcal{T}))\Big)$$
$$+\mathcal{E}_2\Big(\epsilon\,{}^{\rho}\mathcal{I}^{\zeta+\varsigma}_{0^+}g(\varpi,p(\varpi),q(\varpi)) - {}^{\rho}\mathcal{I}^{\xi}_{0+}f(\mathcal{T},p(\mathcal{T}),q(\mathcal{T}))\Big)\Bigg]. \tag{34}$$

For brevity's sake, we'll use the following notations:

$$\mathcal{J}_1 = \frac{\Big(\mathcal{T}^{\rho\xi}(1+|\delta||\widehat{\mathcal{E}}|)\Big)}{\rho^{\xi}\Gamma(\xi+1)} + \frac{|\delta||\pi||\mathcal{E}_1|\sigma^{\rho(\xi+\varrho)}}{\rho^{\xi+\varrho}\Gamma(\xi+\varrho+1)}, \tag{35}$$

$$\mathcal{K}_1 = |\delta|\left(\frac{|\mathcal{E}_1|\mathcal{T}^{\rho\zeta}}{\rho^{\zeta}\Gamma(\zeta+1)} + \frac{|\widehat{\mathcal{E}}||\epsilon|\varpi^{\rho(\zeta+\varsigma)}}{\rho^{\zeta+\varsigma}\Gamma(\zeta+\varsigma+1)}\right), \tag{36}$$

$$\mathcal{J}_2 = |\delta|\left(\frac{\mathcal{T}^{\rho\xi}|\mathcal{E}_2|}{\rho^{\xi}\Gamma(\xi+1)} + \frac{|\pi||\widehat{\mathcal{E}}|\sigma^{\rho(\xi+\varrho)}}{\rho^{\xi+\varrho}\Gamma(\xi+\varrho+1)}\right), \tag{37}$$

$$\mathcal{K}_2 = \frac{\Big(\mathcal{T}^{\rho\zeta}(1+|\delta||\widehat{\mathcal{E}}|)\Big)}{\rho^{\zeta}\Gamma(\zeta+1)} + \frac{|\delta||\epsilon||\mathcal{E}_2|\varpi^{\rho(\zeta+\varsigma)}}{\rho^{\zeta+\varsigma}\Gamma(\zeta+\varsigma+1)}, \tag{38}$$

$$\Phi = \min\{1-[\psi_1(\mathcal{J}_1+\mathcal{J}_2)+\hat{\psi_1}(\mathcal{K}_1+\mathcal{K}_2)], 1-[\psi_2(\mathcal{J}_1+\mathcal{J}_2)+\hat{\psi_2}(\mathcal{K}_1+\mathcal{K}_2)]\}. \tag{39}$$

Theorem 1. *Assume that $f, g : \mathcal{E}\times\mathbb{R}\times\mathbb{R}\to\mathbb{R}$ are continuous functions satisfying the condition: $(\mathcal{A}_1)$ there exists constants $\psi_m, \hat{\psi_m} \geq 0 (m = 1,2)$ and $\psi_0, \hat{\psi_0} > 0$ such that*

$$|f(\tau,o_1,o_2)| \leq \psi_0 + \psi_1|o_1| + \psi_2|o_2|,$$
$$|g(\tau,o_1,o_2)| \leq \hat{\psi_0} + \hat{\psi_1}|o_1| + \hat{\psi_2}|o_2|, \forall o_m \in \mathbb{R}, m = 1,2.$$

If $\psi_1(\mathcal{J}_1+\mathcal{J}_2)+\hat{\psi_1}(\mathcal{K}_1+\mathcal{K}_2) < 1, \psi_2(\mathcal{J}_1+\mathcal{J}_2)+\hat{\psi_2}(\mathcal{K}_1+\mathcal{K}_2) < 1$. Then $\exists$ at least one solution for the BVP (1) and (2) on $\mathcal{E}$, where $\mathcal{J}_1, \mathcal{K}_1, \mathcal{J}_2, \mathcal{K}_2$ are given by (35)–(38) respectively.

Proof. We define operator $\Delta : \mathcal{P}\times\mathcal{Q}\to\mathcal{P}\times\mathcal{Q}$ as being completely continuous in the first step. The continuity of the functions f and g implies that the operators Δ_1 and Δ_2 are continuous. As a result, the operator Δ is continuous. Let $\Psi\subset\mathcal{P}\times\mathcal{Q}$ be a bounded set to demonstrate the uniformly bounded operator Δ. Then $\hat{\mathcal{N}_1}$ and $\hat{\mathcal{N}_2}$ are positive constants such that $|f(\tau,p(\tau),q(\tau))| \leq \hat{\mathcal{N}_1}, |g(\tau,p(\tau),q(\tau))| \leq \hat{\mathcal{N}_2}, \forall(p,q)\in\Psi$. Then we have

$$
\begin{aligned}
|\Delta_1(p,q)(\tau)| \leq & {}^{\rho}\mathcal{I}^{\xi}_{0+}|f(\tau,p(\tau),q(\tau))| + |\delta(\tau)| \Bigg[|\widehat{\mathcal{E}}| \Big(|\epsilon| \, {}^{\rho}\mathcal{I}^{\zeta+\varsigma}_{0+} |g(\varpi,p(\varpi),q(\varpi))| + {}^{\rho}\mathcal{I}^{\xi}_{0+}|f(\mathcal{T},p(\mathcal{T}),q(\mathcal{T}))| \Big) \\
& + |\mathcal{E}_1| \Big(|\pi| \, {}^{\rho}\mathcal{I}^{\xi+\varrho}_{0+} |f(\sigma,p(\sigma),q(\sigma))| + {}^{\rho}\mathcal{I}^{\zeta}_{0+}|g(\mathcal{T},p(\mathcal{T}),q(\mathcal{T}))| \Big) \Bigg] \\
& \leq \hat{\mathcal{N}_1} \left\{ \frac{|\delta||\pi||\mathcal{E}_1|\sigma^{\rho(\xi+\varrho)}}{\rho^{\xi+\varrho}\Gamma(\xi+\varrho+1)} + \frac{\left(\mathcal{T}^{\rho\xi}(1+|\delta||\widehat{\mathcal{E}}|)\right)}{\rho^{\xi}\Gamma(\xi+1)} \right\} \\
& + \hat{\mathcal{N}_2} \left\{ \left(\frac{|\widehat{\mathcal{E}}||\epsilon|\varpi^{\rho(\zeta+\varsigma)}}{\rho^{\zeta+\varsigma}\Gamma(\zeta+\varsigma+1)} + \frac{|\mathcal{E}_1|\mathcal{T}^{\rho\zeta}}{\rho^{\zeta}\Gamma(\zeta+1)} \right) |\delta| \right\},
\end{aligned}
$$

when taking the norm and using (35) and (36), that yields for $(p,q) \in \Psi$,

$$
||\Delta_1(p,q)|| \leq \mathcal{J}_1\hat{\mathcal{N}_1} + \mathcal{K}_1\hat{\mathcal{N}_2}. \tag{40}
$$

Likewise, we obtain

$$
\begin{aligned}
||\Delta_2(p,q)|| \leq & \hat{\mathcal{N}_2} \left\{ \frac{|\delta||\epsilon||\mathcal{E}_2|\varpi^{\rho(\zeta+\varsigma)}}{\rho^{\zeta+\varsigma}\Gamma(\zeta+\varsigma+1)} + \frac{\left(\mathcal{T}^{\rho\zeta}(1+|\delta||\widehat{\mathcal{E}}|)\right)}{\rho^{\zeta}\Gamma(\zeta+1)} \right\} \\
& + \hat{\mathcal{N}_1} \left\{ |\delta| \left(\frac{|\pi||\widehat{\mathcal{E}}|\sigma^{\rho(\xi+\varrho)}}{\rho^{\xi+\varrho}\Gamma(\xi+\varrho+1)} + \frac{\mathcal{T}^{\rho\xi}|\mathcal{E}_2|}{\rho^{\xi}\Gamma(\xi+1)} \right) \right\} \\
\leq & \mathcal{J}_2\hat{\mathcal{N}_1} + \mathcal{K}_2\hat{\mathcal{N}_2},
\end{aligned} \tag{41}
$$

using (37) and (38). Based on the inequalities (40) and (41), we can conclude that Δ_1 and Δ_2 are uniformly bounded, which indicates that the operator Δ is uniformly bounded. Next, we show that Δ is equicontinuous. Let $\tau_1, \tau_2 \in \mathcal{E}$ with $\tau_1 < \tau_2$. Then we have

$$
\begin{aligned}
& |\Delta_1(p,q)(\tau_2) - \Delta_1(p,q)(\tau_1)| \\
& \leq |{}^{\rho}\mathcal{I}^{\xi}_{0+}f(\tau_2,p(\tau_2),q(\tau_2)) - {}^{\rho}\mathcal{I}^{\xi}_{0+}f(\tau_1,p(\tau_1),q(\tau_1))| \\
& + |\delta(\tau_2) - \delta(\tau_1)| \Bigg[\widehat{\mathcal{E}} \Big(|\epsilon| \, {}^{\rho}\mathcal{I}^{\zeta+\varsigma}_{0+} |g(\varpi,p(\varpi),q(\varpi))| + {}^{\rho}\mathcal{I}^{\xi}_{0+}|f(\mathcal{T},p(\mathcal{T}),q(\mathcal{T}))| \Big) \\
& + \mathcal{E}_1 \Big(|\pi| \, {}^{\rho}\mathcal{I}^{\xi+\varrho}_{0+} |f(\sigma,p(\sigma),q(\sigma))| + {}^{\rho}\mathcal{I}^{\zeta}_{0+}|g(\mathcal{T},p(\mathcal{T}),q(\mathcal{T}))| \Big) \Bigg] \\
& \leq \frac{\rho^{1-\xi}\hat{\mathcal{N}_1}}{\Gamma(\xi)} \left| \int_0^{\tau_1} \left[\frac{\theta^{\rho-1}}{(\tau_2^{\rho}-\theta^{\rho})^{1-\xi}} - \frac{\theta^{\rho-1}}{(\tau_1^{\rho}-\theta^{\rho})^{1-\xi}} \right] d\theta + \int_{\tau_1}^{\tau_2} \frac{\theta^{\rho-1}}{(\tau_2^{\rho}-\theta^{\rho})^{1-\xi}} d\theta \right| \\
& + |\delta(\tau_2) - \delta(\tau_1)| \left[|\widehat{\mathcal{E}}| \left(\frac{\hat{\mathcal{N}_2}|\epsilon|\varpi^{\rho\zeta+\varsigma}}{\rho^{\zeta+\varsigma}\Gamma(\zeta+\varsigma+1)} + \frac{\hat{\mathcal{N}_1}\mathcal{T}^{\rho\xi}}{\rho^{\xi}\Gamma(\xi+1)} \right) \right] \\
& + |\delta(\tau_2) - \delta(\tau_1)| \left[|\mathcal{E}_1| \left(\frac{\hat{\mathcal{N}_1}|\pi|\sigma^{\rho\xi+\varrho}}{\rho^{\xi+\varrho}\Gamma(\xi+\varrho+1)} + \frac{\hat{\mathcal{N}_2}\mathcal{T}^{\rho\zeta}}{\rho^{\zeta}\Gamma(\zeta+1)} \right) \right] \\
& \to 0 \text{ as } \tau_2 \to \tau_1.
\end{aligned} \tag{42}
$$

independent of (p,q) with respect to $|f(\tau,p(\tau_1),q(\tau_1))| \leq \hat{\mathcal{N}_1}$ and $|g(\tau,p(\tau_1),q(\tau_1))| \leq \hat{\mathcal{N}_2}$. Similarly, we can express $|\Delta_2(p,q)(\tau_2) - \Delta_2(p,q)(\tau_1)| \to 0$ as $\tau_2 \to \tau_1$ independent of (p,q) in terms of the boundedness of f and g. As a result of the equicontinuity of Δ_1 and Δ_2, operator Δ is equicontinuous. As a result of the Arzela–Ascoli theorem, the operator is compact. Finally, we demonstrate that the set $\Pi(\Delta) = \{(p,q) \in \mathcal{P} \times \mathcal{Q} : \lambda\Delta(p,q);$

$0 < \lambda < 1\}$ is bounded. Let $(p,q) \in \Pi(\Delta)$.Then $(p,q) = \lambda\Delta(p,q)$. For any $\tau \in \mathcal{E}$, we have $p(\tau) = \lambda\Delta_1(p,q)(\tau), q(\tau) = \lambda\Delta_2(p,q)(\tau)$. By utilizing $(\mathcal{A}_1)$ in (33), we obtain

$$\begin{aligned}
|p(\tau)| \leq & {}^{\rho}\mathcal{I}^{\xi}_{0+}(\psi_0, \psi_1|p(\tau)|, \psi_2|q(\tau)|) \\
& + |\delta(\tau)| \Bigg(|\widehat{\mathcal{E}}| \Big(|\epsilon| {}^{\rho}\mathcal{I}^{\zeta+\varsigma}_{0+}(\hat{\psi}_0 + \hat{\psi}_1|p(\varpi)| + \hat{\psi}_2|q(\varpi)|) + {}^{\rho}\mathcal{I}^{\xi}_{0+}(\psi_0 + \psi_0|p(\mathcal{T})| + \psi_2|q(\mathcal{T})|) \Big) \\
& + |\mathcal{E}_1| \Big(|\pi| {}^{\rho}\mathcal{I}^{\xi+\varrho}_{0+}(\psi_0 + \psi_1|p(\sigma)| + \psi_2|q(\sigma)|) + {}^{\rho}\mathcal{I}^{\zeta}_{0+}(\hat{\psi}_0 + \hat{\psi}_1|p(\mathcal{T})| + \hat{\psi}_2|q(\mathcal{T})|) \Big) \Bigg),
\end{aligned}$$

which results when taking the norm for $\tau \in \mathcal{E}$,

$$||p|| \leq (\psi_0 + \psi_1||p|| + \psi_2||q||)\mathcal{J}_1 + (\hat{\psi}_0 + \hat{\psi}_1||p|| + \hat{\psi}_2||q||)\mathcal{K}_1. \tag{43}$$

Similarly, we are capable of obtaining that

$$||q|| \leq (\hat{\psi}_0 + \hat{\psi}_1||p|| + \hat{\psi}_2||q||)\mathcal{K}_2 + (\psi_0 + \psi_1||p|| + \psi_2||q||)\mathcal{J}_2. \tag{44}$$

From (43) and (44), we get

$$\begin{aligned}
||p|| + ||q|| =& \psi_0(\mathcal{J}_1 + \mathcal{J}_2) + \hat{\psi}_0(\mathcal{K}_1 + \mathcal{K}_2) + ||p||[\psi_1(\mathcal{J}_1 + \mathcal{J}_2) + \hat{\psi}_1(\mathcal{K}_1 + \mathcal{K}_2)] \\
& + ||q||[\psi_1(\mathcal{J}_1 + \mathcal{J}_2) + \hat{\psi}_1(\mathcal{K}_1 + \mathcal{K}_2)],
\end{aligned}$$

which results, with $||(p,q)|| = ||p|| + ||q||$,

$$||(p,q)|| \leq \frac{\psi_0(\mathcal{J}_1 + \mathcal{J}_2) + \hat{\psi}_0(\mathcal{K}_1 + \mathcal{K}_2)}{\Phi}.$$

As a result, $\Pi(\Delta)$ is bounded. Thus, the nonlinear alternative of Leray–Schauder [26] is valid and the operator Δ has at least one fixed point. It implies that the BVP (1) and (2) contain at least one solution on $\mathcal{E}$. □

Theorem 2. *Assume that $f, g : \mathcal{E} \times \mathbb{R} \times \mathbb{R} \to \mathbb{R}$ are continuous functions satisfying the condition: $(\mathcal{A}_2)$ there exists constants $\phi_m, \hat{\phi}_m \geq 0 (m = 1,2)$ such that*

$$\begin{aligned}
|f(\tau, o_1, o_2) - f(\tau, \hat{o}_1, \hat{o}_2)| \leq& \phi_1|o_1 - \hat{o}_1| + \phi_2|o_2 - \hat{o}_2|, \\
|g(\tau, o_1, o_2) - g(\tau, \hat{o}_1, \hat{o}_2)| \leq& \hat{\phi}_1|o_1 - \hat{o}_1| + \hat{\phi}_2|o_2 - \hat{o}_2|, \forall o_m, \hat{o}_m \in \mathbb{R}, m = 1,2.
\end{aligned}$$

Furthermore, there exist $\mathcal{S}_1, \mathcal{S}_2 > 0$ such that $|f(\tau,0,0)| \leq \mathcal{S}_1, |g(\tau,0,0)| \leq \mathcal{S}_2$, Then, given that

$$(\mathcal{J}_1 + \mathcal{J}_2)(\phi_1 + \phi_2) + (\mathcal{K}_1 + \mathcal{K}_2)(\hat{\phi}_1 + \hat{\phi}_2) < 1, \tag{45}$$

the BVP (1) and (2) has a unique solution on $\mathcal{E}$, where $\mathcal{J}_1, \mathcal{K}_1, \mathcal{J}_2, \mathcal{K}_2$ are given by (35)–(38) respectively.

Proof. Let us fix $\varphi \leq \frac{(\mathcal{J}_1+\mathcal{J}_2)\mathcal{S}_1+(\mathcal{K}_1+\mathcal{K}_2)\mathcal{S}_2}{1-((\mathcal{J}_1+\mathcal{J}_2)(\phi_1+\phi_2)+(\mathcal{K}_1+\mathcal{K}_2)(\hat{\phi}_1+\hat{\phi}_2))}$ and demonstrate that $\Delta\mathcal{B}_\varphi \subset \mathcal{B}_\varphi$ when operator Δ is given by (32) and $\mathcal{B}_\varphi = \{(p,q) \in \mathcal{P} \times \mathcal{Q} : ||(p,q)|| \leq \varphi\}$. For $(p,q) \in \mathcal{B}_\varphi, \tau \in \mathcal{E}$

$$\begin{aligned}
|f(\tau, p(\tau), q(\tau))| \leq& \phi_1|p(\tau)| + \phi_2|q(\tau)| + \mathcal{S}_1 \\
\leq& \phi_1||p|| + \phi_2||q|| + \mathcal{S}_1,
\end{aligned}$$

and

$$|g(\tau,p(\tau),q(\tau))| \leq \hat{\phi}_1|p(\tau)| + \hat{\phi}_2|q(\tau)| + \mathcal{S}_2$$
$$\leq \hat{\phi}_1||p|| + \hat{\phi}_2||q|| + \mathcal{S}_2. \tag{46}$$

This guides to

$$|\Delta_1(p,q)(\tau)| \leq {}^{\rho}\mathcal{I}^{\xi}_{0+}\Big[|f(\tau,p(\tau),q(\tau)) - f(\tau,0,0)| + |f(\tau,0,0)|\Big]$$
$$+ |\delta(\tau)|\Bigg(|\widehat{\mathcal{E}}|\Big(|\epsilon|^{\rho}\mathcal{I}^{\zeta+\varsigma}_{0+}g[(\varpi,p(\varpi),q(\varpi)) - g(\varpi,0,0)| + |g(\varpi,0,0)|]$$
$$+^{\rho}\mathcal{I}^{\xi}_{0+}f[(\mathcal{T},p(\mathcal{T}),q(\mathcal{T})) - f(\mathcal{T},0,0)| + |f(\mathcal{T},0,0)]|\Big)$$
$$+ |\mathcal{E}_1|\Big(|\pi|^{\rho}\mathcal{I}^{\xi+\varrho}_{0+}f[f(\sigma,p(\sigma),q(\sigma)) - f(\sigma,0,0)| + |f(\sigma,0,0)|]$$
$$+^{\rho}\mathcal{I}^{\zeta}_{0+}[|g(\mathcal{T},p(\mathcal{T}),q(\mathcal{T})) - g(\mathcal{T},0,0)| + |g(\mathcal{T},0,0)|]\Big)\Bigg)$$

$$\leq(\phi_1||p|| + \phi_2||q|| + \mathcal{S}_1)\left\{\frac{\Big(\mathcal{T}^{\rho\xi}(1+|\delta||\widehat{\mathcal{E}}|)\Big)}{\rho^{\xi}\Gamma(\xi+1)} + \frac{|\delta||\pi||\mathcal{E}_1|\sigma^{\rho(\xi+\varrho)}}{\rho^{\xi+\varrho}\Gamma(\xi+\varrho+1)}\right\}$$
$$+ (\hat{\phi}_1||p|| + \hat{\phi}_2||q|| + \mathcal{S}_2)\left\{|\delta|\left(\frac{|\mathcal{E}_1|\mathcal{T}^{\rho\zeta}}{\rho^{\zeta}\Gamma(\zeta+1)} + \frac{|\widehat{\mathcal{E}}||\epsilon|\varpi^{\rho(\zeta+\varsigma)}}{\rho^{\zeta+\varsigma}\Gamma(\zeta+\varsigma+1)}\right)\right\}$$

$$||\Delta_1(p,q)|| \leq (\phi_1||p|| + \phi_2||q|| + \mathcal{S}_1)\mathcal{J}_1 + (\hat{\phi}_1||p|| + \hat{\phi}_2||q|| + \mathcal{S}_2)\mathcal{K}_1. \tag{47}$$

Similarly, we obtain

$$|\Delta_2(p,q)(\tau)| \leq(\hat{\phi}_1||p|| + \hat{\phi}_2||q|| + \mathcal{S}_2)\left\{\frac{\Big(\mathcal{T}^{\rho\zeta}(1+|\delta||\widehat{\mathcal{E}}|)\Big)}{\rho^{\zeta}\Gamma(\zeta+1)} + \frac{|\delta||\epsilon||\mathcal{E}_2|\varpi^{\rho(\zeta+\varsigma)}}{\rho^{\zeta+\varsigma}\Gamma(\zeta+\varsigma+1)}\right\}$$
$$+ (\phi_1||p|| + \phi_2||q|| + \mathcal{S}_1)\left\{|\delta|\left(\frac{\mathcal{T}^{\rho\xi}|\mathcal{E}_2|}{\rho^{\xi}\Gamma(\xi+1)} + \frac{|\pi||\widehat{\mathcal{E}}|\sigma^{\rho(\xi+\varrho)}}{\rho^{\xi+\varrho}\Gamma(\xi+\varrho+1)}\right)\right\}$$
$$||\Delta_2(p,q)|| \leq(\hat{\phi}_1||p|| + \hat{\phi}_2||q|| + \mathcal{S}_2)\mathcal{K}_2 + (\phi_1||p|| + \phi_2||q|| + \mathcal{S}_1)\mathcal{J}_2. \tag{48}$$

As a result, (47) and (48) follow $||\Delta(p,q)|| \leq \varphi$, and thus $\Delta\mathcal{B}_{\varphi} \subset \mathcal{B}_{\varphi}$. Now, for $(p_1,q_1),(p_2,q_2) \in \mathcal{P} \times \mathcal{Q}$ and any $\tau \in \mathcal{E}$, we get

$$|\Delta_1(p_1,q_1)(\tau) - \Delta_1(p_2,q_2)(\tau)|$$
$$\leq^{\rho}\mathcal{I}^{\xi}_{0+}|f(\tau,p_1(\tau),q_1(\tau)) - f(\tau,p_2(\tau),q_2(\tau))|$$
$$+ |\delta(\tau)|\Bigg(|\widehat{\mathcal{E}}|\Big(|\epsilon|\ {}^{\rho}\mathcal{I}^{\zeta+\varsigma}_{0+}|g(\varpi,p_1(\varpi),q_1(\varpi)) - g(\varpi,p_2(\varpi),q_2(\varpi))|$$
$$+^{\rho}\mathcal{I}^{\xi}_{0+}|f(\mathcal{T},p_1(\mathcal{T}),q_1(\mathcal{T})) - f(\mathcal{T},p_2(\mathcal{T}),q_2(\mathcal{T}))|\Big)$$
$$+ |\mathcal{E}_1|\Big(|\pi|\ {}^{\rho}\mathcal{I}^{\xi+\varrho}_{0+}|f(\sigma,p_1(\sigma),q_1(\sigma)) - f(\sigma,p_2(\sigma),q_2(\sigma))|$$
$$+^{\rho}\mathcal{I}^{\zeta}_{0+}|g(\mathcal{T},p_1(\mathcal{T}),q_1(\mathcal{T})) - g(\mathcal{T},p_2(\mathcal{T}),q_2(\mathcal{T}))|\Big)\Bigg)$$

$$\leq (\phi_1 ||p_1 - p_2|| + \phi_2 ||q_1 - q_2||) \left\{ \frac{\left(\mathcal{T}^{\rho\xi}(1 + |\delta||\widehat{\mathcal{E}}|) \right)}{\rho^{\xi}\Gamma(\xi+1)} + \frac{|\delta||\pi||\mathcal{E}_1|\sigma^{\rho(\xi+\varrho)}}{\rho^{\xi+\varrho}\Gamma(\xi+\varrho+1)} \right\}$$

$$+ (\hat{\phi}_1 ||p_1 - p_2|| + \hat{\phi}_2 ||q_1 - q_2||) \left\{ |\delta| \left(\frac{|\mathcal{E}_1|\mathcal{T}^{\rho\zeta}}{\rho^{\zeta}\Gamma(\zeta+1)} + \frac{|\widehat{\mathcal{E}}||\epsilon|\varpi^{\rho(\zeta+\varsigma)}}{\rho^{\zeta+\varsigma}\Gamma(\zeta+\varsigma+1)} \right) \right\}$$

$$\leq (\mathcal{J}_1(\phi_1 + \phi_2) + \mathcal{K}_1(\hat{\phi}_1 + \hat{\phi}_2))(||p_1 - p_2|| + ||q_1 - q_2||).$$

Similarly, we obtain

$$|\Delta_2(p_1, q_1)(\tau) - \Delta_2(p_2, q_2)(\tau)|$$

$$\leq (\hat{\phi}_1 ||p_1 - p_2|| + \hat{\phi}_2 ||q_1 - q_2||) \left\{ \frac{\left(\mathcal{T}^{\rho\zeta}(1 + |\delta||\widehat{\mathcal{E}}|) \right)}{\rho^{\zeta}\Gamma(\zeta+1)} + \frac{|\delta||\epsilon||\mathcal{E}_2|\varpi^{\rho(\zeta+\varsigma)}}{\rho^{\zeta+\varsigma}\Gamma(\zeta+\varsigma+1)} \right\}$$

$$+ (\phi_1 ||p_1 - p_2|| + \phi_2 ||q_1 - q_2||) \left\{ |\delta| \left(\frac{\mathcal{T}^{\rho\xi}|\mathcal{E}_2|}{\rho^{\xi}\Gamma(\xi+1)} + \frac{|\pi||\widehat{\mathcal{E}}|\sigma^{\rho(\xi+\varrho)}}{\rho^{\xi+\varrho}\Gamma(\xi+\varrho+1)} \right) \right\}$$

$$\leq (\mathcal{J}_2(\phi_1 + \phi_2) + \mathcal{K}_2(\hat{\phi}_1 + \hat{\phi}_2))(||p_1 - p_2|| + ||q_1 - q_2||)).$$

Thus we obtain

$$||\Delta_1(p_1, q_1)(\tau) - \Delta_1(p_2, q_2)(\tau)|| \leq (\mathcal{J}_1(\phi_1 + \phi_2) + \mathcal{K}_1(\hat{\phi}_1 + \hat{\phi}_2))(||p_1 - p_2|| + ||q_1 - q_2||). \tag{49}$$

In a similar manner,

$$||\Delta_2(p_1, q_1)(\tau) - \Delta_2(p_2, q_2)(\tau)|| \leq (\mathcal{J}_2(\phi_1 + \phi_2) + \mathcal{K}_2(\hat{\phi}_1 + \hat{\phi}_2))(||p_1 - p_2|| + ||q_1 - q_2||). \tag{50}$$

Hence, using (49) and (50) we can get

$$||\Delta(p_1, q_1)(\tau) - \Delta(p_2, q_2)(\tau)|| \leq ((\mathcal{J}_1 + \mathcal{J}_2)(\phi_1 + \phi_2) + (\mathcal{K}_1 + \mathcal{K}_2)(\hat{\phi}_1 + \hat{\phi}_2))$$
$$(||p_1 - p_2|| + ||q_1 - q_2||).$$

As a consequence of condition $((\mathcal{J}_1 + \mathcal{J}_2)(\phi_1 + \phi_2) + (\mathcal{K}_1 + \mathcal{K}_2)(\hat{\phi}_1 + \hat{\phi}_2)) < 1$, Δ is a contraction operator. As an outcome of the Banach fixed point theorem, we can conclude that operator has a unique fixed point, which is the unique solution of the problem (1), and (2). □

For brevity's sake, we'll use the following notations:

$$\hat{\Omega}_1 = \mathcal{J}_1 - \frac{\mathcal{T}^{\rho\xi}}{\rho^{\xi}\Gamma(\xi+1)} + \mathcal{K}_1, \tag{51}$$

$$\hat{\Omega}_2 = \mathcal{J}_2 - \frac{\mathcal{T}^{\rho\zeta}}{\rho^{\zeta}\Gamma(\zeta+1)} + \mathcal{K}_2. \tag{52}$$

Theorem 3. *Assume that $f, g : \mathcal{E} \times \mathbb{R} \times \mathbb{R} \to \mathbb{R}$ are continuous functions satisfying the assumption $(\mathcal{A}_2)$ in Theorem 2. Furthermore, there exist positive constants $\mathcal{U}_1, \mathcal{U}_2$ such that $\forall \tau \in \mathcal{E}$ and $r_i \in \mathbb{R}, i = 1, 2$.*

$$|f(\tau, r_1, r_2)| \leq \mathcal{U}_1, \qquad |g(\tau, r_1, r_2)| \leq \mathcal{U}_2. \tag{53}$$

If

$$\frac{\mathcal{T}^{\rho\xi}(\phi_1 + \phi_2)}{\rho^{\xi}\Gamma(\xi+1)} + \frac{\mathcal{T}^{\rho\zeta}(\hat{\phi}_1 + \hat{\phi}_2)}{\rho^{\zeta}\Gamma(\zeta+1)} < 1, \tag{54}$$

then the BVP (1), and (2) has at least one solution on $\mathcal{E}$.

Proof. Let us define a closed ball $\mathcal{B}_\varphi = \{(p,q) \in \mathcal{P} \times \mathcal{Q} : ||(p,q)|| \leq \varphi\}$ as follows and split Δ_1, Δ_2 as:

$$\Delta_{1,1}(p,q)(\tau) = \delta(\tau)\Bigg(\widehat{\mathcal{E}}\Big(\epsilon\, {}^\rho\mathcal{I}_{0+}^{\zeta+\varsigma} g(\varpi, p(\varpi), q(\varpi)) - {}^\rho\mathcal{I}_{0+}^{\xi} f(\mathcal{T}, p(\mathcal{T}), q(\mathcal{T}))\Big) + \mathcal{E}_1\Big(\pi\, {}^\rho\mathcal{I}_{0+}^{\xi+\varrho} f(\sigma, p(\sigma), q(\sigma)) - {}^\rho\mathcal{I}_{0+}^{\zeta} g(\mathcal{T}, p(\mathcal{T}), q(\mathcal{T}))\Big)\Bigg), \quad (55)$$

$$\Delta_{1,1}(p,q)(\tau) = {}^\rho\mathcal{I}_{0+}^{\xi} f(\tau, p(\tau), q(\tau)), \quad (56)$$

$$\Delta_{2,1}(p,q)(\tau) = \delta(\tau)\Bigg(\widehat{\mathcal{E}}\Big(\pi\, {}^\rho\mathcal{I}_{0+}^{\xi+\varrho} f(\sigma, p(\sigma), q(\sigma)) - {}^\rho\mathcal{I}_{0+}^{\zeta} g(\mathcal{T}, p(\mathcal{T}), q(\mathcal{T}))\Big) + \mathcal{E}_2\Big(\epsilon\, {}^\rho\mathcal{I}_{0+}^{\zeta+\varsigma} g(\varpi, p(\varpi), q(\varpi)) - {}^\rho\mathcal{I}_{0+}^{\xi} f(\mathcal{T}, p(\mathcal{T}), q(\mathcal{T}))\Big)\Bigg), \quad (57)$$

$$\Delta_{2,2}(p,q)(\tau) = {}^\rho\mathcal{I}_{0+}^{\zeta} g(\tau, p(\tau), q(\tau)). \quad (58)$$

In the Banach space $\mathcal{P} \times \mathcal{Q}, \Delta_1(p,q)(\tau) = \Delta_{1,1}(p,q)(\tau) + \Delta_{1,2}(p,q)(\tau)$, and $\Delta_2(p,q)$ $(\tau) = \Delta_{2,1}(p,q)(\tau) + \Delta_{2,2}(p,q)(\tau)$ on $\mathcal{B}_\varphi$ are closed, bounded and convex subsets of $\mathcal{P} \times \mathcal{Q}$. Let us fix $\varphi \leq \max\{\mathcal{J}_1\mathcal{U}_1 + \mathcal{K}_1\mathcal{U}_2, \mathcal{J}_2\mathcal{U}_1 + \mathcal{K}_2\mathcal{U}_2\}$ and show that $\Delta\mathcal{B}_\varphi \subset \mathcal{B}_\varphi$ to verify Krasnoselskii's theorem [27] condition (i), If we choose $p = (p_1, p_2), q = (q_1, q_2) \in \mathcal{B}_\varphi$, and utilizing condition (53), we obtain

$$\begin{aligned}
&|\Delta_{1,1}(p,q)(\tau) + \Delta_{1,2}(p,q)(\tau)| \\
\leq& {}^\rho\mathcal{I}_{0+}^{\xi} |f(\tau, p(\tau), q(\tau))| \\
&+ |\delta(\tau)|\Bigg(|\widehat{\mathcal{E}}|\Big(|\epsilon|\, {}^\rho\mathcal{I}_{0+}^{\zeta+\varsigma} |g(\varpi, p(\varpi), q(\varpi))| + {}^\rho\mathcal{I}_{0+}^{\xi} |f(\mathcal{T}, p(\mathcal{T}), q(\mathcal{T}))|\Big) \\
&+ |\mathcal{E}_1|\Big(|\pi|\, {}^\rho\mathcal{I}_{0+}^{\xi+\varrho} |f(\sigma, p(\sigma), q(\sigma))| + {}^\rho\mathcal{I}_{0+}^{\zeta} |g(\mathcal{T}, p(\mathcal{T}), q(\mathcal{T}))|\Big)\Bigg) \\
\leq& \mathcal{U}_1\left\{\frac{\left(\mathcal{T}^{\rho\xi}(1 + |\delta||\widehat{\mathcal{E}}|)\right)}{\rho^\xi\Gamma(\xi+1)} + \frac{|\delta||\pi||\mathcal{E}_1|\sigma^{\rho(\xi+\varrho)}}{\rho^{\xi+\varrho}\Gamma(\xi+\varrho+1)}\right\} \\
&+ \mathcal{U}_2\left\{|\delta|\left(\frac{|\mathcal{E}_1|\mathcal{T}^{\rho\zeta}}{\rho^\zeta\Gamma(\zeta+1)} + \frac{|\widehat{\mathcal{E}}||\epsilon|\varpi^{\rho(\zeta+\varsigma)}}{\rho^{\zeta+\varsigma}\Gamma(\zeta+\varsigma+1)}\right)\right\} \\
\leq& \mathcal{U}_1\mathcal{J}_1 + \mathcal{U}_2\mathcal{K}_1 \leq \varphi.
\end{aligned}$$

In a similar manner, we can find that

$$|\Delta_{2,1}(p,q)(\tau) + \Delta_{2,2}(p,q)(\tau)| \leq \mathcal{U}_1\mathcal{J}_2 + \mathcal{U}_2\mathcal{K}_2 \leq \varphi.$$

Clearly the above two inequalities lead to the fact that $\Delta_1(p,q) + \Delta_2(p,q) \in \mathcal{B}_\varphi$. Thus, we define operator $(\Delta_{1,2}, \Delta_{2,2})$ as a contraction-satisfying condition (iii) of Krasnoselskii's theorem [27]. For $(p_1, q_1), (p_2, q_2) \in \mathcal{B}_\varphi$, we have

$$\begin{aligned}|\Delta_{1,2}(p_1,q_1)(\tau)-\Delta_{1,2}(p_2,q_2)(\tau)| \leq& \frac{\rho^{1-\xi}}{\Gamma(\xi)}\int_0^{\tau}\frac{\theta^{\rho-1}}{(\tau^{\rho}-\theta^{\rho})^{1-\xi}} \\ &\times |f(\theta,p_1(\theta),q_1(\theta))-f(\theta,p_2(\theta),q_2(\theta))|d\theta \\ \leq& \frac{\mathcal{T}^{\rho\xi}}{\rho^{\xi}\Gamma(\xi+1)}(\phi_1||p_1-p_2||+\phi_2||q_1-q_2||) \end{aligned} \quad (59)$$

and

$$\begin{aligned}|\Delta_{2,1}(p_1,q_1)(\tau)-\Delta_{2,1}(p_2,q_2)(\tau)| \leq& \frac{\rho^{1-\zeta}}{\Gamma(\zeta)}\int_0^{\tau}\frac{\theta^{\rho-1}}{(\tau^{\rho}-\theta^{\rho})^{1-\zeta}} \\ &\times |g(\theta,p_1(\theta),q_1(\theta))-g(\theta,p_2(\theta),q_2(\theta))|d\theta \\ \leq& \frac{\mathcal{T}^{\rho\zeta}}{\rho^{\zeta}\Gamma(\zeta+1)}(\hat{\phi}_1||p_1-p_2||+\hat{\phi}_2||q_1-q_2||). \end{aligned} \quad (60)$$

As a result (59) and (60),

$$\begin{aligned}&|(\Delta_{1,2},\Delta_{2,2})(p_1,q_1)(\tau)-(\Delta_{1,2},\Delta_{2,2})(p_2,q_2)(\tau)| \\ \leq& \frac{\mathcal{T}^{\rho\xi}(\phi_1+\phi_2)}{\rho^{\xi}\Gamma(\xi+1)}+\frac{\mathcal{T}^{\rho\zeta}(\hat{\phi}_1+\hat{\phi}_2)}{\rho^{\zeta}\Gamma(\zeta+1)}(||p_1-p_2||+||q_1-q_2||),\end{aligned}$$

is a contraction by (54). Therefore, condition (iii) of the Theorem is satisfied. Following that, we can establish that the operator $(\Delta_{1,1},\Delta_{2,1})$ satisfies the Krasnoselskii theorem's [27] condition (ii). We can infer the continuous existence of the $(\Delta_{1,1},\Delta_{2,1})$ operator by examining the continuity of the f,g functions. For each $(p,q)\in\mathcal{B}_{\varphi}$ we have

$$\begin{aligned}&|\Delta_{1,1}(p,q)(\tau)| \\ \leq& |\delta(\tau)|\Bigg(|\widehat{\mathcal{E}}|\Big(|\epsilon|\,{}^{\rho}\mathcal{I}_{0+}^{\zeta+\varsigma}|g(\varpi,p(\varpi),q(\varpi))|+{}^{\rho}\mathcal{I}_{0+}^{\xi}|f(\mathcal{T},p(\mathcal{T}),q(\mathcal{T}))|\Big) \\ &+|\mathcal{E}_1|\Big(|\pi|\,{}^{\rho}\mathcal{I}_{0+}^{\xi+\varrho}|f(\sigma,p(\sigma),q(\sigma))|+{}^{\rho}\mathcal{I}_{0+}^{\zeta}|g(\mathcal{T},p(\mathcal{T}),q(\mathcal{T}))|\Big)\Bigg)\end{aligned}$$

$$\begin{aligned}\leq& \mathcal{U}_1\left\{\frac{\left(\mathcal{T}^{\rho\xi}(|\delta||\widehat{\mathcal{E}}|)\right)}{\rho^{\xi}\Gamma(\xi+1)}+\frac{|\delta||\pi||\mathcal{E}_1|\sigma^{\rho(\xi+\varrho)}}{\rho^{\xi+\varrho}\Gamma(\xi+\varrho+1)}\right\} \\ &+\mathcal{U}_2\left\{|\delta|\left(\frac{|\mathcal{E}_1|\mathcal{T}^{\rho\zeta}}{\rho^{\zeta}\Gamma(\zeta+1)}+\frac{|\widehat{\mathcal{E}}||\epsilon|\varpi^{\rho(\zeta+\varsigma)}}{\rho^{\zeta+\varsigma}\Gamma(\zeta+\varsigma+1)}\right)\right\} \\ =& \hat{\Omega}_1,\end{aligned}$$

$$\begin{aligned}|\Delta_{2,1}(p,q)(\tau)| \leq& \mathcal{U}_2\left\{\frac{\left(\mathcal{T}^{\rho\zeta}(|\delta||\widehat{\mathcal{E}}|)\right)}{\rho^{\zeta}\Gamma(\zeta+1)}+\frac{|\delta||\epsilon||\mathcal{E}_2|\varpi^{\rho(\zeta+\varsigma)}}{\rho^{\zeta+\varsigma}\Gamma(\zeta+\varsigma+1)}\right\} \\ &+\mathcal{U}_1\left\{|\delta|\left(\frac{\mathcal{T}^{\rho\xi}|\mathcal{E}_2|}{\rho^{\xi}\Gamma(\xi+1)}+\frac{|\pi||\widehat{\mathcal{E}}|\sigma^{\rho(\xi+\varrho)}}{\rho^{\xi+\varrho}\Gamma(\xi+\varrho+1)}\right)\right\} \\ =& \hat{\Omega}_2,\end{aligned}$$

which leads to

$$||(\Delta_{1,1},\Delta_{2,1})(p,q)|| \leq \hat{\Omega}_1+\hat{\Omega}_2.$$

From the above inequalities, the set $(\Delta_{1,1}, \Delta_{2,1})\mathcal{B}_\varphi$ is uniformly bounded. The following step will demonstrate that the set $(\Delta_{1,1}, \Delta_{2,1})\mathcal{B}_\varphi$ is equicontinuous. For $\tau_1, \tau_2 \in \mathcal{E}$ with $\tau_1 < \tau_2$ and for any $(p,q) \in \mathcal{B}_\varphi$ we get

$$|\Delta_{1,1}(p,q)(\tau_2) - \Delta_{1,1}(p,q)(\tau_1)|$$

$$\begin{aligned}
\leq& |\delta(\tau_2) - \delta(\tau_1)| \Bigg(|\widehat{\mathcal{E}}| \Big(|\epsilon|\, {}^{\rho}\mathcal{I}_{0+}^{\zeta+\varsigma} |g(\omega, p(\omega), q(\omega))| + {}^{\rho}\mathcal{I}_{0+}^{\zeta} |f(\mathcal{T}, p(\mathcal{T}), q(\mathcal{T}))| \Big) \\
&+ |\mathcal{E}_1| \Big(|\pi|\, {}^{\rho}\mathcal{I}_{0+}^{\xi+\varrho} |f(\sigma, p(\sigma), q(\sigma))| + {}^{\rho}\mathcal{I}_{0+}^{\zeta} |g(\mathcal{T}, p(\mathcal{T}), q(\mathcal{T}))| \Big) \Bigg) \\
\leq& |\delta(\tau_2) - \delta(\tau_1) \Bigg(\mathcal{U}_1 \Big(\frac{\big(\mathcal{T}^{\rho\xi}(|\delta||\widehat{\mathcal{E}}|)\big)}{\rho^{\xi}\Gamma(\xi+1)} + \frac{|\delta||\pi||\mathcal{E}_1|\sigma^{\rho(\xi+\varrho)}}{\rho^{\xi+\varrho}\Gamma(\xi+\varrho+1)} \Big) \\
&+ \mathcal{U}_2|\delta| \Big(\frac{|\mathcal{E}_1|\mathcal{T}^{\rho\zeta}}{\rho^{\zeta}\Gamma(\zeta+1)} + \frac{|\widehat{\mathcal{E}}||\epsilon|\omega^{\rho(\zeta+\varsigma)}}{\rho^{\zeta+\varsigma}\Gamma(\zeta+\varsigma+1)} \Big) \Bigg).
\end{aligned}$$

Likewise, we obtain

$$\begin{aligned}
&|\Delta_{2,1}(p,q)(\tau_2) - \Delta_{2,1}(p,q)(\tau_1)| \\
\leq& |\delta(\tau_2) - \delta(\tau_1) \Bigg(\mathcal{U}_2 \Big(\frac{\big(\mathcal{T}^{\rho\zeta}(|\delta||\widehat{\mathcal{E}}|)\big)}{\rho^{\zeta}\Gamma(\zeta+1)} + \frac{|\delta||\epsilon||\mathcal{E}_2|\omega^{\rho(\zeta+\varsigma)}}{\rho^{\zeta+\varsigma}\Gamma(\zeta+\varsigma+1)} \Big) \\
&+ \mathcal{U}_1 \Big(|\delta| \Big(\frac{\mathcal{T}^{\rho\xi}|\mathcal{E}_2|}{\rho^{\xi}\Gamma(\xi+1)} + \frac{|\pi||\widehat{\mathcal{E}}|\sigma^{\rho(\xi+\varrho)}}{\rho^{\xi+\varrho}\Gamma(\xi+\varrho+1)} \Big) \Big) \Bigg).
\end{aligned}$$

Therefore $|(\Delta_{1,1}, \Delta_{2,1}(\tau_2)) - (\Delta_{1,1}, \Delta_{2,1}(\tau_1))| \to 0$ as $\tau_2 \to \tau_1$ independent of $(p,q) \in \mathcal{B}_\varphi$ Thus the set $(\Delta_{1,1}, \Delta_{2,1})\mathcal{B}_\varphi$ is equicontinuous. As an outcome, the Arzela–Ascoli theorem implies that the operator $(\Delta_{1,1}, \Delta_{2,1})$ is compact on $\mathcal{B}\varphi$. Krasnoselskii's theorem [27] statement leads us to the conclusion that the problem (1) and (2) has at least one solution on $\mathcal{E}$. □

4. Example

Consider the following Liouville–Caputo type generalized FDEs coupled system:

$$\begin{cases} {}^{\frac{3}{4}}_{C}\mathcal{D}_{0+}^{\frac{5}{4}} p(\tau) = f(\tau, p(\tau), q(\tau)), \tau \in \mathcal{E} := [0,1], \\ {}^{\frac{3}{4}}_{C}\mathcal{D}_{0+}^{\frac{31}{20}} q(\tau) = g(\tau, p(\tau), q(\tau)), \tau \in \mathcal{E} := [0,1], \end{cases} \tag{61}$$

supplemented with boundary conditions:

$$\Big\{ p(0) = 0,\ q(0) = 0,\ p(1) = \tfrac{1}{6}\, {}^{\frac{3}{4}}\mathcal{I}^{\frac{13}{20}} q(\tfrac{7}{10}),\ q(1) = \tfrac{1}{7}\, {}^{\frac{3}{4}}\mathcal{I}^{\frac{17}{20}} p(\tfrac{1}{2}), \tag{62}$$

where $\xi = \frac{5}{4}, \zeta = \frac{31}{20}, \rho = \frac{3}{4}, \mathcal{T} = 1, \epsilon = \frac{1}{6}, \omega = \frac{7}{10}, \pi = \frac{1}{7}, \sigma = \frac{1}{2}, \varsigma = \frac{13}{20}, \varrho = \frac{17}{20}$ and

$$f(\tau, p(\tau), q(\tau)) = \frac{(1+\tau)}{30} \Big(\frac{|p(\tau)|}{1+|p(\tau)|} + \frac{1}{3}\cos(q(\tau)) + 3\tau \Big), \tag{63}$$

$$g(\tau, p(\tau), q(\tau)) = \frac{e^{-\tau}}{25} \Big(\frac{\sqrt{\tau}+1}{5} + \frac{1}{6}\cos(p(\tau)) + \frac{|q(\tau)|}{1+|q(\tau)|} \Big). \tag{64}$$

With $\psi_0 = \frac{1}{10}, \psi_1 = \frac{1}{30}, \psi_2 = \frac{1}{90}, \hat{\psi}_0 = \frac{1}{125}, \hat{\psi}_1 = \frac{1}{25}$, and $\hat{\psi}_2 = \frac{1}{150}$, the functions f and g clearly satisfy the $(\mathcal{A}_1)$ condition. Next, we find that $(\mathcal{J}_1) = 2.5370237266984113$,

$(\mathcal{K}_1) = 0.17111607453629377, \mathcal{J}_2 = 0.0906406939922634, \mathcal{K}_2 = 2.274156747108814, \mathcal{J}_i, \mathcal{K}_i$ $(i = 1,2)$ are respectively given by (35),(36),(37) and (38), based on the data available. Thus $\psi_1(\mathcal{J}_1 + \mathcal{J}_2) + \hat{\psi}_1(\mathcal{K}_1 + \mathcal{K}_2) \approxeq 0.18539972688882678 < 1, \psi_2(\mathcal{J}_1 + \mathcal{J}_2) + \hat{\psi}_2(\mathcal{K}_1 + \mathcal{K}_2) \approxeq$ $0.04549809015197488 < 1$, all the conditions of Theorem 1 are satisfied, and there is at least one solution for problem (61) and (62) on $[0,1]$ with f and g given by (63) and (64) respectively.

In addition, we'll use

$$f(\tau, p(\tau), q(\tau)) = \frac{\tau}{3} + \frac{3}{4(\tau+16)} + \frac{|p(\tau)|}{1+|p(\tau)|} + \frac{2}{75}\cos(q(\tau)), \tag{65}$$

$$g(\tau, p(\tau), q(\tau)) = \frac{(1+e^{-\tau})}{4} + \frac{19}{400}\cos(p(\tau)) + \frac{1}{60}\frac{|q(\tau)|}{1+|q(\tau)|}, \tag{66}$$

to demonstrate Theorem 2. It is simple to demonstrate that f and g are continuous and satisfy the assumption $(\mathcal{A}_2)$ with $\phi_1 = \frac{3}{64}, \phi_2 = \frac{2}{75}, \hat{\phi}_1 = \frac{19}{400}$ and $\hat{\phi}_2 = \frac{1}{60}$. All the assumptions of Theorem 2 are also satisfied with $(\mathcal{J}_1 + \mathcal{J}_2)(\phi_1 + \phi_2) + (\mathcal{K}_1 + \mathcal{K}_2)(\hat{\phi}_1 + \hat{\phi}_2) \approxeq 0.35014782699385444 < 1$. As a result, Theorem 2 holds true, and the problem (61) and (62) with f and g given by (65) and (66) respectively, has a unique solution on [0,1].

5. Ulam–Hyers Stability Results for the Problem (1) and (2)

The U–H stability of the solutions to the BVP (1) and (2) will be discussed in this section using the integral representation of their solutions defined by

$$p(\tau) = \Delta_1(p,q)(\tau),\ q(\tau) = \Delta_2(p,q)(\tau), \tag{67}$$

where Δ_1 and Δ_2 are given by (33) and (34). Consider the following definitions of nonlinear operators

$$\mathcal{H}_1, \mathcal{H}_2 \in \mathcal{C}(\mathcal{E},\mathbb{R}) \times \mathcal{C}(\mathcal{E},\mathbb{R}) \to \mathcal{C}(\mathcal{E},\mathbb{R}),$$

$$\begin{cases} {}^{\rho}_{C}\mathcal{D}^{\xi}_{0+}p(\tau) - f(\tau,p(\tau),q(\tau)) = \mathcal{H}_1(p,q)(\tau), \tau \in \mathcal{E}, \\ {}^{\rho}_{C}\mathcal{D}^{\zeta}_{0+}q(\tau) - g(\tau,p(\tau),q(\tau)) = \mathcal{H}_1(p,q)(\tau), \tau \in \mathcal{E}. \end{cases}$$

It considered the following inequalities for some $\hat{\lambda}_1, \hat{\lambda}_2 > 0:$

$$||\mathcal{H}_1(p,q)|| \leq \hat{\lambda}_1, ||\mathcal{H}_2(p,q)|| \leq \hat{\lambda}_2. \tag{68}$$

Definition 4. *The coupled system (1) and (2) is said to be U–H stable if $\mathcal{V}_1, \mathcal{V}_2 > 0$ and there exists a unique solution $(p,q) \in \mathcal{C}(\mathcal{E},\mathbb{R})$ of a problem (1) and (2) with*

$$||(p,q) - (p^*,q^*)|| \leq \mathcal{V}_1\hat{\lambda}_1 + \mathcal{V}_2\hat{\lambda}_2,$$

$\forall (p,q) \in \mathcal{C}(\mathcal{E},\mathbb{R})$ of inequality (68).

Theorem 4. *Assume that $(\mathcal{A}_2)$ holds. Then the problem (1) and (2) is U–H stable.*

Proof. Let $(p,q) \in \mathcal{C}(\mathcal{E},\mathbb{R}) \times \mathcal{C}(\mathcal{E},\mathbb{R})$ be the (1)–(2) solution of the problem that satisfies (33) and (34). Let (p,q) be any solution that meets the condition (68):

$$\begin{cases} {}^{\rho}_{C}\mathcal{D}^{\xi}_{0+}p(\tau) = f(\tau,p(\tau),q(\tau)) + \mathcal{H}_1(p,q)(\tau), \tau \in \mathcal{E}, \\ {}^{\rho}_{C}\mathcal{D}^{\zeta}_{0+}q(\tau) = g(\tau,p(\tau),q(\tau)) + \mathcal{H}_1(p,q)(\tau), \tau \in \mathcal{E}, \end{cases}$$

so,

$$\begin{aligned}p^*(\tau) &= \Delta_1(p^*,q^*)(\tau) +^{\rho} \mathcal{I}_{0+}^{\xi}\mathcal{H}_1(p,q)(\tau)\\&+\delta(\tau)\Bigg(\widehat{\mathcal{E}}\Big[\epsilon\,{}^{\rho}\mathcal{I}_{0+}^{\zeta+\varsigma}\mathcal{H}_2(p,q)(\varpi) -^{\rho} \mathcal{I}_{0+}^{\xi}\mathcal{H}_1(p,q)(\mathcal{T})\Big]\\&+\mathcal{E}_1\Big[\pi\,{}^{\rho}\mathcal{I}_{0+}^{\xi+\varrho}\mathcal{H}_1(p,q)(\sigma) -^{\rho} \mathcal{I}_{0+}^{\zeta}\mathcal{H}_2(p,q)(\mathcal{T})\Big]\Bigg).\end{aligned}$$

It follows that

$$\begin{aligned}|\Delta_1(p^*,q^*)(\tau) - p^*(\tau)| &\leq^{\rho}\mathcal{I}_{0+}^{\xi}|\mathcal{H}_1(p,q)(\tau)|\\&+|\delta(\tau)|\Bigg(|\widehat{\mathcal{E}}|\Big[|\epsilon|\,{}^{\rho}\mathcal{I}_{0+}^{\zeta+\varsigma}|\mathcal{H}_2(p,q)(\varpi)| +^{\rho} \mathcal{I}_{0+}^{\xi}|\mathcal{H}_1(p,q)(\mathcal{T})|\Big]\\&+|\mathcal{E}_1|\Big[|\pi|\,{}^{\rho}\mathcal{I}_{0+}^{\xi+\varrho}|\mathcal{H}_1(p,q)(\sigma)| +^{\rho} \mathcal{I}_{0+}^{\zeta}|\mathcal{H}_2(p,q)(\mathcal{T})|\Big]\Bigg)\\&\leq \hat{\lambda}_1\left\{\frac{\left(\mathcal{T}^{\rho\xi}(1+|\delta||\widehat{\mathcal{E}}|)\right)}{\rho^{\xi}\Gamma(\xi+1)}+\frac{|\delta||\pi||\mathcal{E}_1|\sigma^{\rho(\xi+\varrho)}}{\rho^{\xi+\varrho}\Gamma(\xi+\varrho+1)}\right\}\\&+\hat{\lambda}_2\left\{|\delta|\left(\frac{|\mathcal{E}_1|\mathcal{T}^{\rho\zeta}}{\rho^{\zeta}\Gamma(\zeta+1)}+\frac{|\widehat{\mathcal{E}}||\epsilon|\varpi^{\rho(\zeta+\varsigma)}}{\rho^{\zeta+\varsigma}\Gamma(\zeta+\varsigma+1)}\right)\right\}\\&\leq\mathcal{J}_1\hat{\lambda}_1+\mathcal{K}_1\hat{\lambda}_2.\end{aligned}$$

Similarly, we obtain

$$\begin{aligned}|\Delta_2(p^*,q^*)(\tau) - q^*(\tau)| &\leq\hat{\lambda}_2\left\{\frac{\left(\mathcal{T}^{\rho\zeta}(1+|\delta||\widehat{\mathcal{E}}|)\right)}{\rho^{\zeta}\Gamma(\zeta+1)}+\frac{|\delta||\epsilon||\mathcal{E}_2|\varpi^{\rho(\zeta+\varsigma)}}{\rho^{\zeta+\varsigma}\Gamma(\zeta+\varsigma+1)}\right\}\\&+\hat{\lambda}_1\left\{|\delta|\left(\frac{\mathcal{T}^{\rho\xi}|\mathcal{E}_2|}{\rho^{\xi}\Gamma(\xi+1)}+\frac{|\pi||\widehat{\mathcal{E}}|\sigma^{\rho(\xi+\varrho)}}{\rho^{\xi+\varrho}\Gamma(\xi+\varrho+1)}\right)\right\}\\&\leq\mathcal{J}_2\hat{\lambda}_1+\mathcal{K}_2\hat{\lambda}_2,\end{aligned}$$

where $\mathcal{J}_1, \mathcal{K}_1, \mathcal{J}_2$, and $\mathcal{K}_2$ are defined in (35)–(38), respectively. As an outcome, we deduce from operator Δ's fixed-point property, which is defined by (33) and (34),

$$\begin{aligned}|p(\tau) - p^*(\tau)| &=|p(\tau) - \Delta_1(p^*,q^*)(\tau) + \Delta_1(p^*,q^*)(\tau) - p^*(\tau)|\\&\leq|\Delta_1(p,q)(\tau) - \Delta_1(p^*,q^*)(\tau)| + |\Delta_1(p^*,q^*)(\tau) - p^*(\tau)|\\&\leq((\mathcal{J}_1\phi_1+\mathcal{K}_1\phi_1)+(\mathcal{J}_1\phi_2+\mathcal{K}_1\phi_2))\|(p,q)-(p^*,q^*)\|\\&\quad+\mathcal{J}_1\hat{\lambda}_1+\mathcal{K}_1\hat{\lambda}_2.\end{aligned} \tag{69}$$

$$\begin{aligned}|q(\tau) - q^*(\tau)| &=|q(\tau) - \Delta_2(p^*,q^*)(\tau) + \Delta_2(p^*,q^*)(\tau) - q^*(\tau)|\\&\leq|\Delta_2(p,q)(\tau) - \Delta_2(p^*,q^*)(\tau)| + |\Delta_2(p^*,q^*)(\tau) - q^*(\tau)|\\&\leq((\mathcal{J}_2\phi_1+\mathcal{K}_2\phi_1)+(\mathcal{J}_2\phi_2+\mathcal{K}_2\phi_2))\|(p,q)-(p^*,q^*)\|\\&\quad+\mathcal{J}_2\hat{\lambda}_1+\mathcal{K}_2\hat{\lambda}_2.\end{aligned} \tag{70}$$

From the above Equations (69) and (70) it follows that

$$\begin{aligned}||(p,q)-(p^*,q^*)|| \leq & (\mathcal{J}_1+\mathcal{J}_2)\hat{\lambda}_1+(\mathcal{K}_1+\mathcal{K}_2)\hat{\lambda}_2 \\ & +((\mathcal{J}_1+\mathcal{J}_2)(\phi_1+\phi_2)+(\mathcal{K}_1+\mathcal{K}_2)(\hat{\phi}_1+\hat{\phi}_2))||(p,q)-(p^*,q^*)||.\end{aligned}$$

$$\begin{aligned}||(p,q)-(p^*,q^*)|| \leq & \frac{(\mathcal{J}_1+\mathcal{J}_2)\hat{\lambda}_1+(\mathcal{K}_1+\mathcal{K}_2)\hat{\lambda}_2}{1-((\mathcal{J}_1+\mathcal{J}_2)(\phi_1+\phi_2)+(\mathcal{K}_1+\mathcal{K}_2)(\hat{\phi}_1+\hat{\phi}_2))} \\ \leq & \mathcal{V}_1\hat{\lambda}_1+\mathcal{V}_2\hat{\lambda}_2,\end{aligned}$$

with

$$\mathcal{V}_1=\frac{\mathcal{J}_1+\mathcal{J}_2}{1-((\mathcal{J}_1+|\mathcal{J}_2)(\phi_1+\phi_2)+(\mathcal{K}_1+|\mathcal{K}_2)(\hat{\phi}_1+\hat{\phi}_2))},$$

$$\mathcal{V}_2=\frac{\mathcal{K}_1+\mathcal{K}_2}{1-((\mathcal{J}_1+\mathcal{J}_2)(\phi_1+\phi_2)+(\mathcal{K}_1+\mathcal{K}_2)(\hat{\phi}_1+\hat{\phi}_2))}.$$

Hence, the problem (1)–(2) is U–H stable. □

6. Example

Consider the following Liouville–Caputo type generalized FDEs coupled system:

$$\begin{cases} {}^{\frac{19}{20}}_{C}\mathcal{D}^{\frac{5}{4}}_{0+}p(\tau)=\frac{\sqrt{\tau}}{2}+\frac{1}{5(\tau+25)}\frac{|p(\tau)|}{1+|p(\tau)|}+\frac{3}{80}\cos(q(\tau)), \tau\in[0,1], \\ {}^{\frac{19}{20}}_{C}\mathcal{D}^{\frac{31}{20}}_{0+}q(\tau)=\frac{\tau}{5}+\frac{17}{300}\cos(p(\tau))+\frac{1}{70}\frac{|q(\tau)|}{1+|q(\tau)|}, \tau\in[0,1], \end{cases} \tag{71}$$

supplemented with boundary conditions:

$$\left\{p(0)=0,\ q(0)=0,\ p(1)=\frac{5}{6}{}^{\frac{19}{20}}\mathcal{I}^{\frac{13}{20}}q(\tfrac{9}{20}),\ q(1)=\frac{6}{7}{}^{\frac{19}{20}}\mathcal{I}^{\frac{17}{20}}p(\tfrac{13}{20}),\right. \tag{72}$$

where $\xi=\frac{5}{4},\zeta=\frac{31}{20},\rho=\frac{19}{20},\mathcal{T}=1,\epsilon=\frac{5}{6},\varpi=\frac{9}{20},\pi=\frac{6}{7},\sigma=\frac{13}{20},\varsigma=\frac{13}{20},\varrho=\frac{17}{20}$ and

$$|f(\tau,p_1(\tau),q_1(\tau))-f(\tau,p_2(\tau),q_2(\tau))|=\frac{1}{125}|p_1(\tau)-p_2(\tau)|+\frac{3}{80}|q_1(\tau)-q_2(\tau)|, \tag{73}$$

$$|g(\tau,p_1(\tau),q_1(\tau))-g(\tau,p_2(\tau),q_2(\tau))|=\frac{17}{300}|p_1(\tau)-p_2(\tau)|+\frac{1}{70}|q_1(\tau)-q_2(\tau)|. \tag{74}$$

With $\phi_1=\frac{1}{125},\phi_2=\frac{3}{80},\hat{\phi}_1=\frac{17}{300}$, and $\hat{\phi}_2=\frac{1}{70}$, the functions f and g clearly satisfy the $(\mathcal{A}_2)$ condition. Next, we find that $(\mathcal{J}_1)=1.9529307397739033,(\mathcal{K}_1)=0.21135021378560123,\mathcal{J}_2=0.42682560046779994,\mathcal{K}_2=1.6225052940838325,\mathcal{J}_i,\mathcal{K}_i(i=1,2)$ are respectively given by (35),(36),(37) and (38), based on the data available. Thus $((\mathcal{J}_1+\mathcal{J}_2)(\phi_1+\phi_2)+(\mathcal{K}_1+\mathcal{K}_2)(\hat{\phi}_1+\hat{\phi}_2))\approxeq 0.2383953280869716<1$, all the conditions of Theorem 5.2 are satisfied, and there is a unique solution for problem (71) and (72) on $[0,1]$, which is stable for Ulam–Hyers, with f and g given by (73) and (74) respectively.

7. Existence Results for the Problem (1) and (75)

Furthermore, we are investigating the system (1) under the following conditions:

$$\begin{cases} p(0)=0, \quad q(0)=0, \\ p(\mathcal{T})=\epsilon^{\rho}\mathcal{I}^{\varsigma}_{0+}q(\varpi)=\frac{\epsilon\rho^{1-\varsigma}}{\Gamma(\varsigma)}\int_0^{\varpi}\frac{\theta^{\rho-1}}{(\varpi^{\rho}-\theta^{\rho})^{1-\varsigma}}q(\theta)d\theta, \\ q(\mathcal{T})=\pi^{\rho}\mathcal{I}^{\varrho}_{0+}p(\varpi)=\frac{\pi\rho^{1-\varrho}}{\Gamma(\varrho)}\int_0^{\varpi}\frac{\theta^{\rho-1}}{(\varpi^{\rho}-\theta^{\rho})^{1-\varrho}}p(\theta)d\theta, \\ 0<\varpi<\mathcal{T}. \end{cases} \tag{75}$$

Bear in mind that the conditions (2) contain strips of varying lengths, whereas the one in (75) contains only one strip of the same length $(0, \varpi)$. We introduce the following notations for computational ease:

$$\mathcal{E}_1 = \epsilon \frac{\varpi^{\rho(\varsigma+1)}}{\rho^{\varsigma+1}\Gamma(\varsigma+2)}, \ \mathcal{E}_2 = \pi \frac{\varpi^{\rho(\varrho+1)}}{\rho^{\varrho+1}\Gamma(\varrho+2)}, \ \widehat{\mathcal{E}} = \frac{\mathcal{T}^{\rho}}{\rho}, \tag{76}$$

$$\mathcal{G} = \widehat{\mathcal{E}}^2 - \mathcal{E}_1\mathcal{E}_2 \neq 0, \tag{77}$$

$$\delta(\tau) = \left(\frac{\tau^{\rho}}{\rho\mathcal{G}}\right). \tag{78}$$

Lemma 4. *Given the functions* $\hat{f}, \hat{g} \in C(0,\mathcal{T}) \cap \mathcal{L}(0,\mathcal{T}), p, q \in \mathcal{AC}^2_{\gamma}(\mathcal{E})$ *and* $\Lambda \neq 0$. *Then the solution of the coupled BVP:*

$$\begin{cases} {}^{\rho}_{C}\mathcal{D}^{\xi}_{0+}p(\tau) = \hat{f}(\tau), \tau \in \mathcal{E} := [0,\mathcal{T}], \\ {}^{\rho}_{C}\mathcal{D}^{\zeta}_{0+}q(\tau) = \hat{g}(\tau), \tau \in \mathcal{E} := [0,\mathcal{T}], \\ p(0) = 0, \ q(0) = 0, \ p(\mathcal{T}) = \epsilon^{\rho}\mathcal{I}^{\varsigma}_{0+}q(\varpi), \ q(\mathcal{T}) = \pi^{\rho}\mathcal{I}^{\varrho}_{0+}p(\varpi), \ 0 < \varpi < \mathcal{T}, \end{cases} \tag{79}$$

is given by

$$p(\tau) = {}^{\rho}\mathcal{I}^{\xi}_{0+}\hat{f}(\tau) + \delta(\tau)\left(\left[\epsilon\, {}^{\rho}\mathcal{I}^{\zeta+\varsigma}_{0+}\hat{g}(\varpi) - {}^{\rho}\mathcal{I}^{\xi}_{0+}\hat{f}(\mathcal{T})\right] + \left[\pi\, {}^{\rho}\mathcal{I}^{\xi+\varsigma}_{0+}\hat{f}(\varpi) - {}^{\rho}\mathcal{I}^{\zeta}_{0+}\hat{g}(\mathcal{T})\right]\right) \tag{80}$$

and

$$q(\tau) = {}^{\rho}\mathcal{I}^{\zeta}_{0+}\hat{g}(\tau) + \delta(\tau)\left(\left[\pi\, {}^{\rho}\mathcal{I}^{\xi+\varsigma}_{0+}\hat{f}(\varpi) - {}^{\rho}\mathcal{I}^{\zeta}_{0+}\hat{g}(\mathcal{T})\right] + \left[\epsilon\, {}^{\rho}\mathcal{I}^{\zeta+\varsigma}_{0+}\hat{g}(\varpi) - {}^{\rho}\mathcal{I}^{\xi}_{0+}\hat{f}(\mathcal{T})\right]\right). \tag{81}$$

Proof. When ${}^{\rho}\mathcal{I}^{\xi}_{0+}, {}^{\rho}\mathcal{I}^{\zeta}_{0+}$ are applied to the FDEs in (79) and Lemma 4 is used the solution of the FDEs in (79) for $\tau \in \mathcal{E}$ is

$$p(\tau) = {}^{\rho}\mathcal{I}^{\xi}_{0+}\hat{f}(\tau) + a_1 + a_2\frac{\tau^{\rho}}{\rho} = \frac{\rho^{1-\xi}}{\Gamma(\xi)}\int_0^{\tau}\theta^{\rho-1}(\tau^{\rho}-\theta^{\rho})^{\xi-1}\hat{f}(\theta)d\theta + a_1 + a_2\frac{\tau^{\rho}}{\rho}, \tag{82}$$

$$q(\tau) = {}^{\rho}\mathcal{I}^{\zeta}_{0+}\hat{g}(\tau) + b_1 + b_2\frac{\tau^{\rho}}{\rho} = \frac{\rho^{1-\zeta}}{\Gamma(\zeta)}\int_0^{\tau}\theta^{\rho-1}(\tau^{\rho}-\theta^{\rho})^{\zeta-1}\hat{g}(\theta)d\theta + b_1 + b_2\frac{\tau^{\rho}}{\rho}, \tag{83}$$

respectively, for some $a_1, a_2, b_1, b_2 \in \mathcal{R}$. Making use of the boundary conditions $p(0) = q(0) = 0$ in (82) and (83) respectively, we get $a_1 = b_1 = 0$. We obtain by using the generalized integral operators ${}^{\rho}\mathcal{I}^{\varrho}_{0+}, {}^{\rho}\mathcal{I}^{\zeta}_{0+}$ (82) and (83) respectively,

$${}^{\rho}\mathcal{I}^{\varrho}_{0+}p(\tau) = {}^{\rho}\mathcal{I}^{\xi+\varrho}_{0+}\hat{f}(\tau) + a_1\frac{\tau^{\rho\varrho}}{\rho^{\varrho}\Gamma(\varrho+1)} + a_2\frac{\tau^{\rho(\varrho+1)}}{\rho^{\varrho+1}\Gamma(\varrho+2)}, \tag{84}$$

$${}^{\rho}\mathcal{I}^{\varsigma}_{0+}q(\tau) = {}^{\rho}\mathcal{I}^{\zeta+\varsigma}_{0+}\hat{g}(\tau) + b_1\frac{\tau^{\rho\varsigma}}{\rho^{\varsigma}\Gamma(\varsigma+1)} + b_2\frac{\tau^{\rho(\varsigma+1)}}{\rho^{\varsigma+1}\Gamma(\varsigma+2)}, \tag{85}$$

which, when combined with the boundary conditions $p(\mathcal{T})=\epsilon^{\rho}\mathcal{I}^{\varsigma}_{0+}q(\varpi), q(\mathcal{T}) = \pi^{\rho}\mathcal{I}^{\varrho}_{0+}p(\varpi)$, gives the following results:

$${}^{\rho}\mathcal{I}^{\xi}_{0+}\hat{f}(\mathcal{T}) + a_1 + a_2\frac{\mathcal{T}^{\rho}}{\rho} = \epsilon^{\rho}\mathcal{I}^{\zeta+\varsigma}_{0+}\hat{g}(\varpi) + b_1\frac{\epsilon\varpi^{\rho\varsigma}}{\rho^{\varsigma}\Gamma(\varsigma+1)} + b_2\frac{\epsilon\varpi^{\rho(\varsigma+1)}}{\rho^{\varsigma+1}\Gamma(\varsigma+2)}, \tag{86}$$

$$ {}^{\rho}\mathcal{I}^{\zeta}_{0+}\hat{g}(\mathcal{T}) + b_1 + b_2\frac{\mathcal{T}^{\rho}}{\rho} = \pi^{\rho}\mathcal{I}^{\xi+\varrho}_{0+}\hat{f}(\varpi) + a_1\frac{\pi\varpi^{\rho\varrho}}{\rho^{\varrho}\Gamma(\varrho+1)} + a_2\frac{\pi\varpi^{\rho(\varrho+1)}}{\rho^{\varrho+1}\Gamma(\varrho+2)}. \quad (87) $$

Next, we obtain

$$ a_2\widehat{\mathcal{E}} - b_2\mathcal{E}_1 = \epsilon^{\rho}\mathcal{I}^{\zeta+\varsigma}_{0+}\hat{g}(\varpi) -^{\rho}\mathcal{I}^{\xi}_{0+}\hat{f}(\mathcal{T}), \quad (88) $$

$$ b_2\widehat{\mathcal{E}} - a_2\mathcal{E}_2 = \pi^{\rho}\mathcal{I}^{\xi+\varrho}_{0+}\hat{f}(\varpi) -^{\rho}\mathcal{I}^{\zeta}_{0+}\hat{g}(\mathcal{T}), \quad (89) $$

by employing the notations (76)–(78) in (86) and (87) respectively. We find that when we solve the system of Equations (88) and (89) for a_2 and b_2,

$$ a_2 = \frac{1}{\mathcal{G}}\left[\widehat{\mathcal{E}}\left(\epsilon^{\rho}\mathcal{I}^{\zeta+\varsigma}_{0+}\hat{g}(\varpi) -^{\rho}\mathcal{I}^{\xi}_{0+}\hat{f}(\mathcal{T})\right) + \mathcal{E}_1\left(\pi^{\rho}\mathcal{I}^{\xi+\varrho}_{0+}\hat{f}(\varpi) -^{\rho}\mathcal{I}^{\zeta}_{0+}\hat{g}(\mathcal{T})\right)\right], \quad (90) $$

$$ b_2 = \frac{1}{\mathcal{G}}\left[\mathcal{E}_2\left(\epsilon^{\rho}\mathcal{I}^{\zeta+\varsigma}_{0+}\hat{g}(\varpi) -^{\rho}\mathcal{I}^{\xi}_{0+}\hat{f}(\mathcal{T})\right) + \widehat{\mathcal{E}}\left(\pi^{\rho}\mathcal{I}^{\xi+\varrho}_{0+}\hat{f}(\varpi) -^{\rho}\mathcal{I}^{\zeta}_{0+}\hat{g}(\mathcal{T})\right)\right]. \quad (91) $$

Substituting the values of a_1, a_2, b_1, b_2 in (82) and (83) respectively, we get the solution for (79). □

For brevity's sake, we'll use the following notations:

$$ \mathcal{J}_1 = \frac{\left(\mathcal{T}^{\rho\xi}(1+|\delta||\widehat{\mathcal{E}}|)\right)}{\rho^{\xi}\Gamma(\xi+1)} + \frac{|\delta||\pi||\mathcal{E}_1|\varpi^{\rho(\xi+\varrho)}}{\rho^{\xi+\varrho}\Gamma(\xi+\varrho+1)}, \quad (92) $$

$$ \mathcal{K}_1 = |\delta|\left(\frac{|\mathcal{E}_1|\mathcal{T}^{\rho\zeta}}{\rho^{\zeta}\Gamma(\zeta+1)} + \frac{|\widehat{\mathcal{E}}||\epsilon|\varpi^{\rho(\zeta+\varsigma)}}{\rho^{\zeta+\varsigma}\Gamma(\zeta+\varsigma+1)}\right), \quad (93) $$

$$ \mathcal{J}_2 = |\delta|\left(\frac{\mathcal{T}^{\rho\xi}|\mathcal{E}_2|}{\rho^{\xi}\Gamma(\xi+1)} + \frac{|\pi||\widehat{\mathcal{E}}|\varpi^{\rho(\xi+\varrho)}}{\rho^{\xi+\varrho}\Gamma(\xi+\varrho+1)}\right), \quad (94) $$

$$ \mathcal{K}_2 = \frac{\left(\mathcal{T}^{\rho\zeta}(1+|\delta||\widehat{\mathcal{E}}|)\right)}{\rho^{\zeta}\Gamma(\zeta+1)} + \frac{|\delta||\epsilon||\mathcal{E}_2|\varpi^{\rho(\zeta+\varsigma)}}{\rho^{\zeta+\varsigma}\Gamma(\zeta+\varsigma+1)}. \quad (95) $$

To finish up, we will go over the results of existence, uniqueness, and Ulam–Hyers stability for problems (1) and (75), respectively. For reasons that are similar to those in Sections 3–6, we are not providing the proof.

Corollary 1. *Assume that $f, g : \mathcal{E}\times\mathbb{R}\times\mathbb{R}\to\mathbb{R}$ are continuous functions satisfying the condition: $(\mathcal{A}_1)$ there exists constants $\psi_m, \hat{\psi_m} \le 0 (m = 1,2)$ and $\psi_0, \hat{\psi_0} > 0$ such that*

$$ |f(\tau, o_1, o_2)| \le \psi_0 + \psi_1|o_1| + \psi_2|o_2|, $$
$$ |g(\tau, o_1, o_2)| \le \hat{\psi_0} + \hat{\psi_1}|o_1| + \hat{\psi_2}|o_2|, \forall o_m \in \mathbb{R}, m = 1,2. $$

If $\psi_1(\hat{\mathcal{J}_1} + \hat{\mathcal{J}_2}) + \hat{\psi_1}(\hat{\mathcal{K}_1} + \hat{\mathcal{K}_2}) < 1, \psi_2(\hat{\mathcal{J}_1} + \hat{\mathcal{J}_2}) + \hat{\psi_2}(\hat{\mathcal{K}_1} + \hat{\mathcal{K}_2}) < 1$. Then at least one solution for the BVP (1) and (75) on $\mathcal{E}$, where $\hat{\mathcal{J}_1}, \hat{\mathcal{K}_1}, \hat{\mathcal{J}_2}, \hat{\mathcal{K}_2}$ are given by (92)–(95) respectively.

Corollary 2. *Assume that $f, g : \mathcal{E} \times \mathbb{R} \times \mathbb{R} \to \mathbb{R}$ are continuous functions satisfying the condition: $(\mathcal{A}_2)$ there exists constants $\phi_m, \hat{\phi_m} \leq 0(m = 1, 2)$ such that*

$$|f(\tau, o_1, o_2) - f(\tau, \hat{o}_1, \hat{o}_2)| \leq \phi_1|o_1 - \hat{o}_1| + \phi_2|o_2 - \hat{o}_2|,$$
$$|g(\tau, o_1, o_2) - g(\tau, \hat{o}_1, \hat{o}_2)| \leq \hat{\phi}_1|o_1 - \hat{o}_1| + \hat{\phi}_2|o_2 - \hat{o}_2|, \forall o_m, \hat{o}_m \in \mathbb{R}, m = 1, 2.$$

Moreover, there exist $\mathcal{S}_1, \mathcal{S}_2 > 0$ such that $|f(\tau, 0, 0)| \leq \mathcal{S}_1, |f(\tau, 0, 0)| \leq \mathcal{S}_2$, Then, given that

$$(\mathcal{J}_1 + \mathcal{J}_2)(\phi_1 + \phi_2) + (\mathcal{K}_1 + \mathcal{K}_2)(\hat{\phi}_1 + \hat{\phi}_2) < 1, \tag{96}$$

the BVP (1) and (75) has a unique solution on $\mathcal{E}$, where $\hat{\mathcal{J}}_1, \hat{\mathcal{K}}_1, \hat{\mathcal{J}}_2, \hat{\mathcal{K}}_2$ are given by (92)–(95) respectively.

Corollary 3. *Assume that $f, g : \mathcal{E} \times \mathbb{R} \times \mathbb{R} \to \mathbb{R}$ are continuous functions satisfying the assumption $(\mathcal{A}_2)$ in Theorem 2. Further more, there exist positive constants $\mathcal{U}_1, \mathcal{U}_2$ such that $\forall \tau \in \mathcal{E}$ and $r_i \in \mathbb{R}, i = 1, 2$.*

$$|f(\tau, r_1, r_2)| \leq \mathcal{U}_1, \qquad |g(\tau, r_1, r_2)| \leq \mathcal{U}_2. \tag{97}$$

If

$$\frac{\mathcal{T}^{\rho\xi}(\phi_1 + \phi_2)}{\rho^{\xi}\Gamma(\xi + 1)} + \frac{\mathcal{T}^{\rho\zeta}(\hat{\phi}_1 + \hat{\phi}_2)}{\rho^{\zeta}\Gamma(\zeta + 1)} < 1, \tag{98}$$

then the BVP (1), and (75) has at least one solution on $\mathcal{E}$.

Corollary 4. *Assume that (A2) holds. Then the problem (1) and (75) is Ulam–Hyers stable.*

8. Asymmetric Cases

Remark 1. *If $\rho = 1$, the problem (1) generalized Liouville–Caputo type reduces to the classical Caputo form.*

$$\begin{cases} {}^{C}\mathcal{D}^{\xi}_{0+}p(\tau) = f(\tau, p(\tau), q(\tau)), \tau \in \mathcal{G} := [0, \mathcal{T}], \\ {}^{C}\mathcal{D}^{\zeta}_{0+}q(\tau) = g(\tau, p(\tau), q(\tau)), \tau \in \mathcal{G} := [0, \mathcal{T}]. \end{cases} \tag{99}$$

Remark 2. *If $\rho = 1$ in the boundary conditions (2) and (75) generalized Riemann–Liouville integral boundary conditions reduces to the Riemann–Liouville integral conditions respectively.*

$$\begin{cases} p(0) = 0, \quad q(0) = 0, \\ p(\mathcal{T}) = \epsilon\mathcal{I}^{\varsigma}_{0+}q(\varpi) = \frac{\epsilon}{\Gamma(\varsigma)}\int_0^{\varpi}(\varpi - \theta)^{\varsigma-1}q(\theta)d\theta, \\ q(\mathcal{T}) = \pi\mathcal{I}^{\varrho}_{0+}p(\sigma) = \frac{\pi}{\Gamma(\varrho)}\int_0^{\sigma}(\sigma - \theta)^{\varrho-1}p(\theta)d\theta, \\ 0 < \sigma < \varpi < \mathcal{T}, \end{cases} \tag{100}$$

and

$$\begin{cases} p(0) = 0, \quad q(0) = 0, \\ p(\mathcal{T}) = \epsilon\mathcal{I}^{\varsigma}_{0+}q(\varpi) = \frac{\epsilon}{\Gamma(\varsigma)}\int_0^{\varpi}(\varpi - \theta)^{\varsigma-1}q(\theta)d\theta, \\ q(\mathcal{T}) = \pi\mathcal{I}^{\varrho}_{0+}p(\varpi) = \frac{\pi}{\Gamma(\varrho)}\int_0^{\varpi}(\varpi - \theta)^{\varpi-1}p(\theta)d\theta, \\ 0 < \varpi < \mathcal{T}. \end{cases} \tag{101}$$

Remark 3. *If $\rho = 1$ and $\varsigma = \varrho = 1$ in the boundary conditions (2) and (75) generalized Riemann–Liouville integral boundary conditions reduces to the classical integral conditions respectively.*

$$\left\{ p(0) = 0,\ q(0) = 0,\ p(\mathcal{T}) = \epsilon \int_0^{\varpi} q(\theta)d\theta,\ q(\mathcal{T}) = \pi \int_0^{\sigma} p(\theta)d\theta \ 0 < \sigma < \varpi < \mathcal{T} \right. \tag{102}$$

and

$$\left\{ p(0) = 0,\ q(0) = 0,\ p(\mathcal{T}) = \epsilon \int_0^{\varpi} q(\theta)d\theta,\ q(\mathcal{T}) = \pi \int_0^{\varpi} p(\theta)d\theta \ 0 < \varpi < \mathcal{T}. \right. \tag{103}$$

9. Conclusions

This paper employs coupled nonlinear generalized Liouville–Caputo fractional differential equations and Katugampola fractional integral operators to solve a novel class of boundary value problems. Applying the techniques of fixed-point theory to discover the existence criterion for solutions is efficient. While the second outcome provides a sufficient criterion to establish the problem's unique solution, the first and third results define various criteria for the presence of solutions to the given problem. In the fourth section, the Hyers–Ulam stability of the solution was determined. In the remarks, we have shown the asymmetric cases of the assigned problem. Moreover, the form of the solution in these kinds of remarks can be used to study the positive solution and its asymmetry in more depth. We conclude that our results are novel and can be viewed as an expansion of the qualitative analysis of fractional differential equations. Our results are novel in this configuration and add to the literature on nonlinear coupled generalized Liouville–Caputo fractional differential equations with nonlocal boundary conditions utilizing Katugampola-type integral operators. Future research could focus on various conceptions of stability and existence in relation to a Lotka–Volterra prey-predator system/coupled logistic system.

Author Contributions: Conceptualization, M.S.; formal analysis, M.A. and M.S.; methodology, M.A, M.S., K.A. and M.M. All authors have read and agreed to the published version of the manuscript.

Funding: This work was supported by the Deanship of Scientific Research, Vice Presidency for Graduate Studies and Scientific Research, King Faisal University, Saudi Arabia [Grant No.793].

Institutional Review Board Statement: Not applicable.

Informed Consent Statement: Not applicable.

Data Availability Statement: Not applicable.

Conflicts of Interest: The authors declare no conflict of interest.

References

1. Britton, N.F. *Essential Mathematical Biology*; Springer: London, UK, 2003.
2. Ma, Y.; Ji, D. Existence of Solutions to a System of Riemann-Liouville Fractional Differential Equations with Coupled Riemann-Stieltjes Integrals Boundary Conditions. *Fractal Fract.* **2022**, *6*, 543. [CrossRef]
3. Theswan, S.; Ntouyas, S.K.; Ahmad, B.; Tariboon, J. Existence Results for Nonlinear Coupled Hilfer Fractional Differential Equations with Nonlocal Riemann–Liouville and Hadamard-Type Iterated Integral Boundary Conditions. *Symmetry* **2022**, *14*, 1948. [CrossRef]
4. Klafter, J.; Lim, S.; Metzler, R. *Fractional Dynamics: Recent Advances*; World Scientific: Singapore, 2012.
5. Podlubny, I. *Fractional Differential Equations: An Introduction to Fractional Derivatives, Fractional Differential Equations, to Methods of Their Solution and Some of Their Applications*; Elsevier: Amsterdam, The Netherlands, 1998.
6. Valerio, D.; Machado, J.T.; Kiryakova, V. Some pioneers of the applications of fractional calculus. *Fract. Calc. Appl. Anal.* **2014**, *17*, 552–578. [CrossRef]
7. Machado, J.T.; Kiryakova, V.; Mainardi, F. Recent history of fractional calculus. *Commun. Nonlinear Sci. Numer. Simul.* **2011**, *16*, 1140–1153. [CrossRef]
8. Kilbas, A.A.A.; Srivastava, H.M.; Trujillo, J.J. *Theory and Applications of Fractional Differential Equations*; Elsevier Science Limited, North-Holland Mathematics Studies: Amsterdam, The Netherlands, 2006; Volume 204.
9. Bitsadze, A.; Samarskii, A. On some simple generalizations of linear elliptic boundary problems. *Soviet Math. Dokl.* **1969**, *10*, 398–400.

10. Ciegis, R.; Bugajev, A. Numerical approximation of one model of bacterial self-organization. *Nonlinear Anal. Model. Control.* **2012**, *17*, 253–270. [CrossRef]
11. Subramanian, M.; Alzabut, J.; Baleanu, D.; Samei, M.E.; Zada, A. Existence, uniqueness and stability analysis of a coupled fractional-order differential systems involving hadamard derivatives and associated with multi-point boundary conditions. *Adv. Differ. Equ.* **2021**, *2021*, 1–46. [CrossRef]
12. Rahmani, A.; Du, W.S.; Khalladi, M.T.; Kostić, M.; Velinov, D. Proportional Caputo Fractional Differential Inclusions in Banach Spaces. *Symmetry* **2022**, *14*, 1941. [CrossRef]
13. Tudorache, A.; Luca, R. Positive Solutions for a Fractional Differential Equation with Sequential Derivatives and Nonlocal Boundary Conditions. *Symmetry* **2022**, *14*, 1779. [CrossRef]
14. Ahmad, B.; Alghanmi, M.; Alsaedi, A.; Nieto, J.J. Existence and uniqueness results for a nonlinear coupled system involving caputo fractional derivatives with a new kind of coupled boundary conditions. *Appl. Math. Lett.* **2021**, *116*, 107018. [CrossRef]
15. Alsaedi, A.; Alghanmi, M.; Ahmad, B.; Ntouyas, S.K. Generalized liouville–caputo fractional differential equations and inclusions with nonlocal generalized fractional integral and multipoint boundary conditions. *Symmetry* **2018**, *10*, 667. [CrossRef]
16. Boutiara, A.; Etemad, S.; Alzabut, J.; Hussain, A.; Subramanian, M.; Rezapour, S. On a nonlinear sequential four-point fractional q-difference equation involving q-integral operators in boundary conditions along with stability criteria. *Adv. Differ. Equ.* **2021**, *2021*, 1–23. [CrossRef]
17. Baleanu, D.; Alzabut, J.; Jonnalagadda, J.; Adjabi, Y.; Matar, M. A coupled system of generalized sturm–liouville problems and langevin fractional differential equations in the framework of nonlocal and nonsingular derivatives. *Adv. Differ. Equ.* **2020**, *2020*, 1–30. [CrossRef]
18. Muthaiah, S.; Baleanu, D. Existence of solutions for nonlinear fractional differential equations and inclusions depending on lower-order fractional derivatives. *Axioms* **2020**, *9*, 44. [CrossRef]
19. Saeed, A.M.; Abdo, M.S.; Jeelani, M.B. Existence and Ulam–Hyers stability of a fractional-order coupled system in the frame of generalized Hilfer derivatives. *Mathematics* **2021**, *9*, 2543. [CrossRef]
20. Ahmad, D.; Agarwal, R.P.; Rahman, G.U.R. Formulation, Solution's Existence, and Stability Analysis for Multi-Term System of Fractional-Order Differential Equations. *Symmetry* **2022**, *14*, 1342. [CrossRef]
21. Samadi, A.; Ntouyas, S.K.; Tariboon, J. On a nonlocal coupled system of Hilfer generalized proportional fractional differential equations. *Symmetry* **2022**, *14*, 738. [CrossRef]
22. Awadalla, M.; Abuasbeh, K.; Subramanian, M.; Manigandan, M. On a System of ψ-Caputo Hybrid Fractional Differential Equations with Dirichlet Boundary Conditions. *Mathematics* **2022**, *10*, 1681. [CrossRef]
23. Katugampola, U.N. New approach to a generalized fractional integral. *Appl. Math. Comput.* **2011**, *218*, 860–865. [CrossRef]
24. Katugampola, U.N. A new approach to generalized fractional derivatives. *arXiv* **2011**, arXiv:1106.0965.
25. Jarad, F.; Abdeljawad, T.; Baleanu, D. On the generalized fractional derivatives and their caputo modification. *J. Nonlinear Sci. Appl.* **2017**, *10*, 2607–2619. [CrossRef]
26. Granas, A.; Dugundji, J. *Fixed Point Theory*; Springer Science & Business Media: Berlin, Germany, 2013.
27. Krasnoselskii, M. Two remarks on the method of successive approximations, uspehi mat. *Nauk* **1955**, *10*, 123–127.

symmetry

MDPI

Article

The Fractional Hilbert Transform of Generalized Functions

Naheed Abdullah [1,2,*,†] and Saleem Iqbal [2,†]

1 Department of Mathematics, Government Girls PostGraduate College, Quetta 08734, Pakistan
2 Department of Mathematics, University of Balochistan, Quetta 87550, Pakistan
* Correspondence: naheedabdullah13@gmail.com
† These authors contributed equally to this work.

Abstract: The fractional Hilbert transform, a generalization of the Hilbert transform, has been extensively studied in the literature because of its widespread application in optics, engineering, and signal processing. In the present work, we expand the fractional Hilbert transform that displays an odd symmetry to a space of generalized functions known as Boehmians. Moreover, we introduce a new fractional convolutional operator for the fractional Hilbert transform to prove a convolutional theorem similar to the classical Hilbert transform, and also to extend the fractional Hilbert transform to Boehmians. We also produce a suitable Boehmian space on which the fractional Hilbert transform exists. Further, we investigate the convergence of the fractional Hilbert transform for the class of Boehmians and discuss the continuity of the extended fractional Hilbert transform.

Keywords: convolution; Boehmian; fractional Hilbert transform; Hilbert transform; equivalence class; delta sequences; compact support

Citation: Abdullah, N.; Iqbal, S. The Fractional Hilbert Transform of Generalized Functions. *Symmetry* **2022**, *14*, 2096. https://doi.org/10.3390/sym14102096

Academic Editors: Palle E.T. Jorgensen and Alexander Zaslavski

Received: 23 August 2022
Accepted: 21 September 2022
Published: 8 October 2022

1. Introduction

The space of Boehmians is a class of generalized functions that include all regular operators and generalized functions or distributions, and other objects. The theory of Boehmians with two convergences, introduced by Mikusinski [1], broadens the concept of Boehme's regular operators [2]. In contrast to the theory of distributions in which generalized functions are treated as members of the dual space of any space of testing function, the space of Boehmians treats distributions more as algebraic objects. Several integral transforms for various spaces of Boehmians were studied and their properties were investigated in [3–13]. Currently, a large number of studies are available on the extension of classical integral transforms to Boehmians. Karunakaran and Roopkumar introduced the Hilbert transform as continuous linear mapping defined on some space of Boehmians into another space of Boehmians [7]. They also studied the Hilbert transform for the space of ultradistributions [8]. The pioneering work of Zayed [13], Al-Omari, and Agarwal [6] introduced an extension of fractional integral transform to Boehmians by extending the fractional Fourier and Sumudu transforms to the space of integrable Boehmians. The properties and generalizations of various quaternion integral transform [14] and fractional integral transforms were also studied from the perspective of q-calculus analysis [15,16] and rapidly decaying functions [17]. In recent years, the extension of fractional integral transforms to the space of Boehmians has been an active area of research. Many well-known fractional integral transforms have been extended to the space of Boehmians, but an extension of the fractional Hilbert transform (FHT) has not yet been reported. So, the goal of this paper is to extend the FHT to some space of Boehmians. Different definitions of FHT exist in the literature [18–20], but in the generalization of the classical Hilbert transform, it might rightly be said that the fractionalization of Hilbert transform is given by Zayed and is mathematically elaborated in [21]. The fractional Hilbert transform of a function $f(x)$, denoted by $H_\alpha[f(x)]$, is defined as [20]

$$H_\alpha[f(x)] = \frac{1}{\pi}\int_{-\infty}^{\infty} \frac{e^{-i\frac{x^2-t^2}{2}\cot\alpha}}{x-t} f(t)dt \text{ for } \alpha \neq 0,\ \pi/2,\ \pi, \tag{1}$$

where the integral is taken in the sense of the Cauchy principal value. The special case $\alpha = \pi/2$ reduces FHT into the standard Hilbert transform. Indeed, the FHT allows for converting a real signal into a complex signal by suppressing the negative frequency. Such a signal has a wide variety of applications in optics, signal processing, and image processing [22–25]. It also does not flip the domain of the signal—the signal remains in the same domain. However, it lacks detailed mathematical analysis, so we require a thorough mathematical theory of FHT to understand its strengths and limitations. Consequently, we need to extend the existing theory on such a significant transformation in terms of generalized functions. An extension of FHT to some space of Boehmians may have applications in engineering and other sciences, as it may apply in converting functions with discontinuities into smooth functions that consequently lead to the description of various physical occurrences such as point charges [26].

The present paper is organized as follows: Section 1 covers the introduction. Section 2 covers the important definitions and theorems, and we also discuss the abstract construction of Boehmians to render the paper self-contained. Section 3 covers results that comprise a new convolutional operator and a new convolutional theorem for FHT, and proves auxiliary results required for the construction of two Boehmian spaces. Lastly, we extend the FHT to some spaces of Boehmians. Section 4 presents our conclusions.

2. Preliminaries

Let $\mathbb{R}$ be the set of all real numbers, $\mathcal{L}^1(\mathbb{R}) = \mathcal{L}^1$ be the collection of complex-valued measurable functions f defined on $\mathbb{R}$ for which

$$\|f\|_1 = \int_{-\infty}^{\infty} |f(x)|dx < \infty,$$

and $\mathcal{C}^\infty = \mathcal{C}^\infty(\mathbb{R})$ be the set of all infinitely differentiable functions defined on $\mathbb{R}$, such that functions and their derivatives converge uniformly on compact sets in $\mathbb{R}$.

Theorem 1 ([27] Theorem 9.5). *For any function f on $\mathbb{R}$ and for all $t \in \mathbb{R}$, let f_t be defined by*

$$f_t(x) = f(x-t).$$

If $p \geq 1$ and $f \in \mathcal{L}^p$, then mapping $t \to f_t$ is uniformly continuous from $\mathbb{R}$ into $\mathcal{L}^p(\mathbb{R})$.

Definition 1. *Let f and g be any two functions on $\mathbb{R}$; their convolution, denoted by $f * g$, is defined as*

$$f * g = \int_{-\infty}^{\infty} f(t)g(x-t)dt. \tag{2}$$

The Hilbert transform of convolutional operation $*$ is given as follows:

Theorem 2. *If $f,\ g \in \mathcal{L}^1(\mathbb{R})$ with Hilbert transforms Hf, Hg respectively, so that $Hf, Hg \in \mathcal{L}^1(\mathbb{R})$, then*

$$H[f * g] = Hf * g = f * Hg.$$

The FHT may not act as agreeably with the classical convolutional operator as the classical Hilbert transform (Theorem 2).

Boehmian Space

The members of Boehmian spaces are called Boehmians, which are equivalence classes of "quotients of sequences". These equivalence classes are formulated from an integral domain of continuous functions. The integral domain operations for Boehmians are addition and convolution. This convolutional operation may differ from the standard convolutional operation given in Definition 2.

We now present a brief introduction to Boehmians.

Let G be a complex linear space, $(H, .)$ is a commutative semigroup, and let $\otimes : G \times H \to G$, so that the conditions given below hold:

- $(f \otimes \phi) \otimes \psi = f \otimes (\phi.\psi), \quad \forall f \in G, \forall \phi, \psi \in H;$
- $(f + g) \otimes \phi = f \otimes \phi + g \otimes \phi, \quad \forall f, g \in G, \forall \phi \in H;$
- $\lambda(f \otimes \phi) = (\lambda f \otimes \phi) \, \forall f \in G, \quad \forall \phi \in H, \lambda \in \mathbb{C};$
- If $f_n \to f$ as $n \to \infty$ and $\phi \in H$ then $f_n \otimes \phi \to f \otimes \phi$ as $n \to \infty$.

Let Δ be a collection of sequences on H, so that

- If $\{\phi_n\}, \{\psi_n\} \in \Delta$ then $\{\phi_n.\psi_n\} \in \Delta$;
- If $f_n \to f$ as $n \to \infty$ and $\{\phi_n\} \in \Delta$ then $f_n \otimes \phi_n \to f$ as $n \to \infty$.

A pair of sequences $\{f_n, \phi_n\}$ with $f_n \in G$ for all $n \in \mathbb{N}$ and $\{\phi_n\} \in \Delta$ are a quotient of sequences, denoted by $\frac{f_n}{\phi_n}$, if

$$f_n \otimes \phi_m = f_m \otimes \phi_n \;\; \forall m, n \in \mathbb{N}.$$

Two quotients of sequences $\frac{f_n}{\phi_n}$ and $\frac{g_n}{\psi_n}$ are equivalent ($\sim$) if, for every $n \in \mathbb{N}$

$$f_n \otimes \psi_n = g_n \otimes \phi_n.$$

The equivalence class of $\frac{f_n}{\phi_n}$ induced by "$\sim$" is denoted by $\left[\frac{f_n}{\phi_n}\right]$. Every equivalence class is called a Boehmian. The space of all Boehmians is denoted by $\mathcal{B} = \mathcal{B}(G, H, \otimes, \Delta)$. $\mathcal{B}$ is a vector space under the operations of addition and scalar multiplication defined as follows:

- $\lambda\left[\frac{f_n}{\phi_n}\right] = \left[\frac{\lambda f_n}{\phi_n}\right]$;
- $\left[\frac{f_n}{\phi_n}\right] + \left[\frac{g_n}{\psi_n}\right] = \left[\frac{f_n \otimes \phi_n + g_n \otimes \psi_n}{\phi_n \otimes \psi_n}\right]$.

If we define an isomorphism $f \to \left[\frac{f \otimes \phi_n}{\phi_n}\right]$, then G is a subspace of $\mathcal{B}$. Therefore, every element of G can be expressed uniquely as a Boehmian.

3. Results

In this section, we define a new convolutional operation for FHT that yields a generalized result for Theorem 2. Moreover, to extend the FHT to the class of Boehmians, we define two classes of Boehmians. Two convergences of FHT are proved on $\mathcal{C}^\infty$. Lastly, an extension of FHT on Boehmians is introduced.

3.1. Convolutional Structure for Fractional Hilbert Transform

The idea of convolutional operation makes it evident that, given any integral transform, we can associate a convolutional operation to it [28]. So, we introduce a new fractional convolutional operator that helps us in extending FHT to the space of Boehmians.

Definition 2. *Let $f, g \in \mathcal{L}^1(\mathbb{R})$. We define a fractional convolution $(f *_\alpha g)$ as*

$$(f *_\alpha g)(x) = \int_{-\infty}^{\infty} f(x-t) g(t) e^{-it(x-t)\cot\alpha} dt. \tag{3}$$

Lemma 1. *Let $f, g \in \mathcal{L}^1$. Then, $(f *_\alpha g)$ is also in $\mathcal{L}^1$.*

Proof. To prove that $f *_\alpha g \in \mathcal{L}^1$, we consider its $\mathcal{L}^1$ norm.

$$\begin{aligned} \|f *_\alpha g\|_1 &= \int_{-\infty}^{\infty} |f *_\alpha g| dx \\ &\leq \int_{-\infty}^{\infty} \int_{-\infty}^{\infty} |f(x-t)||g(t)| dt dx. \end{aligned}$$

By using Fubini's theorem, we have

$$\|f *_\alpha g\|_1 \leq \int_{-\infty}^{\infty} |f(x-t)| dx \int_{-\infty}^{\infty} |g(t)| dt.$$

Since the $\mathcal{L}^1$ norm is translation invariance, so $\int_{-\infty}^{\infty} |f(x-t)| dx = \|f_t\|_1 = \|f\|_1$. Therefore,

$$\|f *_\alpha g\|_1 \leq \|f\|_1 \|g\|_1.$$

Since $f, g \in \mathcal{L}^1$,

$$\|f *_\alpha g\|_1 \leq \|f\|_1 \|g\|_1 < \infty,$$

which proves that $f *_\alpha g \in \mathcal{L}^1$. □

To extend the FHT to the case of Boehmians, the essential step is to prove the convolutional theorem, and suitable Boehmian spaces can then be constructed by proving the supplementary results. Now, we state and prove the convolutional theorem for FHT.

Theorem 3. *(convolutional Theorem) Assume that $f, g \in \mathcal{L}^1$. Then,*

$$H_\alpha[f *_\alpha g] = H_\alpha[f] *_\alpha g = f *_\alpha H_\alpha[g]. \tag{4}$$

In addition, $(f *_\alpha g) = -(H_\alpha[f] *_\alpha H_\alpha[g])$.

Proof.

$$\begin{aligned} H_\alpha[(f *_\alpha g)(x)] &= \frac{1}{\pi} \int_{-\infty}^{\infty} \frac{e^{-i\frac{x^2-t^2}{2}\cot\alpha}}{x-t} (f *_\alpha g)(t)\, dt \\ &= \frac{1}{\pi} \int_{-\infty}^{\infty} \frac{e^{-i\frac{x^2-t^2}{2}\cot\alpha}}{x-t} \int_{-\infty}^{\infty} f(t-y) g(y) e^{-iy(t-y)\cot\alpha} dy dt. \end{aligned}$$

By changing variables $t - y = v$, the above equation can be simplified to

$$\begin{aligned} H_\alpha[(f *_\alpha g)(x)] &= \frac{1}{\pi} \int_{-\infty}^{\infty} \int_{-\infty}^{\infty} \frac{e^{-i\frac{x^2-2xy+y^2-v^2}{2}\cot\alpha}}{(x-y)-v} f(v) g(y) e^{-i(yx-y^2)\cot\alpha} dv dy \\ &= \int_{-\infty}^{\infty} H_\alpha[f(x-y)] g(y) e^{-iy(x-y)\cot\alpha} dy \\ &= (H_\alpha[f] *_\alpha g)(x). \end{aligned}$$

Similarly,

$$H_\alpha[(f *_\alpha g)(x)] = H_\alpha[(g *_\alpha f)(x)] = (H_\alpha[g] *_\alpha f)(x) = (f *_\alpha H_\alpha[g])(x). \tag{5}$$

If we substitute g by $H_\alpha[g]$ in (4), we can write

$$\begin{aligned} H_\alpha[(f *_\alpha H_\alpha[g])(x)] &= (H_\alpha[f] *_\alpha H_\alpha[g])(x), \\ (f *_\alpha H_\alpha[H_\alpha[g]])(x) &= (H_\alpha[f] *_\alpha H_\alpha[g])(x), \quad \text{(by (5))} \\ f *_\alpha g &= -(H_\alpha[f] *_\alpha H_\alpha[g]), \end{aligned}$$

where $H_\alpha^2 = -I$, and this proves the theorem. □

3.2. Abstract Construction of Boehmians

Now, we construct the Boehmian space required for extending the theory of the fractional Hilbert transform to some space of Boehmians. Here, we refer to only two spaces of Boehmians needed to develop the theory of FHT. Now to define the space of Boehmians, we introduce a class of identities as follows: Let space $\mathcal{D}$ constitute all infinitely differentiable functions with compact support in $\mathbb{R}$. Let

$$S = \{\phi \in \mathcal{D} : \phi \geq 0 \text{ and } \int_{\mathbb{R}} \phi = 1\}.$$

Then, the space of Boehmians is given by

$$\mathcal{B}_1 = \mathcal{B}_1(\mathcal{L}^1(\mathbb{R}), S, *_\alpha, \Delta),$$

where Δ is the collection of all sequences of real-valued functions $\{\phi_n(x)\} \subset S$, such that

1. $\int_{\mathbb{R}} e^{it(x-t)\cot\alpha}\phi_n(x)dx = 1, \forall n \in \mathbb{N}$;
2. $\|\phi_n\|_1 \leq M, \forall n \in \mathbb{N}$ for some $M > 0$;
3. $\lim_{n\to\infty} \int_{|t|>\epsilon} |\phi_n(t)|dt = 0, \epsilon > 0$.

These sequences are *delta sequences*. We now state and prove the results that are needed to build the desired space for Boehmians.

Lemma 2. *The operation $*_\alpha$ is both commutative and associative.*

Proof. To prove that $*_\alpha$ is commutative, consider

$$(f *_\alpha g)(x) = \int_{-\infty}^{\infty} f(x-t)g(t)e^{-i(x-t)\cot\alpha}dt.$$

By changing variable $x - t = \tau$, we can simplify the above equation to

$$(f *_\alpha g)(x) = \int_{-\infty}^{\infty} f(\tau)g(x-\tau)e^{-i(x-\tau)\tau\cot\alpha}d\tau = (g *_\alpha f)(x).$$

To prove the associativity, let us consider

$$\begin{aligned}((f *_\alpha g) *_\alpha h)(x) &= \int_{-\infty}^{\infty} (f *_\alpha g)(x-t)h(t)e^{-i(x-t)\cot\alpha}dt \\ &= \int_{-\infty}^{\infty}\int_{-\infty}^{\infty} f(x-t-u)g(u)h(t)e^{-iu(x-t-u)\cot\alpha}e^{-it(x-t)\cot\alpha}dtdu.\end{aligned}$$

By changing variables $t + u = y$, we can write the above equation as

$$((f *_\alpha g) *_\alpha h)(x) = \int_{-\infty}^{\infty}\int_{-\infty}^{\infty} f(x-y)g(y-t)h(t)e^{-i(y-t)(x-y)\cot\alpha}e^{-it(x-t)\cot\alpha}dtdy.$$

As an application of Fubini's theorem, we have

$$\begin{aligned}((f *_\alpha g) *_\alpha h)(x) &= \int_{-\infty}^{\infty}\int_{-\infty}^{\infty} g(y-t)h(t)e^{-i(-tx+yt+tx-t^2)\cot\alpha}f(x-y)e^{-iy(x-y)\cot\alpha}dtdy \\ &= \int_{-\infty}^{\infty} f(x-y)(g *_\alpha h)(y)e^{-iy(x-y)\cot\alpha}dy \\ &= (f *_\alpha (g *_\alpha h))(x).\end{aligned}$$

Thus, $((f *_\alpha g) *_\alpha h)(x) = (f *_\alpha (g *_\alpha h))(x)$. □

Lemma 3. *Assume that $\{\phi_n\}$ and $\{\psi_n\}$ are in Δ. Then, their convolution $\{\phi_n *_\alpha \psi_n\}$ is also in Δ.*

Proof. To prove that $\{\phi_n *_\alpha \psi_n\} \in \Delta$, we must show that the three conditions for delta sequences are fulfilled.

1. $\int_{\mathbb{R}} e^{it(x-t)\cot\alpha}(\phi_n *_\alpha \psi_n)(x)dx = \int_{\mathbb{R}} e^{it(x-t)\cot\alpha} \int_{-\infty}^{\infty} \left(\phi_n(x-t)\psi_n(t)e^{-it(x-t)\cot\alpha}\right)dtdx.$
   By using Fubini's theorem, we can write
   $$\int_{\mathbb{R}} e^{it(x-t)\cot\alpha}(\phi_n *_\alpha \psi_n)(x)dx = \int_{\mathbb{R}} e^{it(x-t)\cot\alpha} e^{-it(x-t)\cot\alpha}\phi_n(x-t)dx \int_{-\infty}^{\infty} \psi_n(t)dt.$$
   Since $\{\phi_n\}, \{\psi_n\} \in \Delta$, then
   $$\int_{\mathbb{R}} e^{it(x-t)\cot\alpha}(\phi_n *_\alpha \psi_n)(x)dx = 1.$$
2. $$\begin{aligned} \|\phi_n *_\alpha \psi_n\|_1 &= \int_{-\infty}^{\infty} |(\phi_n *_\alpha \psi_n)(x)|dx \\ &= \int_{-\infty}^{\infty} \left| \int_{-\infty}^{\infty} \phi_n(x-t)\psi_n(t)e^{-it(x-t)\cot\alpha}dt \right| dx \\ &\le \int_{-\infty}^{\infty}\int_{-\infty}^{\infty} \left|\phi_n(x-t)\psi_n(t)e^{-it(x-t)\cot\alpha}dt\right| dx \\ &= \|\phi_n\|_1\|\psi_n\|_1 \\ &\le M^2, \quad \forall n \in \mathbb{N}. \end{aligned}$$
   Thus, $\|\phi_n *_\alpha \psi_n\|_1 \le M^2$.
3. $$\begin{aligned} \lim_{n\to\infty}\int_{|t|>\epsilon} |(\phi_n *_\alpha \psi_n)(x)|dx &\le \lim_{n\to\infty}\int_{|t|>\epsilon}\int_{-\infty}^{\infty} |\phi_n(x-t)\psi_n(t)|dtdx \\ &= \|\phi_n\|_1 \lim_{n\to\infty}\int_{|t|>\epsilon} |\psi_n(t)|dt. \end{aligned}$$
   Since $\{\psi_n\} \in \Delta$, then
   $$\lim_{n\to\infty}\int_{|t|>\epsilon} |\psi_n(t)|dt = 0, \qquad \text{for } \epsilon > 0.$$
   Hence,
   $$\int_{|t|>\epsilon} |(\phi_n *_\alpha \psi_n)(x)|dx \to 0 \quad \text{as } n\to\infty, \text{ for } \epsilon > 0.$$
   This completes the proof. □

Lemma 4. *If $f \in \mathcal{L}^1$ and $\phi_n \in \Delta$ then the convolution $f *_\alpha \phi_n \in \mathcal{L}^1$.*

Proof. Let $f \in \mathcal{L}^1$ and $\phi_n \in \Delta$. To show that $f *_\alpha \phi_n \in \mathcal{L}^1$, we consider the $\mathcal{L}^1$-norm.

$$\begin{aligned} \|f *_\alpha \phi_n\|_1 &= \int_{\mathbb{R}} |(f *_\alpha \phi_n)(x)|dx, \\ &= \int_{\mathbb{R}} \left| \int_{-\infty}^{\infty} f(x-t)\phi_n(t)e^{-it(x-t)\cot\alpha}dt \right| dx, \\ &\le \int_{\mathbb{R}}\int_{-\infty}^{\infty} \left| f(x-t)\phi_n(t)e^{-it(x-t)\cot\alpha}\right| dtdx, \\ &= \int_{-\infty}^{\infty} |f(x-t)|dx \int_{-\infty}^{\infty} |\phi_n(t)|dt, \\ &= \|f\|_1\|\phi_n\|_1. \end{aligned}$$

Since $f \in \mathcal{L}^1$ and $\{\phi_n\} \in \Delta$, $\|f *_\alpha \phi_n\|_1 \leq \|f\|_1\|\phi_n\|_1 < \infty$, which proves that $f *_\alpha \phi_n \in \mathcal{L}^1$. □

Lemma 5. *If* $f, g \in \mathcal{L}^1$, $\phi \in S$, *then* $(f+g) *_\alpha \phi = f *_\alpha \phi + g *_\alpha \phi$.

The proof of this lemma is straightforward. Therefore, we omitted the details.

Lemma 6. *Let* $f_n \to f$ *in* $\mathcal{L}^1$ *as* $n \to \infty$ *and* $\phi \in S$. *Then* $f_n *_\alpha \phi \to f *_\alpha \phi$ *in* $\mathcal{L}^1$.

Proof. From Lemma 4, we can write

$$\begin{aligned}\|(f_n *_\alpha \phi) - (f *_\alpha \phi)\|_1 &= \|(f_n - f) *_\alpha \phi\|_1 \\ &\leq \|f_n - f\|_1\|\phi\|_1 \\ &\leq M\|f_n - f\|_1 \to 0 \text{ as } n \to \infty \text{ for } M > 0.\end{aligned}$$

Hence, $f_n *_\alpha \phi \to f *_\alpha \phi$ in $\mathcal{L}^1$ whenever $f_n \to f$ in $\mathcal{L}^1$. □

Lemma 7. *Let* $f_n \to f$ *in* $\mathcal{L}^1$ *and* $\{\phi_n\} \in \Delta$. *Then* $f_n *_\alpha \phi_n \to f$ *in* $\mathcal{L}^1$.

Proof. Let $\{\phi_n\} \in \Delta$ then $\int_{-\infty}^{\infty} \phi_n(t)e^{it(x-t)}dt = 1$; therefore, we can write

$$\begin{aligned}(f_n *_\alpha \phi_n)(x) - f(x) &= \int_{-\infty}^{\infty} f_n(x-t)\phi_n(t)e^{-it(x-t)\cot\alpha}dt - f(x)\int_{-\infty}^{\infty} \phi_n(t)e^{it(x-t)\cot\alpha}dt \\ &= \int_{-\infty}^{\infty}\Big(f_n(x-t)e^{-2it(x-t)\cot\alpha} - f(x)\Big)e^{it(x-t)\cot\alpha}\phi_n(t)dt.\end{aligned}$$

Now, we consider the L^1-norm of the above equation:

$$\begin{aligned}\|f_n *_\alpha \phi_n - f\|_1 &= \int_{-\infty}^{\infty}\left|\int_{-\infty}^{\infty}\Big(f_n(x-t)e^{-2it(x-t)\cot\alpha} - f(x)\Big)e^{it(x-t)\cot\alpha}\phi_n(t)dt\right|dx \\ &\leq \int_{-\infty}^{\infty}\int_{-\infty}^{\infty}|f_n(x-t)e^{-2it(x-t)\cot\alpha} - f(x)||\phi_n(t)|dtdx.\end{aligned}$$

As an application of Fubini's theorem and via Property 2 of delta sequences, we have

$$\begin{aligned}\|f_n *_\alpha \phi_n - f\|_1 &\leq \int_{-\infty}^{\infty}|\phi_n(t)|dt\int_{-\infty}^{\infty}|f_n(x-t)e^{-2it(x-t)\cot\alpha} - f(x)|dx \\ &\leq M\|(f_n)_t e^{-2it(x-t)\cot\alpha} - f\|_1, \quad (M > 0).\end{aligned}$$

Using the triangular inequality of normed spaces,

$$\begin{aligned}\|f_n *_\alpha \phi_n - f\|_1 &\leq M\|(f_n)_t e^{-2it(x-t)\cot\alpha} - f_t e^{-2it(x-t)\cot\alpha}\|_1 + \|f_t e^{-2it(x-t)\cot\alpha} - f\|_1 \\ &\leq M\|(f_n)_t e^{-2it(x-t)\cot\alpha} - f_t e^{-2it(x-t)\cot\alpha}\|_1 + M\|f_t e^{-2it(x-t)\cot\alpha} - f\|_1.\end{aligned}$$

By using the convergence of $f_n \in \mathcal{L}^1$ and Theorem 1, we have

$$\|(f_n)_t e^{-2it(x-t)\cot\alpha} - f_t e^{-2it(x-t)\cot\alpha}\|_1 \to 0 \text{ as } n \to \infty,$$

and

$$\|f_t e^{-2it(x-t)\cot\alpha} - f\|_1 \to 0 \text{ as } t \to 0.$$

Therefore, $\|f_n *_\alpha \phi_n - f\|_1 \to 0$ as $n \to \infty$, hence, $f_n *_\alpha \phi_n \to f$ in $\mathcal{L}^1$. □

In order to extend the FHT to the class of Boehmians, we define another class of Boehmians (as the codomain of the extended fractional Hilbert transform) $\mathcal{B}_2 = \mathcal{B}_2(\mathcal{C}^\infty, S, *_\alpha, \Delta)$ [7]. The notion of delta sequences, quotients, and their equivalence classes remains the same as that in the prior case. We also retain the definitions of addition and scalar multiplication. Now, we define

$$D^m\left[\frac{f_n}{\phi_n}\right]=\left[\frac{D^m f_n}{\phi_n}\right] \text{ for any } \left[\frac{f_n}{\phi_n}\right]\in\mathcal{B}_2.$$

In addition,

$$\left[\frac{f_n}{\phi_n}\right]*_\alpha\left[\frac{g_n}{\psi_n}\right]=\left[\frac{f_n *_\alpha g_n}{\phi_n *_\alpha \psi_n}\right].$$

Since a concept of convergence is required to construct a Boehmian space, we prove two convergences on $\mathcal{C}^\infty$.

Lemma 8. *Let $f_n \to f$ as $n\to\infty$ in $\mathcal{C}^\infty$ then $f_n *_\alpha \phi \to f *_\alpha \phi$ in $\mathcal{C}^\infty$ for all $\phi\in D$; further, for each delta sequence $\{\delta_n\}$, $f_n *_\alpha \delta_n \to f$ as $n\to\infty$ in $\mathcal{C}^\infty$.*

Proof. Let $K\subset\mathbb{R}$ be any compact set, such that $x\in K$. To prove the convergence of a sequence of functions in $\mathcal{C}^\infty$, we must show that the functions and their derivatives converge uniformly on compact sets.

First, we prove that $f_n *_\alpha \phi \to f *_\alpha \phi$ in $\mathcal{C}^\infty$. For this, consider

$$|(f_n *_\alpha \phi - f *_\alpha \phi)(x)| = |((f_n - f) *_\alpha \phi)(x)| \le \int_{-\infty}^{\infty} |(f_n - f)(x-t)|\phi(t)dt.$$

Since t varies over the compact support of ϕ; therefore, $x-t$ also varies over a compact set in $\mathbb{R}$. So, $|((f_n-f)*_\alpha\phi)(x)|\to 0$ as $n\to\infty$ uniformly on compact sets. Then,

$$|(f_n *_\alpha \phi - f *_\alpha \phi)(x)|\to 0 \text{ as } n\to\infty,$$

or we can write

$$f_n *_\alpha \phi \to f *_\alpha \phi \text{ as } n\to\infty, \tag{6}$$

uniformly on compact sets.

In addition,

$$D^m((f_n *_\alpha \phi)-(f *_\alpha \phi)) = (D^m f_n *_\alpha \phi)-(D^m f *_\alpha \phi). \tag{7}$$

Replacing $D^m f_n$ by f_n and $D^m f$ by f in (7), we have

$$D^m((f_n *_\alpha \phi)-(f *_\alpha \phi)) = (f_n *_\alpha \phi)-(f *_\alpha \phi), \tag{8}$$

the right-hand side of (8) approaches zero by (6). Thus,

$$D^m(f_n *_\alpha \phi)\to D^m(f *_\alpha \phi)$$

uniformly on compact sets. Hence, $f_n *_\alpha \phi \to f *_\alpha \phi$ as $n\to\infty$ in $\mathcal{C}^\infty$.

Next, without any loss of generality, let us suppose that $\{\delta_n\}\in\Delta$ is such that it has a compact support. Then,

$$\begin{aligned}|(f_n *_\alpha \delta_n - f)(x)| &= \left|\int_{-\infty}^{\infty} f_n(x-t)\delta_n(t)e^{-it(x-t)\cot\alpha}dt - f(x)\int_{-\infty}^{\infty} e^{it(x-t)\cot\alpha}\delta_n(t)dt\right| \\ &\le \int_{-\infty}^{\infty} |f_n(x-t)e^{-2it(x-t)\cot\alpha} - f(x)|\delta_n(t)dt, \\ &\le \int_{-\infty}^{\infty}\Big(|f_n(x-t)e^{-2it(x-t)\cot\alpha} - f(x-t)e^{-2it(x-t)\cot\alpha}| + |f(x-t)e^{-2it(x-t)\cot\alpha} - f(x)|\Big)\delta_n(t)dt.\end{aligned}$$

Now, both x and t vary over compact sets; therefore, $x-t$ also varies over a compact set. Thus,

$$\int_{-\infty}^{\infty}\Big(|f_n(x-t)e^{-2it(x-t)\cot\alpha} - f(x-t)e^{-2it(x-t)\cot\alpha}| + |f(x-t)e^{-2it(x-t)\cot\alpha} - f(x)|\Big)\delta_n(t)dt \to 0$$

as $n \to \infty$ and $t \to 0$.

We have $f_n *_\alpha \delta_n \to f$ uniformly on compact sets.

Similarly, $D^m(f_n *_\alpha \delta_n) \to D^m(f)$ uniformly on compact sets.

Hence, $f_n *_\alpha \delta_n \to f$ as $n \to \infty$ in $\mathcal{C}^\infty$. □

Lemma 9. *If $f_n \to f$ as $n \to \infty$ in $\mathcal{L}^1$, then $f_n *_\alpha \delta \to f *_\alpha \delta$ as $n \to \infty$ in $\mathcal{C}^\infty$ for every $\delta \in S$.*

Proof. To show the convergence in $\mathcal{C}^\infty$, we assume that x varies over a compact set K.

$$\begin{aligned}
|(f_n *_\alpha \delta - f *_\alpha \delta)(x)| &= |((f_n * -f) *_\alpha \delta)(x)| \\
&= \left|\int_{-\infty}^{\infty}(f_n - f)(x-t)\delta(t)e^{-it(x-t)\cot\alpha}dt\right| \\
&\le \int_{-\infty}^{\infty}|(f_n - f)(x-t)||\delta(t)|dt \\
&\le \|f_n - f\|_1\|\delta\|_\infty.
\end{aligned}$$

Since $f_n \to f$ in $\mathcal{L}^1$ and $\delta \in S$ has a compact support, $x - t$ varies over a compact set, and $|(f_n *_\alpha \delta - f *_\alpha \delta)(x)| \to 0$ as $n \to \infty$ on compact sets. Similarly, we have

$$|D^m[(f_n *_\alpha \delta - f *_\alpha \delta)](x)| \le \|f_n - f\|_1\|D^m\delta\|_\infty.$$

Thus, $D^m(f_n *_\alpha \delta) \to D^m(f *_\alpha \delta)$ on compact sets.

Hence, $f_n *_\alpha \delta \to f *_\alpha \delta$ as $n \to \infty$ in $\mathcal{C}^\infty$. □

3.3. *Fractional Hilbert Transform on Boehmians*

The following result is very important in the aftermath. The proof of the following theorem is similar to the proof of convolution theorem for FHT as in Theorem 2; we omitted the details.

Theorem 4. *If $f \in \mathcal{L}^1$ and $\delta \in \Delta$, then $H_\alpha[f *_\alpha \delta] = H_\alpha[f] *_\alpha \delta$.*

Definition 3. *The fractional Hilbert transform $\mathcal{H}_\alpha : \mathcal{B}_1 \to \mathcal{B}_2$ on Boehmians is defined by*

$$\mathcal{H}_\alpha\left[\frac{f_n}{\phi_n}\right] = \left[\frac{\mathcal{H}_\alpha f_n}{\phi_n}\right],$$

where $\frac{f_n}{\phi_n}$ is an arbitrary representative of any given Boehmian $B \in \mathcal{B}_1$. Since

$$f_n *_\alpha \phi_m = f_m *_\alpha \phi_n \quad \forall m, n \in \mathbb{N}.$$

By Theorem 4, we can write $\mathcal{H}_\alpha[f_n] *_\alpha \phi_m = \mathcal{H}_\alpha[f_m] *_\alpha \phi_n \quad \forall m, n \in \mathbb{N}$.

Therefore, $\frac{\mathcal{H}_\alpha[f_n]}{\phi_n}$ represents a Boehmian in $\mathcal{B}_2$. In a similar manner, let $\frac{g_n}{\psi_n}$ be another representative of B; then, again, with an application of Theorem 4,

$$\frac{\mathcal{H}_\alpha[f_n]}{\phi_n} \sim \frac{\mathcal{H}_\alpha[g_n]}{\psi_n},$$

thus the extended FHT on Boehmians $\mathcal{H}_\alpha : \mathcal{B}_1 \to \mathcal{B}_2$ is well-defined.

Theorem 5. *Let $\mathcal{H}_\alpha : \mathcal{B}_1 \to \mathcal{B}_2$ be the extended FHT; then,*

1. *If $\frac{f_n}{\phi_n} \in \mathcal{B}_1$ then $\frac{\mathcal{H}_\alpha f_n}{\phi_n} \in \mathcal{B}_2$.*

2. $\mathcal{H}_\alpha$ *is well-defined.*
3. $\mathcal{H}_\alpha$ *is a continuous linear map.*
4. $\mathcal{H}_\alpha$ *is an injective map.*

Proof. The proof of the above theorem is similar to those of Hilbert transform on Boehmians; we omitted the details. For details, the reader is referred to [7]. □

4. Conclusions

This paper gave an extension of the fractional Hilbert transform to a class of generalized functions known as Boehmians. It introduces a new convolutional operator, and the consequent convolutional theorem was also presented. In addition, the extended fractional Hilbert transform is a well-defined map between the spaces of Boehmians having properties, such as continuity and linearity, identical to the classical properties of their corresponding classical versions. Lastly, convergence concerning δ and Δ was also examined.

The methods of this paper can also be utilized to extend FHT to the space of ultradistributions. We suggest that readers consider the expansion of the fractional Hilbert transform to q-calculus and develop the theory of the quaternion fractional Hilbert transform.

Author Contributions: Both authors contributed equally to this work. All authors have read and agreed to the published version of the manuscript.

Funding: This research received no external funding.

Data Availability Statement: Not applicable.

Acknowledgments: The authors would like to thank the reviewers for taking the time to read and improve the present article.

Conflicts of Interest: The authors declare no conflict of interest.

Abbreviations

The following abbreviation is used in this manuscript:

FHT Fractional Hilbert transform

References

1. Mikusinski, P. Convergence of Boehmians. *Jpn. J. Math. New Ser.* **1983**, *9*, 159–179. [CrossRef]
2. Boehme, T.K. The support of Mikusiński operators. *Trans. Am. Math. Soc.* **1973**, *176*, 319–334.
3. Al-Omari, S.K.Q.; Kılıçman, A. Note on Boehmians for class of optical Fresnel wavelet transforms. *J. Funct. Spaces Appl.* **2012**, *2012*, 405368. [CrossRef]
4. Al-Omari, S.K.Q.; Kılıçman, A. An estimate of Sumudu transforms for Boehmians. *Adv. Differ. Equ.* **2013**, *2013*, 77. [CrossRef]
5. Al-Omari, S.K.Q. An extension of certain integral transform to a space of Boehmians. *J. Assoc. Arab. Univ. Basic Appl. Sci.* **2015**, *17*, 36–42. [CrossRef]
6. Al-Omari, S.K.Q.; Agarwal, P. Some general properties of a fractional Sumudu transform in the class of Boehmians. *Kuwait J. Sci.* **2016**, *43*, 16–30.
7. Karunakaran, V.; Kalpakam, N.V. Hilbert transform for Boehmians. *Integral Transform. Spec. Funct.* **2000**, *9*, 19–36. [CrossRef]
8. Karunakaran, V.; Roopkumar, R. Boehmians and their Hilbert transforms. *Integral Transform. Spec. Funct.* **2002**, *13*, 131–141. [CrossRef]
9. Loonker, D.; Banerji, P.K. Natural transform for distribution and Boehmian spaces. *Math. Eng. Sci. Aerosp. (MESA)* **2013**, *4*, 69–76.
10. Roopkumar, R. Stieltjes transform for Boehmians. *Integral Transform. Spec. Funct.* **2007**, *18*, 845–853. [CrossRef]
11. Roopkumar, R. Mellin transform for Boehmians. *Bull. Inst. Math. New Ser.* **2009**, *4*, 75–96.
12. Roopkumar, R. Quaternionic Fractional Fourier Transform for Boehmians. *Ukr. Math. J.* **2020**, *72*, 942–952. [CrossRef]
13. Zayed, A.I. Fractional Fourier transform of generalized functions. *Integral Transform. Spec. Funct.* **1998**, *7*, 299–312.
14. Al-Omari, S.K.Q.; Baleanu, D. Quaternion Fourier integral operators for spaces of generalized quaternions. *Math. Methods Appl. Sci.* **2018**, *41*, 9477–9484. [CrossRef]
15. Al-Omari, S.K.Q. Estimates and properties of certain q-Mellin transform on generalized q-calculus theory. *Adv. Differ. Equ.* **2021**, *2021*, 233.
16. Al-Omari, S.K.Q. q-Analogues and properties of the Laplace-type integral operator in the quantum calculus theory. *J. Inequal. Appl.* **2020**, *2020*, 203. [CrossRef]

17. Al-Omari, S.K.Q. Some remarks on short-time Fourier integral operators and classes of rapidly decaying functions. *Math. Methods Appl. Sci.* **2019**, *42*, 5354–5361. [CrossRef]
18. Cusmariu, A. Fractional analytic signals. *Signal Process.* **2002**, *82*, 267–272. [CrossRef]
19. Lohmann, A.W.; Mendlovic, D.; Zalevsky, Z. Fractional hilbert transform. *Opt. Lett.* **1996**, *21*, 281–283. [CrossRef] [PubMed]
20. Zayed, A.I. Hilbert transform associated with the fractional Fourier transform. *IEEE Signal Process. Lett.* **1998**, *5*, 206–208. [CrossRef]
21. Abdullah, N.; Iqbal, S.; Khalid, A.; Al Johani, A.S.; Khan, I.; Rehman, A.; Andualem, M. The Fractional Hilbert Transform on the Real Line. *Math. Probl. Eng.* **2022**, *2022*, 5027907. [CrossRef]
22. Davis, J.A.; McNamara, D.E.; Cottrell, D.M. Analysis of the fractional Hilbert transform. *Appl. Opt.* **1998**, *37*, 6911–6913. [CrossRef]
23. Deng, L.; Hou, Z.; Liu, H.; Sun, Z. A fractional hilbert transform order optimization algorithm based DE for bearing health monitoring. In Proceedings of the 2019 Chinese Control Conference (CCC), Guangzhou, China, 27–30 July 2019; pp. 2183–2186.
24. Lohmann, A.W.; Tepichin, E.; Ramirez, J.G. Optical implementation of the fractional Hilbert transform for two-dimensional objects. *Appl. Opt.* **1997**, *36*, 6620–6626. [CrossRef]
25. Sharma, N. Design of fractional hilbert transform ($\pi/4$). In Proceedings of the 2019 3rd International Conference on Electronics, Communication and Aerospace Technology (ICECA), Coimbatore, India, 12–14 June 2019; pp. 1182–1184.
26. Al-Omari, S.K.Q. A fractional Fourier integral operator and its extension to classes of function spaces. *Adv. Differ. Equ.* **2018**, *2018*, 195. [CrossRef]
27. Rudin, W. *Real and Complex Analysis*; Mathematics Series; McGraw-Hill: New York, NY, USA, 1987.
28. Zayed, A.I. *Handbook of Function and Generalized Function Transformations*; CRC Press: Boca Raton, FL, USA, 2019.

symmetry

MDPI

Article

On the Composition Structures of Certain Fractional Integral Operators

Min-Jie Luo [1,*,†] and Ravinder Krishna Raina [2,†,‡]

1 Department of Mathematics, College of Science, Donghua University, Shanghai 201620, China
2 Department of Mathematics, College of Technology & Engineering, Maharana Pratap University of Agriculture and Technology, Udaipur 313001, India
* Correspondence: mathwinnie@live.com or mathwinnie@dhu.edu.cn
† These authors contributed equally to this work.
‡ Current address: 10/11, Ganpati Vihar, Opposite Sector 5, Udaipur 313002, India.

Abstract: This paper investigates the composition structures of certain fractional integral operators whose kernels are certain types of generalized hypergeometric functions. It is shown how composition formulas of these operators can be closely related to the various Erdélyi-type hypergeometric integrals. We also derive a derivative formula for the fractional integral operator and some applications of the operator are considered for a certain Volterra-type integral equation, which provide two generalizations to Khudozhnikov's integral equation (see below). Some specific relationships, examples, and some future research problems are also discussed.

Keywords: composition operators; Erdélyi-type integral; fractional integral operator; generalized hypergeometric function

MSC: 26A33; 33C20

Citation: Luo, M.-J.; Raina, R.K. On the Composition Structures of Certain Fractional Integral Operators. *Symmetry* **2022**, *14*, 1845. https://doi.org/10.3390/sym14091845

Academic Editors: Francisco Martínez González and Mohammed K. A. Kaabar

Received: 9 August 2022
Accepted: 2 September 2022
Published: 5 September 2022

Publisher's Note: MDPI stays neutral with regard to jurisdictional claims in published maps and institutional affiliations.

1. Introduction

In 1978, Saigo [1] introduced his widely used fractional integral operators $I^{\alpha,\beta,\eta}$ and $J^{\alpha,\beta,\eta}$ (see Equations (16) and (17) below). Saigo's operators involve the Gauss hypergeometric functions ${}_2F_1$ as kernels and possess many properties (see, for example, Refs. [1–5]). Over the past few decades, Saigo's operators have been applied in various branches of mathematics, especially in the *Geometric Function Theory* (see Refs. [6–8]). The symmetry of parameters of various hypergeometric functions injects more choice and flexibility into the theory of Generalized Fractional Calculus.

A natural question that arises is: *Can an operator involving a generalized hypergeometric function ${}_pF_q$ as kernel have such properties as Saigo's operators?* In this direction, some efforts have been made by some authors to find particular forms of operators. In 1987, Goyal and Jain [9] introduced two fractional integral operators I_α^h and K_β^λ, which involve the generalized hypergeometric functions ${}_pF_q$ as kernels. Later, Goyal et al. [10,11] introduced two more general fractional integral operators involving the generalized hypergeometric function ${}_pF_q$ and Srivastava's polynomial S_n^m.

Although very general in form, the properties of the operators I_α^h and K_β^λ introduced by Goyal et al. are far less succinct than those of Saigo's operators. For Saigo's operators $I^{\alpha,\beta,\eta}$ and $J^{\alpha,\beta,\eta}$, we have the following useful properties (see Refs. [12,13]):

$$I^{\alpha,\beta,\eta}x^\lambda = \frac{\Gamma(\lambda)\Gamma(\lambda-\beta+\eta+1)}{\Gamma(\lambda-\beta+1)\Gamma(\lambda+\alpha+\eta+1)}x^{\lambda-\beta} \tag{1}$$

$$(\Re(\alpha) > 0,\ \Re(\lambda) > \max\{0, \Re(\beta-\eta)\} - 1)$$

and

$$J^{\alpha,\beta,\eta}x^{\lambda} = \frac{\Gamma(\beta-\lambda)\Gamma(\eta-\lambda)}{\Gamma(-\lambda)\Gamma(\alpha+\beta+\eta-\lambda)}x^{\lambda-\beta} \tag{2}$$

$$(\Re(\alpha) > 0,\ \Re(\lambda) < \max\{\Re(\beta), \Re(\eta)\}).$$

Under certain conditions, we also have the following composition properties (see Ref. [1], p. 140, Equations (2.22) and (2.23), see also Ref. [3]):

$$I^{\alpha,\beta,\eta} I^{\gamma,\delta,\alpha+\eta} f = I^{\alpha+\gamma,\beta+\delta,\eta} f, \tag{3}$$

$$I^{\alpha,\beta,\eta} I^{\gamma,\delta,\eta-\beta-\gamma-\delta} f = I^{\alpha+\gamma,\beta+\delta,\eta-\gamma-\delta} f, \tag{4}$$

$$J^{\gamma,\delta,\alpha+\eta} J^{\alpha,\beta,\eta} f = J^{\alpha+\gamma,\beta+\delta,\eta} f \tag{5}$$

and

$$J^{\gamma,\delta,\eta-\beta-\gamma-\delta} J^{\alpha,\beta,\eta} f = J^{\alpha+\gamma,\beta+\delta,\eta-\gamma-\delta} f. \tag{6}$$

However, it seems rather difficult to find properties for the operators I_{α}^{h} and K_{β}^{λ} similar to those given above by (1)–(6). Moreover, it is still unknown whether the corresponding generalized fractional derivatives of the forms (see Ref. [3], Equations (3.2) and (3.4))

$$I^{\alpha,\beta,\eta} f = \frac{\mathrm{d}^n}{\mathrm{d}x^n} I^{\alpha+n,\beta-n,\eta-n} f \quad \text{and} \quad J^{\alpha,\beta,\eta} f = (-1)^n \frac{\mathrm{d}^n}{\mathrm{d}x^n} J^{\alpha+n,\beta-n,\eta-n} f \tag{7}$$

can be defined for the operators I_{α}^{h} and K_{β}^{λ}.

Very recently, the authors [14] introduced two fractional integral operators $\mathcal{I}$ and $\mathcal{J}$ (see below Equations (12) and (13)) whose kernels involve a very special class of generalized hypergeometric function. The authors have to some extent overcome the limitations of the operators I_{α}^{h} and K_{β}^{λ} and obtained results similar to (1) and (2). Subsequently, some further results and applications related to $\mathcal{I}$ and $\mathcal{J}$ were discovered in the papers [15,16].

The aim of the present paper is to first establish for the operators $\mathcal{I}$ and $\mathcal{J}$ some results relating to the composition structures of the defined operators analogous to Formulas (3)–(7). We also consider defining the corresponding fractional derivative operators of these operators $\mathcal{I}$ and $\mathcal{J}$. Finally, we shall consider some connections of our work with Khudozhnikov's work [17] on Volterra-type integral equations.

2. Preliminaries

In this paper, the symbols $\mathbb{N}$, $\mathbb{R}_+$, and $\mathbb{C}$ denote the set of natural, positive real, and complex numbers, respectively. The Pochhammer symbol $(a)_k$ is defined by

$$(a)_k := \frac{\Gamma(a+k)}{\Gamma(a)} = \begin{cases} 1 & (k=0; a \in \mathbb{C} \setminus \{0\}) \\ a(a+1)\cdots(a+k-1) & (k \in \mathbb{N};\ a \in \mathbb{C}). \end{cases}$$

In addition, we shall use the convention of writing the finite sequence of parameters $a_1, \cdots, a_p$ by (a_p) and the product of p Pochhammer symbols by $((a_p))_k \equiv (a_1)_k \cdots (a_p)_k$, where an empty product $p = 0$ is treated as unity.

We are particularly interested in the generalized hypergeometric function ${}_{r+p}F_{r+q}$ of the form

$${}_{r+p}F_{r+q}\left[\begin{matrix}(a_p), & (f_r+m_r) \\ (b_q), & (f_r)\end{matrix}; z\right] := \sum_{k=0}^{\infty} \frac{((a_p))_k}{((b_q))_k} \frac{((f_r+m_r))_k}{((f_r))_k} \frac{z^k}{k!}, \tag{8}$$

where $m_1, \cdots, m_r \in \mathbb{N}$. The conditions of convergence of (8) follow easily from the usual definition of the generalized hypergeometric function; see Ref. [18], p. 62 and Ref. [19], p. 30. Several recent results concerning this particular type of generalized hypergeometric function have been obtained in Ref. [20] (see also Ref. [21]).

For convenience, we put

$$m := m_1 + \cdots + m_r, \tag{9}$$

and let σ_j $(0 \le j \le m)$ be determined by the generating relation

$$\prod_{j=1}^{r}(f_j + x)_{m_j} = \sum_{j=0}^{m}\sigma_{m-j}x^j. \tag{10}$$

Obviously, σ_j's depend *only* on f_j $(1 \le j \le r)$. Additionally, we define A_k $(0 \le k \le m)$ by

$$A_k = \sum_{j=k}^{m}\left\{ {j \atop k} \right\}\sigma_{m-j}, \quad A_0 = (f_1)_{m_1}\cdots(f_r)_{m_r}, \quad A_m = 1, \tag{11}$$

where the notation $\left\{ {j \atop k} \right\}$ denotes the Stirling number of the second kind.

Definition 1 ([14], p. 423, Definition 1.1). *Let $x, h, \nu \in \mathbb{R}_+$, $\delta, a, b, f_1, \cdots, f_r \in \mathbb{C}$ and $m_1, \cdots, m_r \in \mathbb{N}$. Also, let $\Re(\mu) > 0$ and φ be a suitable complex-valued function defined on $\mathbb{R}_+$. Then the fractional integral of the first kind of a function φ is defined by*

$$\begin{aligned}(\mathcal{I}\varphi)(x) &\equiv \left(\mathcal{I}^{\mu;a,b:\ (f_r+m_r)}_{h;\nu,\delta:\quad (f_r)}\varphi\right)(x) \\ &:= \frac{\nu x^{-\delta-\nu(\mu+h)}}{\Gamma(\mu)}\int_0^x (x^\nu - s^\nu)^{\mu-1}\,{}_{r+2}F_{r+1}\left[\begin{matrix}a, b, (f_r+m_r) \\ \mu, \quad (f_r)\end{matrix}; 1 - \frac{s^\nu}{x^\nu}\right]\varphi(s)s^{\nu h+\nu-1}\mathrm{d}s,\end{aligned} \tag{12}$$

and the fractional integral of the second kind of a function $\varphi(x)$ is defined by

$$\begin{aligned}(\mathcal{J}\varphi)(x) &\equiv \left(\mathcal{J}^{\mu;a,b:\ (f_r+m_r)}_{h;\nu,\delta:\quad (f_r)}\varphi\right)(x) \\ &:= \frac{\nu x^{\nu h+\nu-1}}{\Gamma(\mu)}\int_x^\infty (s^\nu - x^\nu)^{\mu-1}\,{}_{r+2}F_{r+1}\left[\begin{matrix}a, b, (f_r+m_r) \\ \mu, \quad (f_r)\end{matrix}; 1 - \frac{x^\nu}{s^\nu}\right]\varphi(s)s^{-\delta-\nu(\mu+h)}\mathrm{d}s.\end{aligned} \tag{13}$$

When $r = 0$, we obtain

$$\left(\mathcal{I}^{\mu;a,b}_{h;\nu,\delta}\varphi\right)(x) = \frac{\nu x^{-\delta-\nu(\mu+h)}}{\Gamma(\mu)}\int_0^x (x^\nu - s^\nu)^{\mu-1}\,{}_2F_1\left[\begin{matrix}a, b \\ \mu\end{matrix}; 1 - \frac{s^\nu}{x^\nu}\right]\varphi(s)s^{\nu h+\nu-1}\mathrm{d}s \tag{14}$$

and

$$\left(\mathcal{J}^{\mu;a,b}_{h;\nu,\delta}\varphi\right)(x) = \frac{\nu x^{\nu h+\nu-1}}{\Gamma(\mu)}\int_x^\infty (s^\nu - x^\nu)^{\mu-1}\,{}_2F_1\left[\begin{matrix}a, b \\ \mu\end{matrix}; 1 - \frac{x^\nu}{s^\nu}\right]\varphi(s)s^{-\delta-\nu(\mu+h)}\mathrm{d}s. \tag{15}$$

Some properties of the operators (12) and (13) have been presented in Refs. [14,16]. Further, the operators $\mathcal{I}^{\mu;a,b}_{h;\nu,\delta}$ and $\mathcal{J}^{\mu;a,b}_{h;\nu,\delta}$ have the following special cases:

(a) For $h = 0$, $\nu = 1$ and $\delta = 0$ in (14) and (15), we obtain

$$\left(\mathcal{I}^{\mu;a,b}_{0;1,0}\varphi\right)(x) = {}_2I^{\mu}_{0+}(a,b)\varphi(x) \quad \text{and} \quad \left(\mathcal{I}^{\mu;a,b}_{0;1,0}\varphi\right)(x) = {}_4I^{\mu}_{-}(a,b)\varphi(x),$$

where ${}_2I^{\mu}_{0+}(a,b)$ and ${}_4I^{\mu}_{-}(a,b)$ are two of the four operators introduced by Grinko and Kilbas [22].

(b) When $h = 0$, $\nu = 1$, $\delta = \beta$, $\mu = \alpha$, $a = \alpha + \beta$ and $b = -\eta$ in (14) and (15), then we obtain Saigo's fractional integral operators

$$\begin{aligned}\left(I^{\alpha,\beta,\eta}\varphi\right)(x) &= \left(\mathcal{I}^{\alpha;\alpha+\beta,-\eta}_{0;1,\beta}\varphi\right)(x) \\ &= \frac{x^{-\beta-\alpha}}{\Gamma(\alpha)}\int_0^x (x-s)^{\alpha-1}\,{}_2F_1\left[\begin{matrix}\alpha+\beta, -\eta \\ \alpha\end{matrix}; 1 - \frac{s}{x}\right]\varphi(s)\mathrm{d}s \quad (\Re(\alpha) > 0)\end{aligned} \tag{16}$$

and

$$\left(J^{\alpha,\beta,\eta}\varphi\right)(x) = \left(\mathcal{J}\,{}^{\alpha;\alpha+\beta,-\eta}_{0;1,\beta}\,\varphi\right)(x)$$
$$= \frac{1}{\Gamma(\alpha)}\int_x^\infty (s-x)^{\alpha-1}{}_2F_1\left[\begin{matrix}\alpha+\beta,-\eta\\ \alpha\end{matrix};1-\frac{x}{s}\right]\varphi(s)s^{-\beta-\alpha}\mathrm{d}s \quad (\Re(\alpha)>0). \tag{17}$$

(c) When $a = b = 0$, it is not difficult to observe that $\mathcal{I}\,{}^{\mu;a,b}_{h;\nu,\delta}$ and $\mathcal{J}\,{}^{\mu;a,b}_{h;\nu,\delta}$ contain the *Erdélyi–Kober operators* (see Ref. [19], p. 105 and Ref. [23], p. 322)

$$\left(I^{\mu}_{+;\nu,h}f\right)(x) = \left(\mathcal{I}\,{}^{\mu;0,0}_{h;\nu,0}f\right)(x)$$
$$= \frac{\nu x^{-\nu(\mu+h)}}{\Gamma(\mu)}\int_0^x (x^\nu - s^\nu)^{\mu-1}f(s)s^{\nu h+\nu-1}\mathrm{d}s$$
$$= \frac{1}{\Gamma(\mu)}\int_0^1 (1-u)^{\mu-1}f(xu^{1/\nu})u^h\mathrm{d}u \quad (\Re(\mu)>0,\ \nu,h\in\mathbb{R}_+) \tag{18}$$

and

$$\left(I^{\mu}_{-;\nu,h}\right)f(x) = \left(\mathcal{J}\,{}^{\mu;0,0}_{h-1+1/\nu;\nu,0}f\right)(x)$$
$$= \frac{\nu x^{\nu h}}{\Gamma(\mu)}\int_x^\infty (s^\nu - x^\nu)^{\mu-1}f(s)s^{\nu(1-\mu-h)-1}\mathrm{d}s$$
$$= \frac{1}{\Gamma(\mu)}\int_1^\infty (u-1)^{\mu-1}f(xu^{1/\nu})u^{-\mu-h}\mathrm{d}u \quad (\Re(\mu)>0,\ \nu,h\in\mathbb{R}_+) \tag{19}$$

as special cases. The operators obtained by letting $\nu = 1$ in (18) and (19) are usually denoted by $I^+_{\eta,\alpha}$ and $K^-_{\eta,\alpha}$, respectively (see Ref. [19], p. 106).

The operators defined above by (12) and (13) were previously studied in Refs. [14,16] in the space X^p_c ($c \in \mathbb{R}$, $1 \le p \le \infty$) of those complex-valued Lebesgue measurable functions φ on $\mathbb{R}_+$ for which $\|\varphi\|_{X^p_c} < \infty$, where

$$\|\varphi\|_{X^p_c} := \left(\int_0^\infty |u^c\varphi(u)|^p\frac{\mathrm{d}u}{u}\right)^{1/p}. \tag{20}$$

It follows at once that $X^p_{1/p} = L^p(\mathbb{R}_+)$. For convenience, we define

$$\mathfrak{c}_1(t) := 1 + h + \frac{t}{\nu} \quad\text{and}\quad \mathfrak{c}_2(t) := \mathfrak{c}_1(\delta-1) - \frac{t}{\nu}.$$

The following lemma gives some useful properties of the operators $\mathcal{I}$ and $\mathcal{J}$ relating to the norm defined in (20).

Lemma 1. *Let $\varphi \in X^p_c$.*

(i) *If $\Re(\mu) > 0$ and $\mathfrak{c}_1(-c) + \min\{0, \Re(\mu - a - b - m)\} > 0$, then the operator $\mathcal{I}$ is bounded from X^p_c into $X_{c+\Re(\delta)}$, and*

$$\|\mathcal{I}\varphi\|_{X^p_{c+\Re(\delta)}} \le C_1\|\varphi\|_{X^p_c}.$$

(ii) *If $\Re(\mu) > 0$ and $\Re(\mathfrak{c}_2(-c)) + \min\{0, \Re(\mu - a - b - m)\} > 0$, then the operator $\mathcal{J}$ is bounded from X^p_c into $X_{c+\Re(\delta)}$, and*

$$\|\mathcal{J}\varphi\|_{X^p_{c+\Re(\delta)}} \le C_2\|\varphi\|_{X^p_c}.$$

(iii) *If $\Re(\mu) > 0$ and $\mathfrak{c}_1(-c) + \min\{0, \Re(\mu - a - b)\} > 0$, then the operator $\mathcal{I}\,{}^{\mu;a,b}_{h;\nu,\delta}$ is bounded from X^p_c into X^p_c, and*

$$\|x^{\delta}\mathcal{I}^{\mu;a,b}_{h;\nu,\delta}\varphi\|_{X^p_c} \leq C_1^*\|\varphi\|_{X^p_c}.$$

(iv) If $\Re(\mu) > 0$ *and* $\Re(\mathfrak{c}_2(-c)) + \min\{0, \Re(\mu - a - b)\} > 0$*, then the operator* $\mathcal{J}^{\mu;a,b}_{h;\nu,\delta}$ *is bounded from* X^p_c *into* X^p_c*, and*

$$\|x^{\delta}\mathcal{J}^{\mu;a,b}_{h;\nu,\delta}\varphi\|_{X^p_c} \leq C_2^*\|\varphi\|_{X^p_c}.$$

(v) If $\Re(\mu) > 0$ *and* $\mathfrak{c}_1(-c) > 0$*, then the operator* $I^{\mu}_{+;\nu,h}$ *is bounded from* X^p_c *into* X^p_c*, and*

$$\|I^{\mu}_{+;\nu,h}\varphi\|_{X^p_c} \leq C_1^{**}\|\varphi\|_{X^p_c}.$$

(vi) If $\Re(\mu) > 0$ *and* $\nu h + c > 0$*, then the operator* $I^{\mu}_{-;\nu,h}$ *is bounded from* X^p_c *into* X^p_c*, and*

$$\|I^{\mu}_{-;\nu,h}\varphi\|_{X^p_c} \leq C_2^{**}\|\varphi\|_{X^p_c}.$$

Proof. The results (i) and (ii) are established in Ref. [14], p. 437, Theorem 3.1.

On the other hand, the results (iii) and (iv) are the corollaries of (i) and (ii) (see also Ref. [16], p. 614).

Finally, the results (v) and (vi) follow immediately from (iii) and (iv). These results are consistent with the classical ones. It may be noted that if we set $c = 1/p$ in (v) and (vi), then the operator $I^{\mu}_{+;\nu,h}$ is bounded in $L_p(\mathbb{R}_+)$ provided that $\Re(\mu) > 0$ and $h > -1 + 1/p\nu$, and the operator $I^{\mu}_{-;\nu,h}$ is bounded in $L_p(\mathbb{R}_+)$ provided that $\Re(\mu) > 0$ and $h > -1/p\nu$ (see Ref. [19], p. 107, Lemma 2.28 and Ref. [23], p. 323). □

It should be particularly emphasized here that the operators $\mathcal{I}$ and $\mathcal{J}$ are quite different from the *multiple Erdélyi–Kober fractional integral operators* (see Ref. [4], p. 11, see also Refs. [24,25]), though some special cases of $\mathcal{I}$ and $\mathcal{J}$ when $r = 0$ (e.g., Saigo's operators) can be expressed as multiple Erdélyi–Kober fractional integral operators. The cases that $r = 0$ are very special because Meijer's G-function $G^{2,0}_{2,2}[\sigma]$ and ${}_2F_1[1-\sigma]$ have the following relationship (see [4], p. 18, Equation (1.1.18))

$$G^{2,0}_{2,2}\left[\sigma\middle|\begin{matrix}\gamma_1+\delta_1, \gamma_2+\delta_2\\ \gamma_1, \gamma_2\end{matrix}\right] = \frac{\sigma^{\gamma_2}(1-\sigma)^{\delta_1+\delta_2-1}}{\Gamma(\delta_1+\delta_2)}\,{}_2F_1\left[\begin{matrix}\gamma_2+\delta_2-\gamma_1, \delta_1\\ \delta_1+\delta_2\end{matrix};1-\sigma\right] \quad (21)$$

for $\sigma < 1$. However, there is *no* such relationship between $G^{m,0}_{m,m}[\sigma]$ and ${}_{r+2}F_{r+1}[1-\sigma]$. A slightly more general case than (21) will lead us to the *Marichev–Saigo–Maeda fractional integral operators* (see Refs. [26,27]), which are also very different from our operators $\mathcal{I}$ and $\mathcal{J}$. In addition, the operators $\mathcal{I}$ and $\mathcal{J}$ *cannot* be regarded as special cases of G-transform studied in Ref. [28]. Since the kernels of $\mathcal{I}$ and $\mathcal{J}$ are not of Sonine's type, they *cannot* be included in the theory developed very recently by Luchko (see Ref. [29]).

3. The Main Results

3.1. Composition Formulas

Theorem 1. *Assume that* $\varphi \in X^p_c$*. Let*

$$\lambda_1 \equiv \lambda - a - m, \quad \lambda_2 \equiv \lambda - b - m \quad \text{and} \quad \mathfrak{p}_m \equiv \lambda - a - b - m, \quad (22)$$

where m is given by (9)*. Let* (ϑ_m) *be the nonvanishing zeros of the parametric polynomial* $\mathfrak{Q}_m(t)$ *defined by*

$$\mathfrak{Q}_m(t) = \sum_{k=0}^{m}(-1)^k A_k(\lambda_1)_k(\lambda_2)_k(t)_k(a+k)_{m-k}(b+k)_{m-k} \cdot {}_3F_2\left[\begin{matrix}k-m, k+t, -\mathfrak{p}_m\\ a+k, b+k\end{matrix};1\right], \quad (23)$$

where A_k $(0 \le k \le m)$ *is defined in* (11). *Then for* $\Re(\gamma) > 0$, $\Re(\mu) > 1/p > 0$,

$$h + \min\{0, \Re(\gamma + \mu - a - b - m)\} > \Re(\gamma + \mathfrak{p}_m + (\rho - c)/\nu)$$

and $h + 1 + \min\{0, \Re(\mu - \lambda - \mathfrak{p}_m)\} > \Re((c + \rho)/\nu)$, *we have*

$$\left(\mathcal{I}^{\mu;\lambda_1,\lambda_2:\ (f_r+m_r)}_{h;\nu,\delta:\ \ \ (f_r)}\left(\mathcal{I}^{\ \gamma;-\mathfrak{p}_m,\lambda-\mu}_{h-\gamma-\mathfrak{p}_m-\rho/\nu;\nu,\rho}\varphi\right)\right)(x) = \left(\mathcal{I}^{\ \gamma+\mu;a,b:\ \ (\vartheta_m+1)}_{h-\gamma-\mathfrak{p}_m-\rho/\nu;\nu,\delta+\rho:\ (\vartheta_m)}\varphi\right)(x) \tag{24}$$

where $\mathcal{I}^{\mu;a,b:\ (f_r+m_r)}_{h;\nu,\delta:\ \ (f_r)}$ *and* $\mathcal{I}^{\mu;a,b}_{h;\nu,\delta}$ *are defined by* (12) *and* (14), *respectively.*

Proof. Denote the left-hand side of (24) by $\Phi(x)$. Then by interchanging the order of integration, we obtain

$$\begin{aligned}\Phi(x) = {} & \frac{\nu x^{-\delta-\nu(\mu+h)}}{\Gamma(\mu)}\int_0^x (x^\nu - s^\nu)^{\mu-1}\,{}_{r+2}F_{r+1}\left[\begin{matrix}\lambda_1, \lambda_2, (f_r+m_r)\\ \mu, \quad (f_r)\end{matrix}; 1 - \frac{s^\nu}{x^\nu}\right] s^{\nu h+\nu-1}\\ & \cdot\left\{\frac{\nu s^{-\nu(h-\mathfrak{p}_m)}}{\Gamma(\gamma)}\int_0^s (s^\nu - t^\nu)^{\gamma-1}\,{}_2F_1\left[\begin{matrix}-\mathfrak{p}_m, \lambda-\mu\\ \gamma\end{matrix}; 1 - \frac{t^\nu}{s^\nu}\right]\varphi(t)t^{\nu(h-\gamma-\mathfrak{p}_m)-\rho+\nu-1}dt\right\}ds\\ = {} & \frac{\nu^2 x^{-\delta-\nu(\mu+h)}}{\Gamma(\mu)\Gamma(\gamma)}\int_0^x \varphi(t)t^{\nu(h-\gamma-\mathfrak{p}_m)-\rho+\nu-1}\Delta_1(t)dt,\end{aligned} \tag{25}$$

where

$$\begin{aligned}\Delta_1(t) := {} & \int_t^x s^{\nu h+\nu-1-\nu h+\nu\mathfrak{p}_m}(x^\nu - s^\nu)^{\mu-1}(s^\nu - t^\nu)^{\gamma-1}\\ & \cdot {}_{r+2}F_{r+1}\left[\begin{matrix}\lambda_1, \lambda_2, (f_r+m_r)\\ \mu, \quad (f_r)\end{matrix}; 1 - \frac{s^\nu}{x^\nu}\right]{}_2F_1\left[\begin{matrix}-\mathfrak{p}_m, \lambda-\mu\\ \gamma\end{matrix}; 1 - \frac{t^\nu}{s^\nu}\right]ds.\end{aligned} \tag{26}$$

We shall tackle Equation (24) and leave the verification of the validity of interchanging the order of integration in (25) at the end of the proof.

Letting $s^\nu = x^\nu - u(x^\nu - t^\nu)$ in (26), we have

$$\begin{aligned}\Delta_1(t) = {} & \frac{1}{\nu}x^{\nu\mathfrak{p}_m}(x^\nu - t^\nu)^{\mu+\gamma-1}\int_0^1 u^{\mu-1}(1-u)^{\gamma-1}\left(1 - \left(1 - \frac{t^\nu}{x^\nu}\right)u\right)^{\mathfrak{p}_m}\\ & \cdot {}_{r+2}F_{r+1}\left[\begin{matrix}\lambda_1, \lambda_2, (f_r+m_r)\\ \mu, \quad (f_r)\end{matrix}; \left(1 - \frac{t^\nu}{x^\nu}\right)u\right]{}_2F_1\left[\begin{matrix}-\mathfrak{p}_m, \lambda-\mu\\ \gamma\end{matrix}; \frac{(1-u)(1-t^\nu/x^\nu)}{1-u(1-t^\nu/x^\nu)}\right]du.\end{aligned} \tag{27}$$

The right-hand side of (27) can be evaluated by using an Erdélyi-type integral established by Luo and Raina [21]. For $\Re(\gamma) > \Re(\mu) > 0$ and $z \in \mathbb{C} \setminus [1, \infty)$, Luo and Raina proved that (Ref. [21], p. 482, Theorem 3.2)

$$\begin{aligned}{}_{m+2}F_{m+1}\left[\begin{matrix}a, b, \ (\vartheta_m+1)\\ \gamma, \quad (\vartheta_m)\end{matrix}; z\right] = {} & \frac{\Gamma(\gamma)}{\Gamma(\mu)\Gamma(\gamma-\mu)}\int_0^1 t^{\mu-1}(1-t)^{\gamma-\mu-1}(1-tz)^{\mathfrak{p}_m}\\ & \cdot {}_{r+2}F_{r+1}\left[\begin{matrix}\lambda_1, \lambda_2, \ (f_r+m_r)\\ \mu, \quad (f_r)\end{matrix}; zt\right]{}_2F_1\left[\begin{matrix}-\mathfrak{p}_m, \lambda-\mu\\ \gamma-\mu\end{matrix}; \frac{(1-t)z}{1-tz}\right]dt,\end{aligned} \tag{28}$$

where λ_1, λ_2 and $\mathfrak{p}_m$ are given by (22) and (ϑ_m) are the nonvanishing zeros of the parametric polynomial defined in (23). We note that the parametric polynomial is independent of parameter γ, and thus we may replace γ by $\gamma + \mu$ (without changing the values of λ_1, λ_2, $\mathfrak{p}_m$ and $\mathfrak{Q}_m(t)$) in (28) to get

$$\begin{aligned}{}_{m+2}F_{m+1}\left[\begin{matrix}a, b, \ (\vartheta_m+1)\\ \gamma+\mu, \quad (\vartheta_m)\end{matrix}; z\right] = {} & \frac{\Gamma(\gamma+\mu)}{\Gamma(\mu)\Gamma(\gamma)}\int_0^1 t^{\mu-1}(1-t)^{\gamma-1}(1-tz)^{\mathfrak{p}_m}\\ & \cdot {}_{r+2}F_{r+1}\left[\begin{matrix}\lambda_1, \lambda_2, \ (f_r+m_r)\\ \mu, \quad (f_r)\end{matrix}; zt\right]{}_2F_1\left[\begin{matrix}-\mathfrak{p}_m, \lambda-\mu\\ \gamma\end{matrix}; \frac{(1-t)z}{1-tz}\right]dt,\end{aligned} \tag{29}$$

where $\min\{\Re(\gamma), \Re(\mu)\} > 0$.

Using the Erdélyi-type integral (29) in (27), we obtain

$$\Delta_1(u) = \frac{\Gamma(\mu)\Gamma(\gamma)}{\Gamma(\gamma+\mu)} x^{\nu \mathfrak{p}_m} (x^\nu - t^\nu)^{\mu+\gamma-1} {}_{m+2}F_{m+1}\left[\begin{matrix} a, b, & (\vartheta_m + 1) \\ \gamma+\mu, & (\vartheta_m) \end{matrix}; 1 - \frac{t^\nu}{x^\nu}\right]. \tag{30}$$

Finally, substituting (30) into (25), we get

$$\begin{aligned}\Phi(x) &= \frac{\nu x^{-\delta-\rho}}{\Gamma(\gamma+\mu)} x^{-\nu(\mu+\gamma+(h-\gamma-\mathfrak{p}_m-\rho/\nu))} \int_0^x (x^\nu - t^\nu)^{\mu+\gamma-1} \\ &\quad \cdot {}_{m+2}F_{m+1}\left[\begin{matrix} a, b, & (\vartheta_m + 1) \\ \gamma+\mu, & (\vartheta_m) \end{matrix}; 1 - \frac{t^\nu}{x^\nu}\right] \varphi(t) t^{\nu(h-\gamma-\mathfrak{p}_m-\rho/\nu)+\nu-1} dt \\ &= \left(\mathcal{I}_{h-\gamma-\mathfrak{p}_m-\rho/\nu;\nu,\delta+\rho:\ (\vartheta_m)}^{\gamma+\mu;a,b:\ \ (\vartheta_m+1)} \varphi\right)(x),\end{aligned}$$

which is the desired right-hand side of (24).

Now, we validate the interchanging of the integration. It is sufficient to show that

$$I = \int_0^x (x^\nu - s^\nu)^{\Re(\mu)-1} \left| {}_{r+2}F_{r+1}\left[\begin{matrix} \lambda_1, \lambda_2, & (f_r + m_r) \\ \mu, & (f_r) \end{matrix}; 1 - \frac{s^\nu}{x^\nu}\right]\right| s^{\nu+\Re(\nu\mathfrak{p}_m)-1} \Delta_2(s) ds < \infty,$$

where

$$\begin{aligned}\Delta_2(s) &= \int_0^s (s^\nu - t^\nu)^{\Re(\gamma)-1} \left| {}_2F_1\left[\begin{matrix} -\mathfrak{p}_m, \lambda - \mu \\ \gamma \end{matrix}; 1 - \frac{t^\nu}{s^\nu}\right]\right| |\varphi(t)| t^{\nu h - \Re(\nu\gamma+\nu\mathfrak{p}_m+\rho)+\nu-1} dt \\ &= \frac{1}{\nu} s^{\nu h - \Re(\nu\mathfrak{p}_m+\rho)} \int_0^1 (1-u)^{\Re(\gamma)-1} \left| {}_2F_1\left[\begin{matrix} -\mathfrak{p}_m, \lambda - \mu \\ \gamma \end{matrix}; 1 - u\right]\right| \\ &\quad \cdot \left|\varphi\left(s u^{1/\nu}\right)\right| u^{h-\Re(\gamma+\mathfrak{p}_m+\rho/\nu)-1} du.\end{aligned}$$

Note that (see Ref. [18], p. 63, Theorem 2.1.3 and [30], p. 387)

$${}_2F_1\left[\begin{matrix} a, b \\ c \end{matrix}; 1 - z\right] = \begin{cases} \mathcal{O}(1), & \Re(c-a-b) > 0; \\ \mathcal{O}\left(z^{\Re(c-a-b)}\right), & \Re(c-a-b) < 0; \\ \mathcal{O}(\log z), & a+b=c; \\ \mathcal{O}\left(z^{\Re(c-a-b)}\right) + \mathcal{O}(1), & \Re(c-a-b) = 0,\ c \neq a+b \end{cases} \tag{31}$$

as $z \to 0^+$, so for each s, we have

$$\begin{aligned}\Delta_2(s) &\leq D_1 \cdot s^{\nu h - \Re(\nu\mathfrak{p}_m+\rho)} \int_0^1 (1-u)^{\Re(\gamma)-1} \\ &\quad \cdot u^{h-\Re(\gamma+\mathfrak{p}_m+\rho/\nu)+\min\{0,\Re(\gamma+\mathfrak{p}_m-\lambda+\mu)\}-1} \left|\varphi\left(s u^{1/\nu}\right)\right| du,\end{aligned}$$

where D_1 is a positive number. In view of the definition of the Erdélyi–Kober operator (18), we have

$$\Delta_2(s) \leq D_2 \cdot s^{\nu h - \Re(\nu\mathfrak{p}_m+\rho)} F(s),$$

where $D_2 := D_1\Gamma(\Re(\gamma))$ ($\Re(\gamma) > 0$) and

$$F(s) := \left(I_{+;\nu,h-\Re(\gamma+\mathfrak{p}_m+\rho/\nu)+\min\{0,\Re(\gamma+\mathfrak{p}_m-\lambda+\mu)\}-1}^{\Re(\gamma)} |\varphi|\right)(s).$$

From Lemma 1, we have $F \in X_c^p$, since $\varphi \in X_c^p$ and

$$h + \min\{0, \Re(\gamma+\mathfrak{p}_m-\lambda+\mu)\} > \Re(\gamma+\mathfrak{p}_m+(\rho-c)/\nu).$$

For the generalized hypergeometric function ${}_{p+1}F_p[z]$, we have (see, for example Ref. [31], p. 149)

$$ {}_{p+1}F_p\left[\begin{matrix} a_1,\cdots,a_{p+1} \\ b_1,\cdots,b_p \end{matrix};1-z\right] = \begin{cases} \mathcal{O}(1), & \Re(\psi_p) > 0; \\ \mathcal{O}\left(z^{\Re(\psi_p)}\right), & \Re(\psi_p) < 0; \\ \mathcal{O}(\log z), & \psi_p = 0, \end{cases} \tag{32} $$

as $z \to 0^+$, where $\psi_p := \sum_{\ell=1}^{p} b_\ell - \sum_{\ell=1}^{p+1} a_\ell$. Therefore, for each $x \in \mathbb{R}_+$, we find that

$$ \begin{aligned} I &\leq D_2 D_3 x^{-\nu\min\{0,\Re(\mu-\lambda_1-\lambda_2)-m\}} \\ &\quad \cdot \int_0^x (x^\nu - s^\nu)^{\Re(\mu)-1} s^{\nu\min\{0,\Re(\mu-\lambda_1-\lambda_2)-m\}+\nu h+\nu-\Re(\rho)} F(s)\frac{\mathrm{d}s}{s} \\ &\leq D_2 D_3 x^{-\nu\min\{0,\Re(\mu-\lambda_1-\lambda_2)-m\}} \|F\|_{X_c^p} \\ &\quad \cdot \left(\int_0^x (x^\nu - s^\nu)^{p'\Re(\mu)-p'} s^{p'\nu\min\{0,\Re(\mu-\lambda_1-\lambda_2)-m\}+p'\nu(h+1)-p'\Re(\rho)-p'c-1}\mathrm{d}s\right)^{1/p'} \\ &\leq D_2 D_3 v^{-1/p'} x^{\nu\Re(\mu-\rho/\nu)+\nu h-c} \|F\|_{X_c^p} \\ &\quad \cdot \left(\int_0^1 (1-u)^{p'\Re(\mu)-p'} u^{p'\min\{0,\Re(\mu-\lambda_1-\lambda_2)-m\}+p'(h+1)-p'\Re(\rho)/\nu-p'c/\nu-1}\mathrm{d}s\right)^{1/p'} \\ &< \infty, \end{aligned} $$

where D_3 is a positive number, $1/p + 1/p' = 1$, $p'\Re(\mu) - p' + 1 > 0$ and

$$ \min\{0, \Re(\mu - \lambda_1 - \lambda_2) - m\} + h + 1 > \Re((c+\rho)/\nu). $$

Thus, Fubini's theorem is applicable and the proof is complete. □

Remark 1. *When $r = 0$, we can set $h = 0$, $\nu = 1$, $\mu = \alpha$, $\delta = \lambda - b - \alpha$ and $\rho = a + b - \lambda - \gamma$ in (24) to get*

$$ \begin{aligned} \left(\mathcal{I}^{\alpha;\lambda-a,\lambda-b}_{0;1,\lambda-b-\alpha}\left(\mathcal{I}^{\gamma;a+b-\lambda,\lambda-\alpha}_{0;1,a+b-\lambda-\gamma}\varphi\right)\right)(x) &= \left(\mathcal{I}^{\alpha;\lambda-b,\lambda-a}_{0;1,\lambda-b-\alpha}\left(\mathcal{I}^{\gamma;a+b-\lambda,\lambda-\alpha}_{0;1,a+b-\lambda-\gamma}\varphi\right)\right)(x) \\ &= \left(\mathcal{I}^{\gamma+\alpha;a,b}_{0;1,a-\alpha-\gamma}\varphi\right)(x). \end{aligned} \tag{33} $$

By comparing it with (16), we find that (33) is equivalent to the identity

$$ \left(I^{\alpha,\lambda-b-\alpha,a-\lambda}\left(I^{\gamma,a+b-\lambda-\gamma,\alpha-\lambda}\varphi\right)\right)(x) = \left(I^{\gamma+\alpha,a-\gamma-\alpha,-b}\varphi\right)(x). \tag{34} $$

If we let further $a = \beta + \gamma + \delta + \alpha$, $b = \gamma + \delta - \eta$ and $\lambda = \beta - \eta + \gamma + \delta + \alpha$, then (34) reduces to (4).

Theorem 2. *Assume that $\varphi \in X_c^p$. Let λ_1, λ_2, and $\mathfrak{p}_m$ be defined in (22). Let (ϑ_m) be the nonvanishing zeros of the parametric polynomial $\mathfrak{Q}_m(t)$ defined in (23). Then for $\Re(\gamma) > 0$, $\Re(\mu) > 1/p > 0$,*

$$ h + 1 + \Re((\rho+\delta)/\nu - \mathfrak{p}_m - \gamma) + \min\{0, \Re(\gamma + \mathfrak{p}_m - \lambda + \mu)\} + (c-1)/\nu > 0 $$

and $1 + h + (1+c)/\nu + \min\{0, \Re(\mu - \lambda - \mathfrak{p}_m)\} + \Re((\rho+\delta)/\nu) > 0$, we have

$$ \left(\mathcal{J}^{\mu;\lambda_1,\lambda_2:\ (f_r+m_r)}_{h;\nu,\delta:\ \ (f_r)}\left(\mathcal{J}^{\gamma;-\mathfrak{p}_m,\lambda-\mu}_{h-\gamma-\mathfrak{p}_m+\delta/\nu;\nu,\rho}\varphi\right)\right)(x) = \left(\mathcal{J}^{\gamma+\mu;a,b:\ \ (\vartheta_m+1)}_{h-\mathfrak{p}_m-\gamma;\nu,\delta+\rho:\ (\vartheta_m)}\varphi\right)(x), \tag{35} $$

where $\mathcal{J}^{\mu;a,b:\ (f_r+m_r)}_{h;\nu,\delta:\ \ (f_r)}$ and $\mathcal{J}^{\mu;a,b}_{h;\nu,\delta}$ are defined by (13) and (15), respectively.

Proof. Denote the left-hand side of (35) by $\Psi(s)$. Then, following a similar procedure as described in the proof of Theorem 1, we have

$$\begin{aligned}\Psi(s) &= \frac{\nu x^{\nu h+\nu-1}}{\Gamma(\mu)} \int_x^\infty (s^\nu - x^\nu)^{\mu-1} {}_{r+2}F_{r+1}\left[\begin{matrix}\lambda_1, \lambda_2, (f_r+m_r) \\ \mu, \quad (f_r)\end{matrix}; 1 - \frac{x^\nu}{s^\nu}\right] s^{-\delta-\nu(\mu+h)} \\ &\quad \cdot \left\{ \frac{\nu s^{\nu h+\nu-1}}{\Gamma(\gamma)} s^{-\nu(\gamma+\mathfrak{p}_m)+\delta} \int_s^\infty (t^\nu - s^\nu)^{\gamma-1} {}_2F_1\left[\begin{matrix}-\mathfrak{p}_m, \lambda-\mu \\ \gamma\end{matrix}; 1 - \frac{s^\nu}{t^\nu}\right] \right. \\ &\quad \left. \cdot \varphi(t) t^{-\rho-\nu(h-\mathfrak{p}_m)-\delta} dt \right\} ds \\ &= \frac{\nu^2 x^{\nu h+\nu-1}}{\Gamma(\mu)\Gamma(\gamma)} \int_x^\infty \varphi(t) t^{-\rho-\nu(h-\mathfrak{p}_m)-\delta} \Delta_3(t) dt, \end{aligned} \tag{36}$$

where

$$\begin{aligned}\Delta_3(t) &= \int_x^t s^{\nu-1-\nu(\mu+\gamma+\mathfrak{p}_m)} (s^\nu - x^\nu)^{\mu-1} (t^\nu - s^\nu)^{\gamma-1} \\ &\quad \cdot {}_{r+2}F_{r+1}\left[\begin{matrix}\lambda_1, \lambda_2, (f_r+m_r) \\ \mu, \quad (f_r)\end{matrix}; 1 - \frac{x^\nu}{s^\nu}\right] {}_2F_1\left[\begin{matrix}-\mathfrak{p}_m, \lambda-\mu \\ \gamma\end{matrix}; 1 - \frac{s^\nu}{t^\nu}\right] ds.\end{aligned}$$

Letting

$$s = \frac{tx}{(t^\nu + (x^\nu - t^\nu)u)^{1/\nu}},$$

so that

$$ds = \frac{1}{\nu} tx(t^\nu - x^\nu)(t^\nu + (x^\nu - t^\nu)u)^{-1-1/\nu} du \quad \text{and} \quad u = \frac{t^\nu(x^\nu - s^\nu)}{s^\nu(x^\nu - t^\nu)} \in (0,1),$$

we have

$$\begin{aligned}\Delta_3(t) &= \frac{1}{\nu}(t^\nu - x^\nu)^{\mu+\gamma-1} t^{-\nu\mu} x^{-\nu(\gamma+\mathfrak{p}_m)} \int_0^1 u^{\mu-1}(1-u)^{\gamma-1} \left(1 - \left(1 - \frac{x^\nu}{t^\nu}\right)u\right)^{\mathfrak{p}_m} \\ &\quad \cdot {}_{r+2}F_{r+1}\left[\begin{matrix}\lambda_1, \lambda_2, (f_r+m_r) \\ \mu, \quad (f_r)\end{matrix}; \left(1 - \frac{x^\nu}{t^\nu}\right)u\right] {}_2F_1\left[\begin{matrix}-\mathfrak{p}_m, \lambda-\mu \\ \gamma\end{matrix}; \frac{(1-u)(1-x^\nu/t^\nu)}{1-u(1-x^\nu/t^\nu)}\right] du.\end{aligned}$$

The use of Erdélyi-type integral (29) gives

$$\Delta_3(t) = \frac{1}{\nu} \frac{\Gamma(\mu)\Gamma(\gamma)}{\Gamma(\gamma+\mu)} (t^\nu - x^\nu)^{\mu+\gamma-1} t^{-\nu\mu} x^{-\nu(\gamma+\mathfrak{p}_m)} {}_{m+2}F_{m+1}\left[\begin{matrix}a, b, \quad (\vartheta_m+1) \\ \gamma+\mu, \quad (\vartheta_m)\end{matrix}; 1 - \frac{x^\nu}{t^\nu}\right],$$

and thus (36) becomes

$$\begin{aligned}\Psi(s) &= \frac{\nu x^{\nu(h-\gamma-\mathfrak{p}_m)+\nu-1}}{\Gamma(\mu+\gamma)} \int_x^\infty (t^\nu - x^\nu)^{\mu+\gamma-1} {}_{m+2}F_{m+1}\left[\begin{matrix}a, b, \quad (\vartheta_m+1) \\ \gamma+\mu, \quad (\vartheta_m)\end{matrix}; 1 - \frac{x^\nu}{t^\nu}\right] \\ &\quad \cdot \varphi(t) t^{-(\rho+\delta)-\nu(\mu+\gamma+h-\gamma-\mathfrak{p}_m)} dt \\ &= \left(\mathcal{J}_{h-\mathfrak{p}_m-\gamma;\nu,\delta+\rho:\ (\vartheta_m)}^{\gamma+\mu;a,b:\ (\vartheta_m+1)} \varphi\right)(x),\end{aligned}$$

where (ϑ_m) are the nonvanishing zeros of the parametric polynomial (23).

As in the proof of Theorem 1, we verify the validity of interchanging the order of integration by checking the finiteness of the integral

$$I = \int_x^\infty (s^\nu - x^\nu)^{\Re(\mu)-1} \left| {}_{r+2}F_{r+1}\left[\begin{matrix}\lambda_1, \lambda_2, (f_r+m_r) \\ \mu, \quad (f_r)\end{matrix}; 1 - \frac{x^\nu}{s^\nu}\right] \right| s^{\nu-1-\Re(\nu\mu+\nu\gamma+\nu\mathfrak{p}_m)} \Delta_4(s) ds,$$

where

$$\Delta_4(s) = \int_s^\infty (t^\nu - s^\nu)^{\Re(\gamma)-1} \left| {}_2F_1\left[\begin{matrix} -\mathfrak{p}_m, \lambda - \mu \\ \gamma \end{matrix}; 1 - \frac{s^\nu}{t^\nu} \right] \right| |\varphi(t)| t^{-\Re(\rho+\delta-\nu\mathfrak{p}_m)-\nu h} dt$$
$$= \frac{1}{\nu} s^{1-\nu-\Re(\rho+\delta-\nu\gamma-\nu\mathfrak{p}_m)-\nu h} \int_1^\infty (u-1)^{\Re(\gamma)-1} \left| {}_2F_1\left[\begin{matrix} -\mathfrak{p}_m, \lambda - \mu \\ \gamma \end{matrix}; 1 - \frac{1}{u} \right] \right|$$
$$\cdot |\varphi(su^{1/\nu})| u^{1/\nu-1-\Re(\rho/\nu+\delta/\nu-\mathfrak{p}_m)-h} du.$$

Using (31) gives

$$\Delta_4(s) \le D_4 \cdot s^{1-\nu-\Re(\rho+\delta-\nu\gamma-\nu\mathfrak{p}_m)-\nu h} \int_1^\infty (u-1)^{\Re(\gamma)-1}$$
$$\cdot \left|\varphi(su^{1/\nu})\right| u^{-\Re(\gamma)-h+\Re(\gamma)+1/\nu-1-\Re(\rho/\nu+\delta/\nu-\mathfrak{p}_m)-\min\{0,\Re(\gamma+\mathfrak{p}_m-\lambda+\mu)\}} du,$$

where D_4 is a positive number. Thus we have

$$\Delta_4(s) \le D_4 \cdot s^{1-\nu-\Re(\rho+\delta-\nu\gamma-\nu\mathfrak{p}_m)-\nu h} G(s),$$

where $D_5 := D_4\Gamma(\Re(\gamma))$ $(\Re(\gamma) > 0)$ and

$$G(s) := \left(I^{\Re(\gamma)}_{-;\nu,h-1/\nu+1+\Re(\rho/\nu+\delta/\nu-\mathfrak{p}_m-\gamma)+\min\{0,\Re(\gamma+\mathfrak{p}_m-\lambda+\mu)\}} |\varphi| \right)(s) \in X_c^p.$$

Then from (32) we have

$$I \le D_5 D_6 x^{\nu \min\{0,\Re(\mu-\lambda_1-\lambda_2)-m\}}$$
$$\cdot \int_x^\infty (s^\nu - x^\nu)^{\Re(\mu)-1} s^{-\nu\min\{0,\Re(\mu-\lambda_1-\lambda_2)-m\}-\nu h-\Re(\rho+\delta+\nu\mu)-1} G(s) \frac{ds}{s}$$
$$\le D_5 D_6 x^{\nu \min\{0,\Re(\mu-\lambda_1-\lambda_2)-m\}} \|G\|_{X_c^p}$$
$$\cdot \left(\int_x^\infty (s^\nu - x^\nu)^{p'\Re(\mu)-p'} s^{-p'\nu\min\{0,\Re(\mu-\lambda_1-\lambda_2)-m\}-p'\nu h-p'\Re(\rho+\delta+\nu\mu)-p'-p'c-1} ds \right)^{1/p'}$$
$$\le D_5 D_6 x^{-\nu-\nu h-\Re(\rho+\delta)-1-c-1/p'} \|G\|_{X_c^p}$$
$$\cdot \left(\int_1^\infty (u-1)^{p'\Re(\mu)-p'} u^{-p'\min\{0,\Re(\mu-\lambda_1-\lambda_2)-m\}-p'h-p'\Re((\rho+\delta)/\nu+\mu)-p'(1+c)/\nu-1} ds \right)^{1/p'}$$
$$< \infty.$$

This completes the proof. □

Remark 2. *When $r = 0$, we can set $h = 0$, $\nu = 1$, $\delta = \gamma + \lambda - a - b$ and $\rho = a + b - \lambda - \gamma$ in (35) to get*

$$\left(\mathcal{J}^{\ \mu;\lambda-b,\lambda-a}_{0;1,\gamma+\lambda-a-b} \left(I^{\gamma,a+b-\lambda-\gamma,\mu-\lambda} \varphi \right) \right)(x) = \left(\mathcal{J}^{\ \gamma+\mu;a,b}_{a+b-\lambda-\gamma;1,0} \varphi \right)(x). \tag{37}$$

Letting further $a = \mu + \gamma$ in (37), we have

$$\left(I^{\mu,\lambda-\mu-b,\gamma+\mu-\lambda} \left(I^{\gamma,\mu+b-\lambda,\mu-\lambda} \varphi \right) \right)(x) = \left(\mathcal{J}^{\ \gamma+\mu;\gamma+\mu,b}_{\mu+b-\lambda;1,0} \varphi \right)(x)$$
$$= \left(I^{\gamma+\mu}_{-;1,\mu-\lambda} \varphi \right)(x) = \left(K^-_{\mu-\lambda,\gamma+\mu} \varphi \right)(x). \tag{38}$$

Additionally, by putting $b = \beta + \lambda - \mu$ in (38) and then letting $\lambda = \mu - \eta$ in the resulting equation we get the following clearer form

$$\left(I^{\mu,-\beta,\gamma+\eta} \left(I^{\gamma,\beta,\eta} \varphi \right) \right)(x) = \left(K^-_{\eta,\gamma+\mu} \varphi \right)(x),$$

which is a special case of (5) *when* $\delta = -\beta$. *It does not seem possible to deduce* (5) *by merely specializing the parameters in* (35). *Therefore, it should be interesting to find a composition formula from* (35) *which may include* (5) *or* (6) *as particular cases.*

As depicted in Theorems 1 and 2, the study of the composition structure of the operators $\mathcal{I}$ and $\mathcal{J}$ rests heavily on the existence of a suitable Erdélyi-type integral, because we derive (24) and (35) from the Erdélyi-type integral (29). However, there may possibly be an alternative approach by which the Erdelyi-type integral may be obtained from a known composition structure [1] (see also Refs. [22,32]). Such an approach may be of special interest since our operators involve the generalized hypergeometric function ${}_{r+2}F_{r+1}$ and the methodology may lead to some new results.

3.2. Derivative Formula

In this section we derive a derivative formula involving the fractional integral operator (12).

We introduce here some notations describing necessary concepts that would be used in the sequel. Let (ξ_m) be the nonvanishing zeros of the parametric polynomial $Q_m(t)$ of degree m defined by

$$Q_m(t) = \sum_{j=0}^{m} \sigma_{m-j} \sum_{k=0}^{j} \left\{ {j \atop k} \right\} (b)_k (t)_k (\mu - b - t)_{m-k}, \tag{39}$$

where the σ_j $(0 \le j \le m)$ are determined by the generating relation (10). We define the parametric polynomial $\tilde{Q}_m(t)$ by

$$\tilde{Q}_m(t) = \sum_{j=0}^{m} \tilde{\sigma}_{m-j} \sum_{k=0}^{j} \left\{ {j \atop k} \right\} (\mu - b - m)_k (t)_k (b + m - n - t)_{m-k}, \tag{40}$$

where $\tilde{\sigma}_j$ $(0 \le j \le m)$ are determined by the generating relation

$$\prod_{j=1}^{m} (\xi_j + x) = \sum_{j=0}^{m} \tilde{\sigma}_{m-j} x^j. \tag{41}$$

Theorem 3. *For* $\Re(\mu) > n$ $(n \in \mathbb{N})$, *we have*

$$\begin{aligned} &\frac{\partial^n}{\partial x^n} \left\{ x^{\delta+\nu(\mu-a+h)} \left(\mathcal{I}^{\mu;a,b:\ (f_r+m_r)}_{h;\nu,\delta:\ \ (f_r)} \varphi \right)(x) \right\} \\ &\quad = \nu^n x^{\delta+\nu(\mu-n-a+h)} \left(\mathcal{I}^{\mu-n;a,b-n:\ (\eta_m+1)}_{h;\nu,\delta:\ \ \ (\eta_m)} \varphi \right)(x), \end{aligned} \tag{42}$$

where (η_m) *are the nonvanishing zeros of the parametric polynomial* $\tilde{Q}_m(t)$ *given by* (40).

Proof. Using the Euler-type transformation due to Miller and Paris [20], p. 305, Theorem 3

$${}_{r+2}F_{r+1}\left[\begin{matrix} a, b, (f_r + m_r) \\ \mu, \quad (f_r) \end{matrix}; x \right] = (1-x)^{-a} {}_{m+2}F_{m+1}\left[\begin{matrix} a, \mu - b - m, (\xi_m + 1) \\ \mu, \quad (\xi_m) \end{matrix}; \frac{x}{x-1} \right], \tag{43}$$

we have

$$\begin{aligned} &x^{\delta+\nu(\mu-a+h)} \left(\mathcal{I}^{\mu;a,b:\ (f_r+m_r)}_{h;\nu,\delta:\ \ (f_r)} \varphi \right)(x) \\ &= \frac{\nu}{\Gamma(\mu)} \int_0^x (x^\nu - s^\nu)^{\mu-1} {}_{m+2}F_{m+1}\left[\begin{matrix} a, \mu - b - m, (\xi_m + 1) \\ \mu, \quad (\xi_m) \end{matrix}; 1 - \frac{x^\nu}{s^\nu} \right] \varphi(s) s^{\nu(h-a)+\nu-1} ds, \end{aligned}$$

where (ξ_m) are the nonvanishing zeros of the parametric polynomial $Q_m(t)$ defined by (39). By making use of the Leibniz integral rule, we obtain

$$\frac{\partial}{\partial x}\left\{x^{\delta+\nu(\mu-a+h)}\left(\mathcal{I}^{\mu;a,b:\,(f_r+m_r)}_{h;\nu,\delta:\quad (f_r)}\varphi\right)(x)\right\}$$
$$=\frac{\nu}{\Gamma(\mu)}\frac{\partial}{\partial x}\int_0^x (x^\nu-s^\nu)^{\mu-1}{}_{m+2}F_{m+1}\left[\begin{matrix}a,\mu-b-m,(\xi_m+1)\\ \mu,\quad (\xi_m)\end{matrix};1-\frac{x^\nu}{s^\nu}\right]\varphi(s)s^{\nu(h-a)+\nu-1}\mathrm{d}s$$
$$=\frac{\nu}{\Gamma(\mu)}\int_0^x \frac{\partial}{\partial x}\left\{(x^\nu-s^\nu)^{\mu-1}{}_{m+2}F_{m+1}\left[\begin{matrix}a,\mu-b-m,(\xi_m+1)\\ \mu,\quad (\xi_m)\end{matrix};1-\frac{s^\nu}{x^\nu}\right]\right\}$$
$$\cdot\,\varphi(s)s^{\nu(h-a)+\nu-1}\mathrm{d}s.$$

Taking into account the formula [33], p. 442, Equation (51)

$$\frac{\partial^n}{\partial z^n}\left\{z^{\sigma-1}{}_pF_q\left[\begin{matrix}(a_p)\\ (b_{q-1}),\sigma\end{matrix};z\right]\right\}=(\sigma-n)_n z^{\sigma-n-1}{}_pF_q\left[\begin{matrix}(a_p)\\ (b_{q-1}),\sigma-n\end{matrix};z\right],\tag{44}$$

we have

$$\frac{\partial}{\partial x}\left\{x^{\delta+\nu(\mu-a+h)}\left(\mathcal{I}^{\mu;a,b:\,(f_r+m_r)}_{h;\nu,\delta:\quad (f_r)}\varphi\right)(x)\right\}=\frac{\nu^2}{\Gamma(\mu-1)}\int_0^x (x^\nu-s^\nu)^{(\mu-1)-1}$$
$$\cdot\,{}_{m+2}F_{m+1}\left[\begin{matrix}a,\mu-b-m,(\xi_m+1)\\ \mu-1,\quad (\xi_m)\end{matrix};1-\frac{x^\nu}{s^\nu}\right]\varphi(s)s^{\nu(h-a)+\nu-1}\mathrm{d}s.$$

Next, differentiating n times, we obtain

$$\frac{\partial^n}{\partial x^n}\left\{x^{\delta+\nu(\mu-a+h)}\left(\mathcal{I}^{\mu;a,b:\,(f_r+m_r)}_{h;\nu,\delta:\quad (f_r)}\varphi\right)(x)\right\}=\frac{\nu^{1+n}}{\Gamma(\mu-n)}\int_0^x (x^\nu-s^\nu)^{\mu-n-1}$$
$$\cdot\,{}_{m+2}F_{m+1}\left[\begin{matrix}a,\mu-b-m,(\xi_m+1)\\ \mu-n,\quad (\xi_m)\end{matrix};1-\frac{x^\nu}{s^\nu}\right]\varphi(s)s^{\nu(h-a)+\nu-1}\mathrm{d}s.$$

By applying the Euler-type transformation (43) again, we get

$$\frac{\partial^n}{\partial x^n}\left\{x^{\delta+\nu(\mu-a+h)}\left(\mathcal{I}^{\mu;a,b:\,(f_r+m_r)}_{h;\nu,\delta:\quad (f_r)}\varphi\right)(x)\right\}$$
$$=\frac{\nu^{1+n}x^{-\nu a}}{\Gamma(\mu-n)}\int_0^x (x^\nu-s^\nu)^{\mu-n-1}{}_{m+2}F_{m+1}\left[\begin{matrix}a,b-n,(\eta_m+1)\\ \mu-n,\quad (\eta_m)\end{matrix};1-\frac{s^\nu}{x^\nu}\right]\varphi(s)s^{\nu h+\nu-1}\mathrm{d}s$$
$$=\nu^n x^{\delta+\nu(\mu-n-a+h)}\left(\mathcal{I}^{\mu-n;a,b-n:\,(\eta_m+1)}_{h;\nu,\delta:\qquad (\eta_m)}\varphi\right)(x),\tag{45}$$

where the sequence of parameters (η_m) are the nonvanishing zeros of the parametric polynomial $\tilde{Q}_m(t)$ of degree m given by (40). This completes the proof of (42). □

Before proceeding further, we consider here a simple example.

Example 1. *When $r=1$ and $m=m_1=1$, $f_1=f$ and $\eta_1=\eta$ in (42), we get*

$$\frac{\partial^n}{\partial x^n}\left\{x^{\delta+\nu(\mu-a+h)}\left(\mathcal{I}^{\mu;a,b:\,f+1}_{h;\nu,\delta:\quad f}\varphi\right)(x)\right\}=\nu^n x^{\delta+\nu(\mu-n-a+h)}\left(\mathcal{I}^{\mu-n;a,b-n:\,\eta+1}_{h;\nu,\delta:\qquad \eta}\varphi\right)(x),\tag{46}$$

where η is the nonvanishing zero of the parametric polynomial

$$\tilde{Q}_1(t)=\sum_{j=0}^{1}\tilde{\sigma}_{1-j}\sum_{k=0}^{j}\left\{\begin{matrix}j\\k\end{matrix}\right\}(\mu-b-1)_k(t)_k(b+1-n-t)_{1-k}$$
$$=\tilde{\sigma}_1\left\{\begin{matrix}0\\0\end{matrix}\right\}(b+1-n-t)+\tilde{\sigma}_0\left\{\begin{matrix}1\\0\end{matrix}\right\}(b+1-n-t)+\tilde{\sigma}_0\left\{\begin{matrix}1\\1\end{matrix}\right\}(\mu-b-1)t$$
$$=\tilde{\sigma}_1(b+1-n)+[\tilde{\sigma}_0(\mu-b-1)-\tilde{\sigma}_1]t.$$

Therefore, η can be expressed as

$$\eta = \frac{\tilde{\sigma}_1(b+1-n)}{\tilde{\sigma}_1 - \tilde{\sigma}_0(\mu - b - 1)}.$$

It follows from (41) that $\tilde{\sigma}_0 = 1$ and $\tilde{\sigma}_1 = \xi$, where ξ is the nonvanishing zero of the parametric polynomial

$$Q_1(t) = \sigma_1(\mu - b) + [\sigma_0 b - \sigma_1]t.$$

From (10), we have $\sigma_0 = 1$ and $\sigma_1 = f$ and thus ξ can be written as $\xi = f(\mu - b)/(f-b)$. Hence,

$$\eta = \frac{f(\mu - b)(b+1-n)}{f + b(\mu - b - 1)},$$

wherein we note that η depends on n.

It may be observed that the Euler-type transformation (43) is used twice, so we need to be careful while finding special cases of Theorem 3.

(i) By letting $b = n$ $(n \in \mathbb{N})$ in (42) and noting that ${}_{m+2}F_{m+1}$-function in (42) reduces to 1, we get

$$\begin{aligned}
&\frac{\partial^n}{\partial x^n}\left\{x^{\delta+\nu(\mu-a+h)}\left(\mathcal{I}^{\mu;a,n:\ (f_r+m_r)}_{h;\nu,\delta:\ \ (f_r)}\ \varphi\right)(x)\right\} \\
&\quad = \frac{\nu^{1+n}x^{-\nu a}}{\Gamma(\mu-n)}\int_0^x (x^\nu - s^\nu)^{\mu-n-1}\varphi(s)s^{\nu h+\nu-1}\mathrm{d}s \\
&\quad = \nu^n x^{\nu(\mu-n-a+h)}\left(I^{\mu-n}_{+;\nu,h}\varphi\right)(x),
\end{aligned} \tag{47}$$

where $I^{\mu-n}_{+;\nu,h}$ denotes the Erdélyi–Kober type fractional integral defined by (18). In fact, letting $b = n$ changes the parametric polynomials $Q_m(t)$ and $\tilde{Q}_m(t)$ defined by (39) and (40), respectively. However, if the new polynomials, say $Q^*_m(t)$ and $\tilde{Q}^*_m(t)$, also have nonvanishing zeros, denoted by (ξ^*_m) and (η^*_m) respectively, then (47) holds true. To illustrate here, let us set $b = n$ in Example 1, then $Q_1(t)$ becomes $Q^*_1(t) = f(\mu - n) + (n - f)t$ with $\xi^* = f(\mu - n)/(f - n)$ its nonvanishing zero and $\tilde{Q}_1(t)$ becomes $\tilde{Q}^*_1(t) = \xi^* + (\mu - n - 1 - \xi^*)t$. The nonvanishing zero of $\tilde{Q}^*_1(t)$ is

$$\eta^* = \frac{f(\mu - n)}{f + n(\mu - n - 1)} \quad (f \neq 0,\ \mu \neq n).$$

Therefore, we obtain from (46) that

$$\begin{aligned}
\frac{\partial^n}{\partial x^n}\left\{x^{\delta+\nu(\mu-a+h)}\left(\mathcal{I}^{\mu;a,n:\ f+1}_{h;\nu,\delta:\ \ f}\ \varphi\right)(x)\right\} &= \nu^n x^{\delta+\nu(\mu-n-a+h)}\left(\mathcal{I}^{\mu-n;a,0:\ \eta^*+1}_{h;\nu,\delta:\ \ \eta^*}\ \varphi\right)(x) \\
&= \nu^n x^{\nu(\mu-n-a+h)}\left(I^{\mu-n}_{+;\nu,h}\varphi\right)(x).
\end{aligned}$$

We also observe that the subsitution $b = n$ may always reduce the right-hand side of (42) to a Erdélyi–Kober type integral.

(ii) When $r = 0$, then in view of (14) and (42), we simply obtain

$$\frac{\partial^n}{\partial x^n}\left\{x^{\delta+\nu(\mu-a+h)}\left(\mathcal{I}^{\mu;a,b}_{h;\nu,\delta}\ \varphi\right)(x)\right\} = \nu^n x^{\delta+\nu(\mu-n-a+h)}\left(\mathcal{I}^{\mu-n;a,b-n}_{h;\nu,\delta}\ \varphi\right)(x). \tag{48}$$

Further, if $h = 0$, $\nu = 1$, $\delta = \beta$, $a = \alpha + \beta$, $b = -\eta + n$ and $\mu = \alpha + n$ in (48), we then have

$$\frac{\partial^n}{\partial x^n}\left\{x^n\left(\mathcal{I}^{\alpha+n;\alpha+\beta,-\eta+n}_{0;1,\beta}\ \varphi\right)(x)\right\} = \left(\mathcal{I}^{\alpha;\alpha+\beta,-\eta}_{0;1,\beta}\ \varphi\right)(x).$$

In addition, in view of (16) and the relation

$$x^n\left(\mathcal{I}^{\alpha+n;\alpha+\beta,-\eta+n}_{0;1,\beta}\varphi\right)(x) = \frac{x^{-\beta-\alpha}}{\Gamma(\alpha+n)}\int_0^x (x-s)^{\alpha+n-1}{}_2F_1\left[\begin{matrix}\alpha+\beta,-\eta+n\\ \alpha+n\end{matrix};1-\frac{s}{x}\right]\varphi(s)ds$$
$$= \left(I^{\alpha+n,\beta-n,\eta-n}_{0,x}\varphi\right)(x),$$

we note the following interesting and remarkable relation:

$$\frac{\partial^n}{\partial x^n}\left(I^{\alpha+n,\beta-n,\eta-n}_{0,x}\varphi\right)(x) = \left(I^{\alpha,\beta,\eta}_{0,x}\varphi\right)(x),$$

which serves as the definition of Saigo's generalized fractional derivative (see Ref. [3], p. 8, Equation (3.2)).

4. Relationship with Khudozhnikov's Work

In a very short paper, Khudozhnikov [17] considered in a certain class of integrable functions the following Volterra-type integral equation

$$\int_a^x \frac{(x-s)^{\gamma-1}}{\Gamma(\gamma)}{}_3F_2\left[\begin{matrix}\alpha,\beta,\varepsilon+m\\ \gamma,\quad\varepsilon\end{matrix};1-\frac{x}{s}\right]\varphi(s)ds = g(x), \tag{49}$$

where $0 < \Re(\gamma) < 1$, $m \in \mathbb{N}$ and $0 < a \le x \le b < +\infty$. By using some known formulas from Ref. [33], Khudozhnikov obtained the following result [17], p. 79, Equation (2).

Theorem 4 (Khudozhnikov). *The Volterra-type integral Equation (49) can be reduced to the following system of differential and integral equations:*

$$\begin{cases}\displaystyle\sum_{k=0}^{m}\binom{m}{k}\frac{(\alpha)_k(\beta)_k}{(\varepsilon)_k(-x)^k}y^{(m-k)}(x) = g(x)x^{\alpha+\beta-\gamma},\\ \displaystyle\int_a^x \frac{(x-s)^{\gamma+m-1}}{\Gamma(\gamma+m)}{}_2F_1\left[\begin{matrix}\gamma-\alpha,\gamma-\beta\\ \gamma+m\end{matrix};1-\frac{x}{s}\right]s^{\alpha+\beta-\gamma}\varphi(s)ds = y(x),\end{cases}$$

with initial conditions $y(a) = y'(a) = \cdots = y^{(m-1)}(a) = 0$.

In Ref. [17], Khudozhnikov briefly mentioned that the result can be generalized to those equations involving the generalized hypergeometric functions ${}_{p+1}F_p$, ${}_pF_p$ and ${}_{p-1}F_p$. However, he did not give possible forms of the generalizations or the formulas to be used. In fact, the most likely generalization requires use of a generalized Euler-type transformation, which is not included in Ref. [33]. Therefore, we think that the question of finding a generalization of Theorem 4 is still open.

In this section, we first propose a generalization of Theorem 4. We then consider a Volterra-type integral equation generated by the operator $\mathcal{I}$ defined by (12) and obtain an analogue of Khudozhnikov's theorem.

4.1. A Generalization of Khudozhnikov's Theorem

Let us consider the Volterra-type integral equation

$$\int_a^x \frac{(x-s)^{\gamma-1}}{\Gamma(\gamma)}{}_{r+2}F_{r+1}\left[\begin{matrix}\alpha,\beta,(f_r+m_r)\\ \gamma,\quad(f_r)\end{matrix};1-\frac{x}{s}\right]\varphi(s)ds = g(x), \tag{50}$$

where $0 < \Re(\gamma) < 1$, $m \in \mathbb{N}$ and $0 < a \le x \le b < +\infty$. Obviously, (50) reduces to (49) when $r = 1$, $f_1 = \varepsilon$ and $m_1 = m$.

By using a lemma due to Miller and Paris [20], p. 298, Lemma 4, and the classical Euler transformation [18], p. 68, Equation (2.2.7), we can express the ${}_{r+2}F_{r+1}$-function as a finite sum of ${}_2F_1$-functions given by

$$
{}_{r+2}F_{r+1}\left[\begin{matrix}\alpha,\beta,(f_r+m_r)\\ \gamma,\quad (f_r)\end{matrix};x\right]=\sum_{k=0}^{m}\frac{A_k}{A_0}\frac{(\alpha)_k(\beta)_k}{(\gamma)_k}{}_2F_1\left[\begin{matrix}\alpha+k,\beta+k,\\ \gamma+k,\end{matrix};x\right]x^k
$$

$$
=(1-x)^{\gamma-\alpha-\beta}\sum_{k=0}^{m}\frac{A_k}{A_0}\frac{(\alpha)_k(\beta)_k}{(\gamma)_k}{}_2F_1\left[\begin{matrix}\gamma-\alpha,\gamma-\beta,\\ \gamma+k,\end{matrix};x\right]\left(\frac{x}{1-x}\right)^k. \tag{51}
$$

Then (50) can be written as

$$
g(x)=\int_a^x\frac{(x-s)^{\gamma-1}}{\Gamma(\gamma)}{}_{r+2}F_{r+1}\left[\begin{matrix}\alpha,\beta,(f_r+m_r)\\ \gamma,\quad (f_r)\end{matrix};1-\frac{x}{s}\right]\varphi(s)ds
$$

$$
=\sum_{k=0}^{m}\frac{A_k}{A_0}\frac{(\alpha)_k(\beta)_k}{(-x)^k}x^{\gamma-\alpha-\beta}\int_a^x\frac{(x-s)^{\gamma+k-1}}{\Gamma(\gamma+k)}{}_2F_1\left[\begin{matrix}\gamma-\alpha,\gamma-\beta\\ \gamma+k\end{matrix};1-\frac{x}{s}\right]s^{\alpha+\beta-\gamma}\varphi(s)ds. \tag{52}
$$

Let

$$
y(x):=\int_a^x\frac{(x-s)^{\gamma+m-1}}{\Gamma(\gamma+m)}{}_2F_1\left[\begin{matrix}\gamma-\alpha,\gamma-\beta\\ \gamma+m\end{matrix};1-\frac{x}{s}\right]s^{\alpha+\beta-\gamma}\varphi(s)ds.
$$

In view of the derivative Formula (44), we have

$$
\frac{\partial^{m-k}}{\partial x^{m-k}}\left\{\frac{(x-s)^{\gamma+m-1}}{\Gamma(\gamma+m)}{}_2F_1\left[\begin{matrix}\gamma-\alpha,\gamma-\beta\\ \gamma+m\end{matrix};1-\frac{x}{s}\right]\right\}=\frac{(x-s)^{\gamma+k-1}}{\Gamma(\gamma+k)}{}_2F_1\left[\begin{matrix}\gamma-\alpha,\gamma-\beta\\ \gamma+k\end{matrix};1-\frac{x}{s}\right],
$$

and therefore

$$
y^{(m-k)}(x)=\int_a^x\frac{(x-s)^{\gamma+k-1}}{\Gamma(\gamma+k)}{}_2F_1\left[\begin{matrix}\gamma-\alpha,\gamma-\beta\\ \gamma+k\end{matrix};1-\frac{x}{s}\right]s^{\alpha+\beta-\gamma}\varphi(s)ds
$$

and $y^{(m-k)}(a)=0$ for $k=1,\cdots,m-1$. Now (52) can be expressed as

$$
\sum_{k=0}^{m}\frac{A_k}{A_0}\frac{(\alpha)_k(\beta)_k}{(-x)^k}y^{(m-k)}(x)=x^{\alpha+\beta-\gamma}g(x).
$$

The above steps concerning the integral Equation (50) therefore yield the following theorem.

Theorem 5. *The Volterra-type integral Equation (50) can be reduced to the following system of differential and integral equations:*

$$
\begin{cases}\displaystyle\sum_{k=0}^{m}\frac{A_k}{A_0}\frac{(\alpha)_k(\beta)_k}{(-x)^k}y^{(m-k)}(x)=g(x)x^{\alpha+\beta-\gamma},\\ \displaystyle\int_a^x\frac{(x-s)^{\gamma+m-1}}{\Gamma(\gamma+m)}{}_2F_1\left[\begin{matrix}\gamma-\alpha,\gamma-\beta\\ \gamma+m\end{matrix};1-\frac{x}{s}\right]s^{\alpha+\beta-\gamma}\varphi(s)ds=y(x),\end{cases}
$$

with initial conditions $y(a)=y'(a)=\cdots=y^{(m-1)}(a)=0$, where A_k $(0\le k\le m)$ is defined in (11).

To show that Theorem 5 contains Khudozhnikov's result as a special case, we only need to prove that

$$
\frac{A_k}{A_0}=\binom{m}{k}\frac{1}{(\varepsilon)_k}. \tag{53}
$$

Our calculations require some basics on the theory of combinatorics.

When $r = 1$, $f_1 = \varepsilon$ and $m_1 = m$, we get

$$A_0 = (\varepsilon)_m \quad \text{and} \quad A_k = \sum_{j=k}^{m} \left\{ {j \atop k} \right\} \hat{\sigma}_{m-j}, \tag{54}$$

where $\hat{\sigma}_{m-j}$ is generated by

$$(\varepsilon + x)_m = \sum_{j=0}^{m} \hat{\sigma}_{m-j} x^j. \tag{55}$$

We need in fact to find an explicit expression for $\hat{\sigma}_{m-j}$. By using the Chu–Vandermonde identity [18], p. 70, we have

$$(\varepsilon + x)_m = \sum_{k=0}^{m} \binom{m}{k} (\varepsilon)_{m-k} (x)_k. \tag{56}$$

Recall that

$$(x)_k = \sum_{j=0}^{k} (-1)^{k-j} s(k,j) x^j = \sum_{j=0}^{k} \left[{k \atop j} \right] x^j, \tag{57}$$

where $s(k,j)$ is the Stirling number of the first kind and the symbol $\left[{k \atop j} \right]$ is usually used to denote the *unsigned* Stirling number of the first kind (see Ref. [34], p. 239). Substituting (57) into (56) and then interchanging the order of summation, we obtain

$$(\varepsilon + x)_m = \sum_{k=0}^{m} \binom{m}{k} (\varepsilon)_{m-k} \sum_{j=0}^{k} \left[{k \atop j} \right] x^j = \sum_{j=0}^{m} \sum_{k=j}^{m} \binom{m}{k} (\varepsilon)_{m-k} \left[{k \atop j} \right] x^j. \tag{58}$$

Comparing (58) with (55), it follows that

$$\hat{\sigma}_{m-j} = \sum_{k=j}^{m} \binom{m}{k} (\varepsilon)_{m-k} \left[{k \atop j} \right], \tag{59}$$

and combining (54) with (59) and taking into account the index factorization

$$[k \le j \le m][j \le \ell \le m] = [k \le j \le \ell \le m] = [k \le \ell \le m][k \le j \le \ell],$$

we obtain

$$\begin{aligned} A_k &= \sum_{j=k}^{m} \left\{ {j \atop k} \right\} \sum_{\ell=j}^{m} \binom{m}{\ell} (\varepsilon)_{m-\ell} \left[{\ell \atop j} \right] = \sum_{\ell=k}^{m} \binom{m}{\ell} (\varepsilon)_{m-\ell} \sum_{j=k}^{\ell} \left\{ {j \atop k} \right\} \left[{\ell \atop j} \right] \\ &= \frac{1}{(k-1)!} \sum_{\ell=k}^{m} \binom{m}{\ell} \binom{\ell}{k} (\varepsilon)_{m-\ell} (1)_{\ell-1} = \frac{1}{(k-1)!} \sum_{\ell=0}^{m-k} \binom{m}{\ell+k} \binom{\ell+k}{k} (\varepsilon)_{m-\ell-k} (1)_{\ell+k-1} \\ &= \binom{m}{k} \sum_{\ell=0}^{m-k} \binom{m-k}{\ell} (\varepsilon)_{m-k-\ell} (k)_\ell = \binom{m}{k} \frac{(\varepsilon)_m}{(\varepsilon)_k}, \end{aligned} \tag{60}$$

where we have used the familiar convoluation identity (see, for example Ref. [34], p. 240)

$$\sum_{j=k}^{\ell} \left\{ {j \atop k} \right\} \left[{\ell \atop j} \right] = \binom{\ell}{k} \frac{(\ell-1)!}{(k-1)!} \quad (\ell \ge k \ge 1).$$

Evidently, (60) is equivalent to (53).

4.2. A Variant of Khudozhnikov's Theorem

A comparison of the fractional integral operator $\mathcal{I}$ with Equations (49) and (50) inspire us to consider the following integral equation

$$\int_{\rho}^{x} \frac{(x^{\nu}-s^{\nu})^{\mu-1}}{\Gamma(\mu)} {}_{r+2}F_{r+1}\left[\begin{matrix} a, b, (f_r+m_r) \\ \mu, \quad (f_r) \end{matrix}; 1-\frac{s^{\nu}}{x^{\nu}}\right] \varphi(s) s^{\nu h+\nu-1} \mathrm{d}s = g(x), \tag{61}$$

where $0 < \Re(\mu) < 1$ and $0 < \rho \leq s \leq x < \infty$.

Using the Euler-type transformation (43), then Equation (61) can be converted into

$$\int_{\rho}^{x} \frac{(x^{\nu}-s^{\nu})^{\mu-1}}{\Gamma(\mu)} {}_{m+2}F_{m+1}\left[\begin{matrix} a, \mu-b-m, (\xi_m+1) \\ \mu, \quad (\xi_m) \end{matrix}; 1-\frac{x^{\nu}}{s^{\nu}}\right] \varphi(s) s^{\nu(h-a)+\nu-1} \mathrm{d}s$$
$$= x^{-\nu a} g(x), \tag{62}$$

where (ξ_m) are nonvanishing zeros of the parametric polynomial $Q_m(t)$ of degree m given by (39).

By using the same lemma of Miller and Paris [20], p. 298, Lemma 4 and the Euler transformation [18], p. 68, Equation (2.2.7) or else using Equation (51), we can express (as in the proof of Theorem 5) the ${}_{m+2}F_{m+1}$-function as a finite sum of ${}_2F_1$-functions given by

$$\begin{aligned} {}_{m+2}F_{m+1}&\left[\begin{matrix} a, \mu-b-m, (\xi_m+1) \\ \mu, \quad (\xi_m) \end{matrix}; x\right] \\ &= \sum_{k=0}^{m} \frac{\mathcal{A}_k}{(\mu)_k} {}_2F_1\left[\begin{matrix} a+k, \mu-b-m+k, \\ \mu+k, \end{matrix}; x\right] x^k \\ &= (1-x)^{b-a+m} \sum_{k=0}^{m} \frac{\mathcal{A}_k}{(\mu)_k} {}_2F_1\left[\begin{matrix} \mu-a, b+m, \\ \mu+k, \end{matrix}; x\right] \left(\frac{x}{1-x}\right)^k, \end{aligned} \tag{63}$$

where

$$\mathcal{A}_k := \frac{(a)_k(\mu-b-m)_k}{\xi_1 \cdots \xi_m} \sum_{j=k}^{m} \left\{\begin{matrix} j \\ k \end{matrix}\right\} \tilde{\sigma}_{m-j} \tag{64}$$

and $\tilde{\sigma}_j$ $(0 \leq j \leq m)$ are generated by (41). With the help of (63), the integral Equation (62) can then be written as

$$\sum_{k=0}^{m} \frac{\mathcal{A}_k}{(-x^{\nu})^k} \int_{\rho}^{x} \frac{(x^{\nu}-s^{\nu})^{\mu+k-1}}{\Gamma(\mu+k)} {}_2F_1\left[\begin{matrix} \mu-a, b+m, \\ \mu+k, \end{matrix}; 1-\frac{x^{\nu}}{s^{\nu}}\right] \varphi(s) s^{\nu(h-b-m)+\nu-1} \mathrm{d}s$$
$$= x^{-\nu(b+m)} g(x). \tag{65}$$

By making use of (44), we obtain

$$\frac{\partial^{m-k}}{\partial z^{m-k}}\left\{z^{\mu+m-1} {}_2F_1\left[\begin{matrix} \mu-a, b+m \\ \mu+m \end{matrix}; z\right]\right\} = \frac{(\mu)_m}{(\mu)_k} z^{\mu+k-1} {}_2F_1\left[\begin{matrix} \mu-a, b+m \\ \mu+k \end{matrix}; z\right],$$

and thus

$$\begin{aligned} &\frac{\partial^{m-k}}{\partial x^{m-k}}\left\{\frac{(x^{\nu}-s^{\nu})^{\mu+m-1}}{\Gamma(\mu+m)} {}_2F_1\left[\begin{matrix} \mu-a, b+m \\ \mu+m \end{matrix}; 1-\frac{x^{\nu}}{s^{\nu}}\right]\right\} \\ &\quad = \nu^{m-k} \frac{(x^{\nu}-s^{\nu})^{\mu+k-1}}{\Gamma(\mu+k)} {}_2F_1\left[\begin{matrix} \mu-a, b+m \\ \mu+k \end{matrix}; 1-\frac{x^{\nu}}{s^{\nu}}\right]. \end{aligned} \tag{66}$$

Substituting (66) into (65), we get

$$\sum_{k=0}^{m}\frac{\mathcal{A}_k\nu^k}{(-x^\nu)^k}\int_\rho^x\frac{\partial^{m-k}}{\partial x^{m-k}}\left\{\frac{(x^\nu-s^\nu)^{\mu+m-1}}{\Gamma(\mu+m)}{}_2F_1\left[\begin{matrix}\mu-a,b+m,\\ \mu+m,\end{matrix};1-\frac{x^\nu}{s^\nu}\right]\right\}\varphi(s)s^{\nu(h-b-m)+\nu-1}ds$$
$$=\nu^m x^{-\nu(b+m)}g(x).$$

Finally, using the Leibniz integral rule and simplifying the resulting formula by the Pfaff transformation [18], p. 68, Equation (2.2.6), we obtain

$$\sum_{k=0}^{m}\frac{\mathcal{A}_k\nu^k}{(-x^\nu)^k}\frac{\partial^{m-k}}{\partial x^{m-k}}\left\{x^{\nu(a-\mu)}\int_\rho^x\frac{(x^\nu-s^\nu)^{\mu+m-1}}{\Gamma(\mu+m)}{}_2F_1\left[\begin{matrix}\mu-a,\mu-b,\\ \mu+m,\end{matrix};1-\frac{s^\nu}{x^\nu}\right]\right.$$
$$\left.\varphi(s)s^{\nu(h+\mu-b-a-m)+\nu-1}ds\right\}=\nu^m x^{-\nu(b+m)}g(x).$$

If

$$y(x)=x^{\nu(a-\mu)}\int_\rho^x\frac{(x^\nu-s^\nu)^{\mu+m-1}}{\Gamma(\mu+m)}{}_2F_1\left[\begin{matrix}\mu-a,\mu-b,\\ \mu+m,\end{matrix};1-\frac{s^\nu}{x^\nu}\right]\varphi(s)s^{\nu(h+\mu-b-a-m)+\nu-1}ds,$$

then the above details concerning the integral equation (61) may be put in the following theorem.

Theorem 6. *The Volerra-type integral Equation (61) can be reduced to the following system of differential and integral equations:*

$$\begin{cases}\displaystyle\sum_{k=0}^{m}\frac{\mathcal{A}_k\nu^k}{(-x^\nu)^k}y^{(m-k)}(x)=\nu^m x^{-\nu(b+m)}g(x),\\ \displaystyle x^{\nu(a-\mu)}\int_\rho^x\frac{(x^\nu-s^\nu)^{\mu+m-1}}{\Gamma(\mu+m)}{}_2F_1\left[\begin{matrix}\mu-a,\mu-b,\\ \mu+m,\end{matrix};1-\frac{s^\nu}{x^\nu}\right]\varphi(s)s^{\nu(h+\mu-b-a-m)+\nu-1}ds=y(x),\end{cases}$$

with initial conditions $y(\rho)=y'(\rho)=\cdots=y^{(m-1)}(\rho)=0$, *where* $\mathcal{A}_k$ $(0\le k\le m)$ *is given by* (64).

5. Conclusions

In this paper, some composition formulas of $\mathcal{I}$ and $\mathcal{J}$ defined by (12) and (13) are obtained by making use of a Erdélyi-type integral. We find a derivative formula, which in the future may enable us to define a new fractional derivative operator. Finally, we generalize Khudozhnikov's work on Volterra-type integral equation and find its relationship with our operator $\mathcal{I}$.

Considering the obtained properties of the operators $\mathcal{I}$ and $\mathcal{J}$, we briefly mention here some problems that deserve further study.

(i) Since only two composition formulas for $\mathcal{I}$ and $\mathcal{J}$ are found in the present work, which is still a very small number compared to the number of the composition formulas of Saigo's operators $I^{\alpha,\beta,\eta}$ and $J^{\alpha,\beta,\eta}$, it may be worthwhile if additional composition structures can be discovered for the operators $\mathcal{I}$ and $\mathcal{J}$. The exploration in this direction may also lead us to new discoveries related to the Erdélyi-type integrals;

(ii) The present work together with our previous papers [14,16] have established many fundamental properties of $\mathcal{I}$ and $\mathcal{J}$. For further possible work, some new properties and problems may be worthy of attention in view of the classical books [4,23] on the subject and some recent review articles contained, for example, in Ref. [35]. In particular, it may be worthwhile to first focus on the problem of finding a reasonable analogue of the well known limit case formula, viz. $\lim_{\alpha\to 0}(I^\alpha_{a+}\varphi)(x)=\varphi(x)$ concerning the Riemann–Liouville fractional integral operator (see Ref. [23], p. 51, Theorem 2.7).

Author Contributions: Conceptualization, M.-J.L. and R.K.R.; methodology, M.-J.L. and R.K.R.; writing—original draft preparation, M.-J.L. and R.K.R.; writing—review and editing, M.-J.L. and R.K.R.; funding acquisition, M.-J.L. All authors have read and agreed to the published version of the manuscript.

Funding: The research of the first author is supported by National Natural Science Foundation of China (No. 12001095).

Institutional Review Board Statement: Not applicable.

Informed Consent Statement: Not applicable.

Data Availability Statement: Not applicable.

Acknowledgments: The authors thank the referees for their comments and suggestions.

Conflicts of Interest: The authors declare no conflict of interest. The funders had no role in the design of the study; in the collection, analyses, or interpretation of data; in the writing of the manuscript; or in the decision to publish the results.

References

1. Saigo, M. A remark on integral operators involving the Gauss hypergeometric functions. *Math. Rep. Coll. Gen. Educ. Kyushu Univ.* **1978**, *11*, 135–143.
2. Saigo, M. On the Hölder continuity of the generalized fractional integrals and derivatives. *Math. Rep. Coll. Gen. Educ. Kyushu Univ.* **1980**, *12*, 55–62.
3. Saigo, M. A generalization of fractional calculus and its applications to Euler-Darboux equation. *RIMS Kokyuroku* **1981**, *412*, 33–56.
4. Kiryakova, V. *Generalized Fractional Calculus and Applications*; Pitman Research Notes in Mathematics Series No. 301; Longman Scientific and Technical: Harlow, UK, 1994.
5. Naheed, S.; Mubeen, S.; Rahman, G.; Khan, A.Z.; Nisar, K.S. Certain integral and differential formulas involving the product of Srivastava's polynomials and extended Wright function. *Fractal Fract.* **2022**, *6*, 93. [CrossRef]
6. Dziok, J.; Raina, R.K.; Sokół, J. Applications of differential subordinations for norm estimates of an integral operator. *Proc. R. Soc. Edinb. Sect. A Math.* **2018**, *148*, 281–291. [CrossRef]
7. Kiryakova, V. On two Saigo's fractional integral operators in the class of univalent functions. *Fract. Calc. Appl. Anal.* **2006**, *9*, 159–176.
8. Saigo, M.; Raina, R.K. On the fractional calculus operator involving Gauss's series and its application to certain statistical distributions. *Rev. Téc. Ing. Univ. Zulia* **1991**, *14*, 53–62.
9. Goyal, S.P.; Jain, R.M. Fractional integral operators and the generalized hypergeometric functions. *Indian J. Pure Appl. Math.* **1987**, *18*, 251–259.
10. Goyal, S.P.; Jain, R.M.; Gaur, N. Fractional integral operators involving a product of generalized hypergeometric functions and a general class of polynomials. *Indian J. Pure Appl. Math.* **1991**, *22*, 403–411.
11. Goyal, S.P.; Jain, R.M.; Gaur, N. Fractional integral operators involving a product of generalized hypergeometric functions and a general class of polynomials. II. *Indian J. Pure Appl. Math.* **1992**, *23*, 121–128.
12. Saigo, M.; Saxena, R.K.; Ram, J. Certain properties of operators of fractional integration associated with Mellin and Laplace transformations. In *Current Topics in Analytic Function Theory*; Srivastava, H.M., Owa, S. Eds.; World Scientific: Singapore, 1992; pp. 291–304.
13. Araci, S.; Rahman, G.; Ghaffar, A.; Azeema; Nisar, K.S. Fractional calculus of extended Mittag-Leffler function and its applications to statistical distribution. *Mathematics* **2019**, *7*, 248. [CrossRef]
14. Luo, M.-J.; Raina, R.K. Fractional integral operators characterized by some new hypergeometric summation formulas. *Fract. Calc. Appl. Anal.* **2017** *20*, 422–446. [CrossRef]
15. Luo, M.-J.; Raina, R.K. On a multiple Čebyšev type functional defined by a generalized fractional integral operator. *Tbil. Math. J.* **2017**, *10*, 161–169. [CrossRef]
16. Luo, M.-J.; Raina, R.K. The decompositional structure of certain fractional integral operators. *Hokkaido Math. J.* **2019**, *48*, 611–650. [CrossRef]
17. Khudozhnikov, V.I. Integration of Volterra-type integral equations of the first kind with kernels containing some generalized hypergeometric functions. *Matem. Mod.* **1995**, *7*, 79.
18. Andrews, G.E.; Askey, R.; Roy, R. *Special Functions*; Cambridge University Press: Cambridge, UK, 1999.
19. Kilbas, A.A.; Srivastava, H.M.; Trujillo, J.J. *Theory and Applications of Fractional Differential Equations*; North-Holland Mathematics Studies; Elsevier Science B.V.: Amsterdam, The Netherlands, 2006; Volume 204.
20. Miller, A.M.; Paris, R.B. Transformation formulas for the generalized hypergeometric function with integral parameter differences. *Rocky Mt. J. Math.* **2013**, *43*, 291–327. [CrossRef]
21. Luo, M.-J.; Raina, R.K. Erdélyi-type integrals for generalized hypergeometric functions with integral parameter differences. *Integral Transform. Spec. Funct.* **2017**, *28*, 476–487. [CrossRef]

22. Grinko, A.P.; Kilbas, A.A. On compositions of generalized fractional integrals. *J. Math. Res. Expo.* **1991**, *11*, 165–171.
23. Samko, S.G.; Kilbas, A.A.; Marichev, O.I. *Fractional Integrals and Derivatives: Theory and Applications*; Gordon and Breach: New York, NY, USA, 1993.
24. Kiryakova, V. On the origins of generalized fractional calculus. *AIP Conf. Proc.* **2015**, *1690*, 050007.
25. Kiryakova, V. Fractional calculus operators of special functions? The result is well predictable! *Chaos Solitons Fractals* **2017**, *102*, 2–15. [CrossRef]
26. Baleanu, D.; Agarwal, P. On generalized fractional integral operators and the generalized Gauss hypergeometric functions. *Abstr. Appl. Anal.* **2014**, *2014*, 630840. [CrossRef]
27. Bansal, M.K.; Kumar, D.; Jain, R. A study of Marichev-Saigo-Maeda fractional integral operators associated with the S-generalized Gauss hypergeometric function. *Kyungpook Math. J.* **2019**, *59*, 433–443.
28. Brychkov, Y.A.; Glaeske, H.-J.; Marichev, O.I. Factorization of integral transformations of convolution type. *J. Math. Sci.* **1985**, *30*, 2071–2094. [CrossRef]
29. Luchko, Y. General fractional integrals and derivatives with the Sonine kernels. *Mathematics* **2021**, *9*, 594. [CrossRef]
30. Olver, F.W.J.; Lozier, D.W.; Boisvert, R.F.; Clark, C.W. (Eds.) *NIST Handbook of Mathematical Functions*; Cambridge University Press: New York, NY, USA, 2010.
31. Bühring, W. Generalized hypergeometric functions at unit argument. *Proc. Am. Math. Soc.* **1992**, *114*, 145–153. [CrossRef]
32. Grinko, A.P.; Kilbas, A.A. On compositions of generalized fractional integrals and evaluation of definite integrals with Gauss hypergeometric functions. *J. Math. Res. Expo.* **1991**, *11*, 443–446.
33. Prudnikov, A.P.; Brychkov, J.A.; Marichev, O.I. *Integrals and Series. Volume 3: More Special Functions*; Gould, G.G., Ed.; Gordon and Breach Science Publishers: New York, NY, USA, 1990.
34. Chu, W. Disjoint convolution sums of Stirling numbers. *Math. Commun.* **2021**, *26*, 239–251.
35. Hilfer, R.; Luchko, Y. *Desiderata* for Fractional Derivatives and Integrals. *Mathematics* **2019**, *7*, 149. [CrossRef]

symmetry

MDPI

Article

Numerical Investigation of Nonlinear Shock Wave Equations with Fractional Order in Propagating Disturbance

Jiahua Fang [1], Muhammad Nadeem [2,*], Mustafa Habib [3] and Ali Akgül [4]

1 Yibin University, Yibin 644000, China; fangjiahua@yibinu.edu.cn
2 Faculty of Science, Yibin University, Yibin 644000, China
3 Department of Mathematics, University of Engineering and Technology, Lahore 54890, Pakistan; mustafa@uet.edu.pk
4 Department of Mathematics, Art and Science Faculty, Siirt University, Siirt 56100, Turkey; aliakgul@siirt.edu.tr
* Correspondence: nadeem@yibinu.edu.cn

Abstract: The symmetry design of the system contains integer partial differential equations and fractional-order partial differential equations with fractional derivative. In this paper, we develop a scheme to examine fractional-order shock wave equations and wave equations occurring in the motion of gases in the Caputo sense. This scheme is formulated using the Mohand transform (MT) and the homotopy perturbation method (HPM), altogether called Mohand homotopy perturbation transform (MHPT). Our main finding in this paper is the handling of the recurrence relation that produces the series solutions after only a few iterations. This approach presents the approximate and precise solutions in the form of convergent results with certain countable elements, without any discretization or slight perturbation theory. The numerical findings and solution graphs attained using the MHPT confirm that this approach is significant and reliable.

Keywords: Mohand transform; homotopy perturbation method; shock wave equation

Citation: Fang, J.; Nadeem, M.; Habib, M.; Akgül, A. Numerical Investigation of Nonlinear Shock Wave Equations with Fractional Order in Propagating Disturbance. *Symmetry* **2022**, *14*, 1179. https://doi.org/10.3390/sym14061179

Academic Editor: Francisco Martínez González

Received: 3 May 2022
Accepted: 4 June 2022
Published: 8 June 2022

1. Introduction

In recent decades, various fractional models in science and technology have been designed in terms of nonlinear partial differential equations (PDEs), such as plasma physics, fluid dynamics, nonlinear optics, quantum mechanics, solid-state physics, mathematical biology and chemical kinetics [1–3]. Fractional differential equations have been widely used to model complex phenomena in various branches of science and engineering, such as wave propagation, lattice vibration, optical fiber, nanotechnology and biology [4,5]. The scientific theory of shock waves played a role in the problems of motion of gases and compressible liquids in the second half of the 19th century. They are described by nonlinear hyperbolic PDEs and can be written in their simplest form as [6]

$$D^{\alpha}_{\wp}\vartheta(\Im,\wp)+f\Big(\vartheta(\Im,\wp)\Big)_{\Im}=0, \qquad \Im\in\mathbb{R},\ \wp>0 \tag{1}$$

with the initial condition

$$\vartheta(\Im,0)=\vartheta_0(\Im), \qquad \Im\in\mathbb{R}. \tag{2}$$

The shock wave equation is a nonlinear PDE and has given an important contribution to various studies, such as those of explosions, traffic flow, glacier waves and airplanes breaking the sound barrier. Goswami et al. [7] used an effective scheme based on the Sumudu transform and the homotopy perturbation method to find the numerical solutions of time fractional Schrodinger equations with harmonic oscillator. Singh and Gupta [8] presented the homotopy perturbation method (HPM) to examine the numerical solution of the time fractional shock wave equation and wave equation. Allan and Khaled [9] employed the Adomian decomposition method to provide the analytical solution of the shock wave equation. Das and Kumar [10] proposed a method for calculating the approximate solution of the shock wave equation and shallow water equation with time derivatives. Later, many researchers [11–14] have developed different strategies to achieve the approximate solution of nonlinear shock wave equations of fractional order.

A differential problem of symmetry is a modification that generates the differential equation continuously in such a way that these symmetries can help to achieve the solution of the differential equation. Solving these equations is sometimes easier than solving the Volterra integro-differential equations [15]. Symmetries can be identified by solving a set of connected ordinary differential equations. PDEs of fractional order are PDEs whose symmetry condition is separated into two segments of integer order and fractional order, and the linear scheme of fractional PDEs reveals a wide dimensional trivial solution continuously. Various numerical and analytical approaches have been demonstrated to attain the semi-analytical solution of nonlinear PDEs, such as the (G′/G)-expansion method [16], the neural network approach [17], the variational iteration method [18], the Exp-function method [19], the homotopy perturbation method [20], the homotopy analysis method [21], residual power series [22], the residual power series method [23], the quasi-wavelet method [24], the Haar wavelet method [25] and the two-scale approach [26]. New developments of the HPM can be found in [27,28].

The aim of this paper is to present the idea of the MT coupled with the HPM for the numerical investigation of nonlinear shock wave equations of fractional order. The obtained results are expressed in terms of series with easily computable components. This series solution converges to the exact solution rapidly. This study is summarized as follows: In Section 2, we demonstrate some basic preliminary concepts. In Section 3, a new strategy is sorted out to handle nonlinear expressions. In Section 4, some numerical examples are demonstrated to determine the competence of the proposed strategy, and at last, some results are discussed with our conclusions in Sections 5 and 6.

2. Preliminary Concepts

Definition 1. *Let $\vartheta(\wp)$ be a function precise for $\wp \geq 0$ [29]; then, we have*

$$\mathscr{L}[\vartheta(\wp)] = V(r) = \int_0^\infty \vartheta(\wp)e^{-r\wp}d\wp,$$

which is said to be a Laplace transform, where $\wp$ is a function (i.e., a function of the time domain), defined on $[0,\infty)$, to a function of r (i.e., of the frequency domain).

Definition 2. *If $V(r)$ symbolizes the Laplace transform of $\vartheta(\wp)$, then*

$$\vartheta(\wp) = \mathscr{L}^{-1}V(r),$$

is termed as the inverse Laplace transform of $V(r)$.

Definition 3. *Mohand and Mahgoub [30,31] developed the MT to facilitate ordinary and PDEs. Let the MT be expressed with the help of operator $\mathscr{M}(.)$. Then $\Longrightarrow$*

$$\mathscr{M}[\vartheta(\wp)] = S(r) = r^2\int_0^\infty \vartheta(\wp)e^{-r\wp}d\wp, k_1 \leq r \leq k_2, \qquad k_1, k_2 \in \mathbb{N}$$

where k_1 and k_2 are constants. On the other hand, if $S(r)$ is the MT of $\vartheta(\wp)$, then $\vartheta(\wp)$ is said to be the inverse of $S(r)$, so

$$\mathscr{M}^{-1}\{S(r)\} = \vartheta(\wp) \qquad \Longrightarrow \quad \mathscr{M}^{-1} \text{ is the inverse MT.}$$

One may see that the Laplace transform and the Mohand transform differ in the function of r (i.e., the frequency domain).

Lemma 1. *The MT of a function of fractional order is [32]*

$$\mathcal{M}\{S^{\alpha}(\wp)\} = r^{\alpha}S(r) - \sum_{k=0}^{n-1}\frac{u^{k}(0)}{r^{k}-(\alpha+1)}, \quad 0<\alpha\leq n$$

Proposition 1. *Let $\mathcal{M}\{\vartheta(\wp)\} = S(r)$; then, the MT of $\vartheta'(\wp)$ has the following properties:*

(a) $\mathcal{M}\{\vartheta'(\wp)\} = rS(r) - r^{2}\vartheta(0)$;
(b) $\mathcal{M}\{\vartheta''(\wp)\} = r^{2}S(r) - r^{3}\vartheta(0) - \vartheta^{2}\vartheta'(0)$;
(c) $\mathcal{M}\{\vartheta^{n}(\wp)\} = r^{n}S(r) - r^{n+1}\vartheta(0) - r^{n}\vartheta'(0) - \cdots - r^{2}\vartheta^{n-1}(0)$.

Definition 4. *The fractional derivative [15] in the Caputo sense is*

$$D_{\tau}^{\alpha}\vartheta(\Im,\wp) = \begin{cases} \dfrac{\partial^{n}\vartheta(\Im,\wp)}{\partial\wp^{n}}, & \alpha\in\mathbb{N} \\ \dfrac{1}{\Gamma(n-\alpha)}\int_{0}^{\wp}(t-\phi)^{n-\alpha-1}\vartheta^{n}(\phi)\partial\phi, & n-1<\alpha<n \end{cases}$$

3. Idea of MHPT

In this section, we construct the idea of the MHPT to find the approximate solution of fractional problems. Therefore, consider a differential equation of fractional order

$$D_{\wp}^{\alpha}\vartheta(\Im,\wp) + R\vartheta(\Im,\wp) + N\vartheta(\Im,\wp) = g(\Im,\wp), \tag{3}$$

$$\vartheta(\Im,0) = h(\Im), \tag{4}$$

where $D_{\wp}^{\alpha} = \dfrac{\partial^{\alpha}}{\partial\wp^{\alpha}}$ is an operator with fractional order α; ϑ is the function in the direction of spital $\Im$ and time $\wp$; R is the linear; N represents the nonlinear differential operator; and $g(\Im,\wp)$ is the source term. Employing the MT in Equation (3), we obtain

$$\mathcal{M}\Big[D_{\wp}^{\alpha}\vartheta(\Im,\wp) + R\vartheta(\Im,\wp) + N\vartheta(\Im,\wp)\Big] = \mathcal{M}\Big[g(\Im,\wp)\Big], \tag{5}$$

using the differentiation property of the MT, we obtain

$$r^{\alpha}\Big[R(r) - r\vartheta(0)\Big] = -\mathcal{M}\Big[R\vartheta(\Im,\wp) + N\vartheta(\Im,\wp)\Big] + \mathcal{M}\Big[g(\Im,\wp)\Big],$$

which leads to

$$R(r) = r\vartheta(0) - \frac{1}{r^{\alpha}}\mathcal{M}\Big[R\vartheta(\Im,\wp) + N\vartheta(\Im,\wp) + g(\Im,\wp)\Big].$$

Using the initial condition (4), we obtain

$$R(r) = rh(\Im) - \frac{1}{r^{\alpha}}\mathcal{M}\Big[R\vartheta(\Im,\wp) + N\vartheta(\Im,\wp) + g(\Im,\wp)\Big],$$

thus, operating the inverse MT, we obtain

$$\vartheta(\Im,\wp) = G(\Im,\wp) - \mathcal{M}^{-1}\left[\frac{1}{r^{\alpha}}\mathcal{M}\Big[R\vartheta(\Im,\wp) + N\vartheta(\Im,\wp)\Big]\right], \tag{6}$$

which is called the recurrence relation of $\vartheta(\Im,\wp)$, where

$$G(\Im, \wp) = \mathscr{M}^{-1}\left[rh(\Im) + \mathscr{M}\left\{g(\Im, \wp)\right\}\right].$$

The approximate solution of Equation (3) can be expressed in terms of the power series

$$\vartheta(\Im, \wp) = \sum_{n=0}^{\infty} p^n \vartheta_n(\Im, \wp), \tag{7}$$

and

$$N\vartheta(\Im, \wp) = \sum_{n=0}^{\infty} p^n H_n \vartheta(\Im, \wp), \tag{8}$$

where $p \in [0,1]$ is an embedding parameter and considered as a small parameter, whereas $\vartheta_0(\Im, \wp)$ is an initial guess of Equation (3). The following strategy can be operated to acquire He's polynomials as

$$H_n(\vartheta_0 + \vartheta_1 + \cdots + \vartheta_n) = \frac{1}{n!}\frac{\partial^n}{\partial p^n}\left(N\left(\sum_{i=0}^{\infty} p^i \vartheta_i\right)\right)_{p=0}. \quad n = 0, 1, 2, \cdots$$

With the help of Equations (7) and (8), we can obtain Equation (6) as

$$\sum_{n=0}^{\infty} p^n \vartheta_n(\Im, \wp) = G(\Im, \wp) - p.\mathscr{M}^{-1}\left[\frac{1}{r^{\alpha}}.\mathscr{M}\left\{R\left(\sum_{n=0}^{\infty} p^n \vartheta_n(\Im, \wp)\right) + \sum_{n=0}^{\infty} p^n H_n \vartheta_n(\Im, \wp)\right\}\right].$$

Equating the similar components of p, we obtain

$$\begin{aligned}
p^0 &: \vartheta_0(\Im, \wp) = G(\Im, \wp),\\
p^1 &: \vartheta_1(\Im, \wp) = -\mathscr{M}^{-1}\left[\frac{1}{r^{\alpha}}.\mathscr{M}\left\{R\vartheta_0(\Im, \wp) + H_0\right\}\right],\\
p^2 &: \vartheta_2(\Im, \wp) = -\mathscr{M}^{-1}\left[\frac{1}{r^{\alpha}}.\mathscr{M}\left\{R\vartheta_1(\Im, \wp) + H_1\right\}\right],\\
p^3 &: \vartheta_3(\Im, \wp) = -\mathscr{M}^{-1}\left[\frac{1}{r^{\alpha}}.\mathscr{M}\left\{R\vartheta_2(\Im, \wp) + H_2\right\}\right],\\
&\vdots
\end{aligned} \tag{9}$$

Thus, we can generate Equation (7) in the collection of orders as

$$\vartheta(\Im, \wp) = \vartheta_0(\Im, \wp) + p^1\vartheta_1(\Im, \wp) + p^2\vartheta_2(\Im, \wp) + +p^3\vartheta_3(\Im, \wp) + \cdots. \tag{10}$$

Let $p = 1$; the analytical solution of Equation (3) is

$$\vartheta(\Im, \wp) = \lim_{N\to\infty} \sum_{n=0}^{N} \vartheta_n(\Im, \wp). \tag{11}$$

We put forward this strategy in the strength of upcoming mathematical applications.

Theorem 1. *Consider that* $\Im$ *and* ζ *are two Banach spaces with* $I : \Im \to \zeta$ *as nonlinear operator, such that* $\vartheta; \vartheta^* \in; \Im, \|I(\vartheta) - I(\vartheta^*)\| \leq K\|\vartheta - \vartheta^*\|, \quad 0 < K < 1$. *According to the Banach contraction theorem, I has a unique fixed point* ϑ, *i.e.,* $I\vartheta = \vartheta$. *Let us recall Equation (11); we have*

$$\vartheta(\Im, \wp) = \lim_{N \to \infty} \sum_{n=0}^{N} \vartheta_n(\Im, \wp), \tag{12}$$

and let us assume that $\Im_0 = \vartheta_0 \in \mathcal{S}_p(\vartheta)$, *where* $\mathcal{S}_p(\vartheta) = \{\vartheta^* \in \Im : \|\vartheta - \vartheta^*\| < p\}$; *then, we have*

$$(B_1)\Im_n \in \mathcal{S}_p(\vartheta),$$
$$(B_2) \lim_{n \to \infty} \Im_n = \vartheta.$$

Proof. (B_1) In view of the mathematical induction for $n = 1$, we have

$$\|\Im_1 - \vartheta_1\| = \|T(\Im_0 - T(\vartheta))\| \le K\|\vartheta_0 - \vartheta\|.$$

Consider that the result is true for $n = 1$, so

$$\|\Im_{n-1} - \vartheta\| \le K^{n-1}\|\vartheta_0 - \vartheta\|.$$

Thus, we have

$$\|\Im_n - \vartheta\| = \|T(\Im_{n-1} - T(\vartheta))\| \le K\|\Im_{n-1} - \vartheta\| \le K^n\|\vartheta_0 - \vartheta\|.$$

Hence, using (B_1), we have

$$\|\Im_n - \vartheta\| \le K^n\|\vartheta_0 - \vartheta\| \le K^n p < p,$$

where p is a contact point of a super norm S, which shows $\Im_n \in \mathcal{S}_p(\vartheta)$.
B_2: Since $\|\Im_n - \vartheta\| \le K^n\|\vartheta_0 - \vartheta\|$ and $\lim_{n\to\infty} K^n = 0$.
Therefore, we have $\lim_{n\to\infty}\|\vartheta_n - \vartheta\| = 0 \Rightarrow \lim_{n\to\infty} \vartheta_n = \vartheta$. □

4. Numerical Examples

In this segment, we deal with the MHPT to present the analytical and numerical solutions of time fractional shock wave equations and time fractional wave equations. The obtained results of these two problems show the performance and high accuracy of the suggested approach. The graphical results declare that this approach has good agreement.

4.1. Example 1

Consider the time fractional shock wave equation

$$D_{\wp}^{\alpha}\vartheta + \Big(\frac{1}{c_0} - \frac{\gamma+1}{2}\frac{\vartheta}{c_0^2}\Big)D_{\Im}\vartheta = 0, \qquad (\Im, \wp) \in R \times [0, T], \qquad 0 < \alpha \le 1, \tag{13}$$

where c_0 and γ are constants, and γ is the specific heat. If $c_0 = 2$, and $\gamma = 1.5$, the study case under consideration relates to the flow of air, as

$$\frac{\partial^{\alpha}\vartheta}{\partial\wp^{\alpha}} + \Big(\frac{1}{2} - \frac{5}{16}\vartheta\Big)\frac{\partial\vartheta}{\partial\Im} = 0, \tag{14}$$

with the initial condition

$$\vartheta(\Im, 0) = e^{-\frac{\Im^2}{2}}. \tag{15}$$

Taking the MT of Equation (14), we obtain

$$\mathscr{M}\left[\frac{\partial^{\alpha}\vartheta}{\partial\wp^{\alpha}} + \Big(\frac{1}{2} - \frac{5}{16}\vartheta\Big)\frac{\partial\vartheta}{\partial\Im}\right] = 0.$$

Using the definition of the MT, we can write it as

$$R(r) = r\vartheta(0) - \frac{1}{r^{\alpha}}\mathcal{M}\left[\left(\frac{1}{2} - \frac{5}{16}\vartheta\right)\frac{\partial\vartheta}{\partial\Im}\right].$$

The inverse MT is

$$\vartheta(\Im,\wp) = \vartheta(\Im,0) - \mathcal{M}^{-1}\left[\frac{1}{r^{\alpha}}\mathcal{M}\left\{\left(\frac{1}{2} - \frac{5}{16}\vartheta\right)\frac{\partial\vartheta}{\partial\Im}\right\}\right],$$

which is the recurrence relation of Equation (14); now, using Equation (7) together with the HPM, we obtain

$$\sum_{n=0}^{\infty} p^n\vartheta_n(\Im,\wp) = \vartheta(\Im,0) - p.\mathcal{M}^{-1}\left[\frac{1}{r^{\alpha}}\mathcal{M}\left\{\frac{1}{2}\sum_{n=0}^{\infty} p^n\frac{\partial\vartheta_n}{\partial\Im} - \frac{5}{16}\sum_{n=0}^{\infty} p^n\vartheta_n\frac{\partial\vartheta_n}{\partial\Im}\right\}\right], \tag{16}$$

by comparing, we can obtain the iterations

$$\begin{aligned}
p^0 &: \vartheta_0(\Im,\wp) = \vartheta(\Im,0),\\
p^1 &: \vartheta_1(\Im,\wp) = -\mathcal{M}^{-1}\left[\frac{1}{r^{\alpha}}\mathcal{M}\left\{\frac{1}{2}\frac{\partial\vartheta_0}{\partial\Im} - \frac{5}{16}\vartheta_0\frac{\partial\vartheta_0}{\partial\Im}\right\}\right],\\
p^2 &: \vartheta_2(\Im,\wp) = -\mathcal{M}^{-1}\left[\frac{1}{r^{\alpha}}\mathcal{M}\left\{\frac{1}{2}\frac{\partial\vartheta_1}{\partial\Im} - \frac{5}{16}\left(\vartheta_0\frac{\partial\vartheta_1}{\partial\Im} + \vartheta_1\frac{\partial\vartheta_0}{\partial\Im}\right)\right\}\right],\\
&\vdots
\end{aligned}$$

which give the solutions

$$\begin{aligned}
\vartheta_0(\Im,\wp) &= e^{-\frac{\Im^2}{2}},\\
\vartheta_1(\Im,\wp) &= \left[\frac{1}{2}xe^{-\frac{\Im^2}{2}} - \frac{5}{16}xe^{-\Im^2}\right]\frac{t^{\alpha}}{\Gamma(\alpha+1)},\\
\vartheta_2(\Im,\wp) &= \frac{1}{256}\left[-25e^{-\frac{3\Im^2}{2}} + 80e^{-\Im^2} - 64e^{-\frac{\Im^2}{2}} + 75\Im^2e^{-\frac{3\Im^2}{2}} - 160\Im^2e^{-\Im^2} - 64\Im^2e^{-\frac{\Im^2}{2}}\right]\frac{t^{2\alpha}}{\Gamma(2\alpha+1)},\\
&\vdots
\end{aligned}$$

Proceeding with a similar process, the other elements of ϑ_n can be calculated, and the series solutions are thus completely obtained. This series converges to the exact solution for high iterations. Finally, the analytical solution of $\vartheta(\Im,t)$ can be obtained by using Equation (10), which is in full agreement with [6,13].

4.2. Example 2

Again, assume the time fractional wave equation

$$D^{\alpha}_{\wp}\vartheta + \vartheta D_{\Im}\vartheta - D_{\Im\Im\wp}\vartheta = 0, \tag{17}$$

with the initial condition

$$\vartheta(\Im,0) = 3\,\mathrm{sech}^2\left(\frac{\Im - 15}{2}\right), \tag{18}$$

According to the HPTM, the recurrence relation of Equation (17) can be written as

$$\vartheta(\Im,\wp)=\vartheta(\Im,0)-\mathscr{M}^{-1}\left[\frac{1}{r^{\alpha}}\mathscr{M}\left\{\vartheta\frac{\partial\vartheta}{\partial\Im}-\frac{\partial}{\partial\wp}\left(\frac{\partial^2\vartheta}{\partial\Im^2}\right)\right\}\right],$$

Now, using Equation (7) together with the HPM, we obtain

$$\sum_{n=0}^{\infty}p^n\vartheta_n(\Im,\wp)=\vartheta(\Im,0)-p\mathscr{M}^{-1}\left[\frac{1}{r^{\alpha}}\mathscr{M}\left\{\sum_{n=0}^{\infty}p^n\vartheta_n\frac{\partial\vartheta_n}{\partial\Im}-\frac{\partial}{\partial\wp}\left(\frac{\partial^2}{\partial\Im^2}\sum_{n=0}^{\infty}p^n\vartheta_n\right)\right\}\right], \quad (19)$$

by comparing, we can obtain the iterations

$$\begin{aligned}
p^0 &= \vartheta_0(\Im,\wp)=\vartheta(\Im,0),\\
p^1 &= \vartheta_1(\Im,\wp)=-\mathscr{M}^{-1}\left[\frac{1}{r^{\alpha}}\mathscr{M}\left\{\vartheta_0\frac{\partial\vartheta_0}{\partial\Im}-\frac{\partial}{\partial\wp}\left(\frac{\partial^2\vartheta_0}{\partial\Im^2}\right)\right\}\right],\\
p^2 &= \vartheta_2(\Im,\wp)=-\mathscr{M}^{-1}\left[\frac{1}{r^{\alpha}}\mathscr{M}\left\{\vartheta_0\frac{\partial\vartheta_1}{\partial\Im}+\vartheta_1\frac{\partial\vartheta_0}{\partial\Im}-\frac{\partial}{\partial\wp}\left(\frac{\partial^2\vartheta_1}{\partial\Im^2}\right)\right\}\right],\\
&\vdots
\end{aligned}$$

which give the solutions

$$\begin{aligned}
\vartheta_0(\Im,\wp) &= 3\,\mathrm{sech}^2\left(\frac{\Im-15}{2}\right),\\
\vartheta_1(\Im,\wp) &= 9\,\mathrm{sech}^2\left(\frac{\Im-15}{2}\right)\tanh\left(\frac{\Im-15}{2}\right)\frac{\wp^{\alpha}}{\Gamma(1+\alpha)},\\
\vartheta_2(\Im,\wp) &= \left[-\frac{27}{2}\mathrm{sech}^8\left(\frac{\Im-15}{2}\right)+81\,\mathrm{sech}^6\left(\frac{\Im-15}{2}\right)\tanh^2\left(\frac{\Im-15}{2}\right)\right]\frac{\wp^{2\alpha}}{\Gamma(1+2\alpha)}\\
&\quad-\left[\frac{63}{2}\mathrm{sech}^6\left(\frac{\Im-15}{2}\right)\tanh\left(\frac{\Im-15}{2}\right)-36\,\mathrm{sech}^4\left(\frac{\Im-15}{2}\right)\tanh^3\left(\frac{\Im-15}{2}\right)\right]\frac{\wp^{2\alpha-1}}{\Gamma(2\alpha)},\\
&\vdots
\end{aligned}$$

Proceeding with a similar process, the other elements of ϑ_n can be calculated, and the series solutions are thus completely obtained. This series converges to the exact solution for high iterations. Finally, the analytical solution of $\vartheta(\Im,\wp)$ can be obtained by using Equation (10) as

$$\vartheta(\Im,0)=3\,\mathrm{sech}^2\left(\frac{\Im-15-\wp}{2}\right), \quad (20)$$

which is in full agreement with [6,13].

5. Results and Discussion

In this segment, we demonstrate the physical interpretations of the illustrated problems. We observe that the HPTM is fully capable of handling time fractional shock wave equations. Figure 1a–d show the surface solutions of $\vartheta(\Im,\wp)$ for various time fractional equations in Brownian motion, and it is observed that $\vartheta(\Im,\wp)$ reduces with the growth of $\Im$ and $\wp$ for $\alpha=0.25, 0.50, 0.75$ and 1. Figure 2a–d show the surface solutions of $\vartheta(\Im,\wp)$ for the analytical solution obtained by the MHPT and the exact solution for various values of $\Im$ and $\wp$, respectively. It is observed that $\vartheta(\Im,\wp)$ increases with the increase in $\Im$ and decreases with the increase in $\wp$ for $\alpha=0.25, 0.50, 0.75$ and 1.

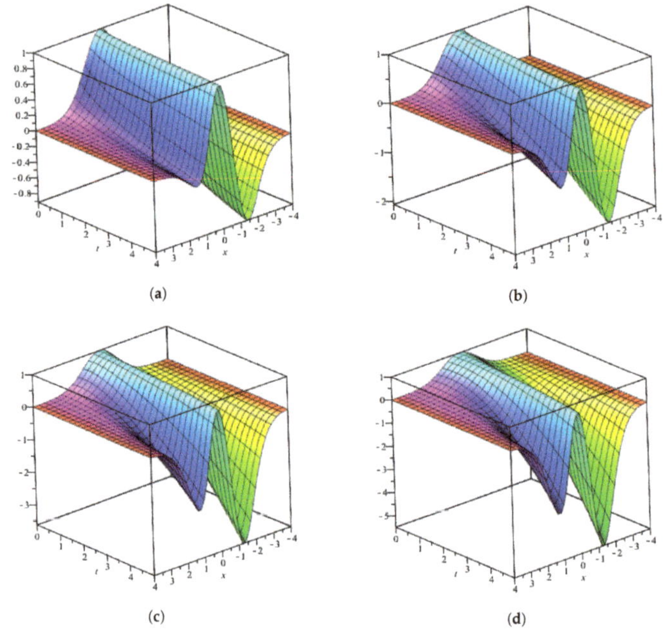

Figure 1. The surface solutions of $u(\Im, \wp)$ with respect to $\Im$ and $\wp$ for distinct values of α. (**a**) Surface solution of $\vartheta(\Im, \wp)$ when $\alpha = 0.25$. (**b**) Surface solution of $\vartheta(\Im, \wp)$ when $\alpha = 0.50$. (**c**) Surface solution of $\vartheta(\Im, \wp)$ when $\alpha = 0.75$. (**d**) Surface solution of $\vartheta(\Im, \wp)$ when $\alpha = 1$.

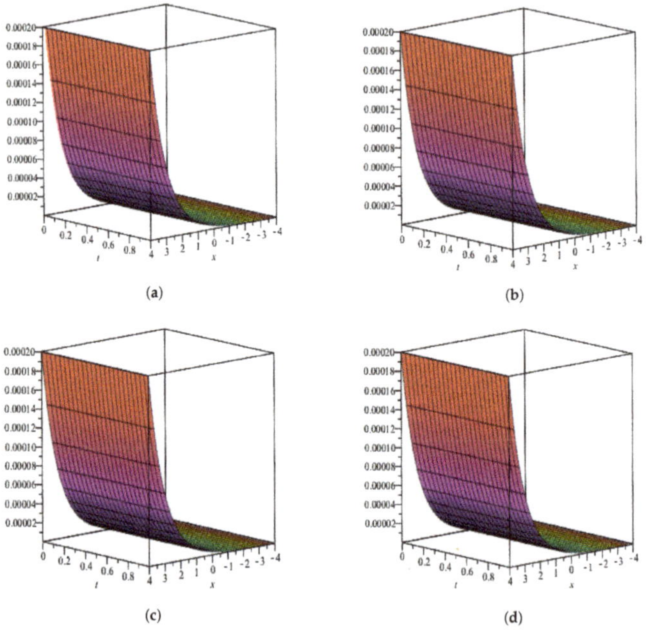

Figure 2. The surface solutions of $\vartheta(\Im, \wp)$ with respect to $\Im$ and $\wp$ for different values of α. (**a**) Surface solution of $\vartheta(\Im, \wp)$ when $\alpha = 0.25$. (**b**) Surface solution of $\vartheta(\Im, \wp)$ when $\alpha = 0.50$. (**c**) Surface solution of $\vartheta(\Im, \wp)$ when $\alpha = 0.75$. (**d**) Surface solution of $\vartheta(\Im, \wp)$ when $\alpha = 1$.

6. Conclusions

In this paper, we successfully apply the HPTM to achieve the approximate and analytical solutions of nonlinear time fractional shock wave and wave equations. This study demonstrates the importance of fractional derivatives and the technique of dealing with the recurrence relation. Since the MT is limited to linear problems only, whereas the HPM is applicable to nonlinear problems, we conclude that the MHPT is the best tool to provide significant results for both linear and nonlinear problems. The MHPT is here directly applied to obtain the series solutions. The present scheme shows higher efficiency and fewer computations than other approaches studied in the literature. All the iterations were calculated with the help of MAPLE Software. The solution graphs show that this approach is suitable for a broad variety of nonlinear fractional differential equations in science and engineering. In future work, this approach could further be extended to solve various nonlinear obstacle problems.

Author Contributions: Conceptualization, formal analysis and funding acquisition, J.F.; investigation, methodology, software and writing—original draft, M.N.; validation and visualization, M.H.; supervision and writing—review and editing, A.A. All authors have read and agreed to the submitted the manuscript.

Funding: This work was supported by the Foundation of Yibin University, China (grant No. 2019QD07).

Institutional Review Board Statement: Not applicable.

Informed Consent Statement: Not applicable.

Data Availability Statement: Not applicable.

Conflicts of Interest: The authors declare no conflict of interest.

References

1. Gharechahi, R.; Arab Ameri, M.; Bisheh-Niasar, M. High order compact finite difference schemes for solving Bratu-type equations. *J. Appl. Comput. Mech.* **2019**, *5*, 91–102.
2. Sharma, J.R.; Kumar, S.; Cesarano, C. An efficient derivative free one-point method with memory for solving nonlinear equations. *Mathematics* **2019**, *7*, 604. [CrossRef]
3. He, J.H.; Wu, X.H. Variational iteration method: New development and applications. *Comput. Math. Appl.* **2007**, *54*, 881–894. [CrossRef]
4. Singh, J. Analysis of fractional blood alcohol model with composite fractional derivative. *Chaos Solitons Fractals* **2020**, *140*, 110127. [CrossRef]
5. Singh, J.; Ganbari, B.; Kumar, D.; Baleanu, D. Analysis of fractional model of guava for biological pest control with memory effect. *J. Adv. Res.* **2021**, *32*, 99–108. [CrossRef]
6. Kumar, D.; Singh, J.; Kumar, S.; Singh, B. Numerical computation of nonlinear shock wave equation of fractional order. *Ain Shams Eng. J.* **2015**, *6*, 605–611. [CrossRef]
7. Goswami, A.; Rathore, S.; Singh, J.; Kumar, D. Analytical study of fractional nonlinear Schrödinger equation with harmonic oscillator. *Discret. Contin. Dyn. Syst.-S* **2021**, *14*, 3589–3610. [CrossRef]
8. Singh, M.; Gupta, P.K. Homotopy perturbation method for time-fractional shock wave equation. *Adv. Appl. Math. Mech.* **2011**, *3*, 774–783. [CrossRef]
9. Allan, F.M.; Al-Khaled, K. An approximation of the analytic solution of the shock wave equation. *J. Comput. Appl. Math.* **2006**, *192*, 301–309. [CrossRef]
10. Das, S.; Kumar, R. Approximate analytical solutions of fractional gas dynamic equations. *Appl. Math. Comput.* **2011**, *217*, 9905–9915. [CrossRef]
11. Khalid, N.; Abbas, M.; Iqbal, M.K.; Singh, J.; Ismail, A.I.M. A computational approach for solving time fractional differential equation via spline functions. *Alex. Eng. J.* **2020**, *59*, 3061–3078. [CrossRef]
12. Khatami, I.; Tolou, N.; Mahmoudi, J.; Rezvani, M. Application of homotopy analysis method and variational iteration method for shock wave equation. *J. Appl. Sci.* **2008**, *8*, 848–853. [CrossRef]
13. Singh, J.; Kumar, D.; Kumar, S. A new fractional model of nonlinear shock wave equation arising in flow of gases. *Nonlinear Eng.* **2014**, *3*, 43–50. [CrossRef]
14. Berberler, M.E.; Yildirim, A. He's homotopy perturbation method for solving the shock wave equation. *Appl. Anal.* **2009**, *88*, 997–1004. [CrossRef]
15. Phuong, N.D.; Tuan, N.A.; Kumar, D.; Tuan, N.H. Initial value problem for fractional Volterra integrodifferential pseudo-parabolic equations. *Math. Model. Nat. Phenom.* **2021**, *16*, 27. [CrossRef]

16. Alam, M.N. Exact solutions to the foam drainage equation by using the new generalized (G′/G)-expansion method. *Results Phys.* **2015**, *5*, 168–177. [CrossRef]
17. Dai, P.; Yu, X. An Artificial Neural Network Approach for Solving Space Fractional Differential Equations. *Symmetry* **2022**, *14*, 535. [CrossRef]
18. Dahmani, Z.; Anber, A. The variational iteration method for solving the fractional foam drainage equation. *Int. J. Nonlinear Sci.* **2010**, *10*, 39–45.
19. Khani, F.; Hamedi-Nezhad, S.; Darvishi, M.; Ryu, S.W. New solitary wave and periodic solutions of the foam drainage equation using the Exp-function method. *Nonlinear Anal. Real World Appl.* **2009**, *10*, 1904–1911. [CrossRef]
20. Gul, H.; Ali, S.; Shah, K.; Muhammad, S.; Sitthiwirattham, T.; Chasreechai, S. Application of Asymptotic Homotopy Perturbation Method to Fractional Order Partial Differential Equation. *Symmetry* **2021**, *13*, 2215. [CrossRef]
21. Sedighi, H.M.; Shirazi, K.H.; Zare, J. An analytic solution of transversal oscillation of quintic non-linear beam with homotopy analysis method. *Int. J. Non-Linear Mech.* **2012**, *47*, 777–784. [CrossRef]
22. Al-qudah, Y.; Alaroud, M.; Qoqazeh, H.; Jaradat, A.; Alhazmi, S.E.; Al-Omari, S. Approximate Analytic–Numeric Fuzzy Solutions of Fuzzy Fractional Equations Using a Residual Power Series Approach. *Symmetry* **2022**, *14*, 804. [CrossRef]
23. Jena, R.M.; Chakraverty, S.; Jena, S.K.; Sedighi, H.M. Analysis of time-fractional fuzzy vibration equation of large membranes using double parametric based Residual power series method. *ZAMM-J. Appl. Math. Mech./Z. für Angew. Math. Und Mech.* **2021**, *101*, e202000165. [CrossRef]
24. Kumar, S.; Gómez-Aguilar, J.F. Numerical solution of Caputo-Fabrizio time fractional distributed order reaction-diffusion equation via quasi wavelet based numerical method. *J. Appl. Comput. Mech.* **2020**, *6*, 848–861.
25. Arbabi, S.; Nazari, A.; Darvishi, M.T. A semi-analytical solution of foam drainage equation by Haar wavelets method. *Optik* **2016**, *127*, 5443–5447. [CrossRef]
26. Habib, S.; Islam, A.; Batool, A.; Sohail, M.U.; Nadeem, M. Numerical solutions of the fractal foam drainage equation. *GEM-Int. J. Geomath.* **2021**, *12*, 1–10. [CrossRef]
27. Nadeem, M.; Yao, S.W. Solving the fractional heat-like and wave-like equations with variable coefficients utilizing the Laplace homotopy method. *Int. J. Numer. Methods Heat Fluid Flow* **2020**, *31*, 273–292. [CrossRef]
28. Gupta, S.; Kumar, D.; Singh, J. Analytical solutions of convection–diffusion problems by combining Laplace transform method and homotopy perturbation method. *Alex. Eng. J.* **2015**, *54*, 645–651. [CrossRef]
29. Nadeem, M.; Li, F. Modified Laplace Variational Iteration Method for Analytical Approach of Klein–Gordon and Sine–Gordon equations. *Iran. J. Sci. Technol. Trans. Sci.* **2019**, *43*, 1933–1940. [CrossRef]
30. Nadeem, M.; He, J.H.; Islam, A. The homotopy perturbation method for fractional differential equations: Part 1 Mohand transform. *Int. J. Numer. Methods Heat Fluid Flow* **2021**, *31*, 3490–3504. [CrossRef]
31. Aggarwal, S.; Chauhan, R. A comparative study of Mohand and Aboodh transforms. *Int. J. Res. Advent Technol.* **2019**, *7*, 520–529. [CrossRef]
32. Shah, R.; Khan, H.; Farooq, U.; Baleanu, D.; Kumam, P.; Arif, M. A New Analytical Technique to Solve System of Fractional-Order Partial Differential Equations. *IEEE Access* **2019**, *7*, 150037–150050. [CrossRef]

symmetry

MDPI

Article

Approximate Solution of Nonlinear Time-Fractional Klein-Gordon Equations Using Yang Transform

Jinxing Liu [1], Muhammad Nadeem [1,*], Mustafa Habib [2] and Ali Akgül [3]

1 Faculty of Science, Yibin University, Yibin 644000, China; liujx03@163.com
2 Department of Mathematics, University of Engineering and Technology, Lahore 54890, Pakistan; mustafa@uet.edu.pk
3 Department of Mathematics, Art and Science Faculty, Siirt University, Siirt 56100, Turkey; aliakgul@siirt.edu.tr
* Correspondence: nadeem@yibinu.edu.cn

Abstract: The algebras of the symmetry operators for the Klein–Gordon equation are important for a charged test particle, moving in an external electromagnetic field in a space time manifold on the isotropic hydrosulphate. In this paper, we develop an analytical and numerical approach for providing the solution to a class of linear and nonlinear fractional Klein–Gordon equations arising in classical relativistic and quantum mechanics. We study the Yang homotopy perturbation transform method (𝕐HPTM),which is associated with the Yang transform (𝕐T) and the homotopy perturbation method (HPM), where the fractional derivative is taken in a Caputo–Fabrizio (CF) sense. This technique provides the solution very accurately and efficiently in the form of a series with easily computable coefficients. The behavior of the approximate series solution for different fractional-order $\wp$ values has been shown graphically. Our numerical investigations indicate that 𝕐HPTM is a simple and powerful mathematical tool to deal with the complexity of such problems.

Keywords: fractional Klein–Gordon equation; Yang transform; homotopy perturbation method; series solution

Citation: Liu, J.; Nadeem, M.; Habib, M.; Akgül, A. Approximate Solution of Nonlinear Time-Fractional Klein-Gordon Equations Using Yang Transform. *Symmetry* **2022**, *14*, 907. https://doi.org/10.3390/sym14050907

Academic Editor: Eduard Marusic-Paloka

Received: 14 April 2022
Accepted: 26 April 2022
Published: 29 April 2022

Publisher's Note: MDPI stays neutral with regard to jurisdictional claims in published maps and institutional affiliations.

1. Introduction

Recently, fractional calculus has grown in popularity due to its significant prospective applications in physics and engineering such as biology, mathematics, chemistry, fluid mechanics, physics, and nonlinear optics [1,2]. Fractional partial differential Equations (FPDEs) are a contemporary tool in calculus that can be used to simulate a wide range of classifications in applied sciences and engineering [3–5].

The Klein–Gordon (KG) equation performs a significant role in mathematical physics and many other scientific studies such as quantum field theory, nonlinear optics, and solid-state physic [6–10]. On the other hand, the fractional-order KG equation is derived from the classical KG equation by substituting the time order derivative with the fractional derivative of order $\wp$. The fractional-order KG equation can be illustrated as below

$$D_q^{\wp}\vartheta(\epsilon,q) - D_\epsilon^2\vartheta(\epsilon,q) + a_1\vartheta(\epsilon,q) + a_2G(\vartheta(\epsilon,q)) = f(\epsilon,q), \quad (1)$$

with initial conditions

$$\vartheta(\epsilon,0) = f_1(\epsilon), \;\; \vartheta_q(\epsilon,0) = f_2(\epsilon), \quad (2)$$

where $D_q^{\wp}$ represents the Caputo fractional time derivative, a_1 and a_2 are real constants, $f(\epsilon,q)$, $f_1(\epsilon)$ and $f_2(\epsilon)$ are known as analytical functions, whereas $G(\vartheta(\epsilon,q))$ is a nonlinear, and ϑ is an unknown function of ϵ and q.

Various authors [11–15] have investigated different analytical and numerical strategies to examine the solution to the KG equation but with some restrictions and lacks. Tamsir and Srivastava [16] used fractional reduced differential transform to obtain the analytical solution of linear and nonlinear KG equation with time-fractional order. Bansu and

Kumar [17] used a radial basis approach, and Kurulay [18] applied the homotopy analysis method to evaluate the numerical solution of the space-time fractional KG equation. Later, Khader and Adel [19] applied a hybridization scheme to achieve the solution of the fractional KG equation. Zhmud and Dimitrov [20] developed the fractional integration method, which is based on extrapolation using a series of integrating and differentiating links with a time constant that changes symmetrically from one step to another. In order to obtain the solution of FPDEs, several valuable strategies have been considered, such as the generalized differential transform method [21], the adomian decomposition method [22], the homotopy analysis method [23], the variational iteration method [24], the homotopy perturbation method [25], the Elzaki transform decomposition method [26], the fractional wavelet method [27,28] and the residual power series method [29,30].

In this paper, we present the Yang homotopy perturbation transform method ($\mathbb{Y}$HPTM), which is a composition of $\mathbb{Y}$T and HPM. The primary objective of this approach is to investigate the approximate solution of fractional KG equations and minimize the computational work that overcomes nonlinear problems easily. Next, this scheme can promptly deal with the nonlinear KG equation. Finally, this method can reduce the range of the computations and generate an approximate solution with elegantly computed expressions, which is its most impressive advantage. The design of this paper is framed as follows. In Section 2, we start with some primary definitions of Caputo–Fabrizio. In Section 3, we formulate the idea of the Yang homotopy perturbation transform method. In Section 4, we perform this scheme on some illustrative examples to show its capability and efficiency. Concluding remarks are given in Section 5.

2. Preliminaries and Concepts

Definition 1. *The CF derivative is described as [31]*

$$ {}^{CF}D_q^{\wp}\vartheta(\epsilon,q) = \frac{S(\wp)}{1-\wp}\int_0^q [Q'(\varrho)K(q,\varrho)]d\varrho, \quad n-1<\wp\leq n \tag{3}$$

$S(\wp)$ *is the normalization function with* $S(0)=S(1)=1$*, and then, Equation (3) becomes as*

$$ {}^{CF}D_q^{\wp}\vartheta(\epsilon,q) = \frac{S(\wp)}{1-\wp}\int_0^q [Q(q)-Q(\varrho)]K(q,\varrho)d\varrho, \quad n-1<\wp\leq n \tag{4}$$

Definition 2. *The fractional CF integral is stated as [32]*

$$ {}^{CF}I_q^{\wp}\vartheta(\epsilon,q) = \frac{1-\wp}{S(\wp)}Q(q) + \frac{\wp}{S(\wp)}\int_0^q Q(\varrho)d\varrho, \quad q\geq 0,\ \wp\epsilon(0,1]. \tag{5}$$

Definition 3. *For* $S(\wp)=1$*, the Laplace transform of the CF derivative is [33]*

$$ L\Big[{}^{CF}D_q^{\wp}Q[(q)]\Big] = \frac{vL[Q(q)-Q(0)]}{v+\wp(1-v)}. \tag{6}$$

Definition 4. *The* $\mathbb{Y}$*T of* $Q(q)$ *is framed as [34]*

$$ \mathbb{Y}[Q(q)] = \chi(v) = \int_0^{\infty} Q(q)e^{-\frac{q}{v}}dq. \ q>0 \tag{7}$$

Remarks

The $\mathbb{Y}$T of some helpful expressions are as follows:

$$\mathbb{Y}[1] = v;$$
$$\mathbb{Y}[q] = v^2;$$
$$\mathbb{Y}[q^i] = \Gamma(i+1)v^{i+1}.$$

Lemma 1. *Let the Laplace transform of $Q(q)$ be $F(v)$, and then $\chi(v) = F(1/v)$ [35].*

Proof. From Equation (7), we can obtain the Yang transform by putting $q/v = \zeta$ as

$$L[Q(q)] = \int_0^\infty Q(v\zeta)e^{\zeta}d\zeta, \ \zeta > 0 \tag{8}$$

since $L[Q(q)] = F(v)$, which implies that

$$F(v) = L[Q(q)] = \int_0^\infty Q(q)e^{-vq}dq. \tag{9}$$

Putting $q = \zeta/v$ in Equation (9), we obtain

$$F(v) = \frac{1}{v}\int_0^\infty Q\Big(\frac{\zeta}{v}\Big)e^{\zeta}d\zeta. \tag{10}$$

Thus, from Equation (8), we obtain:

$$F(v) = \chi\Big(\frac{1}{v}\Big). \tag{11}$$

Furthermore, from Equations (7) and (9), we obtain

$$F\Big(\frac{1}{v}\Big) = \chi(v). \tag{12}$$

The links between Equations (11) and (12) represent the duality connection among the Laplace and Yang transforms. □

Lemma 2. *Let $Q(q)$ be a function, then $\mathbb{Y}T$ of CF derivatives of $Q(q)$ is [35]*

$$\mathbb{Y}\Big[Q(q)\Big] = \frac{\mathbb{Y}\Big[Q(q) - vQ(0)\Big]}{v + \wp(v-1)}. \tag{13}$$

Proof. The fractional Laplace transform of CF is defined as in Equation (13)

$$L\Big[Q(q)\Big] = \frac{L\Big[vQ(q) - Q(0)\Big]}{v + \wp(1-v)}. \tag{14}$$

However, we have a correlation among the $\mathbb{Y}$T and Laplace properties, namely $\chi(v) = F(1/v)$, so put $1/v$ for v in Equation (14), and we obtain

$$\begin{aligned} \mathbb{Y}\Big[Q(q)\Big] &= \frac{\mathbb{Y}\Big[\frac{1}{v}Q(q) - Q(0)\Big]}{\frac{1}{v} + \wp(1-\frac{1}{v})}, \\ \mathbb{Y}\Big[Q(q)\Big] &= \frac{\mathbb{Y}\Big[Q(q) - vQ(0)\Big]}{1 + \wp(v-1)}. \end{aligned} \tag{15}$$

Thus, the proof is satisfied. □

3. Idea of Yang Homotopy Perturbation Transform Method ($\mathbb{Y}$HPTM)

In this part, we will demonstrate the concept of $\mathbb{Y}$HPTM. Let us assume a nonlinear fractional-order PDE, such as

$${}^{CF}D_q^{\wp}\vartheta(\epsilon, q) + R\vartheta(\epsilon, q) + N\vartheta(\epsilon, q) = g(\epsilon, q), \tag{16}$$

$$\vartheta(\epsilon,0) = h(\epsilon), \tag{17}$$

where $g(\epsilon,q)$ is called the source function. Applying the $\mathbb{Y}$T to Equation (16),

$$\frac{1}{v^{\wp}}\mathbb{Y}\Big[\vartheta(\epsilon,q) - v\vartheta(\epsilon,0)\Big] = -\mathbb{Y}[R(\vartheta(\epsilon,q)) + N(\vartheta(\epsilon,q)) + \mathbb{Y}[g(\epsilon,q)]],$$
$$\mathbb{Y}[\vartheta(\epsilon,q)] = vh(\epsilon) - v^{\wp}\Big[\mathbb{Y}[R(\vartheta(\epsilon,q)) + N(\vartheta(\epsilon,q))]\Big] + \mathbb{Y}[g(\epsilon,q)].$$

By using inverse $\mathbb{Y}$T,

$$\vartheta(\epsilon,q) = \vartheta(\epsilon,0) - \mathbb{Y}^{-1}\Big[v^{\wp}\Big[\mathbb{Y}[R(\vartheta(\epsilon,q)) + N(\vartheta(\epsilon,q))]\Big] + \mathbb{Y}[g(\epsilon,q)]\Big]. \tag{18}$$

However, HPM is stated as

$$\vartheta(\epsilon,q) = \sum_{i=0}^{\infty} p^i \vartheta_i(\epsilon,q), \tag{19}$$

and

$$N\vartheta(\epsilon,q) - \sum_{i=0}^{\infty} p^i H_i \vartheta(\epsilon,q). \tag{20}$$

The following strategy can be operated to acquire the He's polynomials,

$$H_i(\vartheta_0 + \vartheta_1 + \cdots + \vartheta_i) = \frac{1}{n!}\frac{\partial^i}{\partial p^i}\left(N\Big(\sum_{i=0}^{\infty} p^i \vartheta_i\Big)\right)_{p=0} . \quad n = 0,1,2,\cdots$$

With the help of Equations (19) and (20), we can obtain Equation (18), such as

$$\sum_{i=0}^{\infty} p^i \vartheta_i(\epsilon,q) = \vartheta(\epsilon,0) - p\mathbb{Y}^{-1}\left[v^{\wp}\mathbb{Y}\left\{R\Big(\sum_{i=0}^{\infty} p^i \vartheta_i(\epsilon,q)\Big) + \sum_{i=0}^{\infty} p^i H_n \vartheta_i(\epsilon,q)\right\}\right].$$

We can obtain the following terms by evaluating the p components:

$$\begin{aligned}
p^0 &= \vartheta_0(\epsilon,q) = \vartheta(\epsilon,0),\\
p^1 &= \vartheta_1(\epsilon,q) = -\mathbb{Y}^{-1}\left[v^{\wp}\mathbb{Y}\left\{R\vartheta_0(\epsilon,q) + H_0(\vartheta)\right\}\right],\\
p^2 &= \vartheta_2(\epsilon,q) = -\mathbb{Y}^{-1}\left[v^{\wp}\mathbb{Y}\left\{R\vartheta_1(\epsilon,q) + H_1(\vartheta)\right\}\right],\\
p^3 &= \vartheta_3(\epsilon,q) = -\mathbb{Y}^{-1}\left[v^{\wp}\mathbb{Y}\left\{R\vartheta_2(\epsilon,q) + H_2(\vartheta)\right\}\right],\\
&\vdots\\
p^i &= \vartheta_i(\epsilon,q) = -\mathbb{Y}^{-1}\left[v^{\wp}\mathbb{Y}\left\{R\vartheta_i(\epsilon,q) + H_i(\vartheta)\right\}\right],
\end{aligned} \tag{21}$$

Thus, we can summarize the set of Equations (21) in the series form, such as

$$\begin{aligned}
\vartheta(\epsilon,q) &= \vartheta_0(\epsilon,q) + \vartheta_1(\epsilon,q) + \vartheta_2(\epsilon,q) + \cdots\\
\vartheta(\epsilon,q) &= \lim_{N\to\infty} \sum_{n=0}^{N} \vartheta_n(\epsilon,q)
\end{aligned} \tag{22}$$

4. Numerical Applications

4.1. Example 1

Consider a linear time-fractional KG problem

$$D_q^{\wp}\vartheta(\epsilon,q) - D_\epsilon^2\vartheta(\epsilon,q) - \vartheta(\epsilon,q) = 0, \tag{23}$$

with the initial condition

$$\vartheta(\epsilon,0) = 1+\sin(\epsilon). \tag{24}$$

Taking $\mathbb{Y}$T of Equation (23), we obtain

$$\mathbb{Y}\Big[\frac{\partial^{\wp}\vartheta}{\partial q^{\wp}}\Big] = \mathbb{Y}\Big[\frac{\partial^2\vartheta}{\partial\epsilon^2}+\vartheta\Big].$$

Executing the differential property of $\mathbb{Y}$T, we obtain

$$\frac{1}{v^{\wp}}\mathbb{Y}\Big[\vartheta(\epsilon,q) - v\vartheta(\epsilon,0)\Big] = \mathbb{Y}\Big[\frac{\partial^2\vartheta}{\partial q^2}+\vartheta\Big],$$
$$\mathbb{Y}\Big[\vartheta(\epsilon,q)\Big] = v\vartheta(\epsilon,0) + v^{\wp}\mathbb{Y}\Big[\frac{\partial^2\vartheta}{\partial q^2}+\vartheta\Big].$$

The inverse $\mathbb{Y}$T indicates

$$\vartheta(\epsilon,q) = \vartheta(\epsilon,0) + \mathbb{Y}^{-1}\Big[v^{\wp}\Big\{\mathbb{Y}\Big(\frac{\partial^2\vartheta}{\partial q^2}+\vartheta\Big)\Big\}\Big].$$

Employing HPM such as

$$\vartheta(\epsilon,q) = \vartheta_0 + p\vartheta_1 + p^2\vartheta_2 + \cdots,$$
$$\sum_{i=0}^{\infty} p^i\vartheta_i(\epsilon,q) = 1+\sin(q) + p\left(\mathbb{Y}^{-1}\Big[v^{\wp}\Big\{\mathbb{Y}\Big(\sum_{i=0}^{\infty} p^i\frac{\partial^2\vartheta_i}{\partial q^2} + \sum_{i=0}^{\infty} p^i\vartheta_i\Big)\Big\}\Big]\right),$$

on comparing the identical of p, we obtain

$$p^0 = \vartheta_0(\epsilon,q) = \vartheta(\epsilon,0),$$
$$p^1 = \vartheta_1(\epsilon,q) = \mathbb{Y}^{-1}\left[v^{\wp}\left\{\mathbb{Y}\Big(\frac{\partial^2\vartheta_0}{\partial q^2}+\vartheta_0\Big)\right\}\right],$$
$$p^2 = \vartheta_2(\epsilon,q) = \mathbb{Y}^{-1}\left[v^{\wp}\left\{\mathbb{Y}\Big(\frac{\partial^2\vartheta_1}{\partial q^2}+\vartheta_1\Big)\right\}\right],$$
$$p^3 = \vartheta_3(\epsilon,q) = \mathbb{Y}^{-1}\left[v^{\wp}\left\{\mathbb{Y}\Big(\frac{\partial^2\vartheta_2}{\partial q^2}+\vartheta_2\Big)\right\}\right],$$
$$\vdots$$

With help of Equation (24), we gain the iterations successively $\vartheta_i(\epsilon)$, $i = 1,2,3,\cdots$, as follows:

$$\vartheta_0(\epsilon,q) = 1+\sin(\epsilon),$$
$$\vartheta_1(\epsilon,q) = \frac{1}{\Gamma(1+\wp)}q^{\wp},$$

$$
\begin{aligned}
\vartheta_2(\epsilon, q) &= \frac{1}{\Gamma(1+2\wp)} q^{2\wp}, \\
\vartheta_3(\epsilon, q) &= \frac{1}{\Gamma(1+3\wp)} q^{3\wp}, \\
&\vdots \\
\vartheta_i(\epsilon, q) &= \frac{1}{\Gamma(1+i\wp)} q^{i\wp},
\end{aligned}
$$

Thus, the approximate solution can be obtained by:

$$
\begin{aligned}
\vartheta(\epsilon, q) &= 1 + \sin(\epsilon) + \frac{1}{\Gamma(1+\wp)} q^{\wp} + \frac{1}{\Gamma(1+2\wp)} q^{2\wp} + \frac{1}{\Gamma(1+3\wp)} q^{3\wp} + \cdots \\
&= 1 + \sin(\epsilon) + \sum_{i=0}^{\infty} p^i \frac{q^{i\wp}}{\Gamma(1+i\wp)},
\end{aligned} \tag{25}
$$

which implies the exact solution of Equation (23), In particular, at $\wp = 1$, we obtain

$$
\vartheta(\epsilon, q) = 1 + \sin(\epsilon), \tag{26}
$$

which is in full agreement.

Figure 1a–d indicate the physical behavior of the obtained solution at $\epsilon \in [0, 4]$ and $q \in [0, 0.8]$. From these figures, it can be observed that the solution graphs of the problem show the friendly touch with each other. Figure 1a–d demonstrate that the solution achieved by $\mathbb{Y}$HPTM approaches the precise solution very rapidly with more iterations. In Figure 2, we have plotted the graph of $\vartheta(\epsilon, q)$ at different fractional order of $\wp = 0.25, 0.50, 0.75, 1$ and $\epsilon \in [0, 2\pi]$ with different values of q.

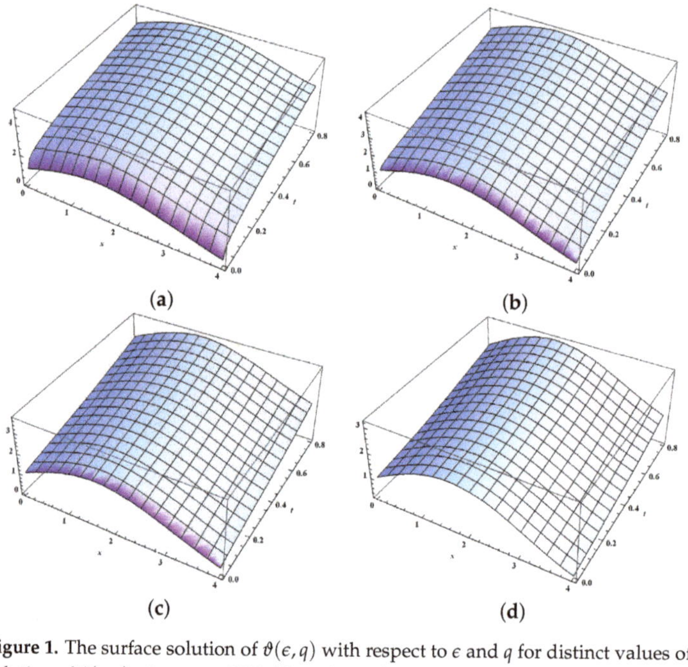

Figure 1. The surface solution of $\vartheta(\epsilon, q)$ with respect to ϵ and q for distinct values of $\wp$: (**a**) surface solution of $\vartheta(\epsilon, q)$ when $\wp = 0.25$; (**b**) surface solution of $\vartheta(\epsilon, q)$ when $\wp = 0.50$; (**c**) surface solution of $\vartheta(\epsilon, q)$ when $\wp = 0.75$; (**d**) surface solution of $\vartheta(\epsilon, q)$ when $\wp = 1$.

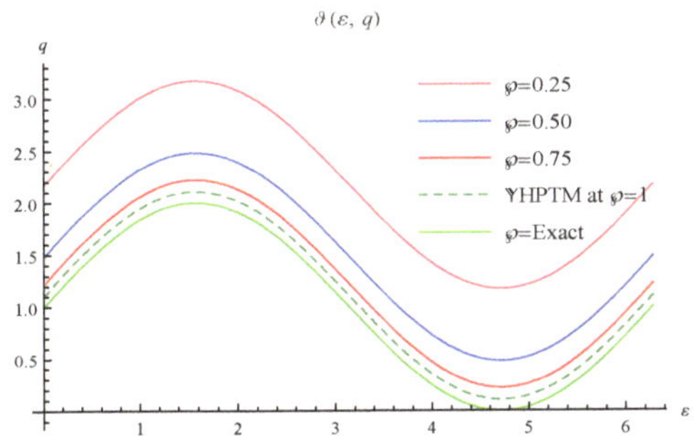

Figure 2. Plot of $\vartheta(\epsilon, q)$ for different values of $\wp$.

4.2. Example 2

Assume a nonlinear time-fractional KG problem

$$D_q^{\wp}\vartheta(\epsilon,q) - D_\epsilon^2\vartheta(\epsilon,q) + \vartheta^2(\epsilon,q) = 0, \tag{27}$$

with the initial condition

$$\vartheta(\epsilon,0) = 1 + \sin(\epsilon). \tag{28}$$

Taking the Yang transform of Equation (27), we obtain

$$\mathbb{Y}\Big[\frac{\partial^{\wp}\vartheta}{\partial q^{\wp}}\Big] = \mathbb{Y}\Big[\frac{\partial^2\vartheta}{\partial\epsilon^2} - \vartheta^2\Big].$$

Executing the differential property of $\mathbb{Y}$T, we obtain

$$\frac{1}{v^{\wp}}\mathbb{Y}\Big[\vartheta(\epsilon,q) - v\vartheta(\epsilon,0)\Big] = \mathbb{Y}\Big[\frac{\partial^2\vartheta}{\partial q^2} - \vartheta^2\Big],$$

$$\mathbb{Y}\Big[\vartheta(\epsilon,q)\Big] = v\vartheta(\epsilon,0) + v^{\wp}\mathbb{Y}\Big[\frac{\partial^2\vartheta}{\partial q^2} - \vartheta^2\Big].$$

The inverse $\mathbb{Y}$T indicates

$$\vartheta(\epsilon,q) = \vartheta(\epsilon,0) + \mathbb{Y}^{-1}\Big[v^{\wp}\Big\{\mathbb{Y}\Big(\frac{\partial^2\vartheta}{\partial q^2} - \vartheta^2\Big)\Big\}\Big].$$

Employing HPM such as

$$\sum_{i=0}^{\infty} p^i\vartheta_i(\epsilon,q) = 1 + \sin(q) + p\Bigg(\mathbb{Y}^{-1}\Big[v^{\wp}\Big\{\mathbb{Y}\Big(\sum_{i=0}^{\infty} p^i\frac{\partial^2\vartheta_i}{\partial q^2} - \sum_{i=0}^{\infty} p^i\vartheta_i^2\Big)\Big\}\Big]\Bigg),$$

on comparing the identical of p, we obtain

$$\begin{aligned}
p^0 &= \vartheta_0(\epsilon,q) = \vartheta(\epsilon,0),\\
p^1 &= \vartheta_1(\epsilon,q) = \mathbb{Y}^{-1}\Bigg[v^{\wp}\Bigg\{\mathbb{Y}\Big(\frac{\partial^2\vartheta_0}{\partial q^2} - \vartheta_0^2\Big)\Bigg\}\Bigg],\\
p^2 &= \vartheta_2(\epsilon,q) = \mathbb{Y}^{-1}\Bigg[v^{\wp}\Bigg\{\mathbb{Y}\Big(\frac{\partial^2\vartheta_1}{\partial q^2} - 2\vartheta_0\vartheta_1\Big)\Bigg\}\Bigg],\\
p^3 &= \vartheta_3(\epsilon,q) = \mathbb{Y}^{-1}\Bigg[v^{\wp}\Bigg\{\mathbb{Y}\Big(\frac{\partial^2\vartheta_2}{\partial q^2} - \vartheta_1^2 - 2\vartheta_0\vartheta_2\Big)\Bigg\}\Bigg],\\
&\vdots
\end{aligned}$$

With help of Equation (28), we gain the iterations successively $\vartheta_i(\epsilon)$, $i = 1, 2, 3, \cdots,$ as follows:

$$\begin{aligned}
\vartheta_0(\epsilon, q) =& 1 + \sin(\epsilon), \\
\vartheta_1(\epsilon, q) =& \frac{-q^{\wp}}{\Gamma(1+\wp)}\Big(1 + 3\sin(\epsilon) + \sin^2(\epsilon)\Big), \\
\vartheta_2(\epsilon, q) =& \frac{q^{2\wp}}{\Gamma(1+2\wp)}\Big(11\sin(\epsilon) + 12\sin^2(\epsilon) + 2\sin^3(\epsilon)\Big), \\
\vartheta_3(\epsilon, q) =& \frac{q^{3\wp}}{\Gamma(1+3\wp)}\Big(18 - 57\sin(\epsilon) - 160\sin^2(\epsilon) - 82\sin^3(\epsilon) - 10\sin(4\epsilon)\Big), \\
&\vdots
\end{aligned}$$

Thus, the approximate solution can be obtained by:

$$\begin{aligned}
\vartheta(\epsilon, q) =& 1 + \sin(\epsilon) - \frac{q^{\wp}}{\Gamma(1+\wp)}\Big(1 + 3\sin(\epsilon) + \sin^2(\epsilon)\Big) + \frac{q^{2\wp}}{\Gamma(1+2\wp)}\Big(11\sin(\epsilon) + 12\sin^2(\epsilon) + 2\sin^3(\epsilon)\Big) \\
&+ \frac{q^{3\wp}}{\Gamma(1+3\wp)}\Big(18 - 57\sin(\epsilon) - 160\sin^2(\epsilon) - 82\sin^3(\epsilon) - 10\sin(4\epsilon)\Big) + \cdots
\end{aligned}$$

Figure 3a–d indicate the physical behavior of the obtained solution at $\epsilon \in [0, 1]$ and $q \in [0, 1]$. From these figures, it can be observed that with the increase in the value of $\wp$, the approximate solution become close to the exact solution at $\wp = 1$. In Figure 4, we have plotted the graph of $\vartheta(\epsilon, q)$ with different fractional orders of $\wp = 0.25, 0.50, 0.75, 1$ at $\epsilon \in [0, 2\pi]$ with different values of q. It is obvious that this approximation can only be employed numerically, even though a closed form solution is not accessible. It can be seen that our approximate solution using YHPTM in Table 1 is more significant than that obtained in [36,37].

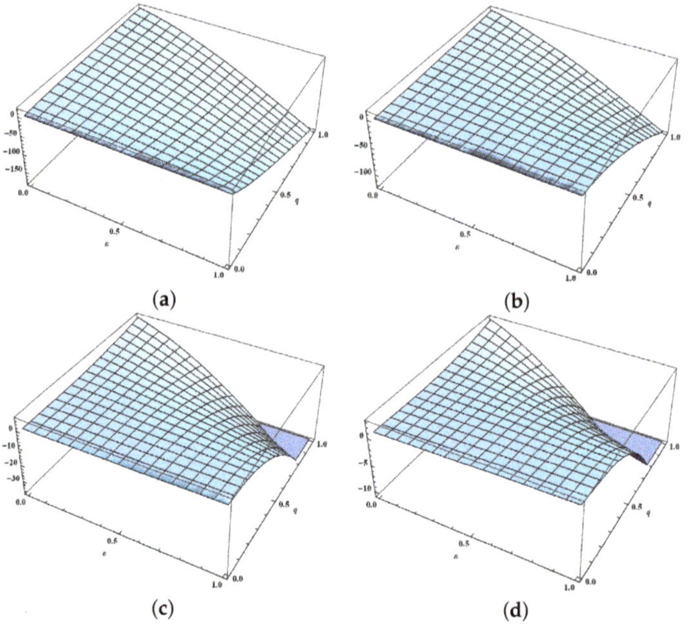

Figure 3. The surface solution of $\vartheta(\epsilon, q)$ with respect to ϵ and q for distinct values of $\wp$: (**a**) surface solution of $\vartheta(\epsilon, q)$ when $\wp = 0.25$; (**b**) surface solution of $\vartheta(\epsilon, q)$ when $\wp = 0.50$; (**c**) surface solution of $\vartheta(\epsilon, q)$ when $\wp = 0.75$; (**d**) surface solution of $\vartheta(\epsilon, q)$ when $\wp = 1$.

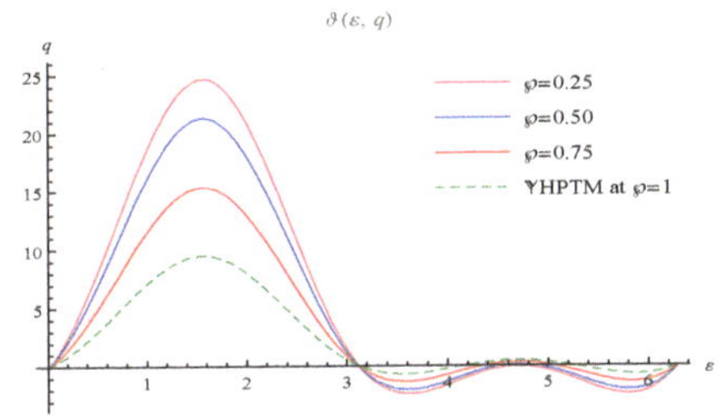

Figure 4. Plot of $\vartheta(\epsilon,q)$ for different values of $\wp$

Table 1. Comparison between the value $\vartheta(\epsilon,q)$ for the solution of the KG equation.

Sr. No.	$q=0.1$			$q=0.2$			$q=0.3$		
ϵ	[36]	[37]	YHPTM	[36]	[37]	YHPTM	[36]	[37]	YHPTM
0.0	0.9949999861	0.9950000249	0.903	0.9799991162	0.9800015775	0.824	0.9549900052	0.9550176534	0.781
0.1	1.093291132	1.093291179	0.976100	1.073723730	1.073726319	0.871321	1.073723730	1.073726319	0.792208
0.2	1.190502988	1.190503087	1.04725	1.166134875	1.166138050	0.915126	1.125945576	1.125974851	0.794835
0.3	1.285668610	1.285668848	1.11584	1.256326130	1.256331032	0.955409	1.208114007	1.208147932	0.789972
0.4	1.377844211	1.377844710	1.18132	1.343423788	1.343432104	0.992136	1.287043874	1.287088824	0.778571
0.5	1.466118315	1.466119219	1.24317	1.426594492	1.426608263	1.0252	1.362025218	1.362089477	0.761295
0.6	1.549620480	1.549621939	1.3009	1.505052082	1.505073495	1.05442	1.432404521	1.432497282	0.738476
0.7	1.627529538	1.627531694	1.35406	1.578063673	1.578094808	1.07951	1.497587424	1.497717706	0.710192
0.8	1.699081273	1.699084244	1.40223	1.644954933	1.644997540	1.0023	1.557040327	1.557215916	0.676451
0.9	1.763575490	1.763579356	1.44504	1.705114628	1.705169916	1.11635	1.610291023	1.610517519	0.63744
1.0	1.820382425	1.820387216	1.48219	1.757998450	1.758066925	1.12781	1.656928567	1.657208637	0.593784

4.3. Example 3

Consider another nonlinear time-fractional KG problem

$$D_q^{\wp}\vartheta(\epsilon,q) - D_{\epsilon}^{2}\vartheta(\epsilon,q) + \vartheta(\epsilon,q) - \vartheta^3(\epsilon,q) = 0, \tag{29}$$

with the initial condition

$$\vartheta(\epsilon,0) = -\operatorname{sech}(\epsilon). \tag{30}$$

According to the idea of $\mathbb{Y}$HPTM, we can obtain the following relation

$$\sum_{i=0}^{\infty} p^i\vartheta_i(\epsilon,q) = \operatorname{sech}(\epsilon) + p\left(\mathbb{Y}^{-1}\left[v^{\wp}\left\{\mathbb{Y}\left(\sum_{i=0}^{\infty} p^i\frac{\partial^2\vartheta_i}{\partial q^2} - \sum_{i=0}^{\infty} p^i\vartheta_i + \sum_{i=0}^{\infty} p^i\vartheta_i^2\right)\right\}\right]\right),$$

when the coefficients of like powers of p are compared, we obtain

$$\begin{aligned}
p^0 &= \vartheta_0(\epsilon,q) = \vartheta(\epsilon,0),\\
p^1 &= \vartheta_1(\epsilon,q) = \mathbb{Y}^{-1}\left[v^{\wp}\left\{\mathbb{Y}\left(\frac{\partial^2\vartheta_0}{\partial q^2} - \vartheta_0 + \vartheta_0^3\right)\right\}\right],\\
p^2 &= \vartheta_2(\epsilon,q) = \mathbb{Y}^{-1}\left[v^{\wp}\left\{\mathbb{Y}\left(\frac{\partial^2\vartheta_1}{\partial q^2} - \vartheta_1 + 3\vartheta_0^2\vartheta_1\right)\right\}\right],\\
p^3 &= \vartheta_3(\epsilon,q) = \mathbb{Y}^{-1}\left[v^{\wp}\left\{\mathbb{Y}\left(\frac{\partial^2\vartheta_2}{\partial q^2} - \vartheta_2 + 3\vartheta_0\vartheta_1^2 + 3\vartheta_0^2\vartheta_2\right)\right\}\right],\\
&\ \,\vdots
\end{aligned}$$

with help of Equation (30), we gain the iterations successively $\vartheta_i(\epsilon), i = 1,2,3,\cdots$, as follows:

$$\begin{aligned}
\vartheta_0(\epsilon,q) &= -\operatorname{sech}(\epsilon),\\
\vartheta_1(\epsilon,q) &= -\frac{q^{\wp}}{\Gamma(1+\wp)}\Big(2\operatorname{sech}(\epsilon) - 3\operatorname{sech}^3(\epsilon)\Big),\\
\vartheta_2(\epsilon,q) &= -\frac{q^{2\wp}}{\Gamma(1+2\wp)}\Big(3\operatorname{sech}(\epsilon) - 34\operatorname{sech}^3(\epsilon) - 18\operatorname{sech}^5(\epsilon)\Big),\\
\vartheta_3(\epsilon,q) &= -\frac{q^{3\wp}}{\Gamma(1+3\wp)}\Big(64\operatorname{sech}^3(x) - 288\operatorname{sech}^5(\epsilon) + 240\operatorname{sech}^7(\epsilon)\Big),\\
&\vdots
\end{aligned}$$

Thus, the approximate solution can be obtained by:

$$\begin{aligned}
\vartheta(\epsilon,q) = &-\operatorname{sech}(\epsilon) - \frac{q^{\wp}}{\Gamma(1+\wp)}\Big(2\operatorname{sech}(\epsilon) - 3\operatorname{sech}^3(\epsilon)\Big) - \frac{q^{2\wp}}{\Gamma(1+2\wp)}\Big(3\operatorname{sech}(\epsilon) - 34\operatorname{sech}^3(\epsilon) - 18\operatorname{sech}^5(\epsilon)\Big)\\
&- \frac{q^{3\wp}}{\Gamma(1+3\wp)}\Big(64\operatorname{sech}^3(\epsilon) - 288\operatorname{sech}^5(\epsilon) + 240\operatorname{sech}^7(\epsilon)\Big) + \cdots
\end{aligned}$$

Figure 5a–d indicates the physical behavior of the obtained solution at $\epsilon \in [-2,2]$ and $q \in [0,0.1]$. From these figures, it can be observed that with increase in the value of $\wp$, the approximate solution graph comes close to the exact exact solution at $\wp = 1$. In Figure 6, we have plotted the graph of $\vartheta(\epsilon,q)$ at different fractional orders of $\wp = 0.25, 0.50, 0.75, 1$ and $\epsilon \in [0,2\pi]$ with different values of q. We compared our graphical results obtained by $\mathbb{Y}$HPTM, which converges to the exact solution very rapidly with a small amount of q compared to [38] at $\wp = 1$.

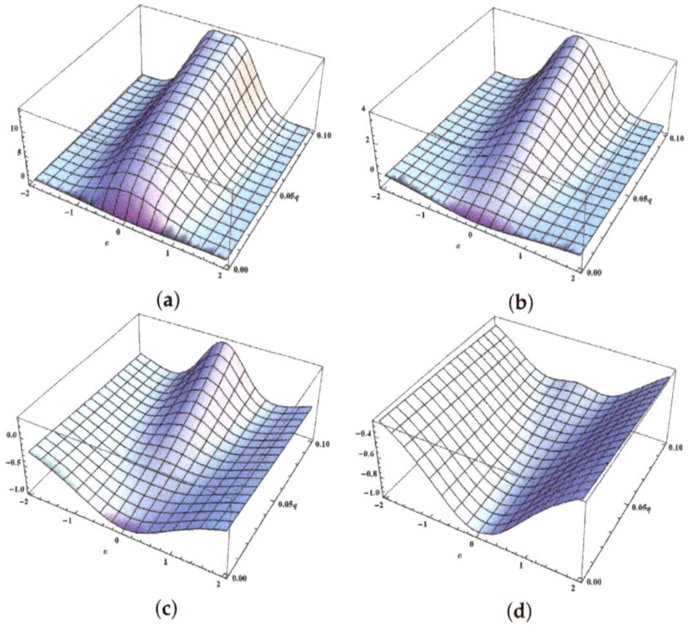

Figure 5. The surface solution of $\vartheta(\epsilon,q)$ with respect to ϵ and q for distinct values of $\wp$. (**a**) surface solution of $\vartheta(\epsilon,q)$ when $\wp = 0.25$; (**b**) surface solution of $\vartheta(\epsilon,q)$ when $\wp = 0.50$; (**c**) surface solution of $\vartheta(\epsilon,q)$ when $\wp = 0.75$; (**d**) surface solution of $\vartheta(\epsilon,q)$ when $\wp = 1$.

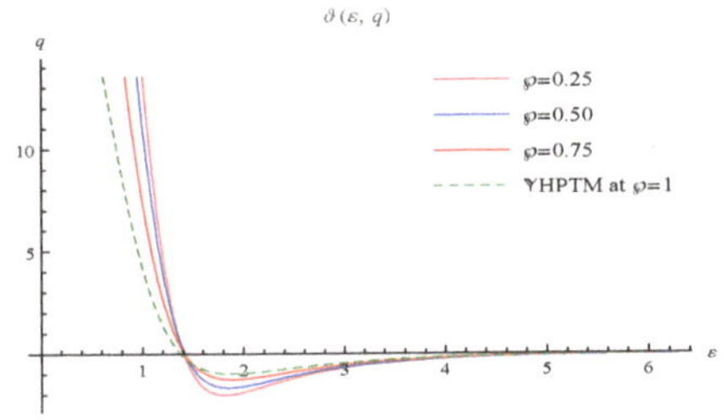

Figure 6. Plot of $\vartheta(\epsilon, q)$ for different values of $\wp$.

5. Conclusions

In this study, 𝕐HPTM has been utilized to achieve an approximate solution of nonlinear time-fractional KG equations. We demonstrate some illustrations to show the validity of the method, which reveals that the obtained findings are satisfactory. It is important to note that in order to improve the accuracy, a greater number of iterations with excessive order of p are required. The best advantage of 𝕐HPTM is that it generates the approximate solution in a quickly convergent power series form. As a result, this strategy can be adopted to elucidate a wide classification of nonlinear challenges in science and engineering with no need for linearization, discretization or perturbation. The proposed strategy has the privilege of being able to tackle linear and nonlinear time-fractional KG problems simultaneously. Mathematica package 11.0.1 has been operated for the graphical analysis as well as for the computations in this paper. We recommend the readers to consider this problem for the Atangana–Baleanu fractional derivative and many others in the place of the Caputo–Fabrizio operator.

Author Contributions: Conceptualization, formal analysis and funding acquisition, J.L.; investigation, methodology, software and writing—original draft, M.N.; validation and visualization, M.H.; writing—review and editing and supervision, A.A. All authors have read and agreed to the published version of the manuscript.

Funding: This work was supported by the Foundation of Yibin University, China (Grant No. 2019QD07).

Institutional Review Board Statement: Not applicable.

Informed Consent Statement: Not applicable.

Data Availability Statement: Not applicable.

Conflicts of Interest: The authors declare that they have no competing of interest.

Abbreviations

The following abbreviations are used in this manuscript:

𝕐HPTM	Yang homotopy perturbation transform method
𝕐T	Yang transform
CF	Caputo–Fabrizio
FPDEs	Fractional partial differential equations
KG	Homotopy perturbation method
HPM	Klein–Gordon

References

1. Miller, K.S.; Ross, B. *An Introduction to the Fractional Calculus and Fractional Differential Equations*; Wiley: New York, NY, USA, 1993.
2. Baleanu, D.; Machado, J.A.T.; Luo, A.C. *Fractional Dynamics and Control*; Springer Science & Business Media: Berlin, Germany, 2011.
3. Owolabi, K.M.; Atangana, A.; Akgul, A. Modelling and analysis of fractal-fractional partial differential equations: Application to reaction-diffusion model. *Alex. Eng. J.* **2020**, *59*, 2477–2490. [CrossRef]
4. Ara, A.; Khan, N.A.; Razzaq, O.A.; Hameed, T.; Raja, M.A.Z. Wavelets optimization method for evaluation of fractional partial differential equations: An application to financial modelling. *Adv. Differ. Equ.* **2018**, *1*, 8. [CrossRef]

5. Wang, G.W.; Xu, T.Z. The improved fractional sub-equation method and its applications to nonlinear fractional partial differential equations. *Rom. Rep. Phys.* **2014**, *66*, 595–602.
6. Almalahi, M.A.; Ibrahim, A.B.; Almutairi, A.; Bazighifan, O.; Aljaaidi, T.A.; Awrejcewicz, J. A qualitative study on second-order nonlinear fractional differential evolution equations with generalized abc operator. *Symmetry* **2022**, *14*, 207. [CrossRef]
7. Dehghan, M.; Mohebbi, A.; Asgari, Z. Fourth-order compact solution of the nonlinear klein-gordon equation. *Numer. Algorithms* **2009**, *52*, 523–540. [CrossRef]
8. Venkatesh, S.; Balachandar, S.R.; Ayyaswamy, S.; Krishnaveni, K. An efficient approach for solving klein-gordon equation arising in quantum field theory using wavelets. *Comput. Appl. Math.* **2018**, *37*, 81–98. [CrossRef]
9. Singh, H.; Kumar, D.; Pandey, R.K. An efficient computational method for the time-space fractional klein-gordon equation. *Front. Phys.* **2020**, *8*, 281. [CrossRef]
10. Amin, M.; Abbas, M.; Iqbal, M.K.; Baleanu, D. Numerical treatment of time-fractional klein–gordon equation using redefined extended cubic b-spline functions. *Front. Phys.* **2020**, *8*, 288. [CrossRef]
11. Khan, H.; Khan, A.; Chen, W.; Shah, K. Stability analysis and a numerical scheme for fractional klein-gordon equations. *Math. Methods Appl. Sci.* **2019**, *42*, 723–732. [CrossRef]
12. Ganji, R.; Jafari, H.; Kgarose, M.; Mohammadi, A. Numerical solutions of time-fractional klein-gordon equations by clique polynomials. *Alex. Eng. J.* **2021**, *60*, 4563–4571. [CrossRef]
13. Belayeh, W.G.; Mussa, Y.O.; Gizaw, A.K. Approximate analytic solutions of two-dimensional nonlinear klein-gordon equation by using the reduced differential transform method. *Math. Probl. Eng.* **2020**, *2020*, 1–12. [CrossRef]
14. Gepreel, K.A.; Mohamed, M.S. Analytical approximate solution for nonlinear space-time fractional klein-gordon equation. *Chin. Phys. B* **2013**, *22*, 010201. [CrossRef]
15. Nadeem, M.; Li, F. Modified laplace variational iteration method for analytical approach of klein-gordon and sine-gordon equations. *Iran. J. Sci. Technol. Trans. A Sci.* **2019**, *43*, 1933–1940. [CrossRef]
16. Tamsir, M.; Srivastava, V.K. Analytical study of time-fractional order klein-gordon equation. *Alex. Eng. J.* **2016**, *55*, 561–567. [CrossRef]
17. Bansu, H.; Kumar, S. Numerical solution of space-time fractional klein-gordon equation by radial basis functions and chebyshev polynomials. *Int. J. Appl. Comput. Math.* **2021**, *7*, 7. [CrossRef]
18. Kurulay, M. Solving the fractional nonlinear klein-gordon equation by means of the homotopy analysis method. *Adv. Differ. Equ.* **2012**, *2012*, 187. [CrossRef]
19. Khader, M.; Adel, M. Analytical and numerical validation for solving the fractional klein-gordon equation using the fractional complex transform and variational iteration methods. *Nonlinear Eng.* **2016**, *5*, 141–145. [CrossRef]
20. Zhmud, V.; Dimitrov, L. Using the fractional differential equation for the control of objects with delay. *Symmetry* **2022**, *14*, 635. [CrossRef]
21. Odibat, Z.; Momani, S. A generalized differential transform method for linear partial differential equations of fractional order. *Appl. Math. Lett.* **2008**, *21*, 194–199. [CrossRef]
22. Ghoreishi, M.; Ismail, A.M.; Ali, N. Adomian decomposition method (adm) for nonlinear wave-like equations with variable coefficient. *Appl. Math. Sci.* **2010**, *4*, 2431–2444.
23. Tan, Y.; Abbasbandy, S. Homotopy analysis method for quadratic riccati differential equation. *Commun. Nonlinear Sci. Numer. Simul.* **2008**, *13*, 539–546. [CrossRef]
24. Batiha, B.; Noorani, M.; Hashim, I. Numerical solution of sine-gordon equation by variational iteration method. *Phys. Lett. A* **2007**, *370*, 437–440. [CrossRef]
25. Qin, Y.; Khan, A.; Ali, I.; Qurashi, M.A.; Khan, H.; Shah, R.; Baleanu, D. An efficient analytical approach for the solution of certain fractional-order dynamical systems. *Energies* **2020**, *13*, 2725. [CrossRef]
26. Khan, H.; Khan, A.; Kumam, P.; Baleanu, D.; Arif, M. An approximate analytical solution of the navier-stokes equations within caputo operator and elzaki transform decomposition method. *Adv. Differ. Equ.* **2020**, *2020*, 622.
27. Yuanlu, L. Solving a nonlinear fractional differential equation using chebyshev wavelets. *Commun. Nonlinear Sci. Numer. Simul.* **2010**, *15*, 2284–2292.
28. Rehman, M.U.; Khan, R.A. The legendre wavelet method for solving fractional differential equations. *Commun. Nonlinear Sci. Numer. Simul.* **2011**, *16*, 4163–4173. [CrossRef]
29. Ali, M.; Jaradat, I.; Alquran, M. New computational method for solving fractional riccati equation. *J. Math. Comput. Sci.* **2017**, *17*, 106–114. [CrossRef]
30. Alquran, M.; Al-Khaled, K.; Ali, M.; Arqub, O.A. Bifurcations of the time-fractional generalized coupled hirota-satsuma kdv system. *Waves Wavelets Fractals* **2017**, *3*, 31–39. [CrossRef]
31. Caputo, M.; Fabrizio, M. On the singular kernels for fractional derivatives. some applications to partial differential equations. *Progr. Fract. Differ. Appl.* **2021**, *7*, 79–82.
32. Alesemi, M.; Iqbal, N.; Abdo, M.S. Novel investigation of fractional-order cauchy-reaction diffusion equation involving caputo-fabrizio operator. *J. Funct. Spaces* **2022**, *2022*, 1–14. [CrossRef]
33. Shah, N.A.; El-Zahar, E.R.; Chung, J.D. Fractional analysis of coupled burgers equations within yang caputo-fabrizio operator. *J. Funct. Spaces* **2022**, *2022*, 1–13. [CrossRef]

34. Yang, X.J. A new integral transform method for solving steady heat-transfer problem. *Therm. Sci.* **2016**, *20* (suppl. 3), 639–642. [CrossRef]
35. Ahmad, S.; Ullah, A.; Akgül, A.; la Sen, M.D. A novel homotopy perturbation method with applications to nonlinear fractional order kdv and burger equation with exponential-decay kernel. *J. Funct. Spaces* **2021**, *2021*, 1–11. [CrossRef]
36. El-Sayed, S.M. The decomposition method for studying the klein-gordon equation. *Chaos Solitons Fractals* **2003**, *18*, 1025–1030. [CrossRef]
37. Yusufoğlu, E. The variational iteration method for studying the klein-gordon equation. *Appl. Math. Lett.* **2008**, *21*, 669–674. [CrossRef]
38. Golmankhaneh, A.K.; Golmankhaneh, A.K.; Baleanu, D. On nonlinear fractional klein-gordon equation. *Signal Process.* **2011**, *91*, 446–451. [CrossRef]

MDPI
St. Alban-Anlage 66
4052 Basel
Switzerland
www.mdpi.com

Symmetry Editorial Office
E-mail: symmetry@mdpi.com
www.mdpi.com/journal/symmetry

www.ingramcontent.com/pod-product-compliance
Lightning Source LLC
LaVergne TN
LVHW070157100526
838202LV00015B/1960

* 9 7 8 3 0 3 6 5 9 4 2 4 8 *